Principles of
3D Image Analysis and Synthesis

THE KLUWER INTERNATIONAL SERIES IN ENGINEERING AND COMPUTER SCIENCE

PRINCIPLES OF 3D IMAGE ANALYSIS AND SYNTHESIS

EDITED BY

BERND GIROD
Information Systems Laboratory
Stanford University
Stanford, CA, USA

GÜNTHER GREINER
Lehrstuhl für Graphische Datenverarbeitung
Universität Erlangen-Nürnberg
Erlangen, Germany

HEINRICH NIEMANN
Lehrstuhl für Mustererkennung
Universität Erlangen-Nürnberg
Erlangen, Germany

Kluwer Academic Publishers
Boston/Dordrecht/London

Distributors for North, Central and South America:
Kluwer Academic Publishers
101 Philip Drive
Assinippi Park
Norwell, Massachusetts 02061 USA
Telephone (781) 871-6600
Fax (781) 871-6528
E-Mail <kluwer@wkap.com>

Distributors for all other countries:
Kluwer Academic Publishers Group
Post Office Box 322
3300 AH Dordrecht, THE NETHERLANDS
Telephone 31 78 6576 000
Fax 31 78 6576 254
E-Mail <orderdept@wkap.nl>

Electronic Services <http://www.wkap.nl>

Library of Congress Cataloging-in-Publication Data

Principles of 3D image analysis and synthesis / edited by Bernd Girod, Günther Greiner, Heinrich Niemann.
p. cm – (The Kluwer international series in engineering and computer science ; SECS 556)
Includes bibliographical references and index.

1. Image processing—digital techniques. 2. Image analysis. I. Girod, Bernd. II. Greiner, Günther. III. Niemann, Heinrich. IV. Series.

TA1637.P75 2000 ISBN 978-1-4419-4982-0
621.36'7—dc21 00-029626

Printed on acid-free paper.

Printed in the United States of America

Contents

Contributing Authors

Michael Breuer
Lehrstuhl für Strömungsmechanik, Universität Erlangen-Nürnberg

Swen Campagna
Lehrstuhl für Graphische Datenverarbeitung, Universität Erlangen-Nürnberg

Katja Daubert
AG 4 Computergraphik, Max-Planck-Institut für Informatik, Saarbrücken

Joachim Denzler
Lehrstuhl für Mustererkennung, Universität Erlangen-Nürnberg

Peter Eisert
Lehrstuhl für Nachrichtentechnik, Universität Erlangen-Nürnberg

Thomas Ertl
Visualisierung und Interaktive Systeme, Universität Stuttgart

Arnd Gebhard
Lehrstuhl für Mustererkennung, Universität Erlangen-Nürnberg

Bernd Girod
Information Systems Laboratory, Stanford University

Sabine Girod
Department of Functional Restoration, Stanford University School of Medicine

Günther Greiner
Lehrstuhl für Graphische Datenverarbeitung, Universität Erlangen-Nürnberg

Peter Hastreiter
Lehrstuhl für Graphische Datenverarbeitung, Universität Erlangen-Nürnberg

Gerd Häusler
Lehrstuhl für Optik, Universität Erlangen-Nürnberg

Wolfgang Heidrich
AG 4 Computergraphik, Max-Planck-Institut für Informatik, Saarbrücken

Benno Heigl
Lehrstuhl für Mustererkennung, Universität Erlangen-Nürnberg

Matthias Hopf
Visualisierung und Interaktive Systeme, Universität Stuttgart

Kai Hormann
Lehrstuhl für Graphische Datenverarbeitung, Universität Erlangen-Nürnberg

Manfred Kaltenbacher
Lehrstuhl für Sensorik, Universität Erlangen-Nürnberg

Stefan Karbacher
Lehrstuhl für Optik, Universität Erlangen-Nürnberg

Xavier Laboureux
Lehrstuhl für Optik, Universität Erlangen-Nürnberg

Hermann Landes
Lehrstuhl für Sensorik, Universität Erlangen-Nürnberg

Reinhard Lerch
Lehrstuhl für Sensorik, Universität Erlangen-Nürnberg

Marcus Magnor
Lehrstuhl für Nachrichtentechnik, Universität Erlangen-Nürnberg

Heinrich Niemann
Lehrstuhl für Mustererkennung, Universität Erlangen-Nürnberg

Dietrich Paulus
Lehrstuhl für Mustererkennung, Universität Erlangen-Nürnberg

Rudolf Rabenstein
Lehrstuhl für Nachrichtentechnik, Universität Erlangen-Nürnberg

Frank Schäfer
Lehrstuhl für Strömungsmechanik, Universität Erlangen-Nürnberg

Annette Scheel
AG 4 Computergraphik, Max-Planck-Institut für Informatik, Saarbrücken

Hartmut Schirmacher
AG 4 Computergraphik, Max-Planck-Institut für Informatik, Saarbrücken

Nikolaus Schön
Lehrstuhl für Optik, Universität Erlangen-Nürnberg

Harald Schönfeld
Lehrstuhl für Optik, Universität Erlangen-Nürnberg

Stephan Seeger
Lehrstuhl für Optik, Universität Erlangen-Nürnberg

Hans-Peter Seidel
AG 4 Computergraphik, Max-Planck-Institut für Informatik, Saarbrücken

Marc Stamminger
AG 4 Computergraphik, Max-Planck-Institut für Informatik, Saarbrücken

Eckehard Steinbach
Lehrstuhl für Nachrichtentechnik, Universität Erlangen-Nürnberg

Norbert Strobel
Lehrstuhl für Nachrichtentechnik, Universität Erlangen-Nürnberg

Matthias Teschner
Lehrstuhl für Nachrichtentechnik, Universität Erlangen-Nürnberg

Christian Teitzel
Lehrstuhl für Graphische Datenverarbeitung, Universität Erlangen-Nürnberg

Lutz Trautmann
Lehrstuhl für Nachrichtentechnik, Universität Erlangen-Nürnberg

Christian Vogelgsang
Lehrstuhl für Graphische Datenverarbeitung, Universität Erlangen-Nürnberg

Rüdiger Westermann
Visualisierung und Interaktive Systeme, Universität Stuttgart

Matthias Zobel
Lehrstuhl für Mustererkennung, Universität Erlangen-Nürnberg

Preface

Traditionally, say 15 years ago, three-dimensional image analysis (aka computer vision) and three-dimensional image synthesis (aka computer graphics) were separate fields. Rarely were experts working in one area interested in and aware of the advances in the other field. Over the last decade, this has changed dramatically. Maybe it is a result of the growing maturity of each of these areas that they are less concerned with themselves. Vision and graphics communities are today engaged in a mutually beneficial exchange, learning from each other and coming up with new ideas and techniques that build on the state-of-the-art in both fields. Many of us today believe that we will ultimately have one unified field, that, besides vision and graphics, also might encompass traditional image processing and image communication, and names such as 'Visual Computing', 'Imaging Sciences', or 'Image Systems Engineering' have been proposed.

Without doubt, three-dimensional image analysis and synthesis is very much an application-driven field. The declining cost of processors, memory, and sensors continues to expand the scope of viable applications at a breath-taking speed. New systems solutions are in reach by combining state-of-the-art techniques from vision and graphics. The thorough scientific treatment of the underlying principles, for example, the limitations of sensors, the reliability, accuracy, and complexity of image processing algorithms, the fidelity of interactive visualization schemes, or the appropriate mathematical formulation of the interaction of diverse methods from computer vision and computer graphics, is the prerequisite for such advanced systems.

This book is the result of a most fruitful collaboration between scientists at the University of Erlangen-Nürnberg, Germany, that, coming from diverse fields, are working together propelled by the vision of a unified area of three-dimensional image analysis and synthesis. As a formal framework for this collaboration, we set up the "Graduiertenkolleg Dreidimensionale Bildanalyse und -synthese" (Graduate Research Center 3D Image Analysis and Synthesis) initially, which is being supported by the Deutsche Forschungsgemeinschaft (German Research Foundation) since January 1996. The Graduiertenkolleg comprises a program of research and advanced studies for doctoral students, with special emphasis on problems of 3D image acquisition, computer vision, 3D graphics, and selected applications ranging from medicine to manufacturing. In January 1998, this program was substantially augmented and broadened by the addition of a "Sonderforschungsbereich (SFB)" (similar to an NSF Center of Excellence in the US). The SFB 603 "Model-based Analysis and Visualization of Complex Scenes and Sensor Data," again funded by the Deutsche Forschungsgemeinschaft, complements the Graduate Research Center through currently 12 long-term collaborative research projects. These SFB projects address much more ambitious goals, often involving implementation and testing of large systems and investigations with a time-line exceeding five years. Within the SFB, a working group was formed that set out to systematically review and compile the state-of-the-art in image analysis and synthesis, and quickly we realized that publishing the results of our effort in the form of a book and thus sharing them with the scientific community might be worthwhile.

It is our long-term vision to develop a theoretically well-founded general methodology for the design of systems for image synthesis and interpretation. This methodology should be based on the formulation of relevant problems as optimization problems, the use of model-based techniques, sensor data fusion, and hierarchical processing algorithms and data structures. Models are the backbone of 3D image analysis and synthesis. Modeling 3D scenes is central to computer graphics, as it is for 3D vision techniques. Often, these models are implicitly built into the flow of an algorithm, but more and more we are seeing techniques that keep the model explicit and separate from the algorithm, thus deserving the label 'model-based'. Initially we considered to organize the contents of this book around the various types of models used, ranging from surface geometry, reflectance and illumination to statistical models or semantic networks. But since many important state-of-the-art techniques are not model-based in the sense of providing and using an explicit model, we decided to follow a more conventional outline which starts out at the image acquisition end of a hypothetical processing chain, proceeds with analysis, recognition and interpretation of images, towards the representation scenes by 3D geometry, then back to images via rendering and visualization techniques.

3D image analysis and synthesis systems can process regular camera images, but often optical 3D sensors are a viable and more powerful alternative that can acquire 3D data directly. Chapter 1 discusses the principles, the potential and the limitations of such sensors. Chapter 2 considers the processing of multiples views of a scenes acquired by several cameras simultaneously or by one camera as a video sequence over time. Discussion includes the famous structure-from-motion problem and approaches to incorporate illumination estimation into motion estimation techniques. Chapter 3 deals with image recognition and interpretation, including segmentation, statistical models for recognition and image understanding, and active vision, where interpretation results are fed back to the sensor. Chapter 4 "Representation and Processing of Surface Data" is a central chapter since it interfaces both to traditional image analysis and image synthesis techniques. The discussion includes polygon meshes and their reduction, splines, and techniques for reverse engineering. The next two chapters address topics from computer graphics. Chapter 5 presents techniques for synthesizing realistic images. The traditional approach based on scene geometric descriptions and on models for reflection and illumination is discussed, including texturing techniques and global illumination. In addition, the recent developments in image-based rendering are presented. Scientific visualization, discussed in Chapter 6, would not be possible without combining image analysis and synthesis techniques, and thus can serve as an excellent example for the spirit of this book. It treats the state-of-the-art in volume visualization and includes a presentation of flow visualization techniques. We have also included Chapter 7 on Acoustic Imaging, Rendering, and Localization, partly because there are many interesting parallels to the material presented in the previous chapters, partly because we expect to see more applications in the future that can benefit from joint audio-video analysis and rendering. The concluding Chapter 8 is a collection of selected applications that some of us are involved in, ranging from automotive design to surgical planning.

We hope that scientists, engineers, graduate students and educators will find this book a useful reference for their work and a readable, but concise introduction to areas that they might be less familiar with. We have included references to the important seminal papers in each area, and thus the book should be a reasonable starting point for in-depth study of a selected problem. There is far too much material to cover in a one- or two-semester course, but individual chapters might be suitable as supplementary reading material in an advanced graduate level course.

The comprehensive treatment of such a variety of research topics was possible only by a close cooperation of many experts. This book is the result of a true team effort. As editors, we express our sincere appreciation to all the contributing authors listed above. In addition, our special thanks go to M. Magnor, K. Hormann and C. Vogelgsang for the technical assistance in preparing the manuscript. Finally we are grateful to the German Research Foundation DFG for financially supporting many of the contributors.

Stanford and Erlangen, February 2000

BERND GIROD, GÜNTHER GREINER, HEINRICH NIEMANN

1 OPTICAL 3D SENSORS

G. HÄUSLER

INTRODUCTION

This chapter is concerned with three-dimensional (3D) image acquisition, including a brief introduction to the problems of two-dimensional image capturing. We emphasize on 3D data because it represents information about the geometrical shape and is therefore often of much greater value than 2D data that represents the local reflectivity only. Geometrical shape is invariant with respect to rotations and shifts, and unaffected by soiling and illumination conditions. Moreover, the shape is the feature that is usually required for inspection purposes. Assuming this to be true, we might ask ourselves why we do not find 3D sensors everywhere? The answer is that video-cameras and computers are seducingly easy to use. Unfortunately, the acquisition of 3D data is much more difficult than the capture of a video image because of deep physical reasons (the optical transfer function of empty space cuts off most of the longitudinal frequencies) and because of technical reasons (the very limited space-bandwidth product of video cameras).

In this chapter we give some concise overview about the potentials and the limitations of optical and X-ray 3D image acquisition, which hopefully motivates the 2D image processing scientists, as well as people looking for suitable sensors, to solve their inspection and metrology problems. Section 1.1 will discuss some of the important problems of 2D image acquisition, because 3D sensors generally need to acquire 2D images. Section 1.2 will discuss the three major 3D sensor principles that cover a large range of applications, and will give some detailed insight into the physics and information theory of optical 3D sensors.

1.1 2D IMAGE ACQUISITION FOR 'PERFECT' 3D SENSORS

G. HÄUSLER

Optical 3D sensors usually work with light sources, lenses, photodiodes and video cameras. We often have to project a pattern onto the object surface (active triangulation), and we have to observe this pattern. This requires some knowledge about two-dimensional imaging and image acquisition. Most of this knowledge can be found in optics textbooks [251, 769]. This section will specifically consider topics that are not commonly mentioned in the literature, but important to build 3D sensors that work at the physical limits. Those topics are: the *sampling theorem*, *linearity*, *depth of field*, and *coherence considerations*.

1.1.1 Sampling Theorem and Linearity

Video cameras spatially sample the image by an array of light sensitive elements. This operation is only reversible, if certain conditions are satisfied: The *sampling theorem* [636] limits the maximum bandwidth of an image that is spatially sampled. Violating the sampling theorem irreversibly generates artifacts ('*aliases*') in the sampled signal. 'Aliasing' displays severe consequences in quantitative image processing. Unfortunately, the careless acquisition of video images is usually connected with violation of the sampling theorem, because the lenses are too good, i. e., they have too large a bandwidth. In daily life, aliasing is not strongly visible—unless the people in the video film do not wear jackets with narrow stripes—but in 3D sensing aliasing is fatal. This will be explained through the example of triangulation. Active triangulation works the following way: A pattern is projected onto the object under test, via a certain 'direction of illumination', and observed from a different 'direction of observation'. The pattern might be a laser spot, for a local distance measurement, or a fringe pattern, for a full field shape measurement. The lateral location of the observed pattern shifts with the distance of the object. To evaluate the distance or the shape we somehow localize the pattern on the video target. From the locus of the image of a laser spot or of a fringe pattern we evaluate via triangulation the distance of the considered object detail. Since standard video can only distinguish about 500 different lateral positions, it appears that only 500 different distances can be distinguished. However, good 3D sensors can distinguish more than 2,000 or 3,000 different distances. They do this by 'subpixel interpolation'. Subpixel interpolation is possible to a large extent, if we have a priori knowledge about the shape of the point spread function (impulse response) of the video output. If we can assume that the sampled version of this point spread function remains the same if the projected spot moves slowly across the video target, we can localize it much better than the given pixel pitch of the target. In the language of system theory, we need *linearity* and *space invariance*. Aliasing destroys spatially invariant imaging. Linearity is usually satisfied, by CCD cameras, but not by the recent CMOS cameras with very high dynamical range. These cameras are not suited for phase measuring triangulation 3D sensors, because this principle requires perfect linearity.

In practice, people do not much care about the sampling theorem, and in fact, the satisfaction of the sampling theorem is difficult: we need to limit the bandwidth of the imaging lens which often contradicts the desire of high light throughput and depth of field requirements. But we have no chance of building 3D sensors that work at the limits of what is physically possible, if we neglect the requirements of linearity and of the sampling theorem.

1.1.2 Depth of Field

The *depth of field* δz of an imaging system depends on the aperture $\sin u$ of the imaging system, and is given by the well known Rayleigh formula:

$$\delta z = \frac{\lambda}{\sin^2 u}, \tag{1.1}$$

where λ is the wavelength of light. In practice, the depth of field can be increased by stopping down the system. However, we sacrifice light throughput and lateral resolution. It is interesting that there is no fundamental law against expansion of depth of field [300], however, such a method requires considerable technical and computational effort. Depth of field problems may occur both in the projection of patterns onto the object under test, as well as in the imaging of the backscattered patterns. In active laser triangulation, the depth of field for the projection of a laser spot can be overcome by application of the '*Scheimpflug condition*' [295]. For phase measuring triangulation, the depth of field problem is somewhat less critical, if we project a sinusoidal pattern with moderate spatial frequency. In this case, a limited depth of field just causes slight decrease of the fringe contrast. Limited depth of field of the observation system afflicts the performance of 3D sensors more seriously. Specifically, the 3D imaging of sharp object details such as edges, is strongly disturbed. Here, as well, stopping down the system is the only measure to increase the depth of field—with the undesired side effects mentioned above. Therefore, a careful design that optimizes the contradicting requirements is necessary for a good 3D sensor.

1.1.3 Coherence Considerations

Although well considered in microscopy [294, 79], the role of spatial coherence in 3D sensing, or generally, in macroscopic optical metrology, is largely neglected. We briefly will summarize some important facts afflicting the performance of 3D sensors:

Coherence of two waves is the ability to perform measurable (stable) interference contrast. We distinguish *temporal coherence* and *spatial coherence*. Temporal coherence essentially means monochromaticity. A Laser displays nearly perfect temporal coherence. But even white light sources such as thermal sources or fluorescent lamps may generate coherence: spatial coherence. The fundamental theorem is the *van Cittert-Zernike theorem* [251, 79]. It basically says, that in the far field of light sources, two points of the illuminated object will scatter coherent waves, if their distance is smaller than some certain 'coherence width'. The fundamental observation is, that the coherence width will increase with larger distance from the (incoherent!) source. The consequences of coherent illumination can be fatal for metrology. If the observation aperture of the imaging system is smaller than the aperture of the illumination, the image of rough (diffusely scattering objects) suffers from strong parasitic interference

Figure 1.1. Interference image of a rough, tilted planar object. This is what we see if we look into a Michelson interferometer, where one mirror is replaced by a ground glass. The illumination is a spatially coherent, broad band light source. Due to the spatial coherence, the entire image displays speckle. In the center, both, reference wave and object wave are even temporally coherent, and cause high speckle contrast. Due to the tilt, outside of the center, there is no temporal coherence between both waves, hence we have incoherent superposition and less contrast.

noise which is called '*speckle noise*' (see Figure 1.1 or [251]). If the observation aperture is larger than the illumination aperture, the speckle contrast decreases as the ratio of the apertures.

The effect of coherent noise is largely neglected, because it is often hardly visible. If the diffraction pattern of the video lens is much smaller than the size of the video pixels, averaging over a number speckles occurs, with a considerable reduction of the speckle contrast. However, although speckles may not be visually detectable, in quantitative image processing they can be extremely disturbing. We just want to mention that in all optical triangulation systems the speckle noise limits the physically achievable depth uncertainty [185].

After these brief remarks about 2D imaging in 3D sensors, we will now proceed to optical 3D sensors.

1.2 3D SENSORS—PRINCIPLES, POTENTIALS AND LIMITATIONS

G. HÄUSLER

Most of the problems of *industrial inspection*, *reverse engineering* and *virtual reality* require data about the geometrical shape of objects in 3D space. Such 3D data offer advantages over 2D data: shape data are invariant against alteration of the illumination, soiling and object motion. Unfortunately, those data are much more difficult to acquire than video data about the two-dimensional local reflectivity of objects. We will discuss the physics of 3D sensing, and will address the following subjects:

- different type of illumination (coherent or incoherent, structured or unstructured),
- interaction of light with matter (coherent or incoherent, at rough surfaces or at smooth surfaces),
- the consequences of *Heisenberg's uncertainty relation.*

From the knowledge of the underlying physical principles that define the limitations of measuring uncertainty, one can design optimal sensors that work just at those limits, as well as judge available sensors. We will show that the vast number of known 3D sensors, is based on only three different principles. The three principles are different in terms of how the measuring uncertainty scales with the object distance. We will further learn that with only two or three different sensors a great majority of problems from automatic inspection or virtual reality can be solved.

We will not discuss many sensors in detail. Quite a couple of sensor principles are explained in [359]. There, the reader can find chapters about the basic principles such as the different triangulation systems (laser triangulation, phase measuring triangulation, confocal microscopy, photogrammetry), time-of-flight measurement, interferometry.

In this article, we will rather discuss the potentials and limitations of the major sensor principles for the physicist, as well as for the benefit of the user of 3D sensors:

- *laser triangulation*,
- *phase measuring triangulation* and
- *white light interferometry* on rough surfaces.

As mentioned above, it turns out that with this set of sensors, 3D data of objects of different kind or material can be acquired. The measuring uncertainty ranges from about 1 nanometer to a few millimeters, depending on the principle and on the measuring range.

We will give examples of the potentials of each sensor, by examples of measured objects and by discussion of the physical and technological drawbacks. We will specifically address the interests of potential users of those sensors concerning the applicability to real problems.

Before we will continue, we briefly explain the three principles. More explanations will be given later. In *laser triangulation* systems we project a laser spot onto the surface under test, from a certain direction of illumination, and we observe the spot by

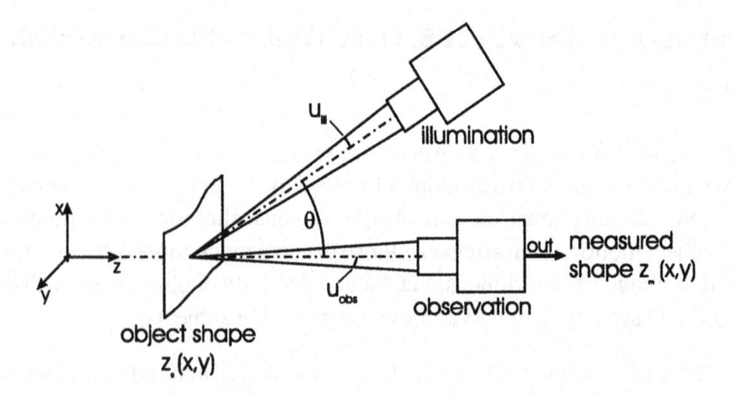

Figure 1.2. Principle of triangulation.

a video line array, from a different direction of observation. The angle between both directions is called the angle of triangulation, see Figure 1.2. If the object distance changes, the lateral position of the spot image changes as well. With simple geometric calculations, we can evaluate the distance of the spot from its lateral position. There is a straight forward improvement of laser-spot-triangulation, by projecting a line, instead of a point. This is sometimes called *laser sectioning*, because an observing video camera sees a profile ('cross section') of the surface under test. To acquire the entire surface, a one-dimensional scan of the laser line over the object is necessary.

With *phase measuring triangulation (pmt)* we can further proceed from a 'line-sensor' to an 'area-sensor' that measures the shape $z(x, y)$ of an entire surface, without any scanning. Such a sensor will be discussed in Section 1.2.3. The basic idea is to project a grid pattern onto the object. If the object surface is curved, the camera observes curved grid lines. If we project sinusoidal patterns with with different phase shifts, it can be shown that from at least three exposures we can derive the local phase of the grid image, and, hence, the distance of each object point.

The two principles discussed above are based on 'triangulation'. The third principle of our list is *white light interferometry*. Interferometry is essentially based on *time-of-flight* measurement, by interference of the object light wave with a reference light wave. Distance variation of the object will cause phase variation of the object wave. Since those phase variations can be measured with extreme accuracy (better than $\frac{\lambda}{1000}$), we can measure shape variations in the sub-nanometer regime. This, however, works only for optically smooth (polished) surfaces. For rough surfaces, a phase evaluation is impossible, since the object wave suffers from *speckle noise*, which means, the phase is arbitrary and does not contain information about the distance. Instead, we use the temporal coherence properties, to detect the time-of-flight of the object wave. We make use of the fact that the reference wave and the object wave display interference contrast only if the path length difference is smaller than the coherence length of the source (which is the case in the center of Figure 1.1). This gives us the possibility to measure the shape of macroscopic objects with an uncertainty of only one micrometer. It should be noted that in interferometric sensors illumination and observation are coaxial. Hence, we can look into narrow holes.

1.2.1 Why 3D Sensors?

Those who investigate or apply computer vision have the long range goal for machines that can 'see' in a technical or less often, natural environment, without aid of an expert. Our practical definition for 'seeing' evolved from many years of problem solving for industrial inspection: for those problems, 'seeing' an object means often

- *segmentation* from other objects,
- *localization* in 6 degrees of freedom,
- *feature detection*, and *template matching*.

If this can be done robust against

- shift, rotation, scale variation, perspective distortion,
- hidden parts, shading, and soiling,

we can solve a majority of problems coming up from industrial inspection, metrology, reverse engineering, and virtual reality. The German VDI/VDE Standard 2628 [724, 293] gives an interesting overview how to define problems in automatic optical inspection.

It is obvious that for many real world problems we cannot achieve the required robustness by just applying more intelligent image processing algorithms. Our approach is that the intelligent algorithms have to be supported by intelligent sensors. To explain this, let us look at Table 1.1, where we compare the visual system and a technical system in terms of its automatic inspection ability. It is important that we separate the system into the 'preprocessing optical hardware' and the 'post-processing computer hardware and algorithms'.

The visual optical system is very poor in terms of space-bandwidth product, the lateral resolution is high only in a field smaller than the angular diameter of the moon. We can see only radiance and color, but not phase or polarization. So we judged the visual optical hardware with '−−'. We want to emphasize that a video image has about the same quality than the visual image (video cameras are built just that way, to satisfy the eye).

This is quite different in modern optical technology: the space-bandwidth product can be extremely large, we can not only see radiance but can take advantage of phase, polarization, coherence. For example, with a differential phase contrast microscope

	visual system	technical system
optical hardware	−−	++
post-processing	++	−

Table 1.1. Visual system vs. technical system.

we can resolve chromosomic structures and with 3D sensors we can resolve 10,000 different depth steps. We judged this category by a '++' (see Table 1.1).

Looking at the post-processing power, the judgement is the other way round: the brain is still unbeatable, for the tasks defined in our definition of 'seeing': we gave a '++', while computers are mildly judged with a '−', due to their serial processing structure.

Now we come to the suggestive point: if we combine a video camera with a computer, we put in series the tools in the main diagonal of our table (three minus signs). That is the wrong diagonal. Unfortunately, we cannot put in series, the elements of the other diagonal (four plus signs), for 'automatic' inspection. But we can take advantage of the high standard of optical sensors to support the weak computer post-processing (right hand column). And we will demonstrate that optical 3D sensors can help to satisfy our requirements above:

3D sensors acquire data about the geometrical 'shape' of the object in 3D space. (Sometimes, for example in tomography, we are interested in 'volume data' [303]). The 'shape'—compared to the 2D video image—is invariant against rotation and shift of the object, against alteration of the illumination, and against soiling. And the shape is that what a majority of tasks in inspection, metrology, localization and recognition, and virtual reality need. By the aid of 3D sensors, we can solve many problems from this list on standard computers. This is mainly possible by relieving the algorithms from the need to consider variations of the image appearance.

If that is so, then why don't we have optical 3D sensors everywhere? There are two answers: nature introduces severe physical limits (still under investigation) to acquire remote 3D information. And our anthropomorphic thinking is seducing us to consider a video camera and computer as a 'technical eye'.

To understand the physical limits of 3D information acquisition gives deep insight in the propagation of information in space, which is interesting by itself. Moreover, the knowledge of those limits enables us to build better sensors (and judge the sensors of the competitors). Eventually, we even might overcome those limits. How could this be possible? We ask the reader for some patience, and stay curious, for a while.

To summarize this section: a video camera and a computer are not like eye and brain. An optical 3D sensor can relieve the computer to consider the variations of the image appearance, since the geometrical shape is invariant. Moreover, to know the geometrical shape is a great value, for many applications. In order to build, to select or to apply good optical 3D sensors we have to understand the potentials as well as the limitations of 3D data acquisition. This will be explained in the next sections.

1.2.2 Some Important Questions about 3D Sensing

Range sensors can measure the distance of stars or measure the thickness of atomic layers on a crystal. What are the principles that give us access to 20 or more orders of magnitude?

There is a confusingly large number of different sensors. Some of them are user friendly described in [176, 293]. A scientific review was given by Besl [58], a more recent review can be found in [359]. Can we put the large number of sensors into only a few categories, to judge their potentials and limitations by basic principles? We will discuss these question in Sections 1.2.3 and 1.2.4, as a summary of results achieved in our group, during the last years.

1.2.3 Triangulation on Optically Rough Surfaces

We will start with a very common method: *triangulation*. We want to acquire the shape of an object surface $z_o(x, y)$, as shown in Figure 1.3. We will evaluate this shape by measuring the distance of one or many object pixels. To be distinguished from the real object shape, the measured shape will be called $z_m(x, y)$.

We will discuss here, only *active triangulation* sensors: there is a source that illuminates the object. Light interacts with the surface of the object. Light is reflected or scattered towards the sensor. The sensor has an aperture to gather the light, and an optical system to image each surface pixel onto a spatially resolving detector (array of photodetectors).

The illumination can be structured or diffuse. It can be spatially coherent or (partially) incoherent. It can be temporally coherent or broad band. It can be polarized or unpolarized. Active triangulation needs structured illumination. Either a small light spot is projected onto the object (we call this a '*point sensor*', since it measures the distance of just one single point). Or we project a narrow line ('*line sensor*', this method known as *light sectioning* [297]. Or we project a grating (*phase measuring triangulation* [276, 466]).

The object surface can be optically smooth, like a mirror, or it can be optically rough, as a ground glass. It is important to note that the attribute smooth or rough depends on the lateral resolution of the observation optics: if we resolve the lateral structure of a ground glass, for example by a high aperture microscope, the surface is smooth, for our purpose. 'Smooth' means for us, that the elementary waves that are collected from the object to form a diffraction limited image spot, contribute only with minor phase variations less than $\pm\frac{\lambda}{4}$, as shown in Figure 1.3. If there are larger phase variations within the elementary waves than we have *diffuse reflection*, or scattering.

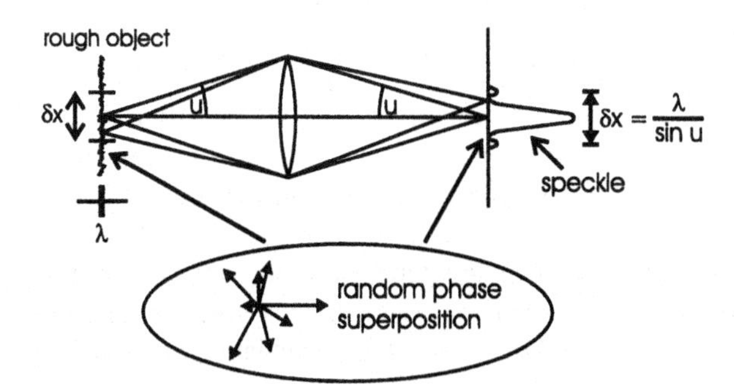

Figure 1.3. For rough surfaces, the accumulation of complex elementary waves within a diffraction limited resolution cell of the sensor leads to the classical random walk problem, since the statistical phase differences are greater than $\pm\pi$. As a result, the lateral location of the spot image will suffer from an uncertainty. This lateral uncertainty leads to a fundamental distance uncertainty δz_m, in triangulation systems.

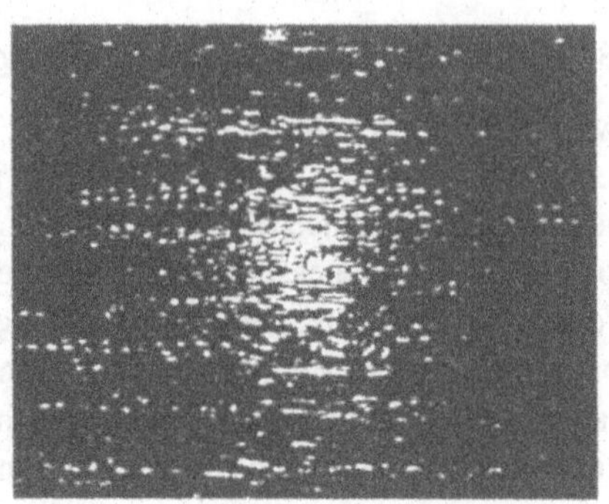

Figure 1.4. Spot image after reflection at a smooth surface (left) and after reflection at a rough surface with speckle noise (right). The localization of the spot image for triangulation is possible only with some uncertainty, introduced by the surface microtopography, we have no access to.

We will see that diffuse reflection in connection with spatial coherence, will introduce a *statistical uncertainty* of the measured shape $z_{\mathrm{m}}(x, y)$. The reason is that the wave scattered from an illuminated object 'point' is not anymore a spherical wave, but suffers from statistical phase variations. These phase errors lead to *speckles* in the image plane of the sensor [250].

For simplicity, let us consider a point sensor. It works by projecting a small light spot onto the object via a certain illumination direction, and observing this spot via a different angle of observation. The angle between these directions will be called '*triangulation angle*' in Figure 1.3. A change of the distance will be encoded as a lateral shift of the spot image on the observing target. From lateral shift, we can evaluate the distance of the object point, or if we do that for many points, we acquire the shape of the object.

Physical Limit of Measuring Uncertainty for Triangulation at Rough Surfaces. The weakness of point triangulation is obvious: it is not robust against shape variation of the spot image. And just such a variation is introduced by speckle, as shown in Figure 1.4.

Since the shape of the spot image depends on the unknown microtopography of the surface, there will be a principal random localization error, theoretically and experimentally determined in [185]. Its standard deviation δz_{m} will be given by:

$$\delta z_{\mathrm{m}} = \frac{c \cdot \lambda}{2\pi \sin u_{\mathrm{obs}} \sin \Theta}. \tag{1.2}$$

With Figure 1.3, θ is the angle of triangulation, $\sin u_{\mathrm{obs}}$ is the aperture of observation, λ is the wavelength of light, and c is the speckle contrast. The speckle contrast is unity for laser illumination. We have to emphasize that it is not the monochromaticity that causes speckle. It is the spatial coherence. And strong spatial coherence is always present, if the aperture of the illumination u_{ill} is smaller than the aperture of observation. With a small light source we can achieve high contrast speckles, even if the source emits white light!

Hence, Equation 1.2 is valid for phase measuring triangulation as well, we just have to use the correct speckle contrast, which is smaller than unity for properly designed phase measuring triangulation (pmt) systems [301].

Equation 1.2 introduces a physical lower limit of the measuring uncertainty of triangulation sensors. To explain this let us measure a macroscopically planar ground glass with a surface roughness of 1 micrometer. We will use a sensor with an aperture of observation of $\frac{1}{100}$, which is realistic for a macroscopic object, an angle of triangulation of 20 degrees, and a wavelength of 0.8 microns, from laser illumination. Then we will find a standard deviation of the measured distance of about 37 microns, which is much larger than the surface roughness. Such a large statistical error is not acceptable, for many applications.

Can we somehow overcome the limitation? Before we start thinking about such a task we should be aware that we fight a deep physical principle: it turns out that the measuring uncertainty δz_m can be calculated by Heisenberg's principle. From the uncertainty δp_z of the photon impulse p_z, in z-direction, introduced by the aperture of observation, we get an uncertainty δz_m of the measured distance z_m. This is a common physical fact, so far. It surprises however, that we have an uncertainty as if we measure with only one single photon—although in fact we measure with billions of photons [302]. Obviously, we cannot profit from statistical averaging over many photons. The deep reason is that each photon is in the same quantum mechanical state, because of the spatial coherence. This is discouraging, because we probably cannot overcome quantum mechanics.

Triangulation beyond the Coherent Limit. Nature does never give presents, but we sometimes can buy something. How can we pay off nature? We have to destroy spatial coherence! For a point sensor this can only be done at the object surface. Figure 1.5 displays the result of an experiment that proves the importance of spatial coherence for the distance uncertainty.

A different method to destroy spatial coherence is to heat the surface up and make it thermally radiating. This happens in laser material processing. We make use of the thermal radiation from the laser induced plasma, to measure the material wear on line, with very low aperture, through the laser beam, with an uncertainty of less than 5 microns [301].

Since the two possibilities above are not generally applicable, the question arises if we can reduce spatial coherence by illumination with a large source. This can principally be done, for phase measuring triangulation. However, for practical reasons, the size of the illumination aperture can not be much larger than that of the observation aperture. Hence, there will always be a residual speckle contrast of $c = 0.1$ or more. Introducing this into Equation 1.2, we will get a reduced measuring uncertainty [301]. It should be mentioned that triangulation is the inherent principle of so called focus sensing as well. This principle is, for example used in the CD player or in the

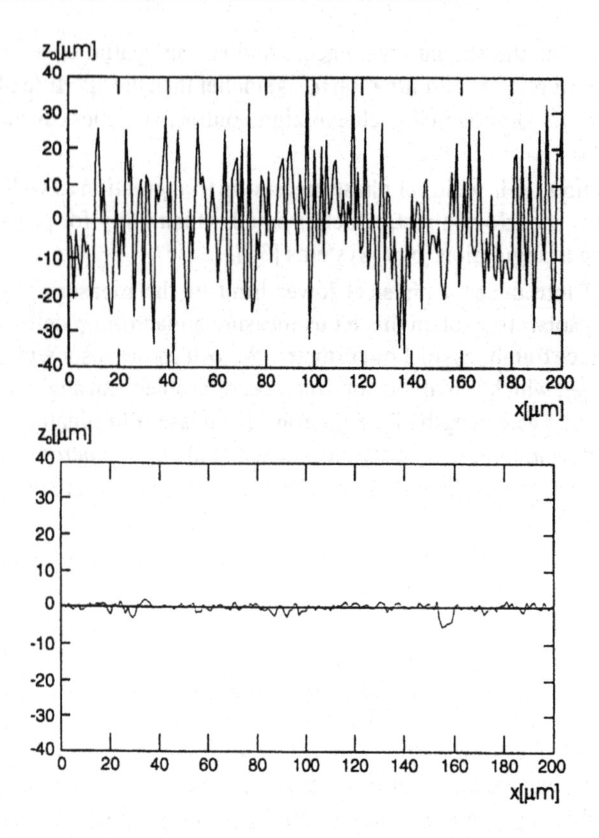

Figure 1.5. Upper: observed image spot from a rough surface, measured with spatially coherent triangulation (laser illumination); lower: the same object measured in fluorescent light. The surface was covered with a very thin fluorescent film. Since fluorescence is perfectly incoherent, the noise is dramatically reduced. This experiment proves the big role of spatial coherence as a limiting factor in triangulation.

confocal scanning microscope [759]. Since the illumination and the observation are implemented through the same aperture u, in these systems Equation 1.2 degenerates to

$$\delta z_{\mathrm{m}} = \frac{c \cdot \lambda}{2\pi \sin u_{\mathrm{obs}}^2}, \tag{1.3}$$

which is, neglecting the factor $c/(2\pi)$, the classical *Rayleigh depth of focus.*

More Triangulation Drawbacks. We have seen that triangulation systems of different kinds cannot overcome the coherent noise limit given by Equation 1.2. This requires some additional remarks: triangulation usually does not even reach the physical limit on real technical surfaces, because the microtopography of the milling or turning process causes errors much larger than that of nice ground surfaces. In Figure 1.6 we see the measuring uncertainty for laser triangulation and different angles

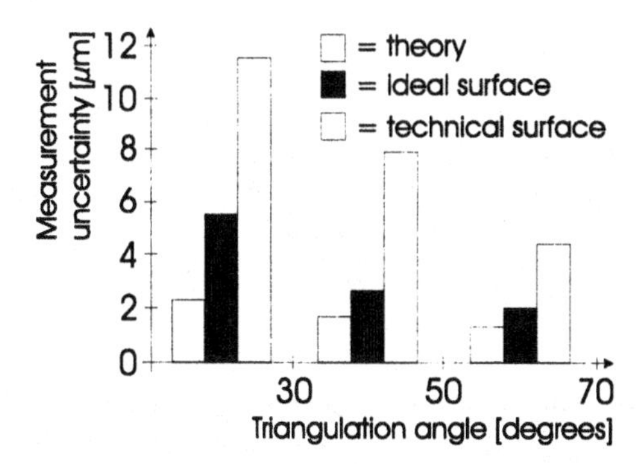

Figure 1.6. Measuring uncertainties for laser triangulation with different angles of triangulation. grey bar: theoretical value according to Equation 1.2, black bar: ideal surface (gypsum), white bar: technical surface (scraped and hardened).

of triangulation: the theoretical value, the experimental value achieved on an 'ideal' surface (gypsum) and on a technical surface (scraped and hardened).

The reason is again the sensitivity of triangulation against shape alterations of the spot image. For real triangulation sensors that can measure macroscopic objects, it turns out that in practice, we cannot get a better uncertainty than about 5 microns—the micrometer regime is until now, the domain of coordinate measuring machines (CMM), that utilize a mechanical probe (which does not suffer from coherent noise). Although CMM technology is slow and touches the object, it is still the proper technology to measure large objects with about 1 micrometer uncertainty.

A further drawback is that in triangulation, illumination and observation are not coaxial. Hence, we cannot avoid shading: some parts of the object are either not illuminated or cannot be seen by the observation system. In spite of these drawbacks, 3D laser scanners are used for many applications.

A 'Real-Time' 3D Camera with Phase Measuring Triangulation. In spite of the drawbacks discussed above, the reduced speckle noise, together with its technical and conceptual robustness, made *phase measuring triangulation* technically and commercially successful, during the last years. Some of the commercially available systems are described in [176]. We will describe in this section one more system, developed in our group. We call it 'real-time 3D camera', because it has the potential to supply 3D data within one video cycle (40 ms, in CCIR standard), or even faster. The system is described in [298].

There are two basic tricks that make the system accurate and fast. A major difficulty for pmt systems is to project a perfect sinusoidal pattern onto the object. Usually, Ronchi gratings are projected and the sinus will be 'somehow' created by defocusing, thus creating higher harmonics on the measured shape. However, it is possible to project a perfect sinusoidal pattern even with a binary mask, by an astigmatic projection lens system [265, 301], see Figure 1.7.

High accuracy needs a precise phase shift between the (usually 4 or 5) exposures. This requires expensive mechanical stages or an LCD projection. The system devel-

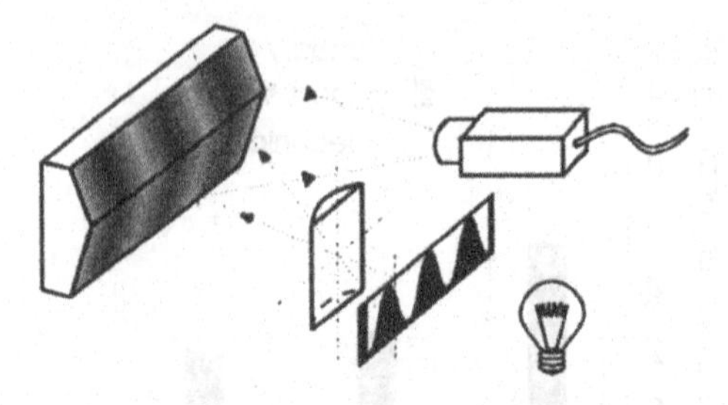

Figure 1.7. Principle of astigmatic projection for phase measuring triangulation.

oped in Erlangen does not use mechanical shift, but an addressable ferroelectric liquid crystal display (FLC). As illustrated in Figure 1.8, with only 4 electrodes we can generate four perfect sinusoidal patterns, each with the proper phase shift. The patterns can be switched within microseconds, due to the fast FLC. The '3D real-time camera' that is realized by these tricks, is shown in Figure 1.9. Phase measuring triangulation systems can easily be scaled, to measure within the mouth, or to measure the hole body

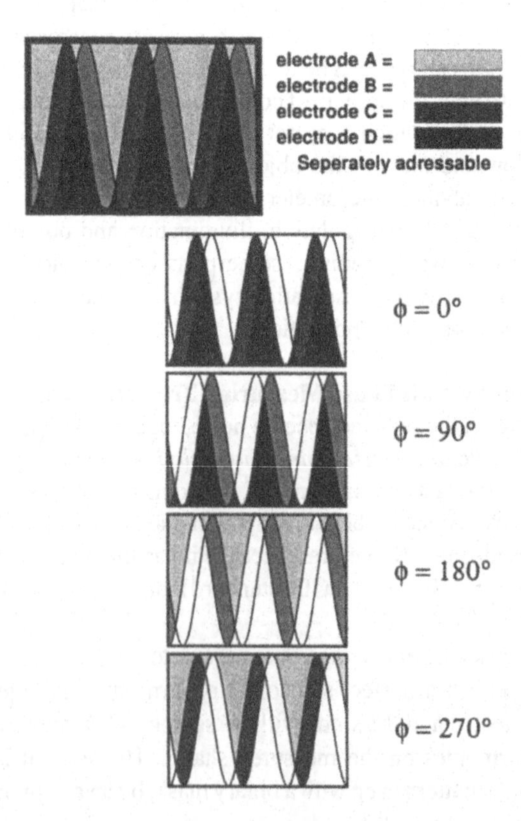

Figure 1.8. With only four electrodes interlaced like a mosaic, we can generate four sinusoidal patterns with a mutual phase shift of 90°.

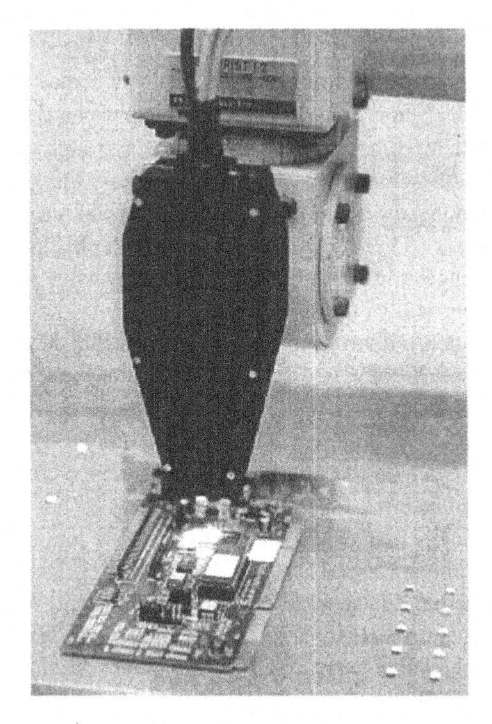

Figure 1.9. 'Real-time 3D camera' for inspection of electronic boards.

of a person. Our system is well adapted to measure people (see Figure 1.10) or being applied in a production line, because it is quite fast.

It should be noted that in case of uncooperative objects, if the camera signal-to-noise ratio is low, sinusoidal patterns are not optimal. For these applications, binary patterns are more appropriate, as described by Malz [466].

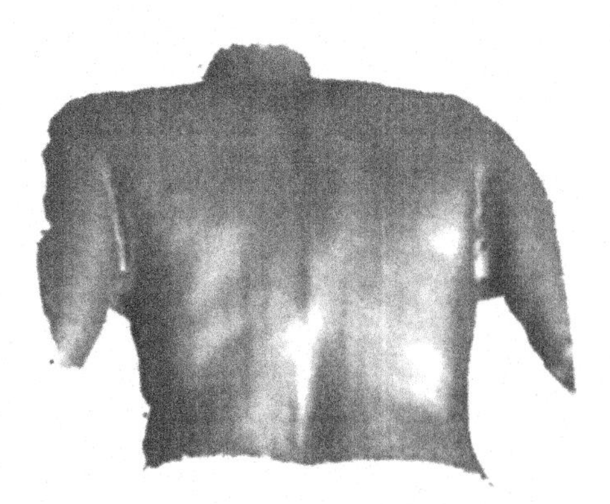

Figure 1.10. Human body, measured by a 'real-time 3D camera'.

1.2.4 White Light Interferometry on Rough Surfaces

Triangulation systems are simple and often useful. But their measuring uncertainty is limited to, say, more than 5 microns, thus leaving a gap to interferometry which is the most precise optical method, but which cannot measure rough objects. Are there other, more advantageous 3D sensing principles that can close this gap?

With classical interferometry we can measure height variations of a mirror, in the range of nanometers, or less. This is possible since smooth objects do not introduce speckle noise. This is suggesting the question, what happens, if we try to measure rough surfaces interferometrically?

Until recently, rough surface interferometry was not possible, because the speckles in the image plane of the interferometer display arbitrary phase, the phase within each speckle independent from the phase in other speckles [250]. So we cannot see any fringes, if we replace one mirror in a Michelson interferometer by the rough object. And it is even useless to evaluate the phase of the interference contrast within each speckle. There is no useful information within that phase

There are two major ideas that yet enable us to measure rough surfaces with an uncertainty in the 1 micrometer regime [189].

The first idea is that within one speckle, the phase is constant, so we can generate interference contrast, in each speckle separately, if we only can generate speckles. And we can do that, as described above, by a sufficiently small aperture of illumination, even with a white, extended light source.

In the second, broad band illumination is used to exploit the limited coherence length of the light. It turns out that interference contrast can be observed only within those speckles that satisfy the equal path length condition: the path length in the object arm of the interferometer has to be approximately the same than that in the reference arm. For a certain object position on the z-axis, we will see interference contrast at one certain contour line of equal distance (or 'height') as shown in Figure 1.1. To acquire the shape of the object, we have to scan the distance (along the z-axis, see Figure 1.11). It should be noted that already Michelson himself used white light interferometry to

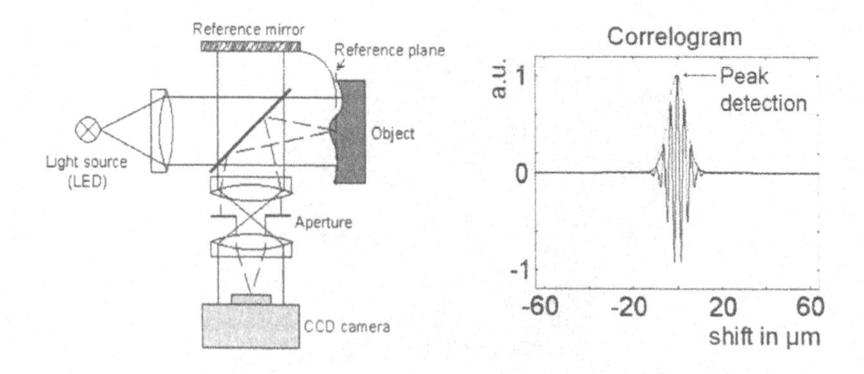

Figure 1.11. Principle of the 'coherence radar' (left). The correlogram shows the (temporal) interference pattern in one single speckle while scanning the object along the z-axis (right).

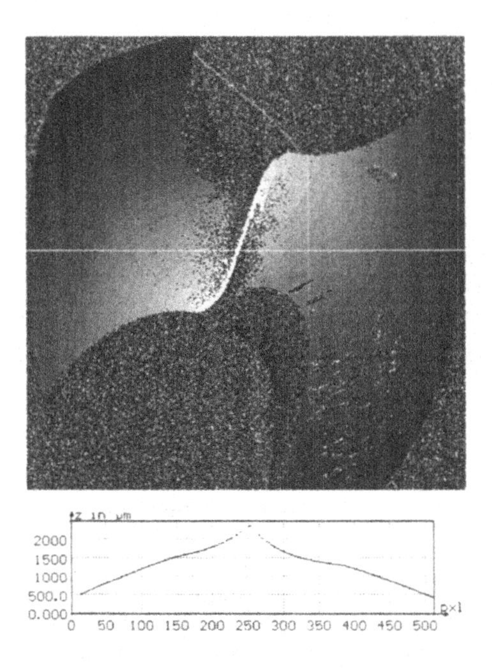

Figure 1.12. Depth map of a drill, seen from the top, and cross section.

measure the length of the (polished!) meter standard. The new ideas enable us to measure optically rough surfaces as well.

While scanning through the z-axis, each pixel of our observation system displays a modulated periodical time signal, which is called 'correlogram'. It is displayed in Figure 1.11, as well. The length of this correlogram signal is about the coherence length, and the time of occurrence, or the position $z_{\mathrm{m}}(x, y)$ of the scanning device at that time, is individual for each pixel: the correlogram has its maximum modulation, if the equal path length condition is satisfied. We store z_{m} for each pixel separately and find the shape of the surface.

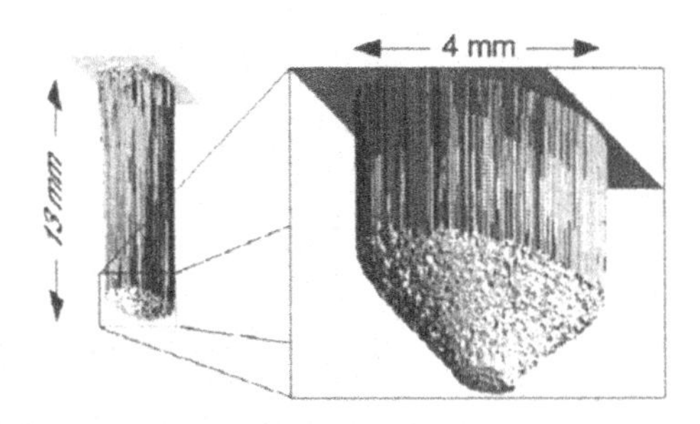

Figure 1.13. Since the measuring uncertainty of the coherence radar does not depend on the observation aperture, we can measure within deep boreholes, with about $1\,\mu$m accuracy.

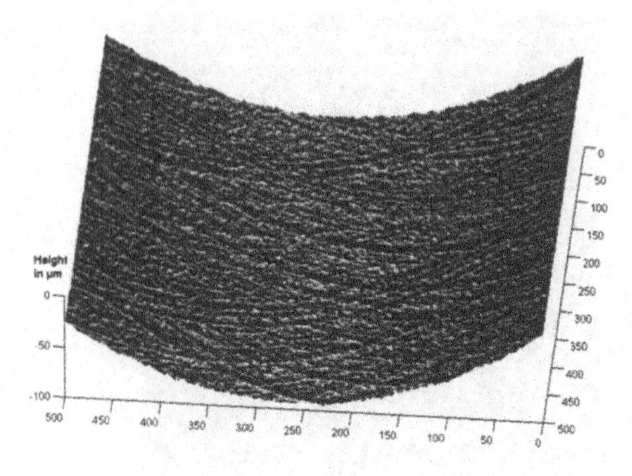

Figure 1.14. Honed cylinder surface.

White light interferometry on rough surfaces, as it is realized in the '*coherence radar*', is extremely powerful. Some measuring examples are displayed in Figures 1.12–1.18. There are unique features, which will be summarized and illustrated by measuring examples:

- The coherence radar is a coaxial method: illumination and observation can be on the same axis. No shading occurs.
- The coherence radar is inherently telecentric, independently from the size of the object. All depths are imaged with the same scale.
- The distance measuring uncertainty on rough surfaces is not given by the apparatus and not limited by the observation aperture. It is only given by the roughness of the surface itself.

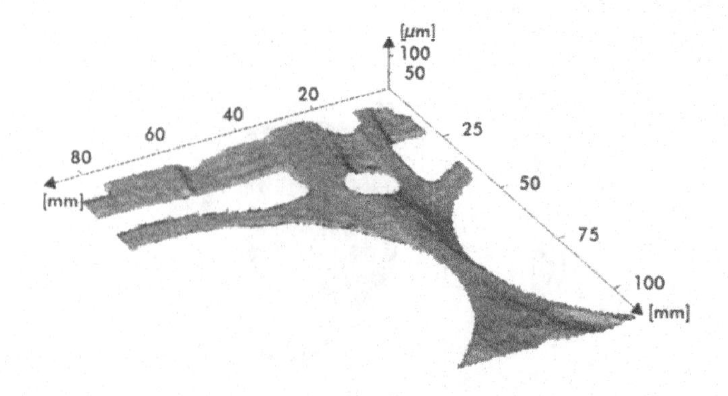

Figure 1.15. Motorblock, measured by large field coherence radar. With the same method we can measure strongly curved polished surfaces [294].

This needs some further explanation: let us consider a macroscopically flat ground glass, where we do not resolve the lateral structure. The standard deviation of the ground glass surface is δz_0. The data $z_m(x, y)$ acquired by the coherence radar will display some statistical variation. Theory and experiments [212] show that the arithmetic value of the magnitude of these variations are equal to δz_0. This is surprising: We can measure the surface roughness although we cannot laterally resolve the microtopography. This remarkable feature is completely different from all other systems for roughness measurements [296].

Since the measuring uncertainty is independent from the aperture, it is independent from the stand off as well. Hence, we can measure distant objects with the same longitudinal accuracy than close objects. In particular, we can measure within deep boreholes, without loss of accuracy (see Figure 1.13).

There are modifications of the coherence radar principle that allow the measurement of objects much larger than the interferometer beam splitter. This is possible by replacing the reference mirror by a groundglass, and utilize divergent illumination in both arms [294].

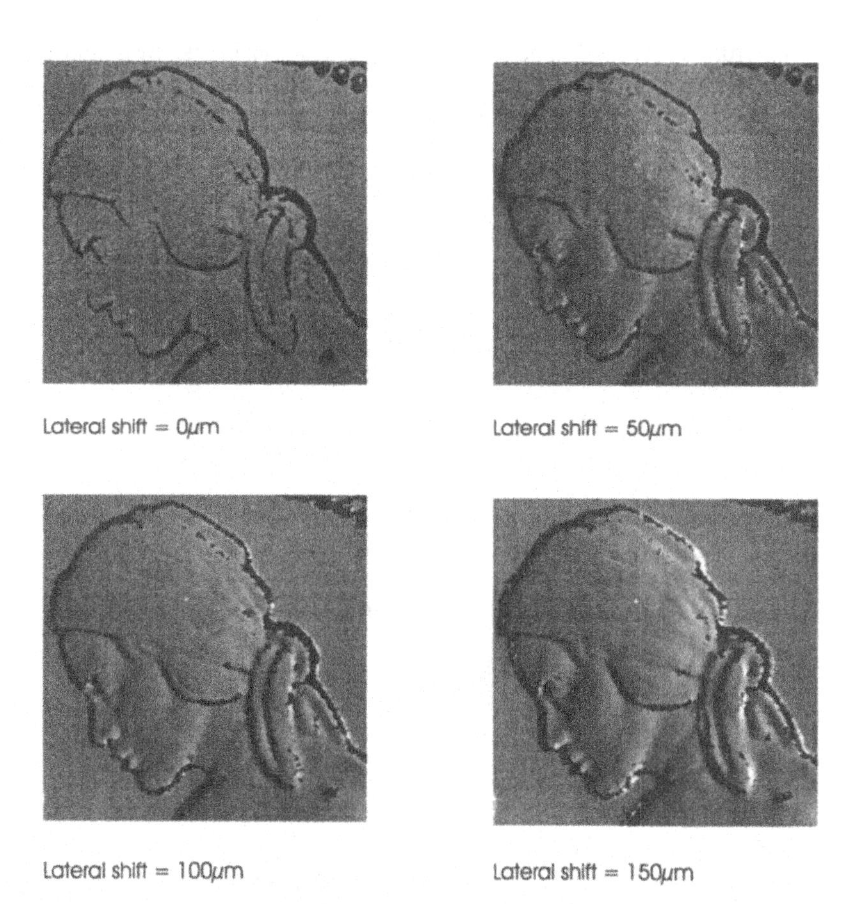

Figure 1.16. Shape difference of two coins (of different age).

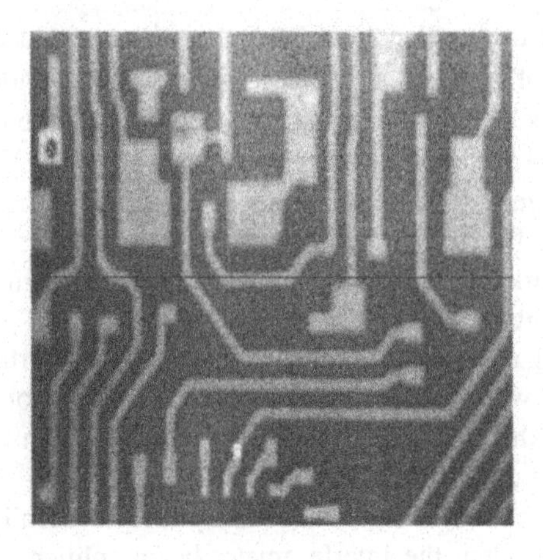

Figure 1.17. Depth map of a hybrid circuit (metal on ceramics), measured by coherence radar. The metal lines have a height of 8 μm.

We can perform the shape comparison of an object against a master, by putting both object and master into the arms of the interferometer. The coherence radar supplies only the difference of the shapes, provided master and object are well adjusted.

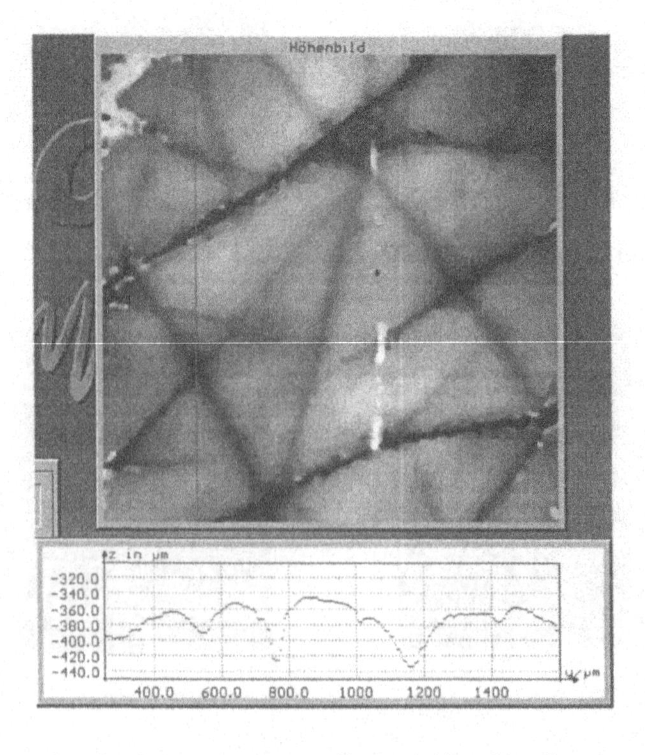

Figure 1.18. In vivo measurement of human skin, with coherence radar.

We can even measure very small deformations in the nanometer regime, provided, there is no decorrelation of the speckle patterns before and after deformation. In case of very large deformation, speckles will be decorrelated. But we still can determine such a deformation absolutely, with a standard deviation slightly larger than the surface roughness [294]. These features cannot be delivered by standard *speckle interferometry* or *ESPI*.

One more feature that cannot be achieved by triangulation, is the ability of the coherence radar to measure translucent objects such as ceramics, paint or even skin. The reason is that we measure essentially the time of flight (with the reference wave as a clock). So we can distinguish the light scattered from the surface, from the light scattered from the bulk of the object.

1.2.5 Summary of the Potentials and Limitations of Optical 3D Sensors

Until now we discussed triangulation and white light interferometry, in terms of potentials and physical limitations and the consequences for application. Of course, it is not only the physical limitations that are important for real problems. There are other conditions such as measuring time, size of the sensor, standoff, costs, availability, sensitivity against temperature, etc. Although these features cannot be discussed here, this section will summarize the results of [298] with the goal to give the reader some feeling, how to select a proper sensor for a given problem.

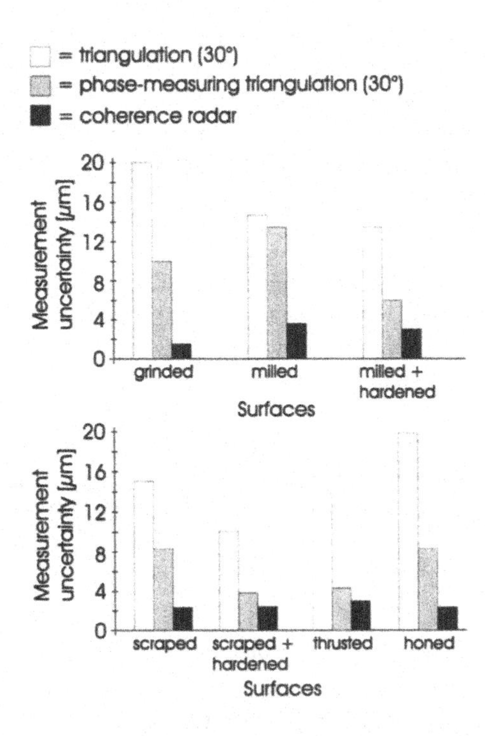

Figure 1.19. Measuring uncertainty, achieved for different sensors and different surface structure.

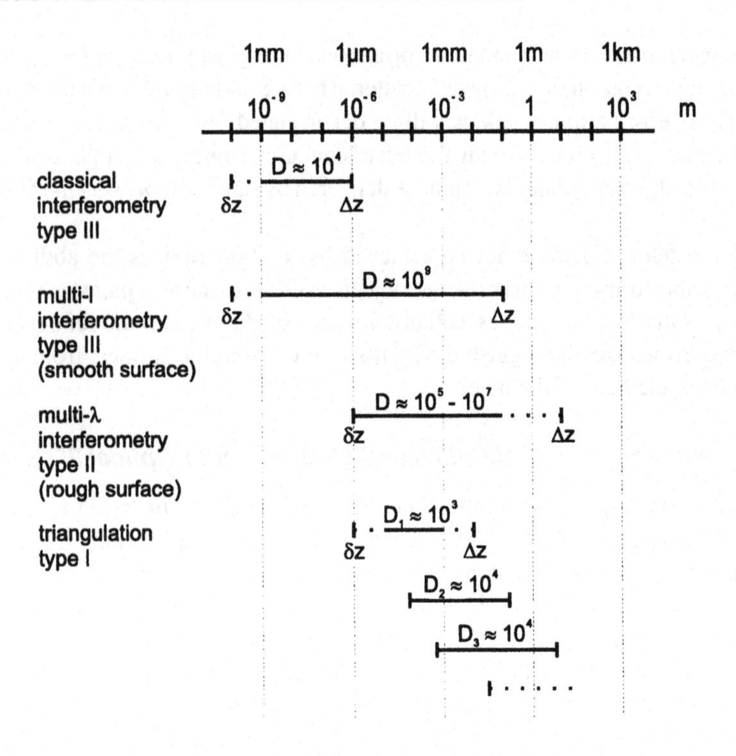

Figure 1.20. The diagram displays the the achievable measuring uncertainty δz for different sensor principles (triangulation (type I), coherence radar (type II), and classical white light interferometry on smooth surfaces (type III)). Δz means the measuring range, D means the dynamical range ($D = \frac{\Delta z}{\delta z}$). It should be noted that the gap between interferometry on smooth surfaces and triangulation, between 1 micron and 5 microns, can be closed by white light rough surface interferometry.

Triangulation and white light interferometry at rough surfaces are based on completely different physical principles. This can be expressed by the fact that the measuring uncertainty δz_{m} scales differently with the standoff distance z_{off}. For triangulation, δz_{m} scales with z_{off}^2. We call this 'type I'. For the coherence radar, δz does not scale at all with the standoff.

It remains to mention that classical interferometry at smooth surfaces has a third type (type III) of scaling behavior [298]. It features optical averaging over the microtopography: δz_{m} is proportional to the inverse off the standoff:

- **type I:** $\delta z_{\text{m}} \sim z_{\text{off}}^2$
- **type II:** δz_{m} independent from z_{off} (coherence radar)
- **type III:** $\delta z_{\text{m}} \sim z_{\text{off}}^{-1}$ (classical interferometry)

We can compress the results in the diagrams of Figures 1.19 and 1.20.

1.2.6 Conclusion

Designing good optical sensors needs understanding of the physical limits. Properly designed, optical 3D sensors supply accurate data about the geometrical shape of objects, as accurate as allowed by physics. The dynamical range allows to distinguish 1,000 – 10,000 different depths. Two main sensor principles, active triangulation and white light interferometry at rough surfaces ('coherence radar') can measure a majority of objects with different surface structure. Once acquired, the geometrical shape can extremely well complement intelligent algorithms, to solve inspection problems, since the algorithms do not have to care about the variable appearance of objects, as necessary in 2D image processing.

Acknowledgement

This section condensed the results of many years of research and of many people, whom I owe to say thank you, for valuable discussion, experiments, software support and new ideas. Those not mentioned in the bibliography I ask to apologize.

I gratefully have to mention, that nearly all projects from our group were funded by the DFG, the BMBF, and the Bayerische Forschungsstiftung, as well as by many companies.

2 MULTIPLE VIEWS AND IMAGE SEQUENCES

E. STEINBACH

INTRODUCTION

Vision is considered to be the most powerful sense that human beings possess. It provides us with a considerable amount of information about our environment and enables us to navigate in the three-dimensional world without physical contact. The two images captured in our visual system produce disparities on the retina which are exploited for 3D measurements of the surroundings. This allows us to learn the position of objects and the relationships between them, even under varying illumination conditions. Taking into account this remarkable performance of the human visual system and the processing unit behind, it is no surprise that technical systems try to imitate the human visual capabilities. The corresponding research area is denoted as *Machine Vision* and has already found implications in a wide variety of technical applications.

This chapter treats several subareas of *Machine Vision* that received considerable attention in the past. The chapter starts with an introduction to the coordinate systems and camera model employed. Three-dimensional motion models as well as the observable projection of 3D motion in the 2D image plane is treated in Section 2.2. The recovery of structure information about the imaged 3D world from two or more camera views is discussed in Section 2.3. Here, the geometrical constraints as well as established solution methods are presented. In Section 2.4 the important problem of object tracking in image sequences is treated. Finally, Section 2.5 presents the real-world problem of illumination estimation that allows a wider applicability of the before-mentioned techniques.

2.1 COORDINATE SYSTEMS AND CAMERA MODELS

B. HEIGL, P. EISERT

In this section we give the theoretical basics of how the position of scene points and cameras as well as rigid transformations and projections are described mathematically.

2.1.1 Rigid Transformation of Coordinates

A cartesian coordinate system in the 3D space is defined by the origin and three orthonormal basis vectors. Suppose that a fixed reference coordinate system is chosen that has three standard unit vectors as a basis. A 3D point in this reference coordinate system is determined by its 3D vector $\boldsymbol{x} = (x, y, z)^T$. Let another coordinate system be identified by the subscript a. The pose of this coordinate system can be described by a translation vector $\boldsymbol{t}_a$ determining the origin, and a rotation matrix $\boldsymbol{R}_a$ consisting of three orthonormal column vectors, which build the new basis: $\boldsymbol{R}_a = (\boldsymbol{r}_{a0}, \boldsymbol{r}_{a1}, \boldsymbol{r}_{a2})$. The matrix $\boldsymbol{R}_a$ is orthonormal, meaning that $\boldsymbol{R}_a^T \boldsymbol{R}_a = \boldsymbol{I}$ and $\det \boldsymbol{R}_a = 1$. If a vector $\boldsymbol{x}_a$ denotes a point given in the coordinate system a, its coordinates can be transformed into the reference coordinate system by

$$\boldsymbol{x} = \boldsymbol{R}_a \boldsymbol{x}_a + \boldsymbol{t}_a \; . \tag{2.1}$$

Vice versa, the coordinates of point $\boldsymbol{x}$ are transformed into the coordinate system a by

$$\boldsymbol{x}_a = \boldsymbol{R}_a^T (\boldsymbol{x} - \boldsymbol{t}_a) \; . \tag{2.2}$$

It should be taken into account that the latter transformation can also be written as $\boldsymbol{x}_a = \widehat{\boldsymbol{R}_a} \boldsymbol{x} + \widehat{\boldsymbol{t}}$, where $\widehat{\boldsymbol{R}_a} = \boldsymbol{R}_a^T$ and $\widehat{\boldsymbol{t}} = -\boldsymbol{R}_a^T \boldsymbol{t}_a$. So, the coordinate system transformation can also be described by $\widehat{\boldsymbol{R}_a}$ and $\widehat{\boldsymbol{t}}_a$, which causes some confusion in the literature.

For two different coordinate systems a and b, a point $\boldsymbol{x}_a$ given in coordinates of system a can be transformed into coordinates of system b by

$$\boldsymbol{x}_b = \boldsymbol{R}_{ab} \boldsymbol{x}_a + \boldsymbol{t}_{ab} \; , \tag{2.3}$$

where $\boldsymbol{R}_{ab} = \boldsymbol{R}_b^T \boldsymbol{R}_a$ and $\boldsymbol{t}_{ab} = \boldsymbol{R}_b^T (\boldsymbol{t}_a - \boldsymbol{t}_b)$.

2.1.2 Parameterization of Rotation Matrices

There are several possibilities to represent the rotation matrix $\boldsymbol{R}$. The most popular parameterization is in terms of Euler angles, where the rotation is composed by three rotations around orthogonal coordinate axes fixed in space. Note, that the order in which the three rotations are applied is essential for the resulting rotation matrix. When rotating first around the z-axis with angle r_z, then around the y-axis (angle r_y), and finally around the x-axis with angle r_x, the following rotation matrix is obtained

$$\boldsymbol{R} = \boldsymbol{R}_x \boldsymbol{R}_y \boldsymbol{R}_z =$$

$$\begin{pmatrix} \cos r_y \cos r_z & -\cos r_y \sin r_z & \sin r_y \\ \sin r_x \sin r_y \cos r_z + \cos r_x \sin r_z & -\sin r_x \sin r_y \sin r_z + \cos r_x \cos r_z & -\sin r_x \cos r_y \\ -\cos r_x \sin r_y \cos r_z + \sin r_x \sin r_z & \cos r_x \sin r_y \sin r_z + \sin r_x \cos r_z & \cos r_x \cos r_y \end{pmatrix} . \tag{2.4}$$

The representation of the rotation with Euler angles, however, has one drawback. For certain configurations one rotational degree of freedom can be lost, the so-called gimbal lock [645] occurs. Consider the case when the object points are first rotated around the x-axis by $-\frac{\pi}{2}$. Then, the y- and the z-axis are aligned and the rotations around the y- and z-axis, respectively, can no longer be distinguished. This is especially unfavorable when interpolating or estimating the 6 motion parameters, since two similar 3D movements can originate from two totally different motion parameter sets.

The gimbal lock can be avoided when using an alternative representation of the rotation matrix. The rotation can be characterized by an arbitrary rotation axis defined by the normalized direction

$$\boldsymbol{n} = [n_x \; n_y \; n_z]^T \quad \text{with} \quad ||\boldsymbol{n}|| = 1 \tag{2.5}$$

and a rotation angle Θ that specifies the amount of rotation around this axis. Note that this description still has three degrees of freedom, two defining the direction of the rotation axis and one for the rotational angle. The corresponding rotation matrix for this case is given by

$$\boldsymbol{R} = \begin{pmatrix} n_x^2+(1-n_x^2)\cos\Theta & n_x n_y(1-\cos\Theta)-n_z\sin\Theta & n_x n_z(1-\cos\Theta)+n_y\sin\Theta \\ n_x n_y(1-\cos\Theta)+n_z\sin\Theta & n_y^2+(1-n_y^2)\cos\Theta & n_y n_z(1-\cos\Theta)-n_x\sin\Theta \\ n_x n_z(1-\cos\Theta)-n_y\sin\Theta & n_y n_z(1-\cos\Theta)+n_x\sin\Theta & n_z^2+(1-n_z^2)\cos\Theta \end{pmatrix}. \tag{2.6}$$

In computer vision, it is often desired to estimate the three degrees of freedom of the rotation. However, this is not an easy problem due to the non-linear relation of the unknown parameters. Therefore, a linearized version of the rotation matrix can be used that is valid under the assumption of small rotation angles ($r_x \ll 1, r_y \ll 1, r_z \ll 1$, or $\Theta \ll 1$). The rotation matrix then reduces to

$$\Delta\boldsymbol{R} \approx \begin{pmatrix} 1 & -r_z & r_y \\ r_z & 1 & -r_x \\ -r_y & r_x & 1 \end{pmatrix} = \begin{pmatrix} 1 & -n_z\Theta & n_y\Theta \\ n_z\Theta & 1 & -n_x\Theta \\ -n_y\Theta & n_x\Theta & 1 \end{pmatrix}. \tag{2.7}$$

For this description, the order in which the three rotations, each corresponding to an Euler angle, are applied is no longer important and the gimbal lock is avoided since the rotation angles are relatively small. In the following, the Euler representation is used. For the linearized rotation matrix, the parameters of one representation can easily be transformed into the other representation using

$$\begin{aligned} \Theta &= \sqrt{r_x^2 + r_y^2 + r_z^2} \\ n_x &= \frac{r_x}{\Theta} \\ n_y &= \frac{r_y}{\Theta} \\ n_z &= \frac{r_z}{\Theta}. \end{aligned} \tag{2.8}$$

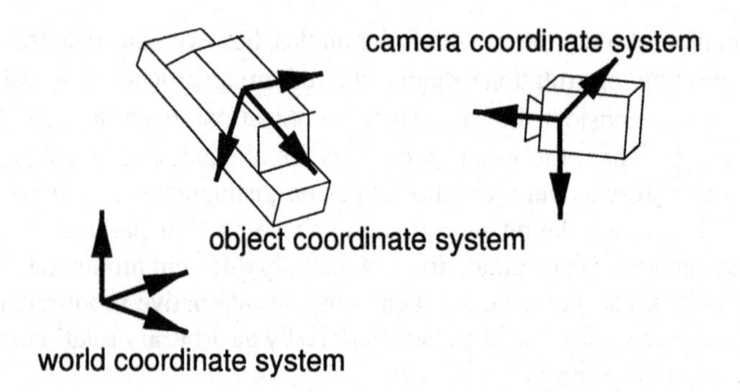

Figure 2.1. Example for the different types of coordinate systems.

2.1.3 Types of Coordinate Systems

In computer vision, often three types of coordinate systems are distinguished for describing a scene with moving objects and cameras:

World coordinate system: it describes a fixed reference frame to which all other coordinate systems and fixed scene points are referred to.

Object coordinate system: it is connected to a given object describing all object points relative to the position of the object. If a rigid object moves, all points belonging to the object are constant with respect to this coordinate system.

Camera coordinate system: it is connected to a given, possibly moving, camera and determines its position and orientation.

Figure 2.1 shows an example for one constellation containing all these types of coordinate systems.

2.1.4 Camera Models

In this section it is described how the projection properties of real cameras can be described mathematically.

Each 3D point is projected along viewing rays onto the image plane. Depending on the chosen camera model, these viewing rays are parallel (parallel projection, or orthographic projection) or they all intersect in a single point, the optical center. The latter projection model is called perspective projection. For the orthographic projection, the image coordinates $\boldsymbol{x} = (x, y)^T$ are directly determined from the 3D object point $\boldsymbol{p} = (x_p, y_p, z_p)^T$ in camera coordinates according to

$$\begin{aligned} x &= x_p \\ y &= y_p. \end{aligned} \tag{2.9}$$

Note, that the depth of the object point has no influence on the resulting image coordinate. All rays coming from the 3D object are assumed to be parallel which is a good approximation for cameras with a large focal length corresponding to a small viewing angle. Due to the linear mapping of the coordinates in Equation 2.9, the use of this projection model leads to simpler algorithms for the estimation of motion parameters from 2D images.

The perspective projection uses a non-linear mapping between 3D and 2D coordinates given by

$$\begin{aligned} x &= f\frac{x_p}{z_p} \\ y &= f\frac{y_p}{z_p} \end{aligned} \tag{2.10}$$

and can model the perspective distortions due to depth variations.

In this chapter we sometimes use homogeneous coordinates to represent image coordinates. The 2D vector $\boldsymbol{x} = (x, y)^T$ is substituted by the vector $\tilde{\boldsymbol{x}} = (\tilde{x}, \tilde{y}, \tilde{w})^T$ where $x = \tilde{x}/\tilde{w}$ and $y = \tilde{y}/\tilde{w}$.

When working with digital images it is more natural to give the position of an image point in *pixel coordinates* $\boldsymbol{n} = (n_x, n_y)^T$. An upper triangular *calibration matrix* $\boldsymbol{K}$ applies the transformation from homogeneous image coordinates $\tilde{\boldsymbol{x}}$ to homogeneous pixel coordinates $\tilde{\boldsymbol{n}}$

$$K = \begin{pmatrix} k_x & s & u \\ 0 & k_y & v \\ 0 & 0 & 1 \end{pmatrix}. \tag{2.11}$$

The values u and v are offsets in pixels of the origin of the image plane, and s indicates the skew of pixels. It is normally assumed to be 0. The values k_x and k_y are the scaled horizontal and vertical focal lengths. They can be expressed in terms of focal length f of the lens and the width d_x and height d_y of each sensor element: $k_x = f/d_x$, $k_y = f/d_y$. The total perspective projection equation can now be written as

$$\rho\tilde{\boldsymbol{n}} = \boldsymbol{K}\boldsymbol{p}. \tag{2.12}$$

The value of the unknown scalar ρ does not affect the 2D image coordinates.

2.1.5 Camera Calibration

The transformation from world to pixel coordinates in Equation 2.12 assumes knowledge about the internal geometric and optical characteristics of the camera (e.g., k_x, k_y). In addition to these two parameters which describe the transformation from metric into pixel coordinates in real images typically a number of real optics distortions are visible which are not covered by the simple pinhole camera perspective projection model. The process of determining the internal parameters and optical characteristics together with the 3D position and orientation of the camera with respect to some reference object coordinate system is typically referred to as *camera calibration*. A standard camera calibration technique that uses a reference object is described in [709]. Figure 2.2 shows an example of a calibration object. The known 3D points are searched in the 2D view and the 3D to 2D point correspondences are used to estimate the internal parameters of the camera which describe the imaging geometry of the camera.

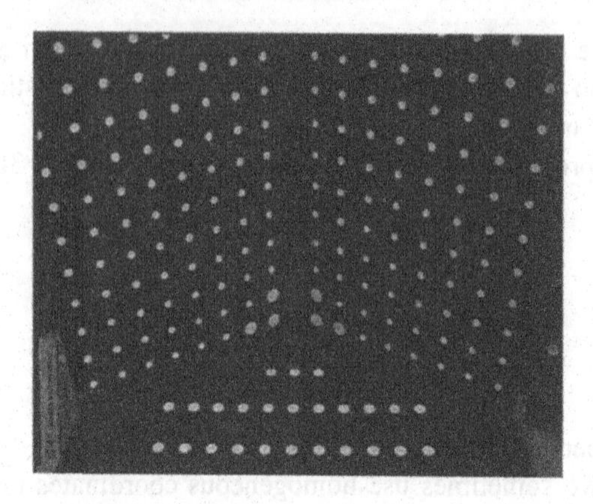

Figure 2.2. Calibration object with ground truth 3D points.

2.2 MOTION MODELS

P. EISERT

This section covers motion and deformation models for the description of the temporal evolution of three-dimensional scenes. In general, the motion characteristics of arbitrary objects can be pretty complex. However, the underlying physical laws also restrict the possible motion or deformation. For a large subclass of objects, the motion and deformation conditions can therefore be represented or at least approximated by a limited set of parameters. This is for example important in computer vision, where the motion and deformation of objects are estimated from multiple images obtained by projecting the real 3D world into the 2D image plane of a camera. To make this ill-conditioned problem mathematically tractable, motion models of the objects are needed that restrict the number of degrees of freedom.

In the following, the rigid body motion is described first which is one of the simplest 3D motion models. Having in mind that the models are used in combination with measurements at discrete time instants, equations for the 3D object point displacements between two time instants are presented. Since many objects in the real world do not follow the rigid body constraint but can also be deformed, the motion model is then extended to describe flexible objects. When viewing a three-dimensional scene with a camera, both the motion of the objects in the scene and the motion of the camera lead to a motion field in the image plane [690, 516]. The 3D and the 2D displacements are coupled via the projection of the camera. In this section, we show the 2D motion fields in the image plane for several types of objects and projections.

2.2.1 Rigid Body Motion

The rigid body motion of an object can be described by a rotation and a translation that is applied to all object points according to the affine transform

$$\boldsymbol{x}_2 = \boldsymbol{R}_{12}(\boldsymbol{x}_1 - \boldsymbol{c}) + \boldsymbol{c} + \boldsymbol{t}_{12}. \tag{2.13}$$

An object point $\boldsymbol{x}_1$ at time t_1 is rotated around an arbitrary center $\boldsymbol{c}$ and translated by $\boldsymbol{t}_{12}$ leading to the object point location $\boldsymbol{x}_2$ at the next time instant t_2. In the rigid body case, the orthonormal rotation matrix $\boldsymbol{R}_{12}$ in Equation 2.4 or Equation 2.6, and the translation vector $\boldsymbol{t}_{12}$ are the same for all object points undergoing the same motion. Six degrees of freedom (three rotational and three translational) are sufficient to describe the rigid body motion.

2.2.2 Deformable Bodies

The rigid body model is well suited to describe the motion of simple objects or the relation between a moving camera and a static scene. However, many real world objects, e.g., human beings or animals, show a more complicated motion characteristic due to deformations of their flexible surface. For the description of such objects, additional degrees of freedom are required. In this context, we focus on parametric deformations [43, 691], where the deformation of an object is represented by a small number of parameters. A small parameter set is of special interest if the current state of an object, given by its pose, velocity, and deformation, should robustly be estimated from sensor data.

Since the deformation of an object should be independent from the object's pose, the deformation is applied in a local object coordinate system. For a rigid body the transformation between the local coordinates $\boldsymbol{x}_0$ specifying the shape of the object and world coordinates $\boldsymbol{x}_1$ in Equation 2.1 corresponding to the current location and pose at time t_1 is given as

$$\boldsymbol{x}_1 = \boldsymbol{R}_1 \boldsymbol{x}_0 + \boldsymbol{c}. \tag{2.14}$$

The local 3D surface points $\boldsymbol{x}_0$ are rotated around the object origin and translated by the vector $\boldsymbol{c}$. Since the rotation defined by $\boldsymbol{R}$ can take on arbitrary values, the linearized version of the rotation matrix as shown in Equation 2.7 cannot be used in this case.

To model a flexible object, we start from a reference shape given by a set of object points $\boldsymbol{x}_0$ and deform it by adding a 3D displacement $\boldsymbol{d}$ to all these points [691]

$$\boldsymbol{x}_1 = \boldsymbol{R}_1(\boldsymbol{x}_0 + \boldsymbol{d}) + \boldsymbol{c}. \tag{2.15}$$

The displacement $\boldsymbol{d}$ is here computed from N deformation parameters $q_0, \ldots, q_{N-1}$ that define the current shape modification as

$$\boldsymbol{d} = D(q_0, q_1, \ldots, q_{N-1}). \tag{2.16}$$

The corresponding 3D motion model between the two time instants t_1 and t_2 can be derived from Equation 2.15 and becomes

$$\boldsymbol{x}_2 = \boldsymbol{R}_{12}(\boldsymbol{x}_1 - \boldsymbol{c}) + \boldsymbol{R}_1 \Delta \boldsymbol{d} + \boldsymbol{c} + \boldsymbol{t}_{12} \tag{2.17}$$

with $\Delta \boldsymbol{d}$ being the change in shape caused by deformation. The additional term $\boldsymbol{R}_1 \Delta \boldsymbol{d}$ compared to the rigid body motion in Equation 2.13 accounts for the deformation. If the transformation D is linear in q_i, it can be represented as

$$D(q_0, q_1, \ldots, q_{N-1}) = \boldsymbol{Q} \cdot \boldsymbol{b}, \tag{2.18}$$

where $\boldsymbol{b}$ is a vector of basis functions and $\boldsymbol{Q}$ a matrix containing the deformation parameters q_i. If these parameters have an influence on all object points, the deformation is called global deformation whereas local deformations occur in the case a parameter q_i affects only a limited region. In both cases the actual shape of the objects is heavily dependent on the transformation D. In the following, different transformations modeling both global and local deformations are presented.

Global Deformation. In 1984, Barr proposed a scheme for global deformation [43] that uses non-linear functions of the coordinates of the reference shape $\boldsymbol{x}_0 = (x_0, y_0, z_0)^T$ to model the deformation given by the transformation D. In that way it is possible to model global deformation like bending, tapering, and twisting. Global tapering along the z-axis can, for example, be modeled as

$$\boldsymbol{d} = \boldsymbol{Q} \cdot \boldsymbol{b} = \begin{pmatrix} q_0 - 1 & 0 & 0 \\ 0 & q_1 - 1 & 0 \\ 0 & 0 & 0 \end{pmatrix} \cdot \begin{pmatrix} x_0 \\ y_0 \\ z_0 \end{pmatrix}. \tag{2.19}$$

In a similar way, twisting around the z-axis can be described by

$$\boldsymbol{d} = \begin{pmatrix} -x_0(1 + \cos\phi) - y_0 \sin\phi \\ -y_0(1 - \cos\phi) + x_0 \sin\phi \\ 0 \end{pmatrix} \tag{2.20}$$

with ϕ being an arbitrary one-dimensional parameterized function of z_0

$$\phi = f(z_0, q_0, q_1, \ldots, q_{N-1}). \tag{2.21}$$

Using polynomials for the non-linear basis functions [761] leads to a different representation of global deformations. The maximum order of the polynomial determines the number of degrees of freedom for the deformation. For second order polynomials the 3D displacement $\boldsymbol{d}$ can be computed as

$$\boldsymbol{d} = \boldsymbol{Q} \cdot \boldsymbol{b} = \begin{pmatrix} q_{00} & q_{01} & q_{02} & q_{03} & q_{04} & q_{05} \\ q_{10} & q_{11} & q_{12} & q_{13} & q_{14} & q_{15} \\ q_{20} & q_{21} & q_{22} & q_{23} & q_{24} & q_{25} \end{pmatrix} \cdot \begin{pmatrix} x_0^2 \\ y_0^2 \\ z_0^2 \\ x_0 y_0 \\ y_0 z_0 \\ y_0 x_0 \end{pmatrix}. \tag{2.22}$$

In spite of the non-linear basis functions, the 3D displacement is linear in the deformation parameters q_{ij} which simplifies their estimation from image data. Figure 2.3 shows different examples of deformations that are created using the global polynomial deformations model. Similar results can also be obtained from a physical based approach [540].

Local Deformation. Global deformation models show the interesting property that a small number of deformation parameters is often sufficient to determine the complete deformation. However, this model type is restricted to simple deformations like

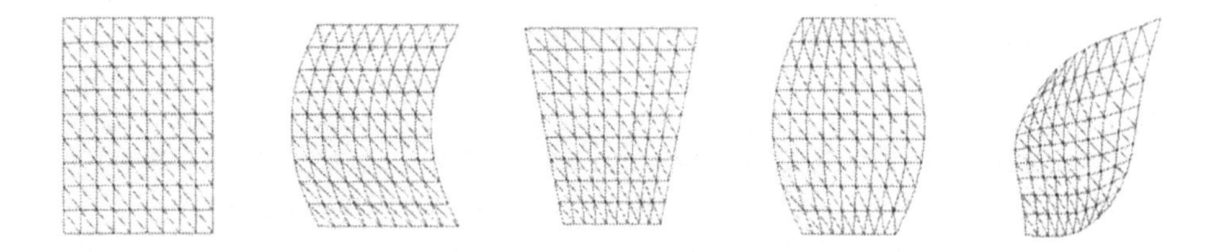

Figure 2.3. Examples of a flexible object which is deformed using global polynomial deformation of third order. From left to right: original object, bending, tapering, pinching, random parameter configuration.

bending, twisting, or pinching. More sophisticated objects often require a deformation characteristic that is locally adapted to the spatially varying properties of the object. Facial expressions of a person, e.g., show locally restricted deformations of the surface since the muscle groups affect only limited regions in the face.

To model such locally changing deformations, the transformation D in Equation 2.16 must also be varied for each reference point $\boldsymbol{x}_0$. In order to restrict the number of degrees of freedom, free form surfaces or splines [227] can be used to model the surface. A set of 3D control points defines the whole shape of the object by adding smoothness constraints to the surface. In contrast to polynomial interpolation, the control points of the spline affect the surface only in a small neighborhood.

The derivation of the local deformation model follows the one of the global deformation by adding 3D displacements to the reference shape

$$\boldsymbol{x}_1 = \boldsymbol{R}_1(\boldsymbol{x}_0 + \boldsymbol{d}) + \boldsymbol{c}. \tag{2.23}$$

The reference points $\boldsymbol{x}_0$ lie on the undeformed spline surface defined by the control points located at a neutral position. The 3D displacement $\boldsymbol{d}$ corresponding to a surface point $\boldsymbol{x}_0$ originates from a change of certain control points. If the relation between a surface point and the control point locations is linear which is, e.g., the case for Bézier- or B-spline surfaces [261] the deformation can be written as

$$\boldsymbol{d}(\boldsymbol{x}_0) = \boldsymbol{Q} \cdot \boldsymbol{b} = \begin{pmatrix} \Delta cp_{x,0} & \Delta cp_{x,1} & \dots & \Delta cp_{x,K-1} \\ \Delta cp_{y,0} & \Delta cp_{y,1} & \dots & \Delta cp_{y,K-1} \\ \Delta cp_{z,0} & \Delta cp_{z,1} & \dots & \Delta cp_{z,K-1} \end{pmatrix} \cdot \begin{pmatrix} b_0(\boldsymbol{x}_0) \\ b_1(\boldsymbol{x}_0) \\ \vdots \\ b_{K-1}(\boldsymbol{x}_0) \end{pmatrix} \tag{2.24}$$

with $\Delta \boldsymbol{cp}_i = (\Delta cp_{x,i}, \Delta cp_{y,i}, \Delta cp_{z,i})^T$ being the change in the control point position of point i. The basis functions $b_i(\boldsymbol{x}_0)$ differ for each point $\boldsymbol{x}_0$ but most of them are equal to zero since a surface point is influenced only by a small number of control points. The actual computation of these basis functions depends on the type of splines. Additional details are, e.g., given in [227].

The concept of using 3D control points to define a 2D surface can easily be extended for the three-dimensional case leading to free form deformations first proposed by Sederberg et al. [628]. The main idea of this approach is to create a cubical bounding box around the reference shape specified by $\boldsymbol{x}_0$. These object points have a unique position in the bounding volume and are moved by displacements $\boldsymbol{d}$ if the surrounding box is deformed. In contrast to free form surfaces the control points of the bounding box are not assigned to the surface of the object but to the volume. They are placed on a cubical grid that is aligned to the cartesian coordinate axes. This 3D lattice of control points defines the complete shape via the basis functions in the same way as the free form surfaces in Equation 2.24. However, the 3D reference points $\boldsymbol{x}_0$ can be located anywhere in the volume and are no longer restricted to lie on a smooth spline surface. To incorporate the 3D structure of the control point grid, the basis functions are computed using trivariate Bernstein polynomials [628] instead of bivariate functions in the case of spline surfaces.

2.2.3 2D Motion Models

Given a projection model (Section 2.1.4), the motion in the image plane can be derived from the motion of the camera and the objects in the scene [179, 690]. Since the aim is to get a description with a small number of motion parameters, assumptions on the projection (perspective or orthographic), the object material (rigid or flexible), and the shape (planar, parabolic, arbitrary) are made. In the following, a 2D motion model, specifying the motion displacements in the image plane between two time instants, is given for each of these classes. First, simple models with a small number of degrees of freedom are presented that are then more and more generalized.

Translational Motion Model. The simplest 2D motion model is a pure translation in the image plane given by

$$\begin{aligned} x_2 &= x_1 + a_0 \\ y_2 &= y_1 + a_1. \end{aligned} \tag{2.25}$$

The 2D displacement between the image points (x_1, y_1) and (x_2, y_2) is constant for all pixels in the image. Such a 2D motion field can be observed under orthographic projection if an arbitrary 3D object is only translated in the 3D space. Observing this 2D motion field under perspective projection is only possible if the object shape is a plane parallel to the image plane in combination with a translation without z-component.

Affine Motion Model. A more complicated 2D motion model is the affine motion model [632, 690]

$$\begin{aligned} x_2 &= a_0 + a_1 x_1 + a_2 y_1 \\ y_2 &= a_3 + a_4 x_1 + a_5 y_1. \end{aligned} \tag{2.26}$$

This model has 6 degrees of freedom and can describe translation, rotation, scaling and shearing in the 2D image plane. The corresponding motion field can be observed under orthographic projection Equation 2.9, if a planar 3D object moves in space. In contrast to the translational motion model both translation and rotation are allowed.

Pure Parameter Model. Adding two more degrees of freedom leads to the pure parameter motion model [710, 179, 690]. This model describes the 2D displacements originating from the perspective projection of the motion trajectories of a plane's 3D surface points. Using the depth information from the plane

$$ax + by + cz = 1 \tag{2.27}$$

and combining it with the motion model given by Equation 2.13 and the perspective projection defined by Equation 2.10 leads to the desired description for the 2D displacements

$$\begin{aligned} x_2 &= \frac{a_0 x_1 + a_1 y_1 + a_2}{a_6 x_1 + a_7 y_1 + 1} \\ y_2 &= \frac{a_3 x_1 + a_4 y_1 + a_5}{a_6 x_1 + a_7 y_1 + 1}. \end{aligned} \tag{2.28}$$

8-Parameter Motion Model. Similar to the pure parameter model, the 8-parameter model has also eight degrees of freedom. However, instead of modeling the displacement field, this model describes the velocity field [690] in the image plane originating from a moving planar patch viewed under perspective projection [677, 448, 632]. The image displacements for this model are given by the following equations which are linear in the model parameters

$$\begin{aligned} x_2 &= a_0 + a_1 x_1 + a_2 y_1 + a_6 x_1^2 + a_7 x_1 y_1 \\ y_2 &= a_3 + a_4 x_1 + a_5 y_1 + a_6 x_1 y_1 + a_7 y_1^2. \end{aligned} \tag{2.29}$$

Bilinear Transformation. Another motion model having 8 degrees of freedom is the bilinear motion model [690]. The 2D displacement field of this model is given by

$$\begin{aligned} x_2 &= a_0 + a_1 x_1 + a_2 y_1 + a_3 x_1 y_1 \\ y_2 &= a_4 + a_5 x_1 + a_6 y_1 + a_7 x_1 y_1. \end{aligned} \tag{2.30}$$

In contrast to the above mentioned models, this description has no direct relation to 3D rigid body motion observed with a specific camera model.

12-Parameter Motion Model. Using all combinations of the image coordinates x_1 and y_1 up to the second order leads to the 12-parameter motion model [179]

$$\begin{aligned} x_2 &= a_0 + a_1 x_1 + a_2 y_1 + a_3 x_1^2 + a_4 y_1^2 + a_5 x_1 y_1 \\ y_2 &= a_6 + a_7 x_1 + a_8 y_1 + a_9 x_1^2 + a_{10} y_1^2 + a_{11} x_1 y_1. \end{aligned} \tag{2.31}$$

In addition to the affine model which can describe the projected motion of a planar object under orthographic projection, this model is able to specify the motion field also for parabolic surfaces.

Rigid Body Motion with Arbitrary Shape. Up to now, the shape of the observed object was constrained to be a plane or a parabolic surface. This is not valid for most real world objects. However, the 2D motion models mentioned so far can be used to approximate the 2D motion of small patches of the 3D object corresponding to small areas in the image plane. The whole displacement field is then obtained by combining several simple motion models each having a well defined set of motion parameters and a region of support.

To describe the complete object motion in the image plane with one set of parameters, the actual shape of the object has to be incorporated to restore the information lost by the projection. In the case of orthographic projection this is straight forward when using the motion Equation 2.13 and the projection (Equation 2.9)

$$\begin{aligned} x_2 &= x_1 + (z(x_1, y_1) - c_z) r_y - (y_1 - c_y) r_z + t_x \\ y_2 &= y_1 - (z(x_1, y_1) - c_z) r_x + (x_1 - c_x) r_z + t_y. \end{aligned} \tag{2.32}$$

To obtain a linear dependency in the motion parameters, the linearized rotation matrix according to Equation 2.7 is used assuming small 3D displacements between two time instants t_1 and t_2. The term $z(x_1, y_1) - c_z$ is related to the object shape and describes the z-component of the 3D surface point corresponding to the pixel at (x_1, y_1). Local object coordinates with the origin located in the object center $\boldsymbol{c} = (c_x, c_y, c_z)^T$ are used to describe this 3D surface point. Note, that a translation component in z-direction cannot be determined from the 2D motion field when orthographic projection is used.

In the case of perspective projection the 2D motion model is no longer linear in the motion parameters which would be desirable to simplify the estimation of the parameter set. However, a linearized model can be derived under the assumption of small motion parameters [448, 500, 738] between two images. This approach makes use of a simplified 3D rigid body motion model with the center of rotation located in the focal point ($\boldsymbol{c} = (0, 0, 0)^T$)

$$\boldsymbol{x}_2 = \boldsymbol{R}_{12} \boldsymbol{x}_1 + \boldsymbol{t}_{12}. \tag{2.33}$$

Projection of this motion model into the image plane using Equation 2.10 and first order approximation leads to the 2D description

$$\begin{aligned} x_2 &= x_1 + x_1 y_1 r_x - (1 + x_1^2) r_y - y_1 r_z - \frac{t_x}{z} - x_1 \frac{t_z}{z} \\ y_2 &= y_1 + (1 + y_1^2) r_x - x_1 y_1 r_y + x_1 r_z - \frac{t_y}{z} - y_1 \frac{t_z}{z} \end{aligned} \tag{2.34}$$

which is linear in the 6 motion parameters r_x, r_y, r_z, t_x, t_y, and t_z.

Since this model is only valid for small motion parameters it is sometimes of advantage to use the motion model given by Equation 2.13. If the object is not rotating around the focal point the translation parameters are much smaller if the information from the rotation center is exploited. Using this motion model we obtain a similar 2D motion description [200, 397]

$$\begin{aligned} x_2 &= x_1 + x_1 (y_1 + \frac{c_y}{z}) r_x - (1 - \frac{c_z}{z} + x_1 (x_1 + \frac{c_x}{z})) r_y - \\ &\quad -(y_1 + \frac{c_y}{z}) r_z - \frac{t_x}{z} - x_1 \frac{t_z}{z} \\ y_2 &= y_1 + (1 - \frac{c_z}{z} + y_1 (y_1 + \frac{c_y}{z})) r_x - y_1 (x_1 + \frac{c_x}{z}) r_y + \\ &\quad +(x_1 + \frac{c_x}{z}) r_z - \frac{t_y}{z} - y_1 \frac{t_z}{z}. \end{aligned} \tag{2.35}$$

Flexible Body Motion with Arbitrary Shape. The more sophisticated motion characteristic of deformable objects is also visible in the displacement field in the 2D image plane. In addition to the rigid body motion parameters that describe rotation and translation, the set of deformation parameters has to be incorporated as well, leading to high parametric 2D motion models of the form

$$\begin{aligned} x_2 &= f(x_1, y_1, z; r_x, r_y, r_z, t_x, t_y, t_z, \Delta q_0, \Delta q_1, \ldots, \Delta q_{N-1}) \\ y_2 &= g(x_1, y_1, z; r_x, r_y, r_z, t_x, t_y, t_z, \Delta q_0, \Delta q_1, \ldots, \Delta q_{N-1}). \end{aligned} \tag{2.36}$$

The functions f and g depend on the actual 3D motion and deformation model and the projection used to model the image formation process. To restrict the complexity of these functions, we focus on 2D models that linearly depend on the motion and deformation parameters. Such models are commonly used, e.g., for the modeling of tissue deformations caused by facial expressions [67, 159, 201, 438].

A comparison of the 3D motion model for rigid bodies as given by Equation 2.13 with the one for flexible bodies in Equation 2.17 shows that the only difference is the additional term $\boldsymbol{R}\Delta\boldsymbol{d}$ accounting for the deformation change. The linear 2D motion models for deformable objects can therefore be derived from the rigid body models by adding the dependency from the deformation parameters.

Assuming orthographic projection and a linear deformation model according to Equation 2.18, the 2D motion model is given by

$$\begin{aligned} x_2 &= x_{2,\text{rigid}} + \sum_{i=0}^{K-1} b_i (r_{11}\Delta q_{1i} + r_{12}\Delta q_{2i} + r_{13}\Delta q_{3i}) \\ y_2 &= y_{2,\text{rigid}} + \sum_{i=0}^{K-1} b_i (r_{21}\Delta q_{1i} + r_{22}\Delta q_{2i} + r_{23}\Delta q_{3i}) \end{aligned} \tag{2.37}$$

with $x_{2,\text{rigid}}$ and $y_{2,\text{rigid}}$ being the corresponding image location of the rigid body model given by Equation 2.32. The parameters r_{mn} are the elements of the rotation matrix $\boldsymbol{R}_1$ in Equation 2.17 and define the current pose of the object.

A similar model is obtained for perspective projection by extending Equation 2.34 with the dependency from the $N = 3K$ deformation parameter changes Δq_{ij}

$$\begin{aligned} x_2 &= x_{2,\text{rigid}} - \sum_{i=0}^{K-1} \frac{b_i}{z} \Big((r_{11} + x_1 r_{31})\Delta q_{1i} + \\ &\quad + (r_{12} + x_1 r_{32})\Delta q_{2i} + (r_{13} + x_1 r_{33})\Delta q_{3i} \Big) \\ y_2 &= y_{2,\text{rigid}} - \sum_{i=0}^{K-1} \frac{b_i}{z} \Big((r_{21} + y_1 r_{31})\Delta q_{1i} + \\ &\quad + (r_{22} + y_1 r_{32})\Delta q_{2i} + (r_{23} + y_1 r_{33})\Delta q_{3i} \Big). \end{aligned} \tag{2.38}$$

This model describes the 2D displacements obtained from the perspective projection of a flexible object moving in the 3D space.

In Section 8.6 this model is used to define the local motion and deformation of a person's skin caused by facial expressions. Exploiting the knowledge of the motion characteristics in the face, parameters defining the mimic are estimated from 2D images. In that way, video sequences showing a talking person can completely be described by these parameters and a 3D model specifying shape and color of the person.

2.3 STRUCTURE FROM MULTIPLE VIEWS

E. STEINBACH, B. HEIGL

A vast variety of applications require the reconstruction of a three-dimensional scene from two or more perspective camera views. Model-based video communication systems, e.g., attempt to extract information about the three-dimensional structure of the scene to be transmitted. Reverse-engineering systems capture a real world object from different viewing positions and fuse the information from all views into a complete 3D description. The control of autonomously moving vehicles requires the reconstruction of the surrounding scene to avoid collisions. Computer-based movie and TV productions reconstruct new views of a scene without additional recording. Augmented reality environments affix virtual objects to real ones in the scene for animation purposes.

Assuming rigid objects, the relative motion between the camera and the objects permits to extract information about the scene structure. This is exploited in *Structure-from-Motion* (SfM) algorithms recovering simultaneously 3D motion parameters and relative depth values. A good overview of the history of SfM and the underlying geometry as well as standard solution techniques can be found in [5, 345, 219].

In the following sections we first present the geometrical constraints that can be derived for two or more views of a scene. Then, standard solution techniques, including linear and non-linear approaches are summarized, followed by brief descriptions of multi-object and multi-view processing. Finally, we show how the relative motion information between two views can be exploited to recover a surface description of the objects under investigation.

2.3.1 Multi-View Relations

A fine introduction to the geometric basics of projective geometry can be found in [487]. In [489], an excellent and detailed description of all following concepts of multi-view relations is given.

Let the pose of a camera with respect to the world coordinate system be given by $\boldsymbol{t}$ and $\boldsymbol{R} = (\boldsymbol{r}_0, \boldsymbol{r}_1, \boldsymbol{r}_2)$ (see Figure 2.4). The vector $\boldsymbol{r}_2$ describes the direction of the so-called optical axis of the camera, which corresponds to its viewing direction. With this parameterization, an arbitrary scene point $\boldsymbol{p}$ given in the world coordinate system is transformed to the camera coordinate system by Equation 2.2.

Using Equation 2.12, the total perspective projection equation can now be written as

$$\rho\tilde{\boldsymbol{n}} = \boldsymbol{K}\boldsymbol{R}^T \left(\boldsymbol{I}_3 \mid -\boldsymbol{t}\right) \tilde{\boldsymbol{p}}, \tag{2.39}$$

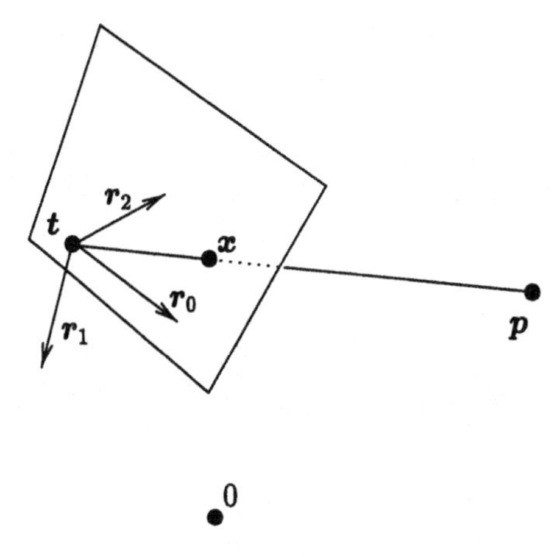

Figure 2.4. Camera geometry and perspective projection.

where $\tilde{p}$ denotes the homogeneous 4-vector corresponding to the 3D point p. Matrix

$$P = KR^T (I_3 \mid -t)$$

contains all camera parameters and is called *projection matrix.*

Epipolar Geometry. The standard epipolar geometry has been introduced and described by [446] and independently by [711]. Figure 2.5 shows the constraints between the projections x_0, x_1 of an unknown scene point p. Please note, that the focal length f is normalized to 1 in the following. If x_0 is known, the scene point p can be situated at any point along the viewing ray of x_0 (see the gray markers in the figure). This means that the projection x_1 of p into the second image has to lie on a line which is determined by intersecting the plane defined by p, t_0 and t_1 with the image plane of the second view. This line is called *epipolar line*. One point of this line can be fixed in

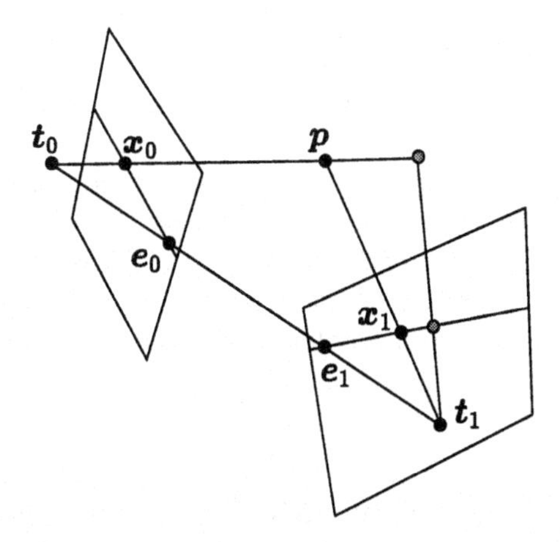

Figure 2.5. Epipolar geometry.

image coordinates by projecting $\boldsymbol{t}_0$ into the second view, yielding the point $\boldsymbol{e}_1$ called *epipole*. Exchanging the two images, similarly an epipole $\boldsymbol{e}_0$ can be defined for the first view.

E-matrix. A formal description of the epipolar geometry results in the essential matrix which contains the unknown motion parameters for two calibrated camera views. It has been introduced by [711].

Having two cameras with parameters $\boldsymbol{t}_0$, $\boldsymbol{R}_0$ and $\boldsymbol{t}_1$, $\boldsymbol{R}_1$, constraints can be defined between the projections $\boldsymbol{x}_0$, $\boldsymbol{x}_1$ of an unknown scene point $\boldsymbol{p}$. Given the known projection $\boldsymbol{x}_0$, the corresponding 3D point $\boldsymbol{p}$ meets the equation

$$\boldsymbol{p} = \boldsymbol{t}_0 + \rho_0 \boldsymbol{R}_0 \tilde{\boldsymbol{x}}. \tag{2.40}$$

Note, that the distance of $\boldsymbol{p}$ is coded in the value ρ_0. Projecting $\boldsymbol{p}$ into the second image, and inserting the upper expression for $\boldsymbol{p}$, we get

$$\rho_1 \tilde{\boldsymbol{x}}_1 = \rho_0 \underbrace{\boldsymbol{R}_1^T \boldsymbol{R}_0}_{\boldsymbol{A}=\Delta \boldsymbol{R}} \tilde{\boldsymbol{x}}_0 + \underbrace{\boldsymbol{R}_1^T (\boldsymbol{t}_0 - \boldsymbol{t}_1)}_{\tilde{e}_1}. \tag{2.41}$$

Therefore, $\tilde{\boldsymbol{x}}_1$ is a linear combination of $\boldsymbol{A}\tilde{\boldsymbol{x}}_0$ and the epipole $\tilde{e}_1$. As known from projective geometry [487], this means that $\boldsymbol{x}_1$ lies on the line containing $\boldsymbol{A}\tilde{\boldsymbol{x}}_0$ and $\tilde{e}_1$, the epipolar line. This line can also be expressed by the 3-vector

$$\boldsymbol{l}_{n_0} = \tilde{e}_1 \times \boldsymbol{A}\tilde{\boldsymbol{x}}_0. \tag{2.42}$$

As a point lies on a line if the scalar product of their homogeneous vectors vanishes, the equation above can also be written as $(\tilde{\boldsymbol{n}}_1)^T \boldsymbol{l}_{n_0} = 0$, or equivalently,

$$\tilde{\boldsymbol{x}}_1^T \underbrace{[\tilde{e}_1]_\times \boldsymbol{A}}_{\boldsymbol{E}} \tilde{\boldsymbol{x}}_0 = 0, \tag{2.43}$$

where $[\tilde{e}_1]_\times$ denotes the antisymmetric matrix performing an outer left multiplication with $\tilde{e}_1$:

$$[\tilde{e}_1]_\times = \begin{pmatrix} 0 & e_3 & -e_2 \\ -e_3 & 0 & e_1 \\ e_2 & -e_1 & 0 \end{pmatrix}. \tag{2.44}$$

Matrix $\boldsymbol{E}$ is called *essential matrix* and describes all motion parameters between the two cameras. The essential matrix can be scaled with an arbitrary factor $\neq 0$ without changing Equation 2.43, also the matrix $[\tilde{e}_1]_\times$ is just determined up to scale. This implies, that if we scale the whole scene by a unique factor, the essential matrix is not affected at all. The translation vector $\boldsymbol{t}_1 - \boldsymbol{t}_0$ is coded in matrix $\boldsymbol{E}$ also just up to scale. In [711] it is proved, that the matrix $\boldsymbol{E}$ determines the camera movement between two views uniquely up to scale. In pixel coordinates, Equation 2.43 can be written as

$$\underbrace{\tilde{\boldsymbol{n}}_1^T \boldsymbol{K}_1^T}_{\tilde{\boldsymbol{x}}_1^T} \boldsymbol{E} \underbrace{\boldsymbol{K}_0 \tilde{\boldsymbol{n}}_0}_{\tilde{\boldsymbol{x}}_0} = 0. \tag{2.45}$$

Matrix $\boldsymbol{E}$ has five degrees of freedom. Since $det([\tilde{e}_1]_\times) = 0$ we can conclude that $det(\boldsymbol{E}) = 0$. Matrix $\boldsymbol{A}$ has full rank, $[\tilde{e}_1]_\times$ is of rank 2; therefore matrix $\boldsymbol{E}$ is of rank 2.

F-matrix. For the case of two uncalibrated camera views a similar, but more general formulation has been introduced in [219, 775]. Here all equations are formulated using pixel coordinates resulting in a mathematical description of epipolar geometry without an explicit use of calibration matrices. Equation 2.41 changes to

$$\rho_1 \tilde{n}_1 = \rho_0 \underbrace{K_1 R_1^T R_0 K_0^{-1}}_{B} \tilde{n}_0 + \underbrace{K_1 R_1^T (t_0 - t_1)}_{\tilde{e}_1}, \tag{2.46}$$

resulting in

$$\tilde{n}_1 \underbrace{[\tilde{e}_1]_\times B}_{F} \tilde{n}_0 = 0. \tag{2.47}$$

Matrix F is called *fundamental matrix* and depends on all calibration parameters of both cameras. F is related to E by $F = K_1^T E K_0$. For the same reason as for the essential matrix, the fundamental matrix is of rank 2 and is defined up to scale. It has seven degrees of freedom.

Trifocal Tensor. Simultaneous processing of three uncalibrated views leads to the trifocal tensor as described in [290, 700]. As for two-view relations, the trifocal tensor can be formulated for the calibrated as well as for the uncalibrated case. We restrict ourselves to the uncalibrated case here. The fundamental trifocal constraint describes the relations of a scene point p in three views. By construction, the constraint is formulated for the projected point n_0 in the first view and 2D lines $l_{1,2}$ in the second and third view containing the projected scene point, respectively. Note, that these lines need not be projections of a unique 3D line. We can express the relation for line l_i containing the projection of scene point p as

$$l_i^T K_i R_i^T (p - t_i) = 0. \tag{2.48}$$

Also regarding the uncalibrated version of Equation 2.40

$$p = t_0 + \rho_0 R_0 K_0^{-1} \tilde{n}_0, \tag{2.49}$$

we get the two equations

$$l_i^T K_i R_i^T (t_0 - t_i) + \rho_0 l_i^T K_i R_i^T R_0 K_0^{-1} \tilde{n}_0 = 0 \quad \text{for} \quad i = 1, 2. \tag{2.50}$$

By eliminating the common scale factor ρ_0 and applying some algebraic transformations the two equations can be combined:

$$l_1^T \cdot T n_0 \cdot l_2 = 0, \tag{2.51}$$

where T is a $3 \times 3 \times 3$ tensor called the *trifocal tensor*. The term $T n_0$ expresses the 3×3 matrix that is built by multiplying each component of n_0 with the corresponding submatrix of T and by summing up these products. Similar to the constraints defined above, the trifocal tensor is defined just up to scale. It has 18 degrees of freedom.

The fundamental trifocal constraint can be transformed to express the relation between three projections of one 3D line:

$$\rho \boldsymbol{l}_0 = \begin{pmatrix} \boldsymbol{l}_1^T \boldsymbol{T}_0 \boldsymbol{l}_2 \\ \boldsymbol{l}_1^T \boldsymbol{T}_1 \boldsymbol{l}_2 \\ \boldsymbol{l}_1^T \boldsymbol{T}_2 \boldsymbol{l}_2 \end{pmatrix}, \tag{2.52}$$

where $\boldsymbol{T}_i$ denotes the i-th submatrix of $\boldsymbol{T}$. Also, the relation between three projections of one 3D point can be expressed:

$$[\boldsymbol{n}_1]_\times [\boldsymbol{T} \boldsymbol{n}_0][\boldsymbol{n}_2]_\times = \boldsymbol{0}. \tag{2.53}$$

More Views. There exist also suggestions to describe the constraints for more than three views. For the case of four views, the concept of the *quadrifocal tensor* is introduced in [705]. It is also proved there, that all relations that exist between image points and lines in five or more views are the epipolar constraints between any pair, the trifocal constraints between any triple and the quadrifocal constraints between any quadruple of views.

2.3.2 Retrieval of Scene and Camera Geometry

The upper concepts describe the constraints for scene projections. In this section we show how the movement of the camera between the recorded images can be calculated from these mathematical terms describing the geometric constraints.

As for each scene the world coordinate system can be chosen arbitrarily, we can suppose without loss of generality, that the world coordinate system coincides with the camera coordinate system of the first of a sequence of camera views. Then, the projection matrix of the first camera is determined by $\boldsymbol{P}_0 = \boldsymbol{K}_0(\boldsymbol{I} \mid \boldsymbol{0})$. Any other projection matrix $\boldsymbol{P}_i$ can be written as $\boldsymbol{P}_i = \boldsymbol{K}_i \boldsymbol{R}_i^T (\boldsymbol{I} \mid -\boldsymbol{t}_i)$. Therefore, if we know the calibration matrices $\boldsymbol{K}_i$ and the projection matrices $\boldsymbol{P}_i$ then we know all about the positions and parameters of each camera. The 3D scene points can be reconstructed from corresponding points by intersecting the viewing rays of each projected point.

In [288] one method is shown how to uniquely calculate the second projection matrix for a pair of cameras with the corresponding essential matrix $\boldsymbol{E}$. The calibration matrices are supposed to be known. One numerical method used is the *singular value decomposition* (SVD) which decomposes each $m \times n$ matrix $\boldsymbol{A}$ with $m \geq n$ into a product of three matrices $\boldsymbol{A} = \boldsymbol{U}\boldsymbol{W}\boldsymbol{V}^T$. The matrix $\boldsymbol{U}$ is a $m \times n$ matrix with orthonormal column vectors, $\boldsymbol{W}$ is a $n \times n$ diagonal matrix, and $\boldsymbol{V}$ is orthonormal with size $n \times n$. If the SVD of $\boldsymbol{E}$ is $\boldsymbol{E} = \boldsymbol{U}\boldsymbol{W}\boldsymbol{V}^T$, there are four solutions for the projection matrix $\boldsymbol{P}_1$:

$$\begin{array}{l} \boldsymbol{K}_1(UZV^T \mid U(0,0,1)^T) \\ \boldsymbol{K}_1(UZV^T \mid -U(0,0,1)^T) \\ \boldsymbol{K}_1(UZ^TV^T \mid U(0,0,1)^T) \\ \boldsymbol{K}_1(UZ^TV^T \mid -U(0,0,1)^T) \end{array} \quad \text{with} \quad \boldsymbol{Z} = \begin{pmatrix} 0 & 1 & 0 \\ -1 & 0 & 0 \\ 0 & 0 & 1 \end{pmatrix}. \tag{2.54}$$

For each solution, each point correspondence can be used to triangulate one 3D scene point $\boldsymbol{p}$. There will be just one solution for $\boldsymbol{P}_1$ such that $\boldsymbol{p}$ will lie in front of both cameras. As the cameras cannot project points lying behind, the true and unique solution of $\boldsymbol{P}_1$ can be recovered. Please note, that knowing the essential matrix and some point correspondences, the position of the cameras as well as the 3D points corresponding to each pair of projected points can be reconstructed up to a unique scale factor.

In the uncalibrated case, when knowing the fundamental matrix $\boldsymbol{F}$ but no calibration matrix $\boldsymbol{K}$, the motion and calibration parameters cannot be determined uniquely as in the calibrated case above. It is well known that from image correspondences alone, the camera projection matrices and the reconstruction of scene points can just be retrieved up to a projective transformation $\boldsymbol{D}$

$$\boldsymbol{P}_i \tilde{\boldsymbol{p}}_i = \boldsymbol{n} = \underbrace{\left(\boldsymbol{P}_i \boldsymbol{D}^{-1}\right)}_{\boldsymbol{P}_i'} \boldsymbol{D} \tilde{\boldsymbol{p}}_i .$$

This relationship can be interpreted as follows: a set of 2D points (the vectors $\boldsymbol{n}_i$) can be projections of 3D scene points (the vectors $\boldsymbol{p}_i$) which are transformed by an arbitrary unknown projective transformation (the 4×4 matrix $\boldsymbol{D}$). Therefore, if we don't know any restrictions for $\boldsymbol{P}_i$, we are not able to reconstruct the vectors $\boldsymbol{p}_i$ uniquely, but up to a unique projective transformation $\boldsymbol{D}$. This means, that from a fundamental matrix $\boldsymbol{F}$ we can retrieve one solution for a pair of projection matrices $\boldsymbol{P}_{0,1}$, but this solution is just a so-called *canonic representation* [458] which has to be transformed using $\boldsymbol{D}$ to yield the true projection matrices:

$$\boldsymbol{P}_0 = (\boldsymbol{I} \mid \boldsymbol{0}) \quad , \quad \boldsymbol{P}_1 = ([e]_\times \boldsymbol{F} \mid e). \tag{2.55}$$

From these projection matrices, we can perform a 3D reconstruction which is related to the choice of $\boldsymbol{D}$. If we know the true 3D positions of n scene points (not all being coplanar) and their corresponding reconstructions, we can retrieve $\boldsymbol{D}$ uniquely [776].

Another way to determine the projective transformation $\boldsymbol{D}$ is to apply so-called *self-calibration* methods. In this case, restrictions are made to the calibration matrices $\boldsymbol{K}_i$ (e.g., the skew parameter s in most cases can be supposed to be 0). Details are beyond the scope of this contribution and can be found, e.g., in [706, 550].

2.3.3 Solution Methods

The previous section gives algebraic descriptions of the geometric constraints when dealing with two or more views of a rigid scene. This section gives an overview of several methods that can be used to estimate the unknown parameters.

Three types of approaches, discrete, continuous, and direct, have been pursued in the computer vision literature. In the discrete approach, information about the displacements of a finite number of discrete points in the image is used to reconstruct the motion. To do this, one has to identify and match feature points in a sequence of images. The minimum number of points required depends on the number of images. In the continuous approach, the optical flow, that is the apparent velocity of image brightness patterns is used. In direct methods the motion parameters are recovered without the explicit establishment of point correspondences or the estimation of the optical flow [334].

In much of the work on recovering surface structure and motion, it is assumed that either a correspondence between a sufficient number of feature points in successive frames has been established or that a reasonable estimate of the full optical flow field is available. In general, identifying features involves determining gray-level corner points. For images of smooth objects, it is difficult to find good features or corners. Further, the correspondence problem has to be solved, that is, feature points from consecutive frames have to be matched. An inherent problem of those methods is the separation of feature matching and the computation of the motion parameters. As a consequence the reported results are very sensitive to feature correspondence errors [711, 5, 745]. The computation of the local flow field exploits a constraint equation between the local intensity changes and the two components of the optical flow. This only gives the component of flow in the direction of the intensity gradient. To compute the full flow field, one needs additional constraints such as the heuristic assumption that the flow field is locally smooth. This, in many cases, leads to an estimated optical flow field that is not the same as the true motion field.

In most practical situations the motion field is not homogeneous as there may be several objects undergoing different motions. Most of the existing motion analysis methods would fail to perform under these circumstances.

Feature-Based Methods. Scanning the extensive *Structure-from-Motion* literature unveils the common assumption that in a pre-processing step 2D features in the first camera image are extracted and labeled. These features are assumed to be successfully associated with their correspondence in the second frame. Such 2D features are, for instance, points in the image, corners, lines, etc. For a large number of SfM systems these 2D point correspondences are the input to the SfM formulation.

Two basic types of methods have been used for this kind of feature-based 3D motion and structure analysis. The first type is iteratively solving non-linear equations, which can be traced back to 1980 [575] when nonlinear equations were derived to relate 3D motion parameters with the observables in the image plane. Although numerical methods could be applied to these nonlinear equations, the solution is not guaranteed, and as reported by a number of researchers, one may end up with a false solution if the initial guess is not sufficiently near the true value. The second type is based on linear formulations of the SfM problem. Linear formulations provide mathematical elegance as well as direct solution methods.

It has been shown in [746] that for each lateral translation (parallel to the image plane), there exists a corresponding type of rotation such that the displacement field of translation can be interpreted by the rotation without significantly violating the epipolar constraint. In the presence of even small pixel-level noise, the displacement of the translation can be interpreted by appropriate rotation and vice versa. Therefore, small pixel-level errors will cause large errors in the estimated rotation and translation. Worse still, once a lateral translation is mistakenly interpreted as the corresponding pure rotation, the estimated translation direction can be arbitrary since pure rotation is an inherently degenerate case.

One common way to improve the estimated motion parameters is to use a large number of points (or a dense displacement field). If the measurement error is random and has a zero mean, the solution error tends to be overcome in the solution of an over-determined system. This is true for computer simulations where the noise is generated by a zero mean pseudo-random number generator. However, with the displacement vector field (or point correspondences) computed automatically by an algorithm, the measurement errors are often biased, and the amount of bias is usually unknown. This fact makes the over-determination less effective.

The next two paragraphs recall standard linear and non-linear feature-based solutions that are both based on predetermined point correspondences. The former provide mathematical elegance as well as direct solution methods, whereas the latter require iterative optimization and must content with local minima.

Linear Solutions. Equation 2.43 is a linear equation with eight unknowns: the elements of the E-matrix. At least eight equations are necessary to recover an estimate of the E-matrix. To increase stability, more than eight point correspondences can be used to perform a least squares minimization of the form

$$||\boldsymbol{x}_2^T \, \boldsymbol{E} \, \boldsymbol{x}_1||^2 \rightarrow \min , \tag{2.56}$$

where $\boldsymbol{x}_1$ and $\boldsymbol{x}_2$ are the point correspondences in frame $I_1$and I_2, respectively, and matrix $\boldsymbol{E}$ contains the essential parameters. This is the standard technique described in [446] and [711].

Least squares minimization is easy to use, but it requires the variables, i.e., the elements of the matrix **E**, to be independent. Here, this is not the case. The solution that the least squares estimator finds does not represent a matrix **E** that is decomposable into $\boldsymbol{t}$ and $\boldsymbol{r}$. Even if, by solving Equation 2.56, a matrix **E** is found that is close to being decomposable, this might be far from minimizing Equation 2.56 in the sense of finding $\boldsymbol{r}$ and $\boldsymbol{t}$ that do so. Another problem is the physical interpretation of what we minimize. Unless **E** is decomposable to $\boldsymbol{r}$, and $\boldsymbol{t}$, then there is no physical interpretation of the quantity that is minimized. For the actual decomposition of the estimated E-matrix into $\boldsymbol{r}$ and $\boldsymbol{t}$ refer to Section 2.3.2.

The least squares solution to the E-matrix in Equation 2.56 requires a two-step approach. In the first step, discriminant features are extracted from one frame and matched in the second frame leading to the desired point correspondences that can be used in Equation 2.56. There is no feedback from the computation of motion parameters to the matching process and typically the motion parameter estimate is very sensitive to errors in the feature point correspondences [711, 745, 5]. In [289] a normalization step has been proposed that increases the robustness of the motion estimate in many situations but still requires the independent establishment of point correspondences.

Non-Linear Solutions. It has been shown [746] that even small pixel-level perturbation may override the image plane information that is essential for the linear algorithms to distinguish different motions. Since the solution of a linear algorithm is generally suboptimal due to quantization and other errors, they can be further improved through optimization of a non-linear formulation of the problem. The non-linear formulation avoids the mutual dependency of the unknowns that is introduced in the linear formulation. Nonlinear equations generally have to be solved through iterative methods with an initial guess or through global search. Iterative methods may converge to a local minimum or even diverge. Searching in the space of motion parameters is computationally expensive. In the absence of noise, i.e., for ideal image point correspondences, the linear approach is preferable since the exact solution can be determined with lower complexity.

In the presence of noise, the non-linear formulation of the SfM problem will lead to better results than the linear solution. However, the linear solution still plays an important role for initialization of iterative non-linear optimization procedures. The linear algorithm is used to obtain preliminary estimates of the motion parameters by random selection of point correspondences. The next step is minimizing the optimal objective function starting from those preliminary estimates as an initial guess. A remarkable accuracy improvement has been reported by this two-stage approach over using the linear algorithm alone [746].

One common non-linear formulation of the 3D motion estimation problem is to accumulate the Euclidean distances of the feature points in the second image from the estimated epipolar lines

$$\min_{\boldsymbol{r},\boldsymbol{t}} \sum_{i=1}^{F} d(n_i, \boldsymbol{l}_i(\boldsymbol{r}, \boldsymbol{t})) \tag{2.57}$$

with F being the number of feature points with known point correspondence in the second image, $\boldsymbol{l}_i(\boldsymbol{r}, \boldsymbol{t})$ the epipolar line for the feature point n_i in the second image for the motion parameter set $\boldsymbol{r}$, $\boldsymbol{t}$, and $d()$ the distance operator computing the Euclidean distance of an image point to a straight line.

Another non-linear formulation called *Optimal Motion and Structure Estimation* can be found in [746]. This formulation of the SfM problem is based on the observation that in addition to the epipolar constraint also the object depth of the feature points can be exploited for increased estimation accuracy. The authors in [746] derive a cost function that incorporates both, the epipolar constraint and the object depth. From the motion parameter set and the observed projections of a 3D point, the two observed projection lines are determined. These two projection lines do not intersect in general. Using these two projection lines the authors in [746] derive the following objective function that is minimized over the motion parameter space

$$\min_{\boldsymbol{r},\boldsymbol{t}} \sum_{i=1}^{F} \left\{ \|\boldsymbol{n}_{i,1} - \boldsymbol{h}_{i,1}(\boldsymbol{r}, \boldsymbol{t})\|^2 + \|\boldsymbol{n}_{i,2} - \boldsymbol{h}_{i,2}(\boldsymbol{r}, \boldsymbol{t})\|^2 \right\} \tag{2.58}$$

with $\boldsymbol{n}_{i,1}$ the pixel position of feature point i in image I_1 and $\boldsymbol{n}_{i,2}$ the corresponding pixel position of feature point i in image I_2. If the true 3D point is on the observed projection line in the first image, the first term in Equation 2.58 is equal to zero, but the second term is generally not. If the true 3D point is on the projection line of the other view, the second term in Equation 2.58 is equal to zero, but the first is generally not. The optimal solution is obtained when determining the 3D point for a given motion that minimizes both terms in Equation 2.58. A computationally efficient choice of this 3D point is the midpoint of the shortest segment that connects the two observed projection lines for the i-th feature point. The pixel position $\boldsymbol{h}_{i,1}(\boldsymbol{r}, \boldsymbol{t})$ then represents the projection of this midpoint in image I_1. Accordingly, $h_{i,2}(r, t)$ represents the projection of this point into the second image.

Direct Methods. Direct methods typically estimate motion or flow field parameters that minimize the absolute or squared intensity differences between two views without explicit computation of the optical flow field or feature point correspondences [498, 408, 334, 53, 601]. In [334] Horn et al. developed a direct method for recovering 3D motion in the case of pure rotation, pure translation, or arbitrary motion when the rotation is known. Bergen et al. describe in [53] a hierarchical estimation framework for the extraction of motion information where different motion models (affine flow, planar surface flow, rigid body motion, and general flow) are incorporated in a unified manner into the estimation process to constrain the image flow field. Negahdaripour and Horn show in [498] how to recover the motion of an observer relative to a planar surface from image brightness derivatives without intermediate computation of optical flow. The resulting non-linear cost function is efficiently solved using an iterative scheme. Kumar et al. [408] construct a Laplacian pyramid of the input images and estimate the motion parameters in a coarse-to-fine manner where at each resolution level the sum of squared differences (SSD) integrated over a region of interest is used as an error measure. This measure is then minimized with respect to the unknown flow field parameters. Girod and Steinbach describe in [244] a direct method that directly imposes the epipolar constraint into the estimation process. The following paragraph describes the approach in [244] as an example of direct 3D rigid body motion estimation techniques.

Example of a Direct Method for Two-View 3D Rigid Body Motion Estimation. The motion estimation approach in [244, 670] is based on the observation that 3D rigid body motion constrains the point correspondences in two views to lie on a straight line in the image plane, the epipolar line. The mathematical derivation of the straight line equation as a function of the 3D motion parameters r, t, the imaging geometry, and the image plane location (n_x, n_y) is presented in Section 2.3.1. The epipolar line for the motion parameter set r, t computed for the pixel (n_x, n_y) in frame I_1 will be named $l_{n_x n_y}(r, t)$ in the following. If we assume a particular motion r, t we can calculate the equation of the epipolar line $l_{n_x n_y}(r, t)$ in I_2. A maximum horizontal and vertical displacement in the image plane (e.g., ± 15 pixels) defines a 2D search area around the point (n_x, n_y) inside which the point correspondences are assumed to fall. The rigid body constraint now reduces the 2D search area in the vicinity of a measurement point to the intersection of the epipolar line with the 2D search area. The point correspondence problem therefore becomes an one-dimensional search problem. Figure 2.6 illustrates the situation for $F = 9$ measurement points (+) in image I_1 and the corresponding one-dimensional search space for the point correspondences along the epipolar line. Since the epipolar line is a function of the motion parameters and the image plane location, the line typically changes its slope and intercept from measurement point to measurement point. All point correspondences have to lie on the corresponding epipolar lines $l_{n_x n_y}(r, t)$ in view I_2. In order to determine the position of the point correspondence along the epipolar line the mean squared error (MSE)

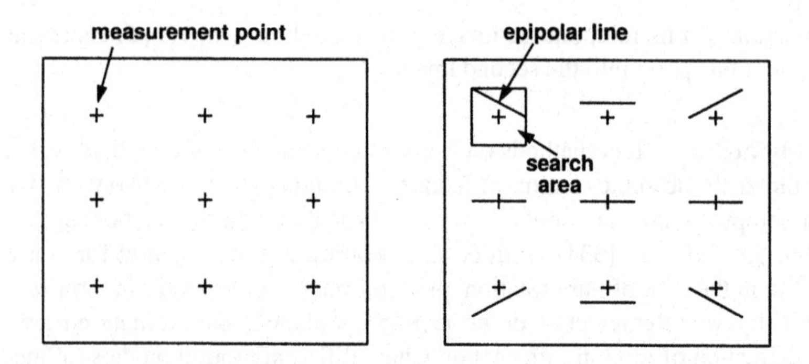

Figure 2.6. Illustration of the rigid body constraint in the image plane. The epipolar line constrains the search space for point correspondences to a one-dimensional search space.

evaluated over a measurement window of size (N, M) which is centered around the measurement point

$$\text{MSE}(n_x, n_y, d_x, d_y) = \tag{2.59}$$

$$\frac{1}{MN} \sum_{r=-\frac{N-1}{2}}^{\frac{N-1}{2}} \sum_{s=-\frac{M-1}{2}}^{\frac{M-1}{2}} (I_1(n_x + r, n_y + s) - I_2(n_x + d_x + r, n_y + d_y + s))^2$$

is evaluated, with $I_1(n_x, n_y)$ and $I_2(n_x, n_y)$ denoting the intensity values in views I_1 and I_2, respectively, and d_x, d_y being the displacements in horizontal and vertical direction. Since the actual motion parameter set $\boldsymbol{r}, \boldsymbol{t}$ is a priori unknown, candidate motion parameter sets have to be selected from the motion parameter space. In order to decide which candidate motion fits best to the actual image data, the following cost function is employed

$$c(\boldsymbol{r}, \boldsymbol{t}) = \sum_{1}^{F} \min_{(n_x + d_x, n_y + d_y) \in l_{n_x n_y}(\boldsymbol{r}, \boldsymbol{t})} \text{MSE}(n_x, n_y, d_x, d_y) \tag{2.60}$$

with F being the total number of measurement locations selected in view I_1. The cost function in Equation 2.60 is the minimum MSE detected along the epipolar lines accumulated over all F measurement locations. The motion parameter set $\boldsymbol{r}, \boldsymbol{t}$ leading to the smallest value for $c(\boldsymbol{r}, \boldsymbol{t})$ in Equation 2.60 is considered as the estimated motion.

Simulation Results. In this section we compare the various techniques described before. The following abbreviations are used throughout the remainder of this section:

TSAI: The linear 8-point algorithm [711].
HART: The normalized linear 8-point algorithm [289].
NP2L: Non-linear minimization of the point to line distance in Equation 2.57.
OSFM: Non-linear minimization of the objective function in Equation 2.58.
DDME: Non-linear minimization of the objective function in Equation 2.60.

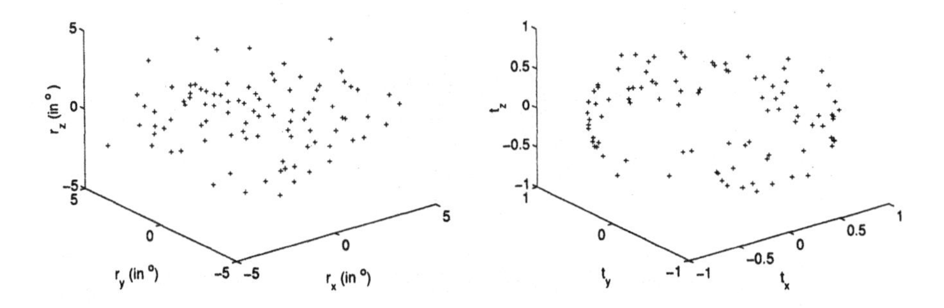

Figure 2.7. Random samples from the rotation space (left). Random samples from the translation space. The translation vector is normalized to unit length (right).

The test sequence consists of an artificial 3D object with a cubic surface in front of an ideal pin-hole camera. The images of the sequence are produced using *Open Inventor* where the objects are captured with a virtual camera. The test scene is smooth and exhibits considerable depth variation. The camera position and pose for the $N = 100$ test views are known and can be compared with the estimated relative motion parameters obtained. Table 2.1 summarizes the experimental setup. The test sequence generation with *Open Inventor* allows to work with an ideal pin-hole camera, a ground truth object and known motion parameters. The motion estimation results therefore can be given as the average deviation of the estimated motion parameters from the original motion parameters which are known from the experimental setup. The left hand side of Figure 2.7 shows a 3D plot of the randomly selected rotation parameters for the N views. The right hand side shows the same plot for the translation parameters. The translation vector in Figure 2.7 is normalized to length 1. The texture that is

Table 2.1. Experimental setup for the test sequence.

symbol	*description*	*value*
N_x	image size (horizontal) in pixel	512
N_y	image size (vertical) in pixel	512
k_x	scaled focal length (horizontal)	512
k_y	scaled focal length (vertical)	512
h	height angle	53.13^o
ar	aspect ratio	1.0
r_x	rotation around the x-axis	$-5^o \leq r_x \leq 5^o$
r_y	rotation around the y-axis	$-5^o \leq r_y \leq 5^o$
r_z	rotation around the z-axis	$-5^o \leq r_z \leq 5^o$
t	translation vector	$t_x^2 + t_y^2 + t_z^2 = 1$

Figure 2.8. Artificial texture that is mapped on the object surface.

mapped on the 3D surface is an artificial three channel random texture reproduced in Figure 2.8. The random texture has two components. A slow varying intensity profile (cosine progression) combined with an additive high frequency texture that has been generated using a random number generator. This texture remains the same for all test frames. Figure 2.9 and Figure 2.10 present histograms of the absolute rotational and translational motion estimation error for the 5 motion estimation techniques. It can be observed from Figures 2.9 and 2.10 that the non-linear methods significantly outperform the linear methods.

Multiple Views. It is well known [746], that for a narrow field of view small rotation and translation produce almost identical motion fields. Therefore it is not possible to separate reliably those types of motion. Processing more than two frames simultaneously can help to overcome this problem. Recently, the analysis of multi-frame sequences had gained considerable attention in the literature. A detailed survey of multi-frame 3D motion and structure estimation can be found in [362].

The recovery of the structure of a surface of an object from a long sequence of data was proposed by Baker, under the assumption that the camera follows a linear trajectory [33]. The notion of path coherence was introduced by Sethi et al. to determine point correspondences between successive time frames [631, 592]. Shariat and Price [637] have introduced a method for recovering a constant rigid motion by considering a single point over five consecutive frames. Broida and Chellappa have explored a possible approach to analyzing such a motion using a Kalman filtering formulation [84]. Chaudhuri et al. recently presented a robust and recursive method using a total least squares approach [115]. Common to all these methods is the establishment of feature point correspondences as a preprocessing step. In practice, however, explicit and exact tracking of features over multiple frames remains difficult due to a variety of reasons. In real world sequences we often lack significant features that can be tracked reliably. Features may appear and disappear due to occlusion and have to be treated accordingly. Wang and Duncan give a formulation of the multi-frame motion estimation problem for binocular image sequences in [736] exploiting the image flow instead of pre-computed feature point correspondences.

Multiple Objects and Robust Algorithms. In real world scenes one has to deal with a varying number of objects in front of the camera. In the general case all objects

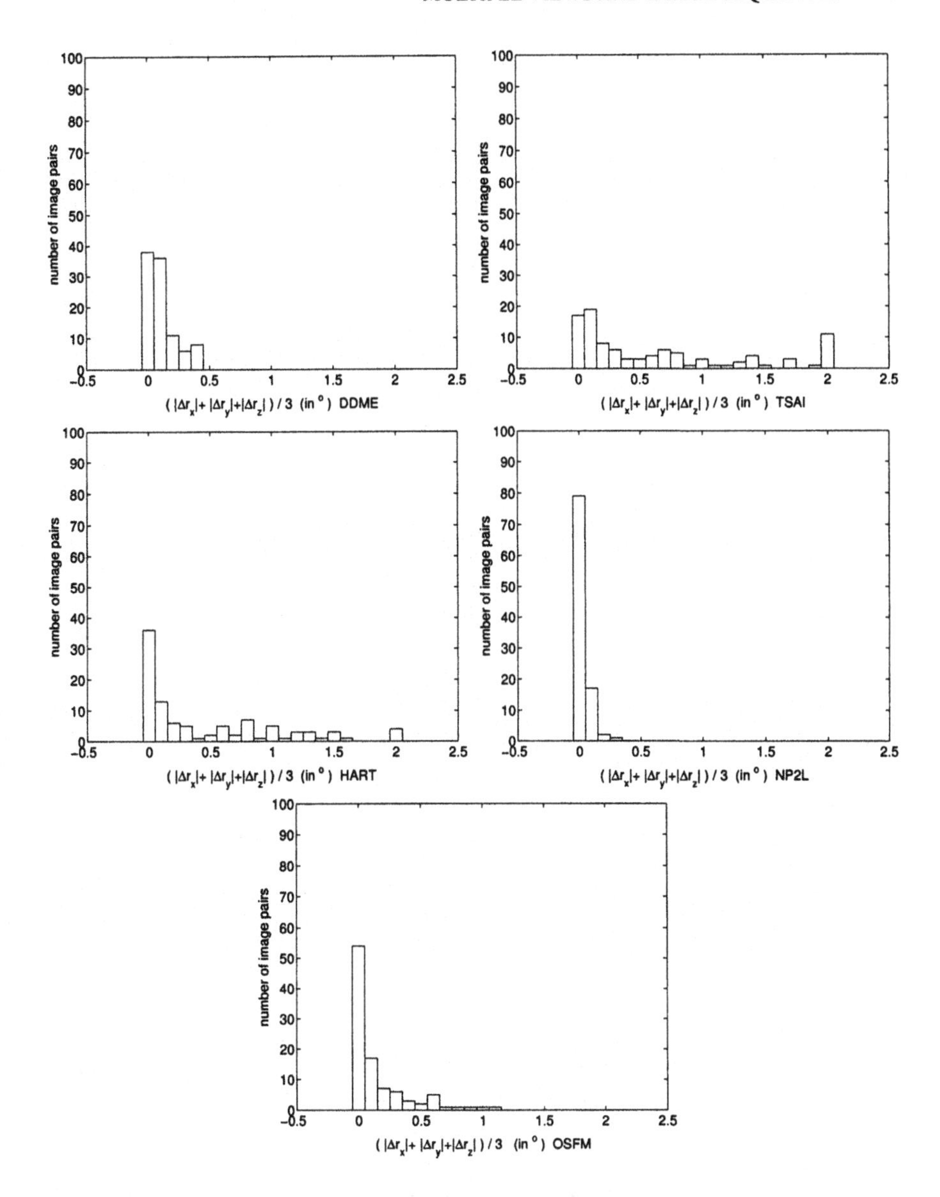

Figure 2.9. Histograms of the absolute rotational estimation error for the synthetic test sequence. The MSE-surfaces are computed with half-pel accuracy. The location of the minima of the MSE-surfaces are used as point correspondences for the feature-based methods TSAI, HART, NP2L, and OSFM. A total number of 2162 measurement windows of size 9×9 are used.

are moving independently which causes different motion fields for different regions of the images. The algorithms from the previous section have to be robust with respect to outliers. The treatment of those outliers is often performed using methods that stem from the field of robust statistics. Description of multiple-motion estimation, the use

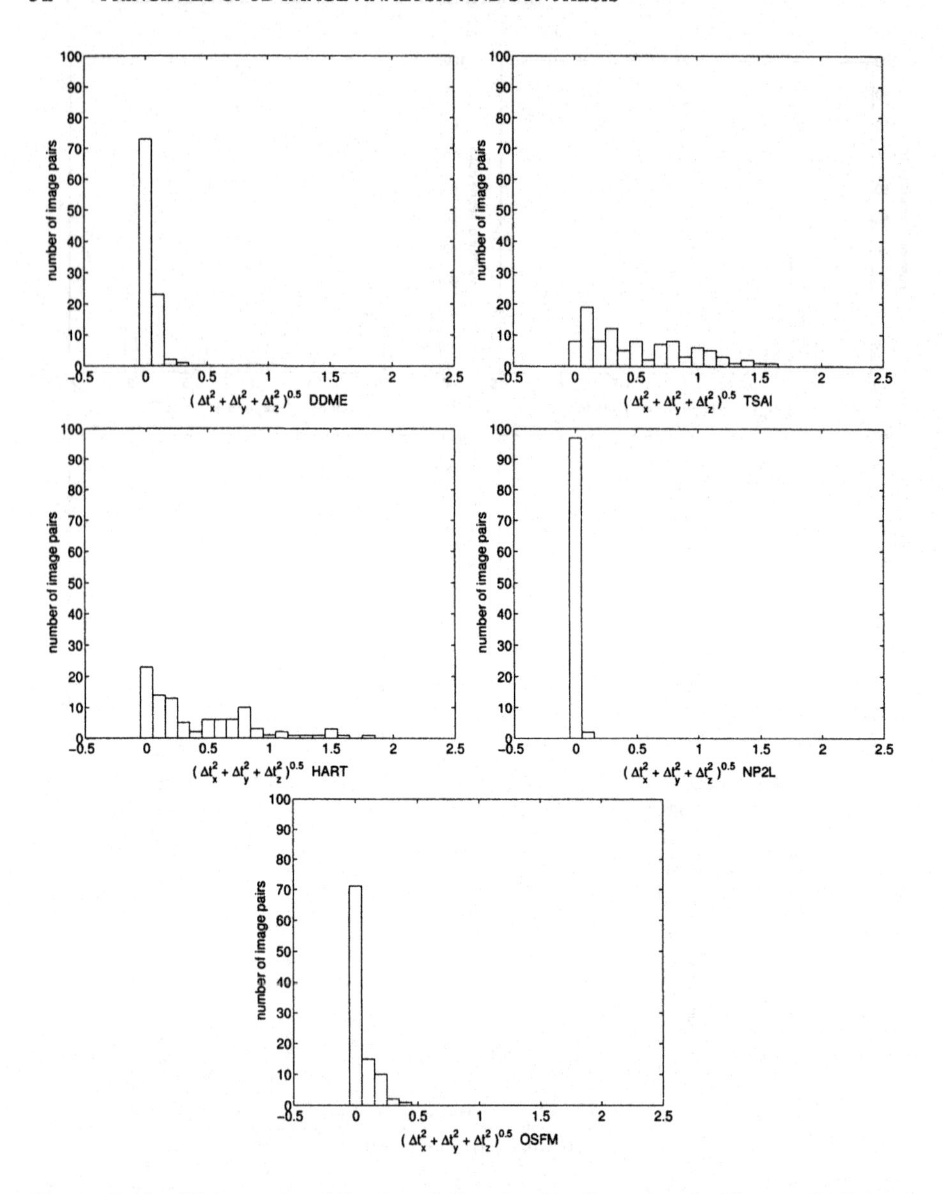

Figure 2.10. Histograms of the translational estimation error for the synthetic test sequence.

of robust estimators and their application to computer vision problems can be found in [582, 592, 668, 66, 577, 669, 700]. As a widely used example for a robust estimator in computer vision the Least Median of Squares (LMedS) estimator will be described in the following.

The LMedS estimator is a well known concept in statistics [582] and has been used in various computer vision applications, including motion analysis [114]. Kumar and Hansen use the LMedS method to obtain a robust estimate of the camera orientation

[409]. Tirumalai et al. have used the LMedS estimator for the recovery of the focus of expansion (FOE) while developing a scheme for estimation of motion from a long sequence of stereo images [696]. An extremely important property is that the LMedS estimator can tolerate up to 50% data contamination by outliers [582]. The linear solution in Equation 2.56 leads to the least squares solution of the estimation of the essential matrix. Outlier lead to very large estimation errors. The corresponding formulation using the LMedS estimator becomes

$$\text{median}\ \left\{ \left| \boldsymbol{x}_2^T\ \mathbf{E}\ \boldsymbol{x}_1 \right| \right\} \quad \rightarrow \quad min. \tag{2.61}$$

There is no analytical solution for this minimization problem. A standard technique to solve the minimization problem is based on Monte-Carlo sampling. J randomly picked K-tuples of data points are chosen from the data set. For each K-tuple the linear set of equations in Equation 2.43 is solved leading to a candidate motion estimate $\boldsymbol{r}_j, \boldsymbol{t}_j$. For each $\boldsymbol{r}_j, \boldsymbol{t}_j$ the corresponding residuals of all feature points are computed and their median is determined. The K-tuple leading to the smallest median is considered to be the solution. An exhaustive search would require the evaluation of $\binom{N}{K}$ combinations, involving high computational complexity. However, with the Monte-Carlo sampling technique only a small probability of not finding a good initial estimate remains. A correct initial estimate can be recovered from the trial solutions only if at least one out of J chosen K-tuples is free from outliers.

2.3.4 Depth Recovery

Once the motion parameters between several views are estimated, the next step is to determine a scene description in terms of scene depth. Given the motion parameters $\boldsymbol{r}, \boldsymbol{t}$ from view I_1 to I_2, a dense map of depth values can be recovered using the structure-from-motion paradigm which says that the depth at pixel position (n_x, n_y) is a function of the estimated motion parameters $\boldsymbol{r}, \boldsymbol{t}$, the measured image plane displacements for this point (d_x, d_y) and the imaging geometry of the camera described by K in Equation 2.11

$$z(n_x, n_y) = f(\boldsymbol{r}, \boldsymbol{t}, n_x, n_y, d_x, d_y, K)\ , \tag{2.62}$$

with

$$(n_x + d_x, n_y + d_y) \in \boldsymbol{l}_{n_x n_y}(\boldsymbol{r}, \boldsymbol{t})\ . \tag{2.63}$$

The standard technique to recover the dense map of depth values is to compute and evaluate the disparity space image (DSI). The disparity space image [354] is defined for a corresponding pair of lines from two rectified views [29, 583] or a stereoscopic picture pair. I_1 will be called the left image and I_2 the right image as typically done for a stereoscopic setup. Because disparity can be modelled as one-dimensional it is possible for a single frame-line to visualize the match error for the various possible disparities in the disparity space image. The information in the DSI one is interested in is the path of true disparity which will be described later together with a couple of characteristics of this path. The picture has the dimensions N_x columns by N_y rows.

The intensities for the N_x pixels of the i-th line of the left image are given by

$$I_L(n_x, i) \quad \text{where} \quad 0 \leq n_x \leq N_x - 1 \quad \text{and} \quad 0 \leq i \leq N_y - 1. \tag{2.64}$$

Analogously, the intensities for the N_x pixel of the i-th line of the right image are given by

$$I_R(n_x, i) \quad \text{where} \quad 0 \leq n_x \leq N_x - 1 \quad \text{and} \quad 0 \leq i \leq N_y - 1. \tag{2.65}$$

The basic form of the disparity space image, the left DSI is defined by

$$DSI_i^L(n_x, d) = \begin{cases} |I_L(n_x, i) - I_R(n_x + d, i)| & 0 \leq (n_x + d) < N_x \\ \text{not defined} & \text{elsewhere} \end{cases} . \tag{2.66}$$

The right DSI is defined by

$$DSI_i^R(n_x, d) = \begin{cases} |I_R(n_x, i) - I_L(n_x - d, i)| & 0 \leq (n_x - d) < N_x \\ \text{not defined} & \text{elsewhere} \end{cases} . \tag{2.67}$$

Comparison of the equation for DSI_i^L and DSI_i^R reveals that DSI_i^L contains the same information as DSI_i^R except that it is skewed along the $x = -d$ axis.

Because practical digital imagery is affected by noise it is often necessary to calculate differences averaged over a window instead of for a single pixel. Now the expression for the filtered disparity space image becomes

$$DSI_i^L(n_x, d) = \frac{1}{M_x M_y} \sum_{n_c = n_x - m_x}^{n_x + m_x} \sum_{i_c = i - m_y}^{i + m_y} (I_L(n_c, i_c) - I_R(n_c + d, i_c)) \tag{2.68}$$

with $M_x = (2m_x + 1)$ and $M_y = (2m_y + 1)$ the width and height of the measurement windows. Ideally, where d corresponds to the true disparity for pixel n_x of line i, the following equality should hold

$$DSI_i^L(n_x, d) = 0 . \tag{2.69}$$

Finding the true disparity d for each pixel is difficult because of several reasons:

- there are matches $DSI_i^L(n_x, d) = 0$ (or very close to zero) for values of d which do not correspond to the true disparity,
- due to occlusion there is sometimes no true match for a particular pixel,
- in practice, image pairs are not noiseless which means that it is not possible to rely upon $DSI_i^L(n_x, d)$ equalling zero for a true match or even on a $DSI_i^L(n_x, d)$ value of zero implying any match at all.

True matches of pixels to their disparity shifted counterparts result in (near) zero disparity space image values. These true matches are visible as horizontal dark lines at appropriate disparity levels in the DSI. There are discontinuities in this horizontal pattern. The nature of this discontinuity is different when going from a low disparity

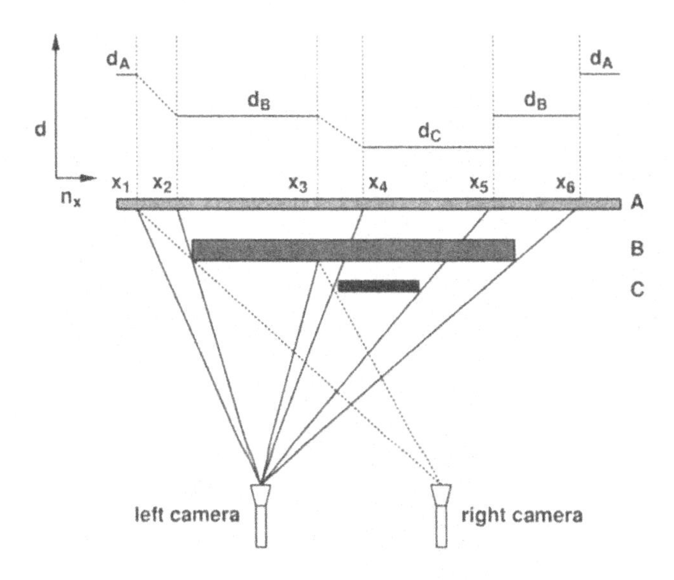

Figure 2.11. Example scene with 3 objects A, B, and C with corresponding disparity path.

to a high one when compared to the discontinuity when going from a high disparity to low one. Figure 2.11 shows an example scene with three objects A, B, and C for illustration of the disparity path determination.

Initially, at the extreme left of the scene, the points of object A observed by the left camera can be matched with the same points as observed by the right camera leading to a disparity d_A. At a point x_1 it is no longer possible to find a match for A because the right camera's view is occluded by B. As the scanning goes on, the left camera encounters the left edge of B at point x_2. As the right camera can also see this point the disparity continues after x_2 at a level of d_B. Between x_1 and x_2 the disparity is indeterminate as indicated by the dotted line in Figure 2.11. The same effect of disparity transition occurs where the left edge of C occludes parts of B to the right camera this time with occlusion between x_3 and x_4. Just after having passed x_4 the left camera sees C which is also visible along its complete length to the right camera so that it may be matched at a disparity of d_C up to x_5. After x_5 B again becomes visible to the left eye. The first point on B that becomes visible to the left camera is also visible to the right camera so that a match at a disparity of d_B can immediately be made from x_5 onwards. The same argument holds for the transition from B to A at point x_6.

There are two types of disparity discontinuity in this left disparity path. The transition from high to low disparity means that the left eye encounters the left edge of (relative) foreground while scanning along the line. A part of the relative background has been occluded to the right camera. The transition from low to high disparity means that the left camera encounters a background object after a foreground object. The right camera notices no occlusion and therefore the disparity transition takes the form of a clean jump.

Figure 2.12. Two example frames of the Flowergarden sequence and the corresponding dense map of depth values. The depth value are grey value coded. The brighter an image point, the closer to the sensor it is.

Similar observations can be made for the right disparity path which are excluded here due to space limitation. Finding a good disparity path corresponding to good matches and obeying the structures which the ordering constraint imposes on a disparity path is a problem closely related to classic combinatorial problems from the field of operations research like the travelling salesman problem where a minimum cost path of some kind is the objective.

A standard technique for finding the disparity path in disparity space images uses dynamic programming [354]. The basic idea is that dynamic programming starts by examining a small part of the problem and finding an optimal solution for this part of the problem. The problem is then extended a little bit and the previously found solution for the sub-problem is extended to a solution which covers the current, slightly extended, view of the problem and so on and so forth.

One significant advantage of the dynamic programming approach is that additional constraints like ordering, neighborhood and smoothness can be incorporated elegantly by simply assigning corresponding costs to the path transitions along the scan line. Details about these constraints and the corresponding cost formulations can be found in [354, 214, 659].

In order to illustrate the observations made in this section, Figure 2.12 finally shows an example of two frames of the *Flowergarden* sequence and the corresponding dense map of depth computed using the disparity space image approach described in this section.

2.4 OBJECT TRACKING IN IMAGE SEQUENCES

J. DENZLER, M. ZOBEL

Motion detection and object tracking from image sequences are two of the most important problems in computer vision, especially when dealing with dynamic environments. Tasks which have to treat such dynamic settings typically arise for autonomous mobile systems, in analyzing traffic scenes, or in surveillance systems. Since it is impossible and not useful, especially in real time applications, to process each frame of the image sequence completely, one has to focus on important events in the video stream (cf. Section 3.6), i.e. one has to track them. Tracking is defined here, to keep track of the movement path of an object either in 2D or in 3D, which might be followed by a movement of the camera or the whole observer (e.g. a moving platform), to keep the object in the center of the camera image.

One can observe three main problems in tracking. The first problem is, how to detect moving objects. If the object to track is known in advance, and there is a model available, object detection can be seen as object recognition (cf. Section 3.4). But then, the detected object has not necessarily to be in motion. Without having a certain object in mind, when looking for moving objects, motion detection becomes more complicated. The key motivation behind motion detection is then, to detect changes in the image from one or several frames to another one. Then the problem arises, to decide whether the change is caused by an independently moving object, by a change in scene illumination, or by egomotion of the observer.

After the detection of pixels in the image with changed value, related pixels have to be identified and combined to regions of motion, i.e. some kind of motion segmentation has to be initiated. The difficulty as well as the quality of the motion segmentation strongly depends on the features used for identifying the motion. The range is from binary images, marking areas in the image where something has changed with respect to a certain threshold, up to optical flow, where each pixel has two attributes, namely the motion in x and y direction. The segmentation in the first case is simple but with low quality, the latter one normally results in high quality but difficult segmentation. Different methods that reside between these two extremes will be discussed later.

Having successfully detected a moving object, appropriate features that are tried to be found again in the next frame have to be extracted. The success of finding the best mapping of the features between certain image frames is strongly influenced by the correspondence problem. It is the most serious problem in motion and object tracking, because normally for each object feature in one frame there is more than one matchable feature in the next frame. Dependent on the number of features the 'combinatory explosion' makes it impossible to evaluate every possible feature match. In addition to that, missing features in one frame, the actual or previous one, result in a further increased complexity.

Finally, when one object could be detected and tracked over a couple of frames, the acquired knowledge about the object's trajectory should be used to improve the quality of tracking, especially in the case of uncertain measurements, sensor noise, or partial occlusion. The solution to this is to provide the tracker with a prediction about the position and appearance of the object in the next frame, given a couple of previous frames. Therefore a certain model of the object's motion must be assumed and the parameters of the model must be determined or estimated by learning from sample data.

2.4.1 Motion Detection

As it was outlined in the previous subsection, the task of motion detection in an image sequence consists of two parts: first, gather information on the changes between succeeding frames (detection), and second, process this information appropriately to obtain an image region that belongs to a moving object in the scene (segmentation). The methods described in this section for detection of change can be distinguished by the kind of changes they take into consideration. In general, they all compute the temporal derivative between images but the more sophisticated and computational more expensive methods take the spatial derivative into account, too. The latter ones are also known by the term *flow based* methods. In the next step, maybe after a necessary preprocessing, some kind of structure has to be brought into the data provided by the change detection. The aim of this process is to identify those image areas that belong to a moving object. According to classical region segmentation, this step is called *motion segmentation*. One major problem for motion detection in general is, that image changes can also be caused by the motion of the camera itself. In very restricted cases, this kind of changes can be handled analytically, or a model of the camera's motion is necessary, as explained below.

In the following, the case of a motion detection based on a model based object localization step with which a known object is localized in the scene, is neglected. In principle, this can be seen as a motion detection step, when applying iteratively for each frame and estimating the displacement in 2D or 3D to decide for a motion. But in general, nothing new is done compared to a model based classification and localization (cf. Section 3.4). One example for such an initialization can be found in [752] in the area of traffic scene analysis.

Change Detection. In the following it is assumed, that no dedicated model for the moving object is used for motion detection. Then features must be defined, which identify motion in the scene. The main feature, which is used as a strong hint to a independently moving object, is a change in the 2D image plane.

Although it is possible to detect changes based on complex features like points, lines, corners, etc., the most straight forward approach is to detect changes in the image at the pixel level. To be more concrete, for each pixel, one needs to decide, whether the pixel has changed its value (gray or color value), whether this change is based on change of illumination or independent motion, and sometimes, in which direction the pixel has moved. If the camera is static, the most simplest approach is to compute a *difference image* from two frames. This corresponds to the computation of the temporal derivative of the image sequences. Taking into account the spatial derivative, one can compute the so called *normal flow*. This results in additional information for each pixel about the direction of the change, which is perpendicular to the edge direction at the pixel location. The most sophisticated approach, but the most computational expensive one is to compute the *optical flow* (cf. Section 2.4.2), for which it is assumed that illumination remains constant over time. A smoothness condition forces neighboring pixel elements to have similar direction of motion.

Before it is discussed how objects are identified from such areas, a look has to be taken at the case of a moving observer. In such a case there is the problem that at the pixel level everything is moving in the scene, the static background as well as moving objects. Especially, in the case that the moving object is correctly tracked, the moving object itself remains static in the image. Without any modification, the difference image algorithm fails in such a case, unless knowledge about the egomotion is available. If the egomotion is restricted to pure rotation around the optical center of the camera, the position of a projected scene point after rotation can be calculated independent of its depth. Then, for each pixel in the image plane the corresponding pixel in the next frame can be computed. One says, the difference image algorithms works now on a so called *motion compensated* difference image [495].

Flow based approaches needs a so called motion model of the camera, if the camera performs egomotion while motion detection. This might also include translational components, in opposite to the restricted case for the difference image. The idea for detecting independently moving objects is based on a so called *outlier detection* [699]. The computed flow vectors are tried to fit into a general motion model. Each flow vector which does not correspond with the motion model, i.e. which is called an outlier, is assumed to belong to an independently moving object.

Beside this approach work has been done on motion detection that make use of statistical classifiers and Markov random fields [314, 403].

Motion Segmentation. Having computed features concerning changes in the image, the features need to be segmented into separate moving objects. For the difference image algorithm typically a threshold is used to eliminate false alarms due to sensor

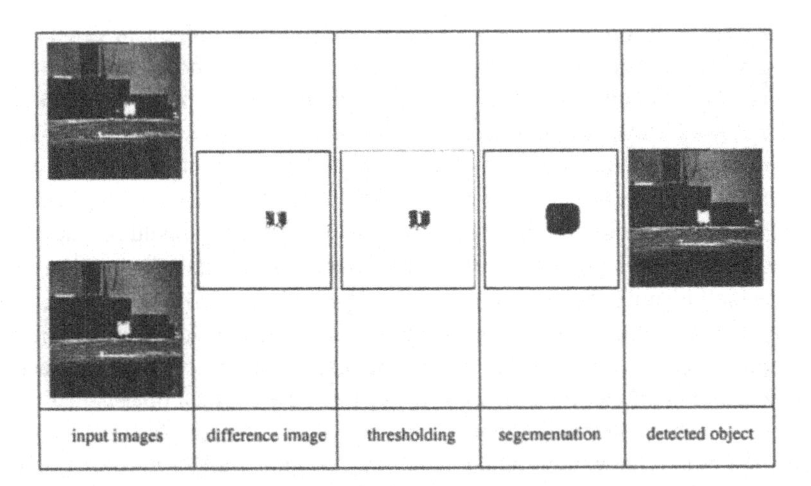

Figure 2.13. An example for motion detection based on a difference image with a static observer, resulting in an initial region for the moving object in the scene (from [167]).

noise or moderate change in illumination. Morphological operations are applied to fuse several small binary areas, which occur at the borders of the moving object, to one large region, covering the moving object completely. Problems occur in such an approach for very slow and very fast moving objects. The former results in an misdetection of the object, the later one in an over-segmentation. To fix such problems, one can try to estimate the static background over time, for example using a Kalman filter [570]. In Figure 2.13 an example for a complete motion detection based on difference images with intermediate results of each step is depicted.

For optical flow based motion detection a complete segmentation step is necessary to extract the moving object. This can be based on areas, in which the smoothness constraint is violated. Typically this occurs at the border of the moving object to the background.

Model based approaches are well suited for motion detection, too. Motion vectors can be used to identify positions in the image, where a model based approach is initialized to match model primitives with image features. Such an approach in the area of car tracking can be found in [402].

2.4.2 Tracking of Moving Objects

After having identified a moving object by means of the previous subsection, the question to be answered now is, where the object has been moved to in the next frame. From the displacement between the two positions the camera may be controlled appropriately to keep the object in the image center. It is clear, that this task could be solved by iteratively applying a motion detection step to each frame in the sequence. Therefore every frame has to be processed completely, and there is no concentration on the most important parts of it, e.g. an appropriate region around the last position of the moving object. Optical flow based methods are something particular, as it will be described below.

Hence, one main problem making motion computation difficult in general, and being the main problem to be solved for tracking is the *correspondence problem.* It summarizes the difficulty for the decision, whether or not two points in two images correspond to the same 3D point in the world. The same problem can be observed in structure from stereo. The other main problem that needs to be solved is the so called *data association problem*, since one is interested in associating image points to physical objects to decide which set of points, i.e. which object should be tracked.

In general, for solving both problems mentioned above, certain features has to be defined, which identify the appearance of 3D objects in the 2D image plane. Such features might be velocity vectors in the image, simple tokens like edges, corners or lines, or more complex ones like regions or contours. These features have to be defined in advance, taking into account the knowledge about the physical objects that should be tracked. Also a kinematic model of the evolution of theses features—constant velocity, constant acceleration, affine motion (cf. Section 2.2.3)—must be provided. This is omitted here. Finally in the case of a model based tracking, we need a detailed idea, how these features are related to the object model.

In the following we will concentrate on different classes of features and the possible approaches to establish correspondence and association.

Optical Flow. The idea of using optical flow is to compute the projected velocity vector for each visible scene point in the image. Main source of information is the spatio-temporal brightness pattern recorded by the camera. To estimate optical flow robustly and consistently a set of constraints must be defined, like the hypothesis that image brightness is invariant along image motion trajectories. To get a smooth optical flow field regularization terms are defined.

The computation of optical flow for object tracking is only the first step, since a second one, the motion segmentation must follow. This second step is difficult, especially when not only the object but also the camera is moving. Then everything in the scene is in motion relative to the camera and a foreground/background segmentation must be done, similar to motion detection.

A more detailed discussion of optical flow based motion computation, which can be separated into the classes gradient-based, correspondence-based and transform methods, is beyond the scope of this section and therefore omitted here. A very deep synopsis can be found, for example, in [483].

For tracking using optical flow as the main source of information a couple of work exists. Assuming a static camera, work on so called temporal integration exists [355], which smoothes difference images over time to increase robustness against background noise. A moving camera is used by [152], who applies optical flow to the problem of tracking moving objects in real time. A stereo camera system with known geometry is used. The image motion between two frames induced by a known camera motion is eliminated. For this, a special configuration of the camera and the rotation axis is needed to eliminate the dependency of the induced optical flow by the depth in the scene. A similar approach based on pure gray value changes can be found in [495]. To compute the difference image under egomotion of the camera for each pixel the corresponding pixel in the previous frame is computed. This also needs a camera/rotation axis configuration, where pure rotation of the camera induces no flow in the image, which is dependent on the depth in the scene. Both approaches are of course not valid for translational motion of the camera unless scene geometry is known. A histogram over the optical flow vectors is used in the work of [697]. In addition to the flow vectors their positions are stored in the histogram, which are needed to locate the moving object in the image. In this biologically motivated approach both the smooth pursuit as well as fast saccades could be demonstrated.

Correlation. Another way to establish correspondence is based on correlation. The complete object or parts of the object are modeled in the image by pixel blocks, i.e. areas of certain size, which are also called templates. Therefore these approaches are sometimes called *template matching*. The position of a template in an image is found by computing the correlation of the pixel block with the image data. When modelling the complete object by one template (for example, deformable templates of [273]) the association problem is solved, too. Otherwise this must be done in an extra step.

In [273], the template identifying the moving objects is not only adopted to changes in illumination, but also to changes in size and 3D position of the objects. The authors in [273] report an improvement in robustness for tracking. Tracking articulated objects by template matching has been proposed in [233]. Dedicated positions of a human body are marked manually in the first frame. These positions (for example, the knee) are then robustly tracked over time by an adaptive template, comparable with the approach of [273].

Points, Corner and Lines. The motivation for using tokens like points, corners or lines as features is, that these features are not as sensible to changes in illumination as gray value based approaches, like optical flow or template matching. This leads to the assumption, that the correspondence problem seems to be easier to solve, but in general this depends strongly on the chosen token and the probability to observe such a token in the image. The more tokens of a given class can be found the more ambiguities will occur, which makes the correspondence problem more difficult. In the case of lines as tokens the association problem becomes more difficult, too.

Tokens are well suited for model based tracking, since the features in most cases directly correspond to model primitives in the 3D world. Thus, it is natural that most token based methods for establishing correspondence are combined with a 2D or 3D model of the moving objects. Examples for tracking rigid objects using a 3D model can be found in [239, 455, 286]. 2D models of the objects are used in [752] in a traffic scene application. Establishing correspondence, also in combination with a prediction step as described below (cf. Section 2.4.3) results not only in a localization of the moving object, but also inherently in a 2D or 3D pose estimation.

Without having an object model, the data association step is very difficult to solve. Therefore, some constraints, like rigid motion of the object must be applied. Often, another constraint is, that only one moving object is allowed. Having a moving camera it is nearly impossible to track objects without having any model.

An interesting approach which uses both, the optical flow based approach as well a a token based approach in combination with a 3D model has been presented by [402]. Assuming a static camera a sparse optical flow field is estimated which defines an area in the image, where a model based matching of line segments is initialized.

Region and Contour Based Tracking. Using image regions as tokens to identify moving objects has the advantage, that the correspondence problem can be solved easier, but the feature extraction task is more difficult and expensive. Again, correspondence must be established first, and then the association of regions to moving objects. The assumption is, that the moving object is represented by its contour, which can be computed by a region segmentation step.

One possible approach for such a region based tracking approach can be found in [192], where the region is not only estimated by information within one image but also using motion information across frames. The motion is again estimated using the optical flow, assuming a static camera. The motion segmentation results in a binary image, marking regions in the image, where motion has occurred. A color region segmentation is fused with this binary motion image to identify such color regions, which have moved over time.

Similar to template matching the idea behind a contour based tracking is to solve the correspondence and association problem within one step. For this, a lot of interest has been spent on so called active, dynamic, elastic contours, sometime also called snakes

[378]. The idea is, that the object is represented by a parametric function in the 2D image plane, that describes the contour of the object. Two energy terms are defined, which describe the deformation ability of the contour over time (internal energy) and identify features in the image corresponding to contour points (external energy). Examples are the spatial gradient or textural features [358, 578]. The localization, i.e. the estimation of the position in the image plane, as well as the extraction, i.e. the computation of the shape of the 2D contour, is done within one step, minimizing the sum of internal and external energy. The advantages of active contours are threefold:

- The image needs only be processed locally around the active contour; this reduces the computation time, an important aspect for real-time applications.
- No distinction is necessary between foreground and background moving objects, since an active contour is attracted by one object, independently of other motion in the scene; this is advantageous in the case of a moving camera and multiple objects in motion.
- Due to the deformation ability of an active contour, this approach is well suited for a purely data driven tracking, where no a priori knowledge of the objects is available.

There exist a couple of promising extensions to the original approach, presented by [378]. In [656] the external energy also contains motion information from optical flow computation, [54] uses the normal flow along the 2D contour to predict the motion of the active contour. In [171, 173] a radial representation of the contour instead of the representation by a parametric function in 2D is introduced. Contour point localization for so called *active rays* is then reduced to a contour point search on 1D rays, which are the gray values that are sampled along a straight line originating from a common reference point lying inside the contour. This leads to an enormous reduction of the computation time. Also, a 3D radial model representation is provided, which can be efficiently combined with the 2D rays, to predict the change of the 2D contour in the case that a 3D model is available [168].

The approach of [68, 69] tackles also the problem of implementing a model based component for the data driven active contours. The changes over time in the shape of the contour, modelled as a 2D spline, as well as the motion is learned from observations. This knowledge impressively increases the tracking result in the case of cluttered background, in combination with a prediction step, described in the next subsection.

2.4.3 Improved Tracking: Prediction

Tracking a moving object by one of the methods described in the previous paragraph results in an increase of knowledge about the object's dynamic. Thus, it is natural to make use of that knowledge to improve tracking. It is inevitable, especially to make tracking more robust against noisy measurements, missing features, or partial occlusions. To apply prediction, it is important to have a model of the object's dynamic behavior. Normally the model is given a priori, and the parameters are obtained and refined while tracking proceeds, or they are learned from image sequences [69].

The aim now is, to estimate the object's state, or its trajectory in 2D or 3D over time to get constraints about the position of the whole object or at least of its features in the next frame. Having such constraints makes the correspondence problem as well as the data association problem much easier.

Crucial for the prediction step is the kind of available knowledge about the object at hand. Having a 3D model of the object, prediction is straight forward: the observations must be taken to estimate the parameters of the object's model, for example its 3D pose. Applying this step iteratively over time, the parameters of the motion model can be estimated, too. This can be used to predict the position of the object in 3D and consequently its appearance in the image.

Having no knowledge about the object prediction in 3D is quite complicated (cf. Section 2.3). One example is the 8-point algorithm to estimate the motion from eight correspondences [447]. Restricting the possible objects' motions, maybe only to motions in one plane, or only scaling, the estimation of the object's state becomes possible from simple 2D models of the object's appearance in the image.

In the unrestricted case, data driven approaches usually define a general motion model, like constant acceleration, and try to estimate at least the motion parameters of the object in the image plane. Then of course only the displacement of the object from one image to the next can be predicted, not the change in the object's appearance.

The problem of predicting the object's state are solved by so called *state estimators*. In the following the Kalman filter [238] and some of it's derivatives are shortly presented. Also a new approach, the so called CONDENSATION algorithm [357], will be described.

Kalman Filter. The Kalman filter is a well established method to estimate the unknown state of a dynamic system [370]. The Kalman filter can be applied in the case, that a state vector shall be estimated from a sequence of erroneous measurements or observations. The goal is to construct an optimal estimator, which minimizes a defined estimation error. A Kalman filter can be used to

- filter, i.e. estimate the actual state using all previous measurements,
- predict, i.e. estimate a state lying in future,
- smooth, i.e. estimate a previous state by use of all the measurements actually done.

There exist two classes of Kalman filters, the *linear* and the *nonlinear* one.

In the linear case, the dynamics of the system as well as the measurements are described by linear systems disturbed by noise terms. It can be shown, that for Gaussian noise the Kalman filter is the best estimator, for arbitrary noise the Kalman filter is the best linear estimator [370]. In the nonlinear case, the matrices are replaced by arbitrary functions and now the nonlinear Kalman filter needs to be used.

The Kalman filter is not restricted to discrete dynamic systems, continuous systems are possible, too. In that case, the state estimation is done by a minimum variance estimation and the Kalman filter is then called *extended* Kalman filter. A linearization of the state transition equation leads to the *linearized* Kalman filter, a linearization around the actual state estimation is called the *iterative extended* Kalman filter. The latter one is more accurate in the estimation but also more computational expensive [401, 238].

The problem in applying Kalman filters to prediction for object tracking lies in the modelling of the system's dynamics and the mapping from state to observation. Such a problem is commonly referred to as *system identification*. For real world problems in computer vision in most cases it is at least impossible to cover the dynamics of the systems exactly. Additionally, the statistical properties of the noise terms can only be approximated. Both influence the quality and success of state estimation and therefore prediction, and must be chosen very carefully.

A more detailed description of the theory of Kalman filter is given in Section 3.5.4. The reader is referred to a very practical guide to the use of Kalman filter in [238]. The original work of Kalman can be found in [370].

Handling Multiple Hypotheses. When applying the Kalman filter to prediction in object tracking one has to decide which measurement in the image plane is caused by the dynamic system. This measurement is compared with the predicted measurement, which is based on the estimated state and leads to one new state estimation. Now, in the case of uncertainty in measurement, which is mostly the case for real problems in computer vision, it would be better to have the chance for computing rather more than one hypothesis for the state than only one estimate. This is impossible in the Kalman filter framework. On the contrary, the application of the Kalman filter is equal to the use of a unimodal density for the state distribution. Thus, only *one* correct state is assumed, multiple hypotheses are not modelled.

To overcome such problems, in [40] several algorithms are described, for example the JPDAF (joint probability data association filter) or the multiple hypotheses filter, whose idea can be mainly summarized as follows: instead of using one Kalman filter, a tree of Kalman filter, that grows in time is processed over time. Each node of the tree corresponds to one hypothesis of the estimated state. The disadvantage is the computational complexity, which also makes it necessary to prune this tree heuristically.

CONDENSATION. The *conditional density propagation algorithm* (CONDENSATION) presented by [357] tackles exactly the disadvantages of the Kalman filter framework. Instead of assuming a unimodal density for the state, a multi modal density can be used. Thus, one gets a belief over all possible states instead of one state estimation. Wrong associations of image features to the model result in a shift of the density toward a wrong state estimate, but the true state will not disappear, i.e. its probability is not zero.

The main technique in the CONDENSATION algorithm is factored sampling. A set is created which includes several possible state estimations. Each element of the state set is associated with the probability of being in this state based on the observation. For the next time step, elements are drawn from this set, depending on their probability of being the correct state. The drawn states are then diffused and propagated over time, using the dynamics of the system. Finally, for each propagated state its probability is computed again, by comparing the predicted observation with the measured image data. The principle of this algorithm is shown in Figure 2.14.

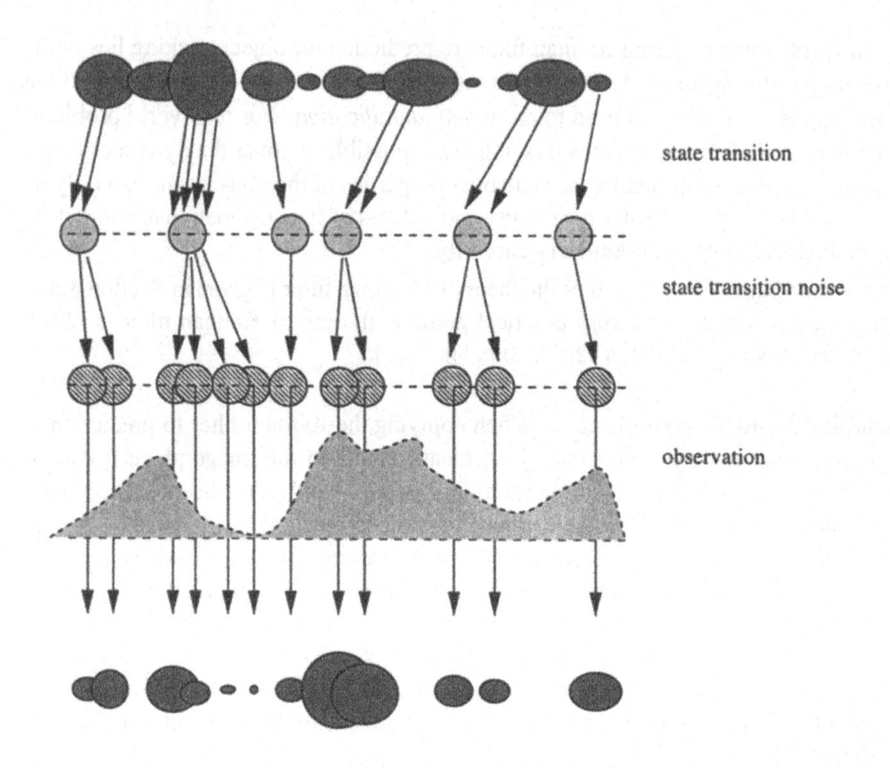

Figure 2.14. Principle of the CONDENSATION algorithm (after [356]).

2.4.4 Concluding Example

In the following an example of a complete system for real time object tracking is shortly discussed to show the principle and realization of the different steps described in this section. The system is called COBOLT (COntour Based Object Localization and Tracking) [172]. Tracking itself is done based on active rays for contour based object tracking. In Figure 3.23 in Section 3.6 the different modules of the system can be seen. The motion detection stage assumes a static camera. Thus, a difference image algorithm can be applied to detect moving objects in the scene. Then tracking starts by extracting the contour of the moving object and sending its position to the camera control module, which moves the camera by reducing the distance between the object center and the image center. A prediction step is applied based on Kalman filter to estimate the position of the contour in 2D for the next image. The system provides also a knowledge based component, which allows for the estimation and prediction of a known object in 3D. To increase robustness in natural scenes, for example outdoor tracking of moving pedestrians, a principle of active vision (cf. Section 3.6), an attention mechanism, is applied. This mechanism detects errors in contour extraction based on too large changes of the contour between successive frames (data driven tracking) or based on the error covariance matrix of the Kalman filter (model driven tracking).

The system has been intensively tested in real scenes and runs in real time on general purpose architectures.

2.5 ILLUMINATION ESTIMATION

N. SCHÖN, P. EISERT

When analyzing multiple views or image sequences for object tracking or recovering geometrical structure, one has to deal with the dependence of 2D images on the illumination conditions of the observed scene. Object tracking relies on the fact that the object looks similar in successive views. However, a change of its position relative to the illuminant or varying illuminants cause effects like glossy regions appearing on different parts of the surface and a change in the observed brightness and color of other surface elements. This is also problematic for motion estimation, because intensity variations caused by a change of illumination conditions have to be distinguished from those variations caused by object movement. There are other applications that require the compensation of illumination effects. In Computer Graphics, visualization techniques require the essential colors of a surface, that depend only on the reflection characteristic of the object, not on the illumination conditions during their measurement. Another example is the search of 2D images in a database that contains only datasets acquired under definite illumination conditions. Illumination effects have to be removed from a new image to make it comparable to the ones in the database.

2.5.1 Introduction

Illumination estimation means the estimation of various parameters of models that consist of three essential parts [544, 497, 614]: illumination of the scene by a light source, reflection of the light by the objects' surfaces and acquisition of an image by a camera. Since models for illumination and reflection are described in detail in Section 5.1 and camera models in Section 2.1, here only the simplifications of these models, that are made for illumination estimation, will be explained. For the definition of the terms *irradiance*, *radiance* and *intensity* see also Section 5.1.

Illumination Model. Most approaches assume a single distant point light source illuminating the scene. This means that the unit vector $\boldsymbol{l}$ which points from a surface element (surfel) in the illumination direction is constant for every surfel. It is given by the tilt angle ϕ_l, which is the angle against the x/z-plane of the world coordinate system (see Section 2.1), and the slant angle θ_l, which is the angle against the z-direction. The relationship between cartesian coordinates and tilt/slant angle of the light source is

$$\begin{aligned} l_x &= \cos\phi_l \sin\theta_l, \\ l_y &= \sin\phi_l \sin\theta_l, \\ l_z &= \cos\theta_l. \end{aligned} \tag{2.70}$$

For the coordinates a_x, a_y, a_z of the surface normal $\boldsymbol{a}$ and its tilt and slant angles θ_a, ϕ_a analogous equations hold.

Since spectral dependencies play an important role in illumination estimation, the following quantities are formulated with respect to the wavelength λ of light. If the spectral composition of the illuminant's intensity I_l is $I_{l,\lambda} := \frac{dI_l}{d\lambda}$, then the spectral irradiation reaching a surface point $\boldsymbol{x}$ with normal vector $\boldsymbol{a}$ in distance r to the light source is given by

$$E_\lambda(\boldsymbol{x}, \boldsymbol{l}) = \frac{I_{l,\lambda} \max\{< \boldsymbol{a}(\boldsymbol{x}), \boldsymbol{l} >, 0\}}{r^2} sr \tag{2.71}$$

and the total irradiance at $\boldsymbol{x}$ is given by

$$E(\boldsymbol{x}, \boldsymbol{l}) = \int E_\lambda(\boldsymbol{x}, \boldsymbol{l}) d\lambda. \tag{2.72}$$

Note that irradiance is zero for surface elements (*surfels*) where the angle between $\boldsymbol{a}$ and $\boldsymbol{l}$ is greater than $\frac{\pi}{2}$.

Reflection Model. The interaction between incoming light and the surface with its specific geometry and reflectance properties is described by reflection models. The relationship between $dL_\lambda(\boldsymbol{x}, \boldsymbol{v})$ leaving a surface point $\boldsymbol{x}$ into the solid angle $d\Omega_o$ around direction $\boldsymbol{v}$ and the irradiance $dE_\lambda(\boldsymbol{x}, \boldsymbol{l})$ arriving at that surface point from the solid angle $d\Omega_i$ around direction $\boldsymbol{l}$ can generally be described by a function called *BRDF* (bidirectional reflectance distribution function) $f_r(\boldsymbol{x}, \boldsymbol{l} \rightarrow \boldsymbol{v}, \lambda)$. By common reflection models which approximate the physical BRDF, the surface radiance is assumed to consist of two components.

1. The Lambertian or diffusely reflected spectral radiance is given by

$$L_{d,\lambda}(\boldsymbol{x}, \boldsymbol{l}) = k_d(\boldsymbol{x}, \lambda) E_\lambda(\boldsymbol{x}, \boldsymbol{l}). \tag{2.73}$$

   $L_{d,\lambda}(\boldsymbol{x}, \boldsymbol{l})$ depends through $E_\lambda(\boldsymbol{x}, \boldsymbol{l})$ on the angle $\theta_{a,l} = \arccos(< \boldsymbol{a}, \boldsymbol{l} >)$ between surface normal $\boldsymbol{a}$ ($\|\boldsymbol{a}\| = 1$) and illumination direction $\boldsymbol{l}$, but not on the viewing direction $\boldsymbol{v}$ ($\|\boldsymbol{v}\| = 1$) (see Figure 2.15). The spectral diffuse reflectance $k_d(\boldsymbol{x}, \lambda)$ contains information about the 'essential' object color [772]. If the light source would be an equal-energy illuminant, i.e. $c_I := E_\lambda(\boldsymbol{x}) = const.$, the Lambertian spectral radiance would be proportional to the spectral reflectance of the surface.

2. The glossily reflected spectral radiance is given by a simplification of the Torrance-Sparrow reflectance model [701](see Section 5.1.3). The glossy reflectance component is modeled only by the microfacet distribution coefficient D, while the Fresnel term F and the geometric attenuation factor G are regarded as constants. Expressing all factors in the equation by $k_g(\boldsymbol{x})$, that are independent of $\boldsymbol{l}$ and $\boldsymbol{v}$ the glossy spectral radiance is

$$L_{g,\lambda}(\boldsymbol{x}, \boldsymbol{l}, \boldsymbol{v}) = k_g(\boldsymbol{x}) \frac{\exp(-c_g(\boldsymbol{x})^2 \alpha(\boldsymbol{x}, \boldsymbol{l}, \boldsymbol{v})^2)}{< \boldsymbol{a}(\boldsymbol{x}), \boldsymbol{l} >< \boldsymbol{a}(\boldsymbol{x}), \boldsymbol{v} >} E_\lambda(\boldsymbol{x}, \boldsymbol{l}) \tag{2.74}$$

   with $\alpha(\boldsymbol{x}, \boldsymbol{l}, \boldsymbol{v}) = \arccos(\max\{< \frac{\boldsymbol{l}+\boldsymbol{v}}{\|\boldsymbol{l}+\boldsymbol{v}\|}, \boldsymbol{a}(\boldsymbol{x}) >, 0\})$.

$k_g(\boldsymbol{x})$ is called the glossy reflection coefficient and specifies the strength of the glossily reflected component. It is assumed to be independent of the wavelength λ of light.

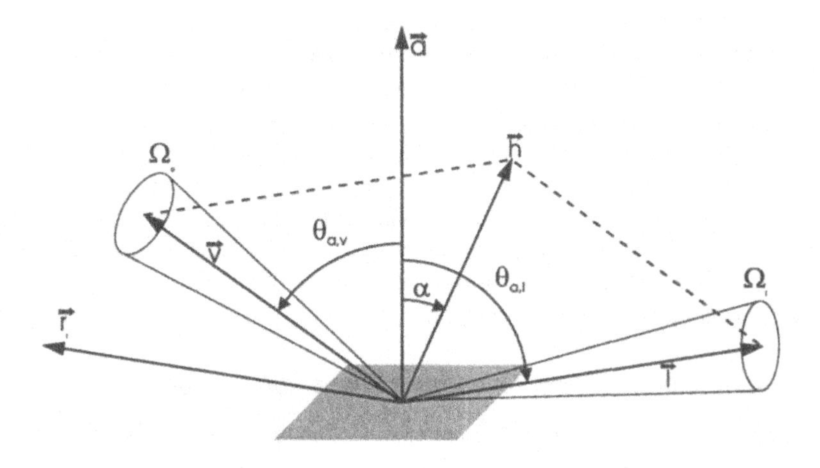

Figure 2.15. Illumination geometry.

$\alpha(\boldsymbol{x}, \boldsymbol{l}, \boldsymbol{v})$ is the angle between the surface normal $\boldsymbol{a}$ and the normal, for which mirror reflection of incident light from direction $\boldsymbol{l}$ into direction $\boldsymbol{v}$ would occur. This mirror normal is given by the halfway vector $\boldsymbol{h} := \frac{\boldsymbol{v}+\boldsymbol{l}}{\|\boldsymbol{v}+\boldsymbol{l}\|}$ between illumination direction and viewing direction (see Figure 2.15). $c_g(\boldsymbol{x})$ is determined by the standard deviation of the Gaussian distribution of microfacet slopes and is a measure for the surface roughness. Increasing surface roughness means a wider standard deviation and therefore a smaller value of the surface roughness coefficient $c_g(\boldsymbol{x})$. Gloss is only visible, if the viewing direction vector $\boldsymbol{v}$ is in a narrow vicinity of the reflection $r_l := 2 < \boldsymbol{l}, \boldsymbol{a} > \boldsymbol{a} - \boldsymbol{l}$ of the illumination vector $\boldsymbol{l}$ on the surface normal $\boldsymbol{a}$.

Combining both reflection components, the spectral radiance emitted by the surface is given by

$$L_{o,\lambda}(\boldsymbol{x}, \boldsymbol{l}, \boldsymbol{v}) = L_{d,\lambda}(\boldsymbol{x}, \boldsymbol{l}) + L_{g,\lambda}(\boldsymbol{x}, \boldsymbol{l}, \boldsymbol{v}). \tag{2.75}$$

Since the Lambertian component depends both on the chromaticities of illuminant and surface reflectance while the chromaticities of the glossy reflectance are equal to those of the illuminant, such models are called 'Dichromatic Reflection Models (DRM)' [635].

Camera Model. For the camera, in most cases one can use the model of a pinhole camera projecting a point $\boldsymbol{x} = (x, y, z)^T$ in the world coordinate system into the point $\boldsymbol{n} = (n_x, n_y)^T$ on the camera's image plane (see Section 2.1). The pixel intensity $I_c(\boldsymbol{n})$ in the image plane comes from the radiance $L_{o,\lambda}$ emitted by the corresponding surface point according to

$$I_c(\boldsymbol{n}) = \int s(\lambda) L_{o,\lambda}(\boldsymbol{x}(\boldsymbol{n}), \boldsymbol{l}, \boldsymbol{v}) d\lambda. \tag{2.76}$$

For the camera and light source being not too close to the object, v and l are approximately constant over the image. $s(\lambda)$ is the spectral camera luminance responsitivity.

In a color camera, it is adapted to the absorption spectra of the cones in the human eye using color filters. The three filters are then represented by the spectral responsitivity functions $s_r(\lambda)$, $s_g(\lambda)$, $s_b(\lambda)$ for the red, green and blue color receptors, respectively. The spectral responsitivity of a B/W camera without color filters be $s_w(\lambda)$.

Which of the parameters and quantities that appear in the model are given and which are to be estimated, depends on the kind of problem in context of which illumination estimation is applied. Some of them are demonstrated in the following subsections. The field of those problems can be coarsely divided into two categories: Illumination estimation *without* a priori information about the shape of the object and illumination estimation *with* a priori shape information.

2.5.2 Illumination Estimation without Shape Information

In some applications, illumination estimation is used as a preprocessing step to increase the robustness or the accuracy of algorithms. The purpose of such algorithms may be e.g. the computation of the three-dimensional shape of objects from one or several intensity images, like in Shape-from-Shading or Photometric Stereo [14]. Other algorithms try to find corresponding views of the same object in different images. The intention hereby is to remove illumination effects from input images which those algorithms are not able to deal with.

Shape from Shading [732, 584] means the reconstruction of object shape from the intensities in a single image. Basically this means the inversion of the pixel intensity equation for a pure Lambertian surface, which one gets by combining Equations 2.71, 2.73 and 2.76:

$$\begin{aligned} I_c(\boldsymbol{n}) &= \int s_x(\lambda) k_d(\boldsymbol{x}(\boldsymbol{n}), \lambda) I_{l,\lambda} \cdot < \boldsymbol{a}(\boldsymbol{n}), \boldsymbol{l} > d\lambda \\ &= I_0(\boldsymbol{n}) \cdot < \boldsymbol{a}(\boldsymbol{n}), \boldsymbol{l} > \end{aligned} \tag{2.77}$$

with $I_0(\boldsymbol{n}) = \int s_x(\lambda) k_d(\boldsymbol{x}(\boldsymbol{n}), \lambda) I_{l,\lambda} d\lambda$ and x=r,g,b,w. If $I_0(\boldsymbol{n})$ and $\boldsymbol{l}$ are known, this allows to compute the slopes of the surface normals from the pixel intensities and to integrate the shape under the assumption that the surface is continuous [732]. $I_0(\boldsymbol{n})$ depends on the reflection characteristic (e.g. the color) of the surface and introduces varations of pixel intensities that do not result from normal slope variations. From a single intensity image they cannot be estimated in advance, so that the reflection characteristic of the surface has to be homogeneous. In most cases, the glossy reflection component is not taken into consideration by Shape from Shading algorithms. To avoid computation of wrong normal slopes, glossy regions have to be identified. In the context of Shape from Shading, Illumination Estimation is used to achieve the light direction $\boldsymbol{l}$, because normals' slopes can only be computed relative to $\boldsymbol{l}$ from the image intensities. For this purpose statistical or feature based methods, as shown below, are used.

In *Photometric Stereo* [333] at least three views with different known illumination directions are used for shape reconstruction. Vice versa it can be shown, that for three intensity values obtained for a pixel with three light sources in non-coplanar positions, the triad of the illumination directions can be determined up to an unknown rotation [765]. The more views and illumination directions are used, the more conditions are given to estimate the parameters in Equation 2.77 by an optimization approach or other standard solution methods. In this sense, normal directions are just additional dimensions in the parameter space of the cost function. Thus, geometrical and illumination parameters can be estimated simultaneously [381]. This can be done separately for every pixel or surfel, respectively, so that the restriction to homogeneous surfaces can be overcome. However, robustness of Photometric Stereo algorithms increases, if illumination direction is estimated in advance by separate methods.

Another task where illumination estimation is necessary is the *tracking of objects in 2D image sequences*. The correspondences of regions in different images that belong to the same object have to be found. In natural scenes with moving objects the illumination direction varies relative to the objects over time. Even the light sources with their spectral characteristics can change. As Equation 2.77 shows, these changes have significant influence on the appearance of a 3D surface in an image. An additional problem are glossy regions wandering over the object while it is moving.

Similar difficulties arise from tasks like motion estimation or finding images, e.g. faces, with arbitrary light conditions in databases with fixed light conditions. Illumination estimation is used here to determine chromaticities and direction of the illumination. The intention is to achieve more comparable input data that are independent of their original light conditions.

Statistical Methods. If parameters of models are not known explicitly, it is sometimes possible to approximate them from their statistical distributions. In the following paragraph it is shown how assumptions about the distribution of normals on a surface are used to estimate *illumination direction*. Then it is explained, how *illumination color* can be computed from the distribution of reflection colors.

Estimation of Illumination Direction. One approach to obtain $\boldsymbol{l}$ works with the statistical assumption that the first derivatives of surface normals $\boldsymbol{a}(n_x, n_y)$, corresponding to pixel indices n_x and n_y, are distributed isotropically along a specific direction $\boldsymbol{n_i} = (n_{x,i}, n_{y,i}), (||\boldsymbol{n_i}|| = 1)$ with index i in the image plane [538], so that their sum across the whole image has to be estimated as zero,

$$E\Big\{\sum_j d\boldsymbol{a}(j \cdot \Delta n \cdot \boldsymbol{n_i})\Big\} = 0. \tag{2.78}$$

with the step size Δn along the chosen direction $\boldsymbol{n_i}$ and the summation index j. The reflection characteristic factor I_0 is implied to be uniform over the whole object. If illumination direction and viewing direction are identical, the average intensity derivative along a particular image direction is zero. Then, a deviation of the average intensity from zero must result from a deviation of the illumination direction from the viewing direction. The first derivation dI_c of intensity in direction $d\boldsymbol{n_i} = (dn_{x,i}, dn_{y,i})$ can be expressed as

$$\begin{aligned} dI_c &= I_0 \cdot (d\boldsymbol{a} \cdot \boldsymbol{l} + \boldsymbol{a} \cdot d\boldsymbol{l}) \\ &= I_0 \cdot (d\boldsymbol{a} \cdot \boldsymbol{l}). \end{aligned} \tag{2.79}$$

Thus the change in image intensity dI_c depends on the change in the surface normal $\boldsymbol{da}$ relative to the illuminant. Along N image directions $\boldsymbol{n}_i$ the average value $\overline{dI}_i$ over image intensities can be computed. With $\hat{l}_x = I_0 \cdot \Delta n l_x$ and $\hat{l}_y = I_0 \cdot \Delta n l_y$ the regression model is

$$\begin{pmatrix} \overline{dI}_1 \\ \overline{dI}_2 \\ \vdots \\ \overline{dI}_N \end{pmatrix} = \begin{pmatrix} dn_{x,1} & dn_{y,1} \\ dn_{x,2} & dn_{y,2} \\ \vdots & \vdots \\ dn_{x,N} & dn_{y,N} \end{pmatrix} \begin{pmatrix} \hat{l}_x \\ \hat{l}_y \end{pmatrix}. \tag{2.80}$$

From this, a maximum likelihood estimate for $\hat{l}_x$ and $\hat{l}_y$ can be computed using a least squares regression. This gives the tilt of the illuminant

$$\phi_l = \arctan\left(\frac{\hat{l}_x}{\hat{l}_y}\right). \tag{2.81}$$

The slant angle θ_l follows from Equation 2.70 and the correlation between (l_x, l_y) and $(\hat{l}_x, \hat{l}_y)$:

$$\theta_l = \arcsin\left(\frac{\sqrt{\hat{l}_x^2 + \hat{l}_y^2}}{I_0 \cdot \Delta n}\right). \tag{2.82}$$

$I_0 \cdot \Delta n$ depends on the variance of $\overline{dI}_i$:

$$I_0 \cdot \Delta n = \sqrt{\overline{(dI^2)} - \overline{dI}^2}. \tag{2.83}$$

with $\overline{(dI^2)} = \sum_i (dI_i)^2$ and $\overline{dI}^2 = (\sum_i dI_i)^2$. This is expression for the fact that the brighter a surface is (and the larger I_0), the stronger fluctuate the derivatives of the observed intensities according to normal variations.

The condition of isotropically distributed surface normals is met only by a few real objects, like a sphere, an ellipsoid or smooth rocks. However, it can be a good *local* approximation for small patches of a surface.

Let us consider a surface point $(x, y, z(x, y))$ and its neighbors, $(x_i, y_i, z_i(x_i, y_i))$, indexed with i. We assume that the surface region can be approximated by a spherical patch. For every point (x_i, y_i, z_i) in the neighborhood of (x, y, z) we get a pair of coordinate increments $(dn_{x,i}, dn_{y,i})$ in the image and the corresponding increment in image intensity dI_i. Using the fact that all neighbors lie on local spherical patches, it can be shown that again a local estimate for the tilt ϕ_l of the illuminant can be computed according to Equation 2.80 and 2.81. This can be done for every pixel in the image and its neighbourhood. One achieves the final estimate by averaging over all local estimations. For cylindrical objects the assumption of local spherical patches is wrong. In average, however, their local estimates cancel each other, due to their symmetry. For a local computation of the slant angle, the variance of intensity derivatives cannot be used, because there are too few pixels to estimate it reliably. But on a spherical patch the average intensity and average squared intensity due to symmetry are functions of I_0 and θ_l only,

$$\overline{I}_{\text{sphere}} = I_0 f_1(\theta_l), \quad \overline{I^2}_{\text{sphere}} = I_0^2 f_2(\theta_l). \tag{2.84}$$

f_1 and f_2 can be approximated by polynomials computed numerically from the intensity distributions on a sphere under various illumination slant angles θ_l. These intensity distributions can be generated synthetically by means of Computer Graphics techniques, measured from a real spherical object or modeled statistically.

Estimation of Illuminant Chromaticities. The chromaticities of the illuminant are $I_{g0,c} = \int s_c(\lambda) \cdot k_g I_{l,\lambda} d\lambda$, $c = r, g, b$ for red, green or blue chromaticity, respectively, and $c = w$ for a B/W camera. According to the Dichromatic Reflectance Model, they are, up to a scaling factor, identical to the chromaticities of the glossy reflection component,

$$I_{g,c}(\boldsymbol{n}) = I_{g0,c}(\boldsymbol{x}) \frac{\exp(-c_g(\boldsymbol{x})^2 \alpha(\boldsymbol{x}, \boldsymbol{l}, \boldsymbol{v})^2)}{< \boldsymbol{a}(\boldsymbol{x}), \boldsymbol{v} > r^2}. \tag{2.85}$$

To use this information, it is necessary to separate glossy and diffuse reflection. Without geometry information it is difficult to compute the positions where glossy reflectance is to be expected. Therefore, properties of the color signal itself have to be exploited. They can be analyzed in RGB color space. If the color appearance of a surface element is scanned under varying illumination directions, the resulting points in RGB-space lie on a plane, the 'color-signal plane' [698], that is defined by both the object surface color and the color of the illuminant. For objects with different colors one gets different color-signal planes, but all these planes intersect in one line. The direction vector of this line represents the color of the illuminant, up to a scaling factor. If multiple objects with different colors are in one image, this provides an estimation for the illuminant color.

If only one object is in an image, diffuse and glossy reflection components have to be separated geometrically in color space. The chromaticities of the reflection components lie on different clusters in RGB space. Within a color-signal plane, the measurements lie on a T-Shape [391, 392], which geometrically separates diffuse and glossy reflection.

To simplify geometrical color analysis and get faster but less accurate algorithms, one can also use a 2D color space [615], like the UV-space derived from the YUV color model.

Feature-Based Methods. Instead of using statistical arguments to approximate the distribution of surface normals for estimating illumination parameters, one can also detect image features for which information about the corresponding surface normals can be presumed. For estimating the light *direction*, especially object boundaries are useful features. Information about illuminant *chromaticities* is provided by glossy regions.

Estimation of Illumination Direction. At *object boundaries* it is reasonable to assume the slant angles of the surface normals to be constant along the boundary and their tilt angles to be just the tilt angles of the boundary contour in the image plane [777]. Image contours can be obtained by edge detectors, e.g. a zero-crossing edge detector. To compute the azimuth angle along a closed boundary contour of N pixels with intensities $I_1 \dots I_N$ and normal tilt angles $\phi_1 \dots \phi_N$ the regression model would be

$$\begin{pmatrix} I_1 \\ I_2 \\ \vdots \\ I_N \end{pmatrix} = \begin{pmatrix} \cos(\phi_1) & \sin(\phi_1) \\ \cos(\phi_2) & \sin(\phi_2) \\ \vdots & \vdots \\ \cos(\phi_N) & \sin(\phi_N) \end{pmatrix} \begin{pmatrix} \hat{l}_x \\ \hat{l}_y \end{pmatrix} \tag{2.86}$$

which can be solved for $\begin{pmatrix} \hat{l}_x \\ \hat{l}_y \end{pmatrix}$ by common regression techniques. The azimuth of $\boldsymbol{l}$ is then

$$\phi_l = \arctan\left(\frac{\hat{l}_x}{\hat{l}_y}\right).$$

Estimation of Illuminant Chromaticities. Illuminant color can be measured from image regions that appear under glossy reflection. They can be found by regarding the brightest points as glossily reflecting (see Figure 2.16) and growing a region around them including all pixels the intensity of which lies over a certain threshold. The pixels surrounding those regions can be regarded as diffusely reflecting. Their average chromaticity values give an estimate for the underground color of the glossy region. The underground chromaticities are to be subtracted from the chromaticities of the glossy regions to get the illuminant chromaticities.

Other approaches operate with moving objects [538, 456, 735]. This means essentially that the normal direction varies. For instance an object stands on a turn table, while viewing and illumination direction are fixed. The normal vectors on the silhouette of the object are assumed to be parallel to the image plane and perpendicular to the

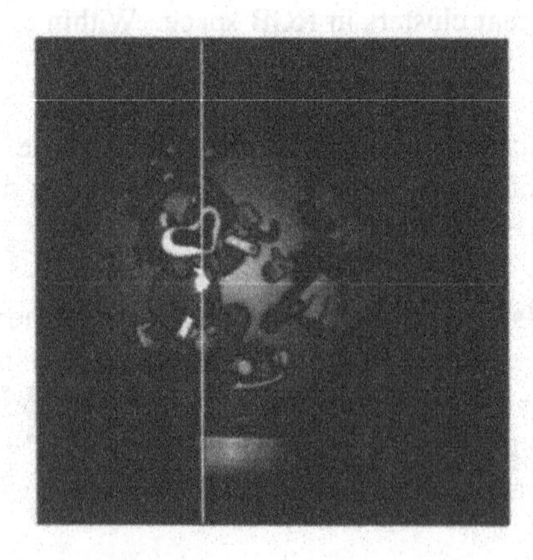

Figure 2.16. Glossy region on a spherical surface. Illuminant color is white.

silhouette boundary. Multiple images are taken which differ only in a known object rotation angle. Keeping track of a particular pixel along the different images gives a set of intensity values for various illumination angles. This process can be regarded as scanning values of the BRDF. From the scanned BRDF profile the reflection components can be separated to estimate illuminant color together with surface roughness and albedo parameters.

2.5.3 Illumination Estimation with Shape Information

Knowing the geometrical shape of an object means to know the normal directions for every point on the surface, which appear in the reflection model (Section 2.5.1). One application where shape is used for the compensation of illumination effects is Coding of Facial Image Sequences [202, 666, 532]. In these model-based approaches, instead of having accurate shape information for every face to be analyzed, a kind of 'medium shape', e.g. the scanned head of a mannequin, is used. Shading and specular highlights are removed from the images as explained below. The aim is to make faces that are illuminated by directional light look as if they were illuminated under ambient light. This is a precondition for the adequate recognition of facial motion and expression.

Modern range sensors allow to determine the shape of objects very accurately. For *Computer Graphics applications* this information can be used to generate complete descriptions of three dimensional objects, including the color structure (*texture*) and surface roughness [351, 41, 598, 597].

Objects with Uniform Reflectance. One may consider a measurement setup where viewing and illumination directions are identical (distant camera and light source, which are close to each other) and thus known. The camera is also part of a light stripe range finder, from which the object's shape is acquired in form of a height map . The normal directions and therefore $< \boldsymbol{a}, \boldsymbol{l} >$ can be computed from shape data. With the assumption of a Lambertian surface with uniform albedo, Equation 2.77 can easily be solved for I_0. If this value is applied for synthesizing a 2D image of the given shape with pure Lambertian reflection, the difference between the generated image and the real image must represent the reflection components ignored by the Lambertian model.

Beside glossy reflection, here also *shadows* and *interreflections* [732] appear, which until now have been completely neglected. Shadows appear when parts of the object keep incident light from reaching other parts of the object. Interreflections result from the fact, that irradiated surface elements themselves illuminate other surfels which are inclined towards them. To get information about the surface roughness parameters only the specular reflection is useful. It can be separated from the others regarding the following facts [351]:

- The mirror normal $\boldsymbol{h}$, i.e. the normal direction for which direct reflection is to be expected, can be computed from viewing and illumination direction (here, it is parallel to the viewing direction). The angles between normals of specular points and the mirror normal have to be small.
- Pixels in shadow regions are darker than predicted by the Lambertian model, which neglects occlusion of surface parts.

- Pixels illuminated by interreflections are brighter than predicted by the Lambertian model, even if the angle between their normal and the mirror normal is large.

Once the glossy regions are separated, their angular extension can be analyzed and the parameters I_g0 and c_g can be estimated.

Objects with Non-Uniform Reflectance. In Computer Graphics, the reflectance properties of objects often are not uniform, but have a fine spatial structure. To produce realistic rendering of such objects, this structure has to be provided. It is computed from color images. If multiple images are given for one 3D view, between which only the illumination direction varies, illumination effects like shading and specular highlights cause the images to appear very different. These effects have to be compensated, because they are not part of the texture itself, but depend on the illumination direction [470] and light spectrum.

Compensation of Shading and Specular Highlights. If color images of an object under equal-energy illumination are given, the texture map can be maintained by computing $I_0(\boldsymbol{n})$ in Equation 2.77 for every pixel in each of the images. This presupposes that pixels appear under diffuse reflection. Then regions that show specular highlights are to be replaced by corresponding regions from one of the other images of the same view. It is important, that in the other image the regions do not appear under specular reflection. On the borders of the replaced regions, however, intensity jumps arise. This can be avoided by building weighted averages of corresponding pixels instead of simply replacing the pixels. The weights for each pixel are related to the probability that they appear under specular reflection. This probability can be modeled by a gaussian function of the form

$$p_{\text{glossy}} = 1 - \exp(-c_g^2 \cdot \alpha^2) \tag{2.87}$$

which decreases with increasing angle $\alpha = \arccos(\max\{< \frac{\boldsymbol{l}+\boldsymbol{v}}{\|\boldsymbol{l}+\boldsymbol{v}\|}, \boldsymbol{a}(\boldsymbol{x}) >, 0\})$ between the surface normal $\boldsymbol{a}$ and the normal $\boldsymbol{h}$ for which the condition for directional reflection is fulfilled. c_g defines the angular width of the glossy highlights and depends on the surface roughness.

Illumination Direction and Shininess. To compute the angular width $1/c_g$, image based highlight segmentation can be used (see Section 2.5.2). The maximum angle between the normals of the pixels at the segment's border provides an estimate for $1/c_g$. Further, if the illumination direction is unknown, the normals of the brightest points in the image are approximations for $\boldsymbol{h}$:

$$E\{\boldsymbol{h}\} = \frac{\sum_i \boldsymbol{a}_i}{\|\sum_i \boldsymbol{a}_i\|}, \tag{2.88}$$

where i is the index of the considered normals $\boldsymbol{a}_i$. Then, $\boldsymbol{l}$ can be computed as

$$\boldsymbol{l} = 2 < \boldsymbol{h}, \boldsymbol{v} > \boldsymbol{h} - \boldsymbol{v}. \tag{2.89}$$

Even if every pixel has been scanned under many illumination directions, the problem can still be ill-conditioned. It is possible that particular surface parts appear in no view under glossy reflection, because for a certain illumination and viewing direction these reflection component can be observed only in very small vicinity of the normal's direction $\boldsymbol{a}$ around $\boldsymbol{h}$. Then, the parameters describing these components, $I_{g0,c}, c = r, g, b$ (see Equation 2.85) and c_g, can't be estimated for these pixels. Therefore, assumptions have to be made how an image can be segmented into regions with small fluctuations of these parameters. If some of the pixels inside such a region allow for an estimation of specular reflectance parameters, these parameters can be assigned to all other pixels in the region. An example are regions with similar hue values in color images [41].

Multiple Views. When compensation of illumination effects is done for various viewing directions of an object, the problem occurs, that the texture parts do not match to each other. The main reason for this is that the applied illumination model is not accurate enough. There are two alternatives to encounter this problem.

One way is to improve the model of illumination, reflection and image acquisition. The fundament for this can be a complete table of acquired camera intensities for all combinations of the parameters v, l, λ, etc. To establish such a table, extended measurements are necessary.

Another way is to blend corresponding texture parts smoothly into each other. This requires the determination of correspondences between pixels in adjacent views. The influence, that a texture element has on the blending operation, should depend on its reliability and resolution. Reliability can be modeled from the knowledge of surface slopes against illumination direction and from noise measurements on geometry and intensity data.

3 RECOGNITION AND INTERPRETATION

H. NIEMANN

INTRODUCTION

Recognizing objects and interpreting the meaning of static or mobile configurations of objects is a problem which has to be solved in different applications, like medical diagnosis, autonomous mobile systems, or remote sensing. Any system for the recognition and interpretation of images, image sequences, or sensor signals in general implicitly or explicitly makes use of a priori knowledge about the origin and properties of the image, about the objects, scenes, and events visible in an image, about the requirements of the user or the application, and about actions and conclusions which may be infered due to the image content. In a *model-based approach* to image recognition and/or interpretation all or at least a significant amount of this a priori knowledge is represented *explicitly* in a *model*. The model may contain *declarative knowledge*, that is, knowledge about structural properties, and *procedural knowledge*, that is, procedures computing certain attributes of the structural components. In the most general case an algorithm is provided which computes an interpretation based on the input image and the available model. Basically, this algorithm specifies which procedural knowledge to activate and which intermediate results to use for further processing. In this sense the algorithm computes a *processing strategy* or *controls* the interpretation process and hence will be referred to as the *control algorithm*. Two approaches to modeling, that is, semantic models and statistical models are treated in Section 3.2 and Section 3.3, respectively.

An interpretation of an image or image sequence should optimally fit to the observed images, should be maximally compatible with the stored model, and should contain most of the information relevant to the task-domain or application scenario. In this setting model-based interpretation of images is an *optimization problem*.

A basic problem occurs if a single static image or a volume-sequence of images is to be interpreted, Examples are single MR or CT images, or volume-equences of MR or CT images. To solve this problem processing often proceeds in three phases. At first, a segmentation of the image into segmentation objects like lines, regions, and vertices is computed. This topic is treated in Section 3.1. Then objects or in general semantically meaningful components in the image are classified as described in Section 3.4. Finally, in the third phase the meaning of configurations of objects is interpreted in the context of the task domain; this is described in Section 3.5.

If a time-sequence of images is to be interpreted, some specializations may be necessary or advisable due to the real-time requirements. In a time-equence of images new objects may become visible, visible objects may move out of the field of vision, the same object may be visible on many image frames, and an object needed to achieve some task may not yet be visible. In such cases it is not advisable to treat every image frame completely independent of the other ones. Rather, one should distinguish the *phases* of

- *initializing* the processing by making a full and detailed analysis of usually a few consecutive images, for example, using the techniques described in Section 3.4 and Section 3.5,
- *tracking* of recognized objects as long as they are visible, see e.g. Section 2.4 and Section 3.2.2,
- *detecting* new objects entering the field of vision, that is, being *attentive* to changes in the image content, see e.g. Section 2.4 or Section 7.4,
- *actively searching* for objects which are important for the fulfilment of the task as described in Section 3.6.

3.1 SEGMENTATION

D. PAULUS

Many applications of image processing and image analysis have been outlined in the previous chapters of this volume. More examples can be found in medical, industrial, military or geographical applications. We saw that simple *classification* problems can be solved with statistical methods and that analysis and understanding of complex patterns generally requires knowledge about the particular task domain.

But even knowledge-based image analysis systems tailored to a certain problem usually have some components which operate on images independently from the specific task. These components are commonly referred to as the image preprocessing stage. Typical operations are a variety of filters, geometric transformations and corrections, extraction of features, like e.g. gray-level histograms.

Preprocessing is not the only part of an image analysis system which can be designed without knowledge about the task domain. The analysis of complex scenes usually requires that the image is split into simpler components.

This process is called the *segmentation* of an image. Segmentation may in some cases be directed by the knowledge-base, but may in many cases as well be initially performed independently of the specific task.

3.1.1 Introduction

The term 'segmentation' is used in many publications with slightly different meaning. For a definition the following two statements are helpful:

Description of an image

"The description of an image is its decomposition or segmentation into simpler constituents (or pattern primitives or segmentation objects) and their relations, the identification of objects, events and situations by symbolic names, and if required, an inference about its implications for the task domain." [508, p.6]

Analysis of an image

"When analyzing (complex) patterns each pattern is given an individual symbolic description." [508, p.5]

Image segmentation thus is a process that transforms the image into a set of simpler constituents. The types of the objects detected depend highly on the algorithm used. Three major classes of algorithms are

- region-based segmentation which searches for objects fulfilling a certain homogenity criterion,
- line-based systems which detect discontinuities in the image function, and
- point-segmentation which identifies interesting points.

The objects detected in the segmentation may be represented in various ways as indicated in Table 3.1. One object in the scene may be represented by more than one representation at the same time. An example is a line which is detected as a chain code and is approximated by a spline.

An initial segmentation requires little if any knowledge about the structural properties of patterns. It results in a *segmentation object* which has certain *attributes*. A segmentation object is an initial symbolic description of an image. Typical attributes of segmentation objects are the location in two-dimensional image coordinates or in three-dimensional world coordinates, gray-level and color, texture, motion, depth, surface normal, shape and reliability or certainty of the detection. Attributes and relations between them are used to represent data driven evidence of parallelism, symmetry, curvature, and collinearity.

Since color images are now provided by most CCD cameras, the ideas of image segmentation which were introduced for gray-level images in the literature, have to be extended to multi-channel images. This is the case for line detection, region segmentation, as well as the identification if points of interest. In the following we will outline the principles of segmentation in their original form. If extensions to color images already exist, we will cite and outline them as well. If no such extension is known yet but is feasible, we will propose it.

Geometric object	Representation
Region	Contour-line
	Characteristic function
	Quad tree
	Run length code
Lines	Chain code
	Polygon
	Spline
Point	Coordinate pair

Table 3.1. Representations of results of the segmentation for 2D images

The organization of the section is as follows: Region segmentation of intensity images is introduced in [338]; we survey this topic in Section 3.1.2. Edge detection and line segmentation is formalized in [469, 106]; we describe these approaches in Section 3.1.3. Detection of corners, vertices, and interesting points is introduced in Section 3.1.4. These segmentation strategies are applied to range images in Section 3.1.5. The problem arises, how to represent these symbolic data; formalisms and data structures for this purpose and implementation issues are covered in Section 3.1.6. In Section 3.1.7 we refer to literature on performance valuation, which is crucial for image segmentation.

Many other ideas exist for image segmentation, e.g., region segmentation using texture information. Due to space limitation, we omit these subjects and refer to the literature, e.g. [190].

3.1.2 Region Segmentation

In the following we describe two region segmentation algorithms and mention a third approach. The first method extends an algorithm which was introduced in [191]. It extends and uses the 'split-and-merge' algorithm of [338], to color differences. The second is based on image morphology.

For the 'split-and-merge' algorithm we assume that we have a quadratic color image with a size which is a power of two; we organize the pixels initially as a quadtree which is a hierarchical data structure for efficient region representation.

The color mean vector $\mu_{i,j}$ in the square window

$$\boldsymbol{W}_{i,j} = [\boldsymbol{w}_{\mu\nu}]_{\mu,\nu=1,\cdots,n} = [\boldsymbol{f}_{a,b}]_{a=i,\cdots,i+n,b=j,\cdots,j+n} \tag{3.1}$$

of size $(n+1) \times (n+1)$ is computed from the color values

$$\boldsymbol{f}_{i,j} = (r_{ab}, g_{ab}, b_{ab})^T \tag{3.2}$$

(a)

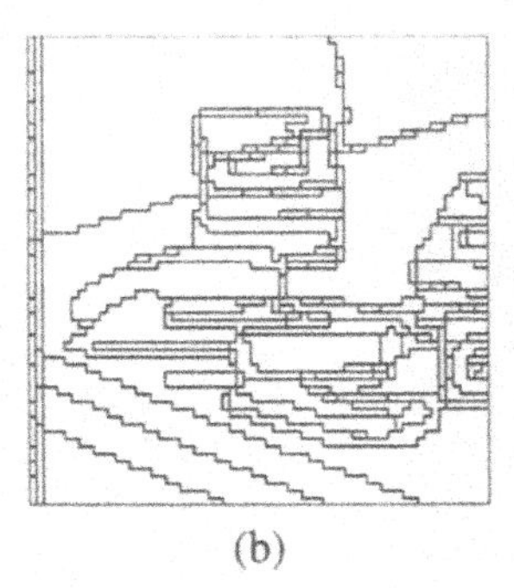
(b)

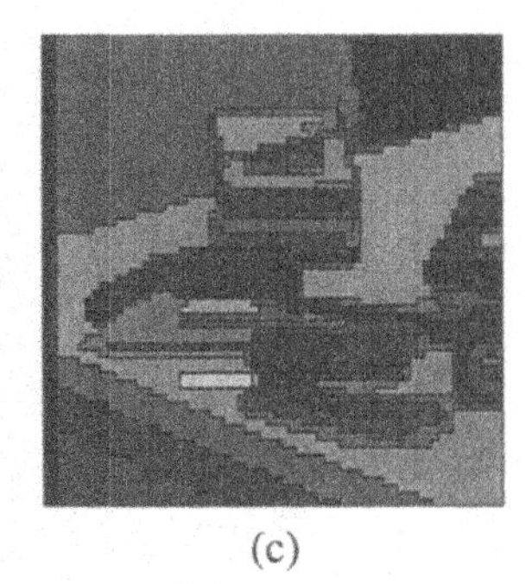
(c)

Figure 3.1. Example of region segmentation: input image (a), contours (b), contours overlayed to regions (c).

at positions $(a,b)^T$ within the window $\boldsymbol{W}_{i,j}$ by an average operation on the color channels:

$$\mu_{i,j} = \frac{1}{n^2} \sum_{a=i}^{a=i+n} \sum_{b=j}^{b=j+n} \boldsymbol{f}_{a,b}. \tag{3.3}$$

The variance is computed as the sum of variances in the color channels in [191]. To generalize this measure to color spaces other than RGB, the variance ${\sigma^2}_{i,j}$ in the square $\boldsymbol{W}_{i,j}$ was defined in [169] as follows, where $d(\mu_1, \mu_2)$ denotes the distance between μ_1 and μ_2

$${\sigma^2}_{i,j} = \frac{1}{n^2} \sum_{a=i}^{a=i+n} \sum_{b=j}^{b=j+n} \left(d(\boldsymbol{f}_{a,b}, \mu_{i,j})\right)^2. \tag{3.4}$$

If the image inside a window $\boldsymbol{W}_{i,j}$ is sufficiently heterogeneous, i.e., if the variance ${\sigma^2}_{i,j}$ is greater than a threshold θ_{v}, then we split the square into four sub-squares. The minimal size of the squares is a parameter of the algorithm which can be used to tune the speed and the desired accuracy and noise sensitivity of the segmentation.

During the subsequent merge phase, four adjacent squares which have a common direct ancestor in the quadtree are merged if their color distance is below a threshold θ_{m}. In a final grouping stage, similar regions are merged if their color distance is below a threshold θ_{g}. The result is no longer a quadtree. A typical sequence of intermediate results of the processing steps is shown in Figure 3.1.

The so-called *watershed transform* [730] has recently gained increasing interest in computer vision. This morphological image segmentation method can in some cases considerably enhance image analysis.

The watershed transform idea can be easily explained by help of a geographical imagination. Suppose a drop of water is falling on a topographic relief—it will run down until it reaches a local minimum. The influence zones of those local minima are called catchment basins. As the water level increases and two basins are going to flow together, a dam has to be built. The dam separates the local minima against each other and can be seen as a watershed (see Figure 3.2 (right)).

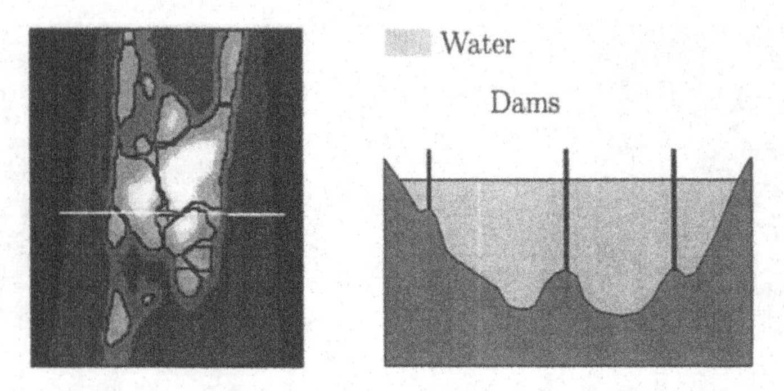

Figure 3.2. One-dimensional effect of the immersion simulation on a gray-level profile (right); cut from a real thermographic image showing the skin temperature on a human forearm (white line on the left), implementation according to Baxes [47]. Immersion depth 33.9^oC.

The two general ideas for the watershed transform are the *global* and the *local approach* to watershed segmentation. In the local approach, decision about further flow of water is based on the neighborhood of a pixel. In the global approach, all pixels are affected by the flooding. First we invert the image and—figuratively spoken—we put holes into each local minimum. Now the mountains are immersed into water which passes through the prepared holes. Whenever two basins unite, a dam is built (see Figure 3.2 (right)). The computed contour lines of these dams surround the areas searched for. During the flooding process, only the points on the border of the basin are required for computation. This contour based algorithm allows for an efficient implementation with linear complexity when these points are queued and inspected in parallel [730]. The immersion can be stopped at a certain level or can be continued until the highest mountain is covered by water and only dams remain visible. Partial flooding can be used to separate objects from the background. This strategy has been applied to segment thermographic images in a medical application [528].

Vincent [730] describes the watershed transform as a special case of a morphological operation. For segmentation purposes, the watershed transform can be treated as a region growing algorithm. Region segmentation requires the definition of a criterion for homogeneity [508]. Catchment basins are homogeneous in the sense that they contain exactly one local minimum together with the related influence zone.

3.1.3 Edge and Line Detection

Literally hundreds of edge detector schemes were proposed in the literature in the past. They can be categorized into three types:

- Gradient based approaches which compute discrete approximations of the partial derivatives f_x and f_y of the image function f.
- Edge masks which filter the image locally by a mask modeling a particular edge type and orientation.

-1	0	1
-2	0	2
-1	0	1

-1	-2	-1
0	0	0
1	2	1

-1	0	1
-1	0	1
-1	0	1

-1	-1	-1
0	0	0
1	1	1

Figure 3.3. Masks for Sobel (left) and Prewitt (right) operator. Masks on the left: f_x, masks on the right: f_y. Note that these masks may be flipped with respect to other literature since we choose the origin of the coordinate system on the left top.

- Local parametric models which approximate the image locally by a function modeling the edge.

We briefly introduce gradient methods and outline an optimal edge detector which uses a parametrized edge model. Edge operators create an edge image in which edges are combined to lines. These lines can be further appoximated to geometric primitives such as straight line segments or circular arcs. We present results of these processing steps.

Parametric models which approximate the image function as well as edge modeles have been proposed for edge detection [508, 529]. In the following we will briefly show methods based in the intensity gradient.

Two of the most common operators for gradient computation in gray-level images are the Sobel operator and the Prewitt operator. These linear operations can be computed by a convolution of the input image with the masks shown in Figure 3.3. In [151] it is shown that the Sobel mask is an approximation of the first derivative. The same idea can be applied to color images in a straight forward manner using color vectors instead of intensity pixels. Edge detectors result in an edge image which contain a matrix of edge elements. These edge elements may be either a representation of the intensity gradient, or—which is more common—an encoding as edge strength f_e and edge orientation f_o; edge orientation is perpendicular to the edge gradient and the strength is a measure computed from the norm of the gradient.

In many cases which are of practical relevance, color information does not considerably increase the quality of edge detection, although color region segmentation has been quite successful e.g. for the segmentation of traffic signs. This is due to the fact that most color edges for real objects are also conceivable in the gray-level equivalent; color regions can use a homogeneity predicate which is not based on intensity, e.g., only chrominance. Results of edge detection on the image Figure 3.1 (a) are shown in Figure 3.4; the result of the Sobel operator on the gray-level input image is shown on the left; the center image shows the edge strength for the color-Sobel operator. On the right we have the edge orientation image computed from the gray-level operator.

Several gradient operators are listed in [508, 529]. We now assume that the horizontal derivative $\boldsymbol{f}_x$ and the vertical derivative $\boldsymbol{f}_y$ can each be computed by a convolution of the image $\boldsymbol{f}$ with a mask:

$$f_x = f \star G_x, \quad (3.5)$$

$$f_y = f \star G_y. \quad (3.6)$$

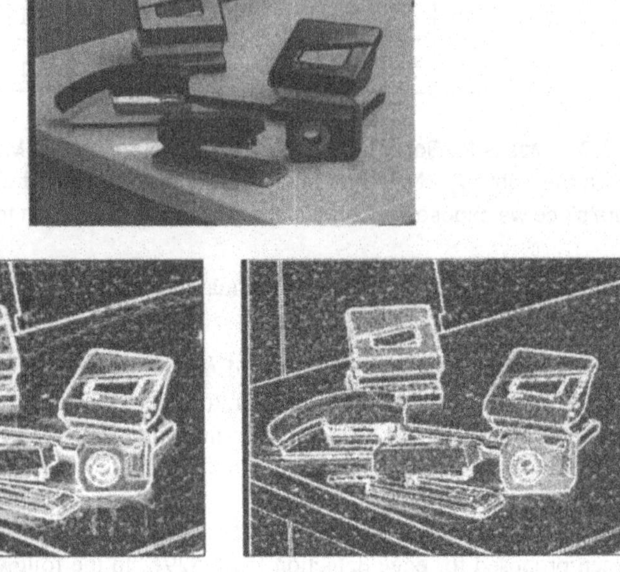

Figure 3.4. Comparison of edge detection in gray-level and color image. Top: original image, middle: edge strength, bottom: edge orientation. Left: gray-level segmentation, right: color segmentation.

Lines can now be computed from edge images tracking the edge elements with high edge strength along the direction indicated by the edge orientation. The resulting lines are often represented as chain codes (cmp. Table 3.2). These chains can be further approximated to find segments which are approximately straight or are circular arcs. Various line following and approximation schemes have been introduced in the literature, e.g. in [284]. A method that computes straight lines directly from edge images is the application of the Hough-transform on edge images. The idea is to represent the straight lines by $y = ax + b$ and to use the quantized parameter space for a, b. An accumulator which holds the quantized parameter values is initialized to 0. For each edge element (f_e, f_o) computed with a gradient operator at a position

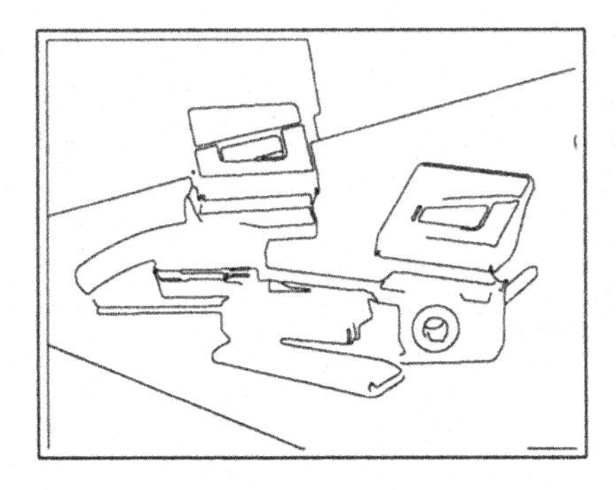

Figure 3.5. Straight lines and circular arcs from line approximation (left), and straight lines computed by the Hough transform (right).

$[i, j]$, we compute the parameter b from i and j and estimate the parameter a from f_o. If the slope a is finite, we increment the accumulator at a, b by f_e. If instead of the parametrization above, a line representation as $r = \cos \alpha x + y$ is chosen, the parameter array has a finite size but the results of edge detection cannot be used as parameters directly. Local maxima in the accumulator are used as indications for straight lines in the image.

Straight lines and circular arcs computed by the line following as described in [284], and straigth lines computed by the Hough-transform are shown in Figure 3.5.

The Canny-operator [106] was shown to be an optimal edge detector under fairly general assumptions, such as the assumption of step edges and the presence of white additive noise. It optimizes three criteria: detection, localization, and uniqueness. The final result of the fairly mathematical derivation in [106] is that the operator is computed from the convolution of the image f with $\partial\phi/\partial\boldsymbol{n}$ where $\phi[x, y] = \exp\left(-\left(x^2 + y^2\right) / \left(2\sigma^2\right)\right)$ is a two-dimensional Gaussian function and $\boldsymbol{n}$ is an estimate of a normal orthogonal to the edge orientation which can be computed form the edge derivatives. The edge orientation estimate as well as the parameters of the Gaussian have to be varied to obtain the optimal result. The result of the convolution is then analyzed to link edge elements to lines; an example is shown in Figure 3.6.

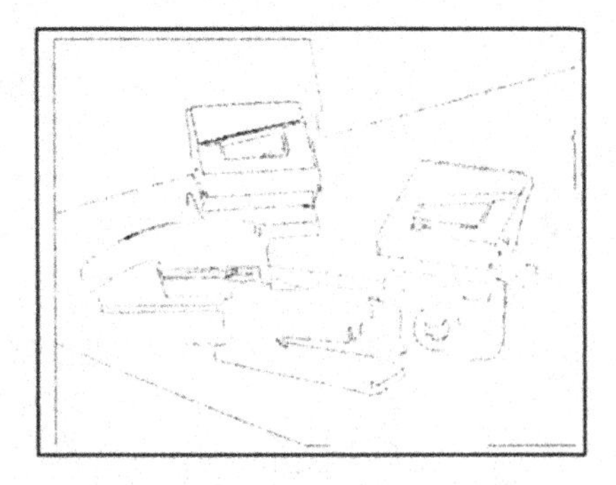

Figure 3.6. Result of Canny operator on the image in Figure 3.1.

3.1.4 Points, Vertices, Corners

A well-known method to find interesting points in an image is due to Moravec and is thus called the *Moravec interest* operator. He extended his own method in [490]; for each pixel in the input image, he uses a local window

$$\boldsymbol{W}_{ij} = [w_{ij,\mu\nu}]_{\mu,\nu=0,\ldots,7} = [f_{a,b}]_{a=i,\ldots,i+7,b=j,\ldots,j+7}\,. \tag{3.7}$$

The following features are computed from these windows:

$$m_{\mathrm{xx}ij} = \sum_{\mu=0}^{6}\sum_{\nu=0}^{6}\left(w_{ij,\mu\nu} - w_{ij,\mu\nu+1}\right)^2, \tag{3.8}$$

$$m_{\mathrm{yy}ij} = \sum_{\mu=0}^{6}\sum_{\nu=0}^{6}\left(w_{ij,\mu\nu} - w_{ij,\mu+1\nu}\right)^2, \tag{3.9}$$

$$m_{\mathrm{xy}ij} = \sum_{\mu=0}^{6}\sum_{\nu=0}^{6}\left(w_{ij,\mu\nu} - w_{ij,\mu+1\nu+1}\right)^2, \tag{3.10}$$

$$m_{\mathrm{yx}ij} = \sum_{\mu=0}^{6}\sum_{\nu=0}^{6}\left(w_{ij,\mu+1\nu} - w_{ij,\mu\nu+1}\right)^2. \tag{3.11}$$

The operator now is

$$H_{ij} = \min\left\{m_{\mathrm{xx}ij}, m_{\mathrm{xy}ij}, m_{\mathrm{yx}ij}, m_{\mathrm{yy}ij}\right\}. \tag{3.12}$$

The application of this operator on an image results in an interest map which is subsequently filtered by a Laplace-operator. The extension to color images is straight forward and uses the window

$$\boldsymbol{W}'_{ij} = [\boldsymbol{w}_{ij,\mu\nu}]_{\mu,\nu=0,\ldots,7} = [\boldsymbol{f}_{a,b}]_{a=i,\ldots,i+7,b=j,\ldots,j+7} \tag{3.13}$$

in a color image, i.e., the scalars $w_{ij,\mu\nu}$ in the Equations 3.8–3.11 are replaced by color vectors and a suitable difference operator. The square of the intensities is replaced by the scalar product of the resulting vectors. Figure 3.7 shows results of the interest operators a gray-level image and a RGB color image where the Euclidian distance of color vectors was used.

Another operator is the Harris corner detector [287]: for an image $\boldsymbol{f}$ we compute a gradient image whose horizontal and vertical components $\boldsymbol{f}_x$ and $\boldsymbol{f}_y$ are convolved with a Gaussian mask $\boldsymbol{G}_\sigma$ of standard derivation σ to obtain two smoothed images $\tilde{\boldsymbol{f}}_x$ and $\tilde{\boldsymbol{f}}_y$:

$$\tilde{\boldsymbol{f}}_x = \boldsymbol{f}_x \star \boldsymbol{G}_\sigma = \boldsymbol{f} \star \boldsymbol{G}_x \star \boldsymbol{G}_\sigma, \tag{3.14}$$

$$\tilde{\boldsymbol{f}}_y = \boldsymbol{f}_y \star \boldsymbol{G}_\sigma = \boldsymbol{f} \star \boldsymbol{G}_y \star \boldsymbol{G}_\sigma. \tag{3.15}$$

The default value for σ is $\sigma = 0.7$. At each position in the image $(i,j)^T$ we compute the matrix

$$\boldsymbol{M}_{ij} = \begin{pmatrix} M_{ij,11} & M_{ij,12} \\ M_{ij,21} & M_{ij,22} \end{pmatrix} \tag{3.16}$$

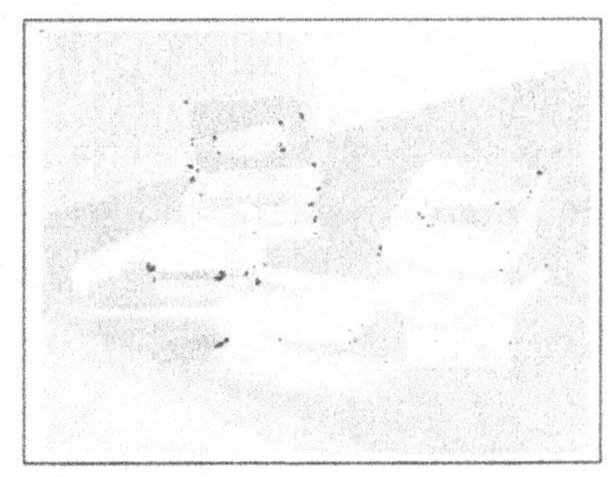

Figure 3.7. Interest map computed by Equation 3.12 for gray-level (left) and color images (right) overlayed to brightened input image from Figure 3.1.

with

$$M_{ij,11} = \tilde{f}^2_{x,ij}, \quad (3.17)$$
$$M_{ij,12} = \tilde{f}_{x,ij} \cdot \tilde{f}_{y,ij}, \quad (3.18)$$
$$M_{ij,21} = \tilde{f}_{x,ij} \cdot \tilde{f}_{y,ij}, \quad (3.19)$$
$$M_{ij,22} = \tilde{f}^2_{y,ij}. \quad (3.20)$$

The result of the Harris operator at position $(i,j)^T$ is now defined as

$$H_{ij} = \det(\boldsymbol{M}_{ij}) - k \operatorname{tr}(\boldsymbol{M}_{ij}), \quad (3.21)$$

where $k = 0.04$ is a fixed value for the computation. Again, the result is an interest map. In [174] it has been generalized to color in as

$$M_{ij,11} = \tilde{\boldsymbol{f}}^2_{x,ij}, \quad (3.22)$$
$$M_{ij,12} = \tilde{\boldsymbol{f}}_{x,ij} \cdot \tilde{\boldsymbol{f}}_{y,ij}, \quad (3.23)$$
$$M_{ij,21} = \tilde{\boldsymbol{f}}_{x,ij} \cdot \tilde{\boldsymbol{f}}_{y,ij}, \quad (3.24)$$
$$M_{ij,22} = \tilde{\boldsymbol{f}}^2_{y,ij}, \quad (3.25)$$

Figure 3.8. Interest map computed by the Harris operator for gray-level (left) and color images (right) overlayed to brightened input image from Figure 3.1.

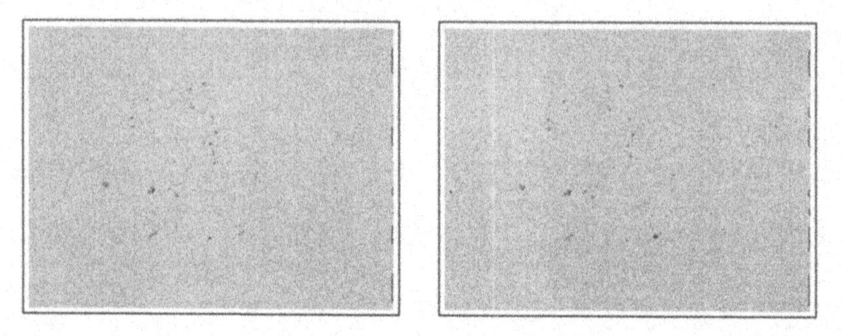

Figure 3.9. Interests map computed for gray-level (left) and color images (right).

where the products and squares are computed by scalar products of the vectors. Smoothing of the color gradient images is done componentwise.

Results of the gray-level and color version of this operator on the test image ot Figure 3.1 are shown in Figure 3.8. Figure 3.9 depicts a comparison of the three interest operators.

In addition to point operators which are mostly useful for point detection in image sequences, we also need detectors for points in single images. The demand for stability and roboustness can be satisfied when a larger context is used for the computation, as it is the case, e.g., by points which lie on lines. Points are detected by intersections or corners on these lines. Corners are identified by the analysis of the curvature of the lines. Thereby interesting points are found by post-processing lines resulting from line segmentation [284]. A result of point detections based on corners and vertices is shown in Figure 3.10.

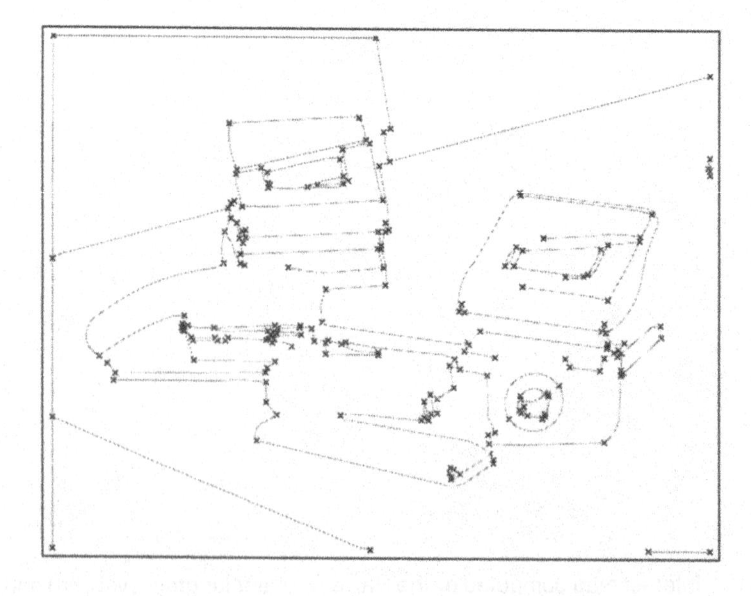

Figure 3.10. Segmented Corners and Vertices indicated by crosses on the lines.

Figure 3.11. Example of range image segmentation. Result of segmentation rendered with different colors corresponding to regions of different curvature characteristics (left). The other two images additionally show the minimal and maximal curvature coded by the brightness.

3.1.5 Segmentation of Range Images

Range images as provided by several sensor types (cf. Section 1.1) are also subject to image segmentation. The principles introduced for intensity images can be applied as well, if the depth information is represented as a matrix of distance values, i.e., as a range *image*. Instead of the intensity gradient, the surface gradient can be used for detection of discontinuities. In order to be less sensitive to noise, the gradient is often computed analytically from a parametric function whose parameters are estimated by an error minimization on a surface patch.

Surface patches can also be detected by region segmentation. The criterion of homogeneity can be, e.g., the minimal and the maximal curvature of the surface or the $H - K$ map (see Section 4.6). These parameters are invariant features with respect to viewpoint changes and can thus be used for object recognition [59] (cmp. Section 3.4). Often, a combination of intensity and range segmentation provides the best results.

An example of a triangle mesh that was reconstructed from multiple range images and segmented using the curvature estimation rule of Section 4.3.1 is shown in in Figure 3.11.

3.1.6 Data Representation

As described in Section 3.5, knowledge-based image analysis consists of a sequence of transformations from the input image over an initial segmentation to a symbolic description. The results of data driven segmentation thus have to be passed to an interface which provides the transfer of the information to the symbolic processing stage.

Data structures for the representation of the results of the segmentation process were presented by various authors as summarized in Table 3.2. A general data structure for the interface must have the expressional power to represent all kinds of information resulting from any segmentation process. Surfaces, lines, regions, vertices and various structural and temporal relations between any two of these objects may be detected during segmentation and have to be stored in these data structures. Naturally, representations for lines use representations of points, representations for regions utilize lines, e.g. for contour representation. The segmentation object [508, 529] has these properties and can be used as a common interface definition. To give an example, regions

Data structure	Author
Primal Sketches	Marr
2.5D Sketches	Marr
Recursive Structure for Line Drawings	Shapiro
Iconic-Symbolic Data Structure	Tanimoto
RSE-Graph	Hanson & Riseman
Line Adjacency Graph	Pavlidis
Region Adjacency Graph	Pavlidis
Region Graph	Fisher
Spatial Data Structure	Shapiro & Haralick
Segmentation Objects	Niemann & Paulus

Table 3.2. Candidates for interfaces from segmentation to knowledge based processing according to [530].

detected with the methods of Section 3.1.2 can be represented as chain codes and are stored as a set of contour lines in a segmentation object; some of the contours may be polylines composed of straight lines; data-driven estimation of parallelism results in relations between some of these lines which are also recorded in the segmentation object. A syntactic description of segmentation objects is given in [508, p. 74–75]; it is also mapped to a syntactic description of concepts of a knowledge base. An object-oriented implementation of segmentation objects can be found in [529]. The system is further extended to contain a hierarchy of image segmentation operators in [529].

3.1.7 Performance Evaluation

Objective and general methods for the evaluation of image processing algorithms, filtes, and segmentation are not available, yet. The major problem is to define the evaluation criterion which often depends on the application, whereas segmentation algorithms were implemented independently from the the application.

In order to compare the results of different algorithms which produce the same type of segmentation results, their parameters have to be known and have to be varied systematically. To give an example, the algorithm in Section 3.1.2 requires the choice of several thresholds θ_{v}, θ_{m}, etc. In most of the cases, the notion of a 'good result' will depend on the application and the task of the overal system. In the context of active vision (Section 3.6), the 'quality' of a result may be chosen sub-optimal, since higher computation speed may be crucial instead of high precision. 'Optimal' solutions such as the Canny operator, may be not the right choice since they require too much computation time or since the underlying assumptions are not valid.

Performance evaluation of computer vision systems is a problem of ongoing research; the sub-task of evaluating segmentation algorithms will be useful for overall evaluation [283]. A survey on evaluation of segmentation methods can be found in [773].

3.2 SEMANTIC MODELS

H. NIEMANN

An image (or observations derived from an image) which belongs to a certain task domain usually can be decomposed into simpler constituents or *segmentation objects*, for example, lines, vertices, or regions, as described in the previous section. Similarly, a time sequence of observations, denoted by a symbolic name, can be decomposed into sub-sequences or phases also denoted by symbolic names. For example, a scintigraphic image sequence showing a 'heart beat' consists mainly of the 'systolic' and 'diastolic' phases. It is a common experience that an image belonging to some task domain has a certain *structure*, and, therefore, not any arrangements of simple constituents result in a valid image belonging to the task domain. This results in spatial, temporal, and/or logical *constraints* which are necessary for image interpretation and may be exploited to reduce the complexity of interpretation.

For the purpose of image interpretation these constraints contain the *knowledge* about the task domain. The relevant knowledge is represented in the computer in a *semantic model*; we will use in this section the term *model* for short. As pointed out in Section 3, in a model-based approach to image interpretation an algorithm is provided which computes an interpretation using the model and the input data. Here we concentrate on the model, not on the algorithm using it. The advantage of a clear separation between the model and the algorithm using it for interpretation is that a change in the model will not affect the algorithm and vice versa. In general a useful model should be as simple as possible in order to facilitate computationally simple and efficient algorithms for its usage, be sufficiently complex in order to adequately represent the relevant aspects of the task domain, and allow at least the automatic estimation of its parameters from given data.

There are different representations of models, in particular we mention the following:

- *Semantic networks*, that is, special directed graphs containing conceptions in nodes and relations in links.

  Examples of this representation are given in [510, 660, 591]. Reasoning in such models or image understanding using such models is mainly based on graph search algorithms and combinatorial optimization, as pointed out in Section 3.5.

- *Spatio-temporal models* of a dynamical system, that is, a system of first-order differential equations.

  Examples of this representation are given in [178, 377]. Its usage mainly amounts to the estimation of a state vector such that an optimal fit between the model and the observations is achieved as pointed out in Section 3.5.4.

- *Bayes networks* (or graphical models), that is, graphs containing causal relations together with probabilities of occurrence.

  This representation is between semantic and statistical models and, therefore, will be treated below in Section 3.3 under statistical models. Examples of its use are given in [101, 56].

- *Rule-based schemes*, that is, the representation of constraints in the form of 'IF <condition>, THEN <conclusion>'. This can be augmented by a judgment (a real number) which gives a measure of confidence in the conclusion or by special conditions which further constrain the applicability of rules as is done in coordinate grammars.

  Examples of this representation are given in [97, 473, 241, 480]. Image understanding using this framework is based on a control stragey for the application of rules to given data as pointed out briefly below.

- *Formal logic*, the 'classical' approach based on well-known predicate calculus. We will not consider this further because of the problem to integrate a judgment of the reliability of results into this formalism. Examples of this representation are given in [585].

3.2.1 Semantic Networks

As mentioned above, knowledge about the task domain is explicitly represented in a model $\mathcal{M}$. In the case of a semantic network this is a special graph whose nodes are concepts C. In order to facilitate the design of a *task-independent* control algorithm which uses the model (and the result of initial segmentation) to interpret an image, all concepts representing whichever item of knowledge have the *same* syntax. There are different definitions of semantic networks; we present the one developed in [510, 591].

Nodes. The designer of a semantic network will start with an analysis of the task domain to determine the relevant items. These may be, for example, conceptions like 'heart', 'systolic phase', 'normal motility', and so on. They may be individuals, e.g. one particular house, or classes of individuals, e.g. houses. A conception is represented internally (in the computer) by a recursive formal data structure which contains the information relevant to the task domain. For example, for autonomous driving of cars on a highway the price of a car will not be relevant, but the size and shape. The internal formal representation is called a *concept*. It is a *node* in the graph which makes up the model of the task domain. The concept is referenced by a name which may be identical to the colloquial conception. For example, the conception 'heart' is represented internally by a concept which may be named 'heart'.

The goal of processing is to associate subsets of pixels with concepts. If some subset of pixels is computed to meet the definition of a concept C with sufficient reliability, an *instance* $I(C)$ of the concept is created, linked to this concept C, and stored in a memory of intermediate results.

The creation of an instance for some subset of pixels may cause restrictions on the possible interpretations of the remaining parts of the image. These restrictions are represented in a *modified concept*. It contains knowledge specialized according to current results of processing. For example, if a subset of pixels was identified as a 'lake', it becomes unlikely to find a 'car' on the lake, but more likely to find a 'boat'.

Only three types of nodes are defined in a semantic network, namely the concept, the instance, and the modified concept as outlined above. In this sense, all semantic networks following the above definition have the *same syntax* regardless of the task domain, but their *content*, i.e., the knowledge represented by them, depends on the task domain.

An example of a semantic network designed for active vision is given in Section 3.6, Figure 3.22.

Links. In a model there will be relations between concepts; they are represented by *links*. Again, only a fixed and small number of link types is defined. In order to allow for a well-structured representation of knowledge, the three links or relations 'part', 'concrete', and 'specialization' are introduced.

The distinction between parts and concretes allows one to represent relations between different *levels of abstraction* or different conceptual systems. For example, a 'left ventricle' is a physical *part* P of a heart attached to the concept 'heart' by the link 'part' which naturally is in the conceptual system of 'hearts'. The same holds for a 'house' and its 'roof'. A left ventricle may 'contract', but a contraction is not a physical part of a heart but a special type of 'motion'. Hence, a left ventricle would be related to a contraction as a *concrete* K by the link 'conc'. Another view of a concretization is that it is closer to the pixel data, for example, a 'boundary line' as a concrete of 'left ventricle' is closer to the pixel data.

In addition, for the parts a distinction is made between context-independent and -dependent parts in order to handle context-dependent relations between objects and their parts. With respect to the usage of a knowledge base this means that a context-independent part may be detected or infered *without* having an instance of the superior concept, whereas a context-dependent part can only be infered *after* having at least a *partial instance* of the superior concept. For example, the 'hood' of a car usually can be detected in an image without having the car, but a 'leg of a chair' usually can be infered from straight lines only if the chair as a context has been established.

The *specialization* link V is introduced in order to have compact knowledge bases and to define hierarchies of conceptions. It is common to imply that all defining elements (parts, concretes, attributes, relations) of a more general concept are handed down to the more special one or are *inherited* by it unless explicitly defined otherwise. For example, an 'contraction' is a special 'motion' related to it by the link 'spec'.

Attributes, Relations, and Judgment. In addition to the links relating a concept to other concepts in the model, a concept C usually needs for its definition attributes A and structural relations S between attributes. Each attribute references a function (or procedure) F which can compute the value of the corresponding attribute. An *attribute* has a type, e.g. length, color, position, and a value which may be a real number (e.g. for length) or a symbol out of a finite set of symbols (e.g. for color).

In general there may be *structural relations* S between the attributes A_C of a concept or between attributes A_P and A_K of its parts and concretes, respectively. Since in image processing noise and processing errors are inevitable, relations usually can only be established with limited precision. Therefore, each relation references a function F computing a measure of the degree of fulfilment of this relation, for example, in the sense of a fuzzy relation. An example is the degree of fulfilment of the relation 'right angle' between two lines.

Finally, every concept C has an attached function F computing a *judgment* $G(I(C))$ of an instance $I(C)$. The judgment may be a goodness or value or alternatively a cost or loss, but obviously this does not make an esential difference. The judgment G may be a scalar or a vector.

The functions associated to a concept are the active elements of the model; they compute values during instantiation. The concepts and links may be viewed as *declarative knowledge*, the functions as *procedural knowledge*. The procedural knowledge is not restricted. It may be, for example, a simple function computing the area or a cricle, a set of rules to compute a diagnostic inference, or a dynamic programming algorithm to match two waveforms.

Concept. In summary, the structure of a concept C is defined by

$$\begin{aligned} C \;=\; & ((D : T_C), && \texttt{name} \\ & (P : C)^*, && \texttt{parts} \\ & (K : C)^*, && \texttt{concretes} \\ & (V : C)^*, && \texttt{specializations} \\ & (L : I)^*, && \texttt{instances} \\ & (A : (T_A \mapsto F))^*, && \texttt{attributes} \\ & (S(A_C, A_P, A_K) \mapsto F)^*, && \texttt{relations} \\ & (G \mapsto F)) && \texttt{judgment} \end{aligned} \tag{3.26}$$

This equation is obvious in view of the above explanations. To obtain compact representations of knowledge it is useful to allow different sets of parts and concretes alternatively for the definition of a concept. For example, this allows one to define chairs having four or five legs in one concept instead of two. Each alternative definition is represented in a *set of modality* H of context-dependent and -independent parts as well as concretes

$$\begin{aligned} H \;=\; & ((P_{\mathrm{ci}} : C^+)^*, && \texttt{contextindep. parts} \\ & (P_{\mathrm{cd}} : C^+)^*, && \texttt{contextdepend. parts} \\ & (K : C^+)^*) && \texttt{concretes} \end{aligned} \tag{3.27}$$

An arbitrary number of sets of modality then will replace the components $(P : C)$ and $(K : C)$ in 3.26. A further generalization is the introduction of obligatory sets of modality H_{obl} which *must* be present in order to compute an instance of a concept and optional sets H_{opt} which *may* be present.

In order to facilitate a task-independent and efficient control algorithm it is useful to request that a model $\mathcal{M}$ of a task domain is *cycle-free*, that is, there is no path in the semantic network leading from a concept C via links of different type back to C, and *consistent*, that is, between two concepts there cannot be links of different type. The designer of a model has to ensure these restrictions which are also meaningful for knowledge representation. A model represented by a semantic network then is a graph having three types of links, the part, the concrete, and the specialization and one type of node, the concept. During processing two more node types are used, the modified concept and the instance, and one more link type, the instance link L.

Instance. In knowledge-based image analysis one tries to determine which objects and so on are actually present in the image. The occurrence of an object is represented by an *instance* $I(C)$ of the corresponding concept C. The relation between the concept and its instance is represented by a link L from C to $I(C)$. An instance is represented by a structure identical to 3.26 except that references to functions are replaced by the actual values computed by those functions from the image. Since due to noise and processing errors the occurrence of an instane can only be infered with limited certainty and precision, a *judgment* G is computed for every instance of a concept as mentioned above.

Due to the hierarchical and cycle-free representation of a knowledge base there are concepts having no parts and no concretes. Such a concept is called a *primitive concept.* The instantiation of primitive concepts is done, by definition, by means of segmentation results. This provides the *interface* between initial segmentation and knowledge-based processing.

The process of instantiation can be defined and implemented by task-independent *rules*. They define the conditions to compute a partial instance (which is necessary to handle context-dependent parts), to instantiate a concept by obligatory components H_{obl}, to enlarge an instance by optional components H_{opt}, to compute a modified concept top-down or bottom-up, or to compute modified concepts from initial segmentation results. Details of the instantiation of concepts have to be omitted here. In principle, in order to compute an instance of a concept, first its obligatory context-independent parts and its obligatory concretes must be instantiated. If this is the case, an instance can be created and its attributes, relations, and judgment computed. An obligatory context-dependent part can only be instantiated if a partial instance of its superior concept is available. The instance may be extended by computing instances of optional parts and concretes. The handling of competing instances in order to compute optimal instances efficiently by graph search is described in Section 3.5.

Function Centered Representation. It is possible to convert a semantic network consisting of linked concepts to a representation based on the functions F in the concept definition 3.26. This is natural since the functions for computing values of attributes, relations, and judgments are the active elements in the model. Each computation by a function needed during instantiation is represented by a node of a directed acyclic graph. Basically, from the arguments of the functions F a dependency structure is created showing which functions may be computed in parallel and which sequentially because their arguments need results provided from other functions. Due to the rigid syntax of a semantic network an automatic conversion of a model to this dependency structure is possible.

The function centered representation is the basis for a control algorithm computing optimal instances by combinatorial optimization as described in Section 3.5. It may also be used for parallel computation of instances [225].

3.2.2 Spatio-Temporal Models

Dynamical System Model. The motion of (rigid) bodies is described by a system of first order differential equations. They define the time change $d\boldsymbol{q}(t)/dt$ of the state vector $q(t)$ depending on the current state, the input $u(t)$ and random noise $w(t)$ in the most general form by a non-linear differential equation

$$\frac{dq(t)}{dt} = \phi\left(q(t), u(t), w(t)\right) . \tag{3.28}$$

In general, the state vector is not directly observable, but only an observation (or measurement) vector $o(t)$, to be considered below. The nonlinear equations are linearized to facilitate analytical treatment

$$\frac{dq(t)}{dt} = A(t)q(t) + B(t)u(t) + w(t) . \tag{3.29}$$

To allow numerical solutions on a (digital) computer, the continuous dynamical system model is converted to a discrete dynamical system model

$$q[k] = \Phi[k-1]q[k-1] + B[k-1]u[k-1] + w[k-1] , \tag{3.30}$$

which is evaluated from time step $k-1$ to time step k.

The explicit form of the equations can be worked out if the motion is specified in detail. Moving vehicles usually are constrained to motion in a plane, often along a road with certain width and curvature. The motion path of the vehicle on the road may be modeled at time k, for example, by a circle with its center in the plane of motion. This way space and time coordinates are related due to the equations of motion.

For example, in [377] different state vectors for ego-motion, vanishing point, and road recognition are used to interpret TV-image sequences of a driving car. The state vector for ego-motion consists of:

- the world coordinates of the origin of the ego-coordinate system in the road plane,
- the direction of the axis of the car in the world system,
- the absolute value of the translational velocity,
- the absolute value of the rotational velocity of the ego-car in the road plane.

Observation Model. As mentioned above, the state vector cannot be observed directly. Rather, it has to be estimated from observations $o(t)$ of object features, for example, prominent lines or corners. The observation will depend on the current state of the moving vehicle corrupted by noise $r(t)$, yielding

$$o(t) = \psi\left(q(t), r(t)\right) . \tag{3.31}$$

Again, the nonlinear dependence is linearized and converted to discrete form

$$o[k] = H[k]q[k] + r[k] . \tag{3.32}$$

If the geometry of the moving vehicle is known, if image features have been selected, and if the parameters of the camera taking an image of the object are known, the dependence of the observation on the state can be computed explicitly. We will not (and, due to space limitations, cannot) treat geometric modeling of a vehicle and feature extraction from images here. The camera projection model is described in Section 2.3.

3.2.3 Rule-Based Systems

Rule-based systems are the basis of the popular expert systems. Detailed treatment is given, for example, in [97, 241]. A rule-based system (or production system) consists of three main components:

- A set of rules (or productions) in the form

$$\text{IF } < \text{condition} >, \text{ THEN } < \text{conclusion} > \quad \text{or}$$
$$\text{condition} \rightarrow \text{conclusion} \, ,$$

  where synonyms for 'condition' are 'premise', 'left side' or 'antecedent' and for 'conclusion' they are 'action', 'right side' or 'consequence'.

- A set of data which are inspected by the rules to check whether the 'condition' part of the rule is fulfilled by a subset of the data. The 'conclusion' part of the rule usually modifies data.

- A control strategy which selects rules for execution. This is necessary because in general several subsets of data may fulfil several 'condition' parts of rules. In this case it must be decided which rule to activate on which subset of data.

The advantages of this approach are that every rule may encode a small and well-defined piece of knowledge, that new knowledge may be added to the system by adding new rules, that the system operation is defined by available data and not by available program code, and that it is straightforward to add a judgment of the reliability of conclusions to the formalism. Problems with this approach result from the fact that large rule bases are difficult to assemble and to maintain, that the effect of competing rules is hard to anticipate, and that the strategy for selecting data and rules for execution is not obvious.

In principle, the format of a rule is arbitrary resulting in a completely general formalism; in practice, the rule format is constrained in implemented systems. An example is a rule format according to

$$\text{IF } A \text{ AND } B \text{ OR } \neg C, \quad \text{THEN } D \, . \tag{3.33}$$

It allows to form the left side from the logical operations AND, OR, and NOT. The right side is one conclusion; if two alternative conclusions can be drawn, for example, D OR E, two rules would be used, one with right side D and one with right side E. Some modifications and generalizations of this simple rule format are to denote a rule by a symbolic name for further reference, to have conditions which consist of attribute-value pairs, to have conclusions of type 'make' a new data element, 'modify' an existing data element, and 'remove' an existing data element, or to allow variables in the condition part which are matched by any data element, to name a few. The coordinate grammars developed in [480] may be viewed as a special rule format. They include, among others, equations constraining the applicability of rules.

The degree of fulfilment of the predicates of the condition part often cannot be assured with certainty or probability one. We denote the degree of fulfilment of a predicate A by $cf(A)$, where cf stands for certainty factor. The certainty factor may be based on probability, basic probability assignments, or on fuzzy logic [558]. All of them provide a formalism for combining evidence. For example, if fuzzy logic is used, the certainty factor of the condition part in 3.33 is

$$cf(A \text{ AND } B \text{ OR } \neg C) = \max\{\min\{cf(A), cf(B)\}, 1 - cf(C)\} . \tag{3.34}$$

A simple, but in many applications sufficient approach to defining the degree of fulfilment of the conclusion is the identitiy mapping from condition to conclusion, that is, in 3.33 one obtains

$$cf(D) = cf(A \text{ AND } B \text{ OR } \neg C) . \tag{3.35}$$

In certain task domains the relevant knowledge is available from experts or text books. This is the case, for example, in cases of medical diagnosis and therapy. A statement may be 'IF the motion of the left ventricle is weak and one of its four segments (i.e., the inferioapical, postero-lateral, basal, or septal segment) shows akinetic behavior, THEN the motional behavior of the left ventricle is akinetic'. The implicit assumption is that there is an expert or that there are algorithms which can determine from suitable sensor data whether 'the motion of the left ventricle is weak' and whether 'one of its four segments shows akinetic behavior'. In the rule format such statements are logical predicates whose truth value has to be provided by functions computing the truth value. The above statement given in natural (English) language can be transformed to the rule

$$\begin{array}{rl} \text{IF} & (\text{lv-weak AND } [\text{ia-akin OR pl-akin OR b-akin OR s-akin}]) \\ \text{THEN} & \text{lv-akin} . \end{array}$$

Integrating a fuzzy judgment according to Equation 3.34 results in

$$\begin{aligned} cf(\text{lv-akin}) &= \min\{cf(\text{lv-weak}), \max\{cf(\text{ia-akin}), cf(\text{pl-akin}), \\ &\qquad cf(\text{b-akin}), cf(\text{s-akin})\}\} . \end{aligned}$$

Details about this example, including the image processing part to determine from an image sequence the relevant predicates, are available in [509]. Another application to image processing is treated in [473].

The control strategy determines pairs of data elements and condition parts of rules matched by the data. A pair [rule name, list of data matched by the condition part of this rule] is called an *instantiation* of the rule. The set of these pairs is the *conflict set*. It is the task of the control strategy to select from the conflict set one instantiation of a rule which is to be executed. The most simple strategy is to order data and rules and execute the first data subset on the first matching rule. However, results of processing then are dependent on the order or rules and data. Another strategy is to execute all rules in the conflict set quasi parallel. This may result in conflicting or contradictory results, for example, 'at image location x is a house' and 'at image location x is a swimming pool'. Conflicting results have to be resolved later. Control strategies are described, for example, in [155, 306].

The straightforward solution to determination of the conflict set is to loop over all data and over all condition parts of rules to find matching pairs. The RETE-algorithm avoids this double loop by checking for changes in the data set and by arranging conditions in a tree-like sorting network in order to check the same condition occurring multiply in several rules only once.

3.3 STATISTICAL MODELS

H. NIEMANN

The interest in statistical models as well as their success in diverse applications mainly arises from the basic result of statistical decision theory that in order to classify observations whith minimal probability of error one has to decide for the alternative having maximal a posteriori probability. In addition, probability theory provides a sound basis of theorems and results, and statistics provides a sound basis for estimating the parameters of statistical models. We denote the observation (or the features) by a vector $\boldsymbol{o}$ of real numbers, and this may be the pixels of an image, the features extracted from an image, a time sequence of signal values, or a time sequence of feature vectors. The set of K alternatives or classes is $\Omega = \{\Omega_1, \ldots, \Omega_K\}$. The posterior probabilities can be computed from

$$p(\Omega_\kappa|\boldsymbol{o}) = \frac{p(\Omega_\kappa)p(\boldsymbol{o}|\Omega_\kappa)}{p(\boldsymbol{o})} \quad \kappa = 1, \ldots, K\,. \tag{3.36}$$

The central problem is to find a suitable *statistical model* of the observation $\boldsymbol{o}$ given the class Ω_κ which is represented by the conditional probability density $p(\boldsymbol{o}|\Omega_\kappa)$.

In general, a statistical model has a certain structure, which usually is derived by the designer, and a set of unknown parameters, which is estimated from given data. Therefore, an essential requirement for statistical modeling is the availability of a sufficient amount of data in order to *learn* (or estimate, or train) parameters with sufficient reliability. Collecting data is cumbersome, but once they are available the parameters can be trained mainly automatically without manual intervention. This is a significant advantage of statistical models.

Initially statistical modeling in pattern recognition started with a kind of elementary and simple distribution for the feature vector $\boldsymbol{o}$ which meanwhile has been generalized in various ways. We consider the following approaches:

- Elementary parametric or nonparametric distributions, usually unimodal, are sufficient in some practically useful applications.

- Mixture distributions or mixture models were introduced to model multimodal distributions.

- Statistical dependencies covering a limited neighborhood can be accounted for in *Markov* random fields.

- Arbitrary, but limited, statistical dependencies can be modeled by *Bayes* networks. These may be viewed as a special type of semantic models, but are treated here, and not in Section 3.2, because we think that statistical inference in *Bayes* networks is the main point.

- The introduction of 'hidden' variables, that is, of variables which are not directly observable, allows, for example, the modeling of time coherence by *Markov* models.
- The (usually unknown) correspondence between image features and model features can be described by a statistical distribution, too. Marginalization over the unknown correspondences eliminates the correspondence problem.
- Coordinate transforms can be introduced into the probability density to account for object translation and rotation.

These generalizations and extensions were accompanied by the development of powerful algorithms for the estimation of the statistical parameters of the models from data which mainly are based on maximum-likelihood estimation combined with the expectation-maximization algorithm.

3.3.1 Elementary Models

A standard elementary statistical model is a normal (or *Gauss*) distribution for the feature vector $\boldsymbol{o}$. It has the advantage of being completely determined by the class conditional mean vector μ_κ and covariance matrix Σ_κ. Another possibility is to represent the probability density by a histogram, which is a nonparametric technique. However, estimating an n-dimensional histogram from data for reasonably large dimension n of the feature vector is clearly infeasible. So this approach is useful, in general, only if in addition class-conditional independence of the features is assumed. In summary we get

$$p(\boldsymbol{o}|\Omega_\kappa) = \begin{cases} p(\boldsymbol{o}|\boldsymbol{a}_\kappa) & : \text{ general case} \\ \mathcal{N}(\boldsymbol{o}|\mu_\kappa, \Sigma_\kappa) & : \text{ normal distr.} \\ \{p(o_{1,i_1}, \ldots, o_{n,i_n})|i_\nu = 0, \ldots, L_{\nu-1}\} & : \text{ histogram} \\ \prod_{\nu=1}^{n} p(o_\nu|\boldsymbol{a}_\kappa) & : \text{ stat. indep.} \end{cases} \tag{3.37}$$

whith the normal distribution defined by

$$\mathcal{N}(\boldsymbol{o}|\mu_\kappa, \Sigma_\kappa) = \frac{1}{\sqrt{|2\pi\Sigma_\kappa|}} \exp\left[-\frac{(\boldsymbol{o}-\mu_\kappa)_t \Sigma_\kappa^{-1}(\boldsymbol{o}-\mu_\kappa)}{2}\right] . \tag{3.38}$$

The unknown parameters $\{\mu_\kappa, \Sigma_\kappa, \ \kappa = 1, \ldots, K\}$ are obtained by maximum-likelihood (ML) estimation from given data. One has a problem of *supervised learning* if the class κ of every observation is known. In this case the k mean vectors and covariance matrices are obtained from k ML-estimations. If the class of an observation is unknown, a problem of *unsupervised learning* results. It still can be solved under general conditions from the ML-approach, but the k estimation problems are not independent. The standard solution is the expectation-maximization algorithm. We consider briefly a special case of it in the next subsection.

3.3.2 Mixture Models

In principle, a mixture distribution can be obtained from an arbitrary family of parametric densities. In practice, only the mixture of normal densities is in use. It is given for every class Ω_κ by

$$p(o|\Omega_\kappa) = \sum_{l=1}^{L_\kappa} p_{\kappa,l}\mathcal{N}(o|\mu_{\kappa,l},\Sigma_{\kappa,l}), \quad \sum_l p_{\kappa,l} = 1 . \tag{3.39}$$

In order to reduce the complexity of computations the covariance matrix is often constrained to a diagonal matrix.

Even if the class of all patterns is known, it is unknown which mixture component l a particular observation should update. Therefore, parameter estimation is not as straightforward as in the supervised learning case, but still can be handled by the ML-approach. The unknown parameters are for every class $\{p_{\kappa,l}, \mu_{\kappa,l}, \Sigma_{\kappa,l} \mid l = 1, \ldots, L_\kappa\}$. The resulting ML equations are

$$p(l|^\varrho o, \kappa) = \frac{\widehat{p}_{\kappa,l}\, p(^\varrho o|\mu_{\kappa,l},\Sigma_{\kappa,l})}{\sum_{l'} \widehat{p}_{\kappa,l'}\, p(^\varrho o|\mu_{\kappa,l'},\Sigma_{\kappa,l'})}, \tag{3.40}$$

$$\widehat{p}_{\kappa,l} = \frac{1}{N_\kappa}\sum_{\varrho=1}^{N_\kappa} p(l|^\varrho o,\kappa), \tag{3.41}$$

$$\widehat{\mu}_{\kappa,l} = \sum_{\varrho=1}^{N_\kappa} \frac{p(l|^\varrho o,\kappa)}{\sum_{\varrho'} p(l|^{\varrho'} o,\kappa)}\, {}^\varrho o, \tag{3.42}$$

$$\widehat{\Sigma}_{\kappa,l} = \sum_{\varrho=1}^{N_\kappa} \frac{p(l|^\varrho o,\kappa)}{\sum_{\varrho'} p(l|^{\varrho'} o,\kappa)}\, ({}^\varrho o - \widehat{\mu}_{\kappa,l})\,({}^\varrho o - \widehat{\mu}_{\kappa,l})^T . \tag{3.43}$$

It is a coupled system of equations which is solved iterartively using a variant of the *expectation-maximization* (EM) algorithm. This algorithm can handle the problem of 'missing information'; in this case the information about the involved mixture component is missing.

The unknown parameters, including L_κ, the number of mixture components, are initialized. One possibility is to obtain L_κ from vector quantization, another is to start with $L_\kappa = 1$ and compute standard ML-estimates of $\mu_\kappa, \Sigma_\kappa$, since we assumed classified observations. Then the mean is split into

$$\mu_{\kappa,l_1} = \mu_{\kappa,l} - \delta, \quad \mu_{\kappa,l_2} = \mu_{\kappa,l} + \delta , \tag{3.44}$$

with δ a small quantity, if

$$\sum_{\varrho=1}^{N_\kappa} p(l|^\varrho o, \kappa, l) > \theta . \tag{3.45}$$

With the new values the above equations are iterated until convergence. This process of splitting the mean and iterating the parameters may be repeated. Mixture densities combined with parameter training by the EM-algorithm has proven to be a powerful approach to stochastic modeling.

A further generalization is achieved by taking into account the observation and its *position* in an image $\boldsymbol{f}$, that is, the triple $[j, k, o_{j,k} \mid j, k = 1, \ldots, M]$, where M is the number of pixels in x and y direction. Summarizing the unknown parameters of the probability density in a vector θ one gets

$$p(\boldsymbol{f} \mid \theta, \kappa) = p(\{[j, k, o_{j,k}] \mid j, k = 1, \ldots, M\} \mid \kappa; \theta) \ . \tag{3.46}$$

To obtain computationally feasible statistical models of this type different specializations were investigated. One example of such a specialization is the use of histograms [610]. Three other types of specialization will be described briefly below.

The first specialization is to consider the probability density of an observation given a grid point. Of course, the grid points used in this modeling may be obtained from a suitable subsampling of the original image grid points, and the observation may be the image intensity itself or some feature vector computed from the intensity values. The *joint probability* of the observation $\boldsymbol{o}_{j,k}$ at the randomly selected image point (j, k) is

$$p([j, k, \boldsymbol{o}_{j,k}]^T | \kappa; \theta) = p(j, k | \kappa) p(\boldsymbol{o}_{j,k} | j, k, \kappa; \theta) \ . \tag{3.47}$$

Assuming mutually independent intensity values and image points results in

$$p(\boldsymbol{f} | \kappa; \theta) = \prod_{j=1}^{M} \prod_{k=1}^{M} p(j, k | \kappa) p(\boldsymbol{o}_{j,k} | j, k, \kappa; \theta) \ . \tag{3.48}$$

This type of density was used in [553] for 2D and 3D object recognition and localization with wavelet coefficients as observation vectors.

The second specialization is to factor the joint probability density (Equation 3.46) just the other way, that is, to consider the density of grid points given the intensity. The resulting joint density of the observation at a grid point then is

$$p([j, k, \boldsymbol{o}_{j,k}]^T | \kappa; \theta) = p(\boldsymbol{o}_{j,k} | \kappa) p(j, k | \boldsymbol{o}_{j,k}, \kappa; \theta) \ . \tag{3.49}$$

Assuming again mutually independent intensity values and image points gives

$$p(\boldsymbol{f} | \kappa; \theta) = \prod_{j=1}^{M} \prod_{k=1}^{M} p(\boldsymbol{o}_{j,k} | \kappa) p(j, k | \boldsymbol{o}_{j,k}, \kappa; \theta) \ . \tag{3.50}$$

The third specialization considers the density of segmentation results. Let preprocessing and segmentation algorithms map the observed image to a set of features, e.g. 2D corner points. Denote the set of corner points by $\boldsymbol{O} = \{\boldsymbol{o}_k | k = 1, 2, \ldots, m\}$, where $\boldsymbol{o}_k \in R^2$; The 2D points of the image belong to 3D points of some 3D object, but the assignment of image and object points is *not* part of the observation. Therefore, we denote the (unknown) assignment of the 2D features $\boldsymbol{o}_k$ in the image to the index i_k of the corresponding feature of the 3D object by ζ_κ. If there are m observed features and n_κ 3D points for class Ω_κ, then the discrete mapping (assignment) ζ_κ is defined by

$$\zeta_\kappa : \left\{ \begin{array}{ccc} \boldsymbol{O} & \rightarrow & \{1, 2, \ldots, n_\kappa\} \\ \boldsymbol{o}_k & \mapsto & i_k \end{array} \right. \ .$$

After some algebraic manipulations we get the model density

$$p(\{o_k|k=1,2,\ldots,m\}|\kappa;\theta) = \sum_{\zeta_\kappa}\prod_{k=1}^{m} p(\zeta_\kappa(o_k))p(o_k|\zeta_\kappa,\kappa;\theta) \tag{3.51}$$

and we observe that the unknown assignments vanished by marginalization of the density. This approach was investigated in [336].

Only the general ideas of the three specializations were given here. The parametric densities have to chosen suitably, usually by assuming a normal density or a mixture of normal densities. It has been shown that not only the problem of object classification can be treated by statistical models of the above type, but also the problem of object localization. This is handled by introducing additional parameters for the location and computing them by ML-estimation. The estimation formula for the unknown parameters have to be worked out, and efficient solutions for the computation of parameters, localization of objects in 2D and 3D, and object classification are to be derived. The details have to left to the above references. These descriptions show the potential of statistical modeling for object recognition in images.

3.3.3 Markov Random Field

A standard probabilistic model in image processing is the *Markov* random field (MRF). We denote by $\widetilde{X} = \{X_i|i=1,2,\ldots,L\}$ a set of random variables. For each random variable X_i we define a *neighborhood*

$$N(X_i) = \{X_j|X_j \text{ is neighbor of } X_i\}.$$

In image processing a random variable X_i is an image pixel f_i and a neighborhood N_i corresponds to pixels 'around' f_i in the image grid, for example, the four- or eight-neighborhood. An MRF is defined to have two basic properties, namely the probability to observe an arbitrary subset of random variables is non-zero (positivity) and the probability of observing a certain random variable $X_i \in X$ depends only on a neighborhood $N(X_i)$ of variables and *not* on all other variables (*Markov* property).

A MRF can be associated with a graph; each random variable X_i corresponds to a node v_i and an undirected edge $e_{i,j}$ links v_i and v_j if X_i and X_j are neighbors, that is, if $X_j \in N(X_i)$. We recall that the cliques of a graph are its fully connected subgraphs.

The basis for computing the joint probability density $p(\widetilde{X})$ is that the joint density is proportional to a *product of real valued functions* associated with the cliques of the MRF. The clique functions are *symmetric* in their arguments, that is, a change of arguments does not change the function. Let $\widetilde{C} = \{C_1,\ldots,C_C\}$ denote the set of cliques, and $X(C_i)$ denote the random variables belonging to C_i. Then there exists a set of *clique functions* $\phi_{C_i}(X_{C_i})$, $i=1,2,\ldots,C$, such that

$$p(\widetilde{X}) = \frac{1}{Z}\prod_{C_i\in\widetilde{C}} \phi_{C_i}(X_{C_i}), \tag{3.52}$$

where Z normalizes the integral over p to one. Due to the positivity constraint the above product can be written as a Gibbs *field*

$$p(\widetilde{X}) = \frac{1}{Z} \exp \left[- \sum_{C_i \in \widetilde{C}} V_{C_i}(X_{C_i}) \right] = \frac{1}{Z} \exp[-U(\widetilde{X})] \, . \tag{3.53}$$

The term V_{C_i} is called a *potential function*; it maps the set of random variables belonging to clique C_i to real values. The sum $U(X)$ of potentials is called the *energy function*. For discrete random variables the normalizing constant Z is

$$Z = \sum_X \exp[-U(X)] \, .$$

MRFs are mostly used for low-level image processing and image modeling [760], applications to object recognition and localization are discussed, e.g., in [484].

3.3.4 Bayes Networks

Estimating and manipulating the joint probability $P(o) = P(o_1, \ldots, o_n)$ of an n-dimensional vector of observations is a complex problem. Therefore, it is desirable to reduce this complexity by limiting the statistical dependencies. This was also the goal of the MRFs mentioned above. The above probability can be factored in various ways, for example,

$$\begin{aligned} P(o) &= P(o_1)P(o_2|o_1)P(o_3|o_2,o_1)\ldots P(o_n|o_{n-1},\ldots,o_2,o_1) \\ &= P(o_2)P(o_3|o_2)P(o_4|o_3,o_2)\ldots P(o_1|o_n,\ldots,o_3,o_2) \\ &= \ldots \end{aligned} \tag{3.54}$$

In an application a random variable will often depend only on a few other variables, but not on all others. This results in limited dependencies, for example, one may have $P(o_3|o_2,o_1) \approx P(o_3|o_2)$. The reduction of complexity is obvious. Instead of dealing with one probability distribution of n-th order one has to deal only with several distributions of much lower order. It is the problem of the designer to determine, either experimentally or from knowledge about the task domain, the relevant dependencies and independencies of the involved observations.

The determined statistical independencies are represented in a *Bayes network*. It is a special graph designed to represent uncertain knowledge and to make inferences based on probability theory. The nodes of the graph represent random variables, in our case observations o_ν, and links reaching a node v_ν from a set Π_ν of direct predecessors represent statistical dependencies The joint probability then is given by

$$\begin{aligned} P(o) &= P(o_1,\ldots,o_n) \\ &= \prod_{\nu=1}^{n} P(o_\nu|\Pi_\nu) \, , \end{aligned} \tag{3.55}$$

which is in direct correspondence to Equation 3.53. Due to its intuitive representation by a graph these models are also termed *graphical models*.

Given the statistical model of Equation 3.55 one can consider the following types of inferences:

- *Associaton.* If a value of an observation is given by $o_j = a_j$, what can be infered about the probability of value a_k of observation o_k; in probabilistic terms one asks for the probability of observing some value of o_k given some value of o_j, where of course, also some subsets of variables may be involved instead of single variables

$$P(o_k = a_k | o_j = a_j) = \frac{P(o_k = a_k, o_j = a_j)}{P(o_j = a_j)} . \tag{3.56}$$

- *Diagnosis.* Given a value $o_n = a_n$ of an observation, what can be said about the values of the other variables; in probabilistic terms one asks for the most probable values

$$(a_1, \ldots, a_{n-1}) = \arg\max P(o_1 = a_1, \ldots, o_{n-1} = a_{n-1}, o_n = a_n) . \tag{3.57}$$

- *Control.* If a certain variable is fixed to some value, $o_n = a_n$, what will be the other values; in probabilistic terms one asks for the most probable values

$$(a_1, \ldots, a_{n-1}) = \arg\max P(o_1 = a_1, \ldots, o_{n-1} = a_{n-1} | o_n = a_n) . \tag{3.58}$$

In order to use the formalism of *Bayes* networks one has to define the random variables (or observations) and their range of values, determine the statistical independencies (or the graphical model), estimate the relevant probabilities, and then make the desired statistical inferences. For further details we refer to [531, 363].

3.3.5 Hidden Markov Model

A hidden *Markov* model generates a stochastic process by two mechanisms. The first mechinisam generates a sequence of states, the second a sequence of observations.

The state sequence of lenght T is $\boldsymbol{s} = [s_n]$. Each state is an element of a finite set S of state symbols $S_i, i = 1, \ldots, I$. At discrete time t_n the model is in state $s_n \in S$. At time t_{n+1} it performs a random transition to a new state $s_{n+1} \in S$. A transition to a new state is done with a certain probability defined by the matrix of state transition probabilities

$$\boldsymbol{A} = [a_{ij}] = P(s_{n+1} = S_j | s_n = S_i) , \quad i, j = 1, 2, \ldots, I .$$

It is assumed that the state transition probability is *independent* of time and dependent on *only one* preceding state. Again, this requirement reduces statistical dependencies and the complexity of computations.

The first state is chosen due to a vector of initial probabilities

$$\pi = [\pi_i] = P(s_1 = S_i) , \quad i = 1, 2, \ldots, I .$$

The states *cannot* be observed, hence the name *hidden Markov* model.

What is observable is the sequence of observations generated by the second mechanism. The sequence of observations is $\boldsymbol{o} = [o_i]$ chosen from a finite set of observation symbols

$$O = \{O_1, O_2, \ldots, O_L\} .$$

An observation $o_i \in O$ is generated when the model is in state S_i. The output symbol is chosen depending on the matrix of output probabilities

$$\begin{aligned} \boldsymbol{B} &= [b_{il}] = P(O_l \text{ emitted in state } S_i | s_n = S_i) \\ & \quad i = 1, \ldots, I, \quad l = 1, \ldots, L . \end{aligned}$$

A hidden *Markov* model is defined by the tripel

$$\text{HMM} = (\pi, \boldsymbol{A}, \boldsymbol{B}) . \tag{3.59}$$

It may be viewed as a stochastic automaton generating a stochastic process. The set of states is chosen according to the task domain. For example, in speech recognition it is a set of subword units of a language, in gesture recognition it may be a set of motion states. For each class, e.g. for each word or each gesture, a separate HMM is defined and its parameters are trained from classified observations. For a new observation the posterior probability is computed that a particular HMM was active when this observation is available. The most probable HMM is chosen according to Equation 3.36. An HMM may be represented by a graph. The nodes are the states, and a link between two nodes is introduced if there is a non-zero transition probability between the two corresponding states.

The three basic computational problems for HMM's are:

- Given an observation sequence $\boldsymbol{o}$ and an HMM with its parameters, what is the *probability* $P(\boldsymbol{o}|\text{HMM})$ of generating $\boldsymbol{o}$ by this HMM?

- Given an $\boldsymbol{o}$ and HMM, what is the most probable *sequence of states* in HMM for generating $\boldsymbol{o}$?

- Given several observations, what are the *parameter values* $\pi, \boldsymbol{A}, \boldsymbol{B}$ maximizing $P(\boldsymbol{o}|\text{HMM})$?

Efficient algorithms to solve these problems are available from the references [564, 625].

3.4 OBJECT RECOGNITION

D. PAULUS

The mapping of real world objects to the image plane including the geometric and the radiometric parts of image formation is basically well understood [766, 219].

From an abstract point of view, a camera is thought of as a geometric engine that projects the 3D world to the 2D image space. The most challenging problems in computer vision which are still not solved entierly are especially related to object recognition [708]. Up to now, there have been no general algorithms that allow the automatic learning of arbitrary 3D objects and their recognition and localization in complex scenes. The term *object recognition* denotes two problems [708]: classification of an object and determination of its pose parameters. By definition, the recognition requires that *knowledge* or *models* of the object are available; formalisms for such knowledge are introduced in Section 3.5. The key idea is to compare the image with a model. The key issues thus are the choice of the representation scheme, of the selection of models, and the method for comparison.

Two major types of objects be distinguished: On the one hand, solid objects have to be recognized. Such problems arise, e.g., in industrial applications where a construction robot assembles objects from several parts. On the other hand, flexible or deformable objects impose further problems on the recognition algorithm. Such 'objects' are, e.g., humans or animals in observation systems.

3.4.1 Introduction

We assume that N_K object classes Ω_κ $(1 \leq \kappa \leq N_K)$ are known and represented as 'knowledge' (i.e., models) in an appropriate manner. The representation of the object can be in two dimensions, it may use a full 3D description, or it can contain a set of 2D views of a 3D object. The object models use an object coordinate system and a reference point (mostly on the object) as its origin.

We also assume that an image is given which may contain data in 2D, 2.5D or 3D. For intensity images in 2D it may be either monochrome, color or a multi-channel image. The object recognition problem can be formalized when the following sub-problems are identified:

The *object classification* problem is to identify an object in the given image $\boldsymbol{f}$. This may be particularily hard if the object is partially occluded, if many objects are in the scene, if the object is small, etc. This part of the problem formally corresponds to a function δ

$$\delta : \boldsymbol{f} \rightarrow \kappa; \kappa \in \{1 \ldots N_K\} \quad . \tag{3.60}$$

The *pose estimate* problem is to compute pose parameters $\boldsymbol{R}$ and $\boldsymbol{t}$ for an object Ω_κ, which determine the transformation of the camera coordinate system to object coordinates (Figure 3.12). Depending on the dimension of the input data and on the object representation, these may be two-dimensional or three-dimensional rotation matrices and translation vectors.

In the following we assume a 3D world containing solid objects and a 2D image. In projective geometry, the transformation parameters can be divided into two groups. There are several representations for the transformation parameters. In [554, 610], in-plane and out-of-plane transformations are distinguished (see Figure 3.12). The rotation matrix $\boldsymbol{R}$ can be described elegantly by the Rodrigues formula, quaternions or by Euler angles (cmp. [219]). Advantages and disadvantages of these representations for object recognition purposes are discussed in [335]

The algorithms for 3D object recognition published so far differ with respect to

- the type and dimension of sensor data,

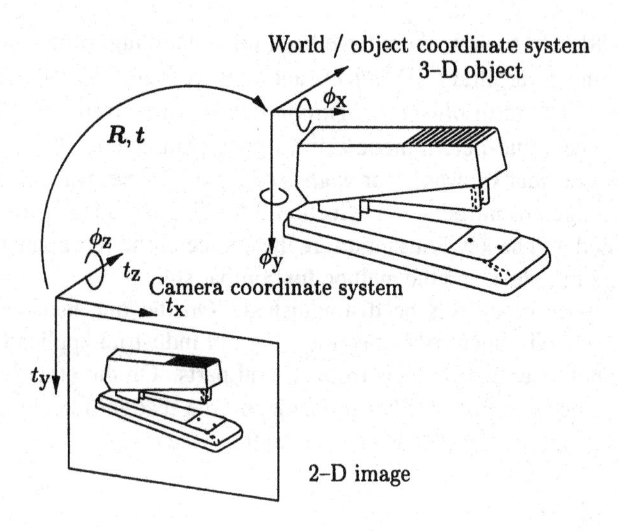

Figure 3.12. Different components of rotation and translation of a 3D object.

- the representation scheme for object models,
- the object classification strategy,
- the strategy for pose recognition, and
- the learning of object models.

It is beyond the scope of this section to provide an exhaustive overview and a lucid discussion of all models successfully applied in computer vision. Instead, we describe the most common and some new approaches and give a general outline of their characteristics.

The organization of the section is as follows: The next section introduces an example of how geometric features can be used for object classification and pose determination. As an example of appearance based methods, we introduce eigenspaces in Section 3.4.3. Arguments for a probabilistic formulation of object recognition modules are listed in Section 3.4.4 and one example of an approach using segmented features as well as one statistical appearance based method is mentioned. The use of invariant features for object recognition is outlined in Section 3.4.5. The architecture of a system for active object recognition is presented in Section 3.4.6. We summarize our review in Section 3.4.7.

3.4.2 Object Recognition Using Geometric Models

State-of-the-art approaches dealing with high-level vision tasks are essentially dominated by model-based object recognition methods [360]. As we outline in the following, such models consist of primitives that can be compared directly to the information found in the segmented images. Whereas the purpose of geometric models is to facilitate the generation of images and models thus had to be similar to the visualization pipeline, the models suitable for object recognition need to have similar structure as segmentation objects (see Section 3.1).

The classical approach is to use geometric models and to represent them explicitly [19]. Generally speaking, segmentation algorithms decompose given images in primitives, reference models are matched to observations, and distance measures are the basis for class decisions and pose estimates. The selection of discriminating image features and the adequate representation of object models define herein the vital problems which have to be solved when designing a vision system. We assume that a segmentation object $\boldsymbol{O}$ is created from an input image $\boldsymbol{f}$ and that an object model C exists which is in a format compatible with the segmentation data. A (partial) match between $\boldsymbol{O}$ and C is found when a subset of the data in the model is found to correspond to a subset of the segmentation object. A function

$$\zeta : C \rightarrow \boldsymbol{O} \tag{3.61}$$

is called an *interpretation*. The information in the model constrains the admissible elements in the segmentation object, e.g., by requirng parallel lines in the three-dimensional object model to be mapped to approximately parallel lines in the segmentation object. Finding an interpretation, generally is a search problem in which data resulting from image segmentation has to be compared to data in models. Since during search this comparison has to be done frequently, it is required for efficiency reasons that segmentation data and model data are compatible in the sense, that a direct comparison of primitives is possible. In particular, efficient and roboust mappings of data represented in model primitives to that represented in segmentation data have to exist. This is not the case, e.g. for CAD models as used in Section 4.3, since these models contain lines which will never be segmented in the image. At least some of the lines prepresented in the model are not present in the object; as an example we can use wire frame models. As another example, the control points in triangulated surfaces will not be detected roboustly in the image data, either. Free form surfaces represented as splines, as proposed in Section 4.6 are parametrized by the spline parameters and their control points, which usually cannot be detected by segmentation, either.

Figure 3.13 shows an example of a surface model $C = \{c_1, \ldots, c_9\}$ and an idealized segmented range image consisting of surface segments in a segmentation object $\boldsymbol{O} =$

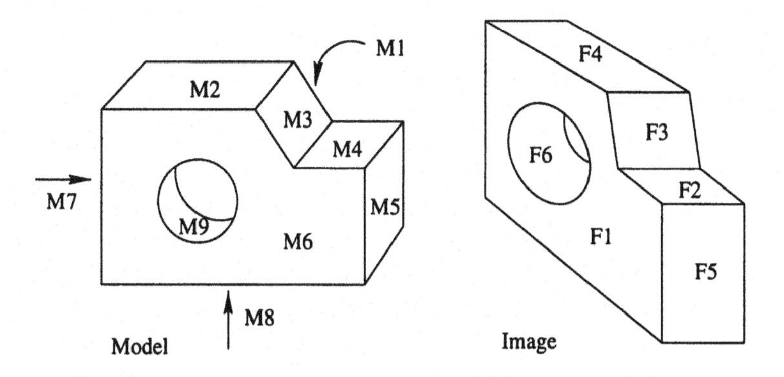

Figure 3.13. Matching of segmented features to a model.

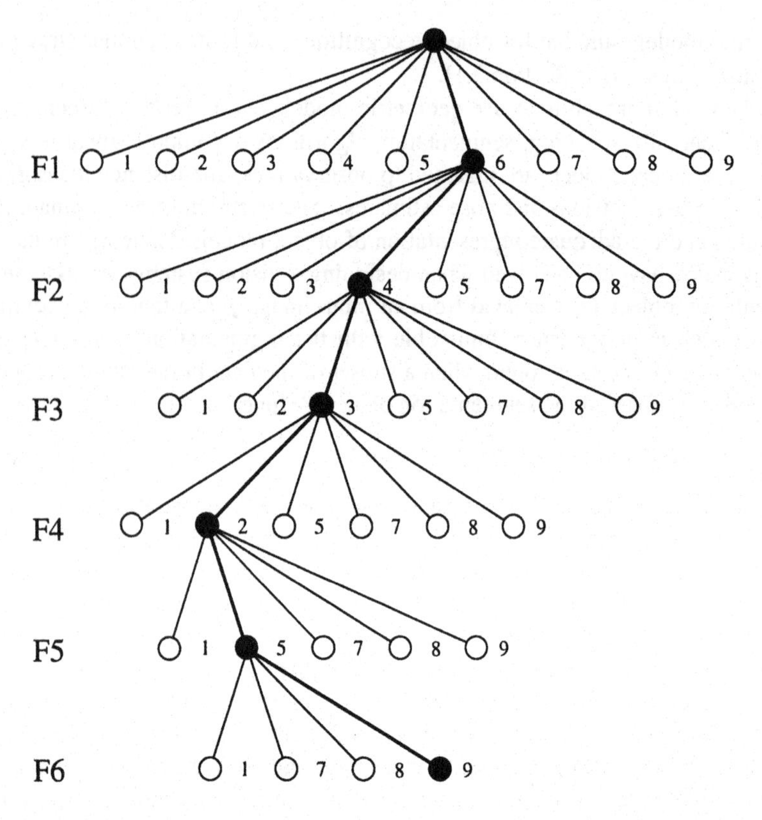

Figure 3.14. Search space for model match for the object in Figure 3.13.

$\{o_1, \ldots, o_6\}$. We constrain the matches such that L-shaped segments may only be matched to the L-shaped model elements $\{c_1, c_6\}$. A depth-first search of a match in Figure 3.14 left results in an incorrect interpretation assignment. In a subsequent verification by backprojection of the hypothesized model parameters to the image and comparison of the features it is required to find the correct match which is shown as a dotted in Figure 3.14 right. The general problem is that features are only compared locally. Since real image segmentation is inaccurate and not complete, a perfect match will almost never be found for real data. One solution is to add 'wildcards' to the segmentation and to the model when processing the interpretation tree. These are nodes which can be matched to any segment or model; a wildcard model is matched to a superfluous segment node; the same technique is used for missing segments [708]. More elaborate matching strategies for semantic models are outlined in Section 3.3.

The pose recognition problem for geometric models and matching of geometric features is strongly related to the problems of 3D reconstruction (see Chapter 4) and camera calibration (see Section 2.1). As noted in Section 3.4.1, the problem is to determine $\boldsymbol{R}$ and $\boldsymbol{t}$ for an object Ω_κ. Assuming a perfect match, we have a set of points

$$\{\boldsymbol{x}_1^{(w)}, \boldsymbol{x}_2^{(w)}, \ldots, \boldsymbol{x}_{N_m}^{(w)}\} \tag{3.62}$$

in world coordinates corresponding to points $\{x_1^{(c)}, x_2^{(c)}, \ldots, x_{N_m}^{(c)}\}$ in camera coordinates related by

$$x_i^{(c)} = Rx_i^{(w)} + t \quad . \tag{3.63}$$

What is observed are the image coordinates of the corresponding projected points $\{x_1^{(I)}, x_2^{(I)}, \ldots, x_{N_m}^{(I)}\}$. Assuming that we are dealing with 3D world coordinates and 2D images, the complete transformation including the projection step can be expressed a 4×3 matrix E if points are written in homogeneous coordinates [219]. Depending on whether the projection parameters are known from calibration or not, and whether the images are registered (see Section 5.1), several strategies exist for the computation of this matrix, see e.g. in [61, 593].

3.4.3 Appearance Based Object Recognition

Appearance based object recognition uses non-geometric models representing the intensities in the projected image. Rather than using an abstract model of geometries and geometric relations, images of an object taken from different viewpoints and under different lighting conditions are used as object representation. Figure 3.15 shows a set of such images. To beat the curse of dimenstionality, the images used for object representation are transformed to lower-dimensional feature vectors. This overcomes several problems related to standard approaches as, for example, the geometric modeling of fairly complex objects and the required feature segmentation. Comparative studies prove the power and the competitiveness of appearance based approaches to solve recognition problems [551]. Well-known and classical pattern recognition algorithms can be used for computer vision purposes, if appearance based methods are applied: feature selection methods [572, 65], feature transforms [572, 708], or even more recent results from statistical learning theory [723].

As the given image and the model share the same representation, the choice of the distance function for matching images with models is simpler than for geometric models. We rearrange the image pixels $f_{i,j}$ in an image vector

$$f' = (f_{1,1}, \ldots, f_{1,N_x}, f_{2,1}, \ldots, f_{2,N_x}, \ldots, f_{N_y,1}, \ldots, f_{N_y,N_x})^T \tag{3.64}$$

The comparison of two images f_1' and f_2' by correlation simply reduces to the dot product of the image vectors f_1' and f_2'

$$s = {f_1'}^T \cdot f_2' \quad ; \tag{3.65}$$

the bigger s gets, the more similar are the images f_1' and f_2'.

Obviously, high dimensional feature vectors such this image vector will not allow the implementation of efficient recognition algorithms [507]. The vectors have to be transformed to lower dimensions. Commonly used transforms are the principal component analysis [494, 376, 120] or in more recent publications the Fisher transform [50]. In the following we motivate a linear tranformation Φ which maps the image vector $f' \in \mathbb{R}^{N_x \cdot N_y}$ to a feature vector $b \in \mathbb{R}^{L_a}$ with $L_a \ll N_x \cdot N_y$ by

$$b = \Phi f' \quad . \tag{3.66}$$

If we choose Φ such that the distance of all features is maximized, this reduces to a problem of eigenvalue computation. From N_a given images written as vectors $f_1', \ldots f_{N_a}'$ of an object we compute the mean vector

$$\mu = \frac{1}{N_a} \sum_{i=1}^{N_a} f_i' \tag{3.67}$$

and from this we create a matrix V whose columns are the image vectors

$$V = \left[(f_1' - \mu) | \ldots | (f_{N_a}' - \mu) \right] \quad . \tag{3.68}$$

Eigenvalue analysis of the matrix $K = V^T V$ yields the eigenvectors $v_1, \ldots v_{N_a}$ sorted by magnitude of the corresponding eigenvalues. A fundamental fact from linear algebra states that an image vector f_j' can be written as a linear combination of the mean image vector and the eigenvectors as

$$f_j' = \mu + \sum_{i=1}^{N_a} b_i^{(j)} v_i \quad . \tag{3.69}$$

An approximation of f' can be obtained if instead of N_a eigenvectors we select only the first $L_a \leq N_a$ vectors. The image vector f_j' is represented by

$$b^{(j)} = \left(b_1^{(j)}, \ldots, b_{L_a}^{(j)}\right)^T = \Phi\left(f_j' - \mu\right) \tag{3.70}$$

and the columns of the matrix Φ are the first L_a eigenvectors $v_1, \ldots v_{L_a}$.

Typically for $N_a = 100$ images we choose only the first $L_a = 15$ eigenvectors. For each object class κ we now record images from different viewpoints and under changing lighting conditions, perform the transformation to eigenspace to obtain

$$\left\{ b^{(\kappa,j)}; j = 1 \ldots N_{a\kappa} \right\} \tag{3.71}$$

for $N_{a\kappa}$ images captured. The vectors $b^{(\kappa,j)}$ of an object of class κ are a manifold in eigenspace. They are used and stored as the object model. The processing steps of this approach are exemplified in Figure 3.15.

The correlation of two normalized images f_i' and f_j' with $||f_i'|| = 1$ can now be approximated by the Euclidian distance of two weight vectors $b^{(i)}$ and $b^{(j)}$ which yields a huge gain in computation speed:

$$||f_i'^T f_j'|| \approx 1 - 0.5||b^{(i)} - b^{(j)}|| \quad . \tag{3.72}$$

For object recognition of a given image we compute its eigenspace representation to create a vector b using Equation 3.70. From the manifolds representing the objects we choose the one which has minimal distance to the computed vector b. Object recognition is thus reduced to the problem of finding the minimum distance between

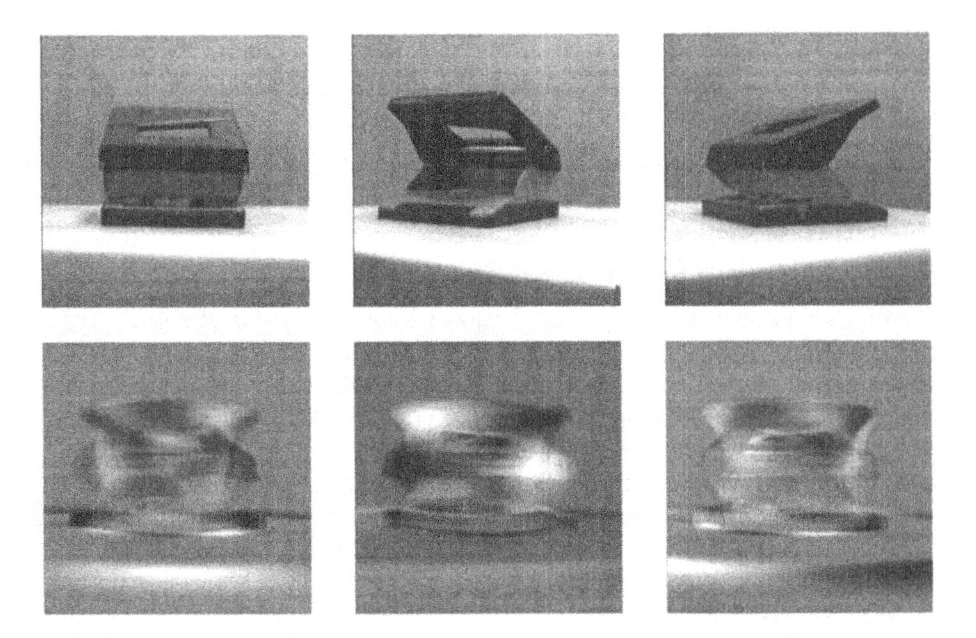

Figure 3.15. Three different views of an object (upper row), mean vector, and eigenvectors v_0, v_{15} (lower row). For the computation 72 views and 360° rotation in 5°-steps were used. The images are taken from the COIL-data set.

an object and a model. In order to generate an image from a vector b, we use the pseudo-inverse Φ^+ of Φ

$$\Phi^+ = \Phi^T \left(\Phi\, \Phi^T\right)^{-1} \tag{3.73}$$

to create

$$f' = \Phi^+ b \quad . \tag{3.74}$$

The key to success in this approach is not to create the matrix

$$K = V\ V^T \tag{3.75}$$

explicitly when the eigenvectors are computed. For a typical image f of size $N_x = 256$ and $N_y = 256$, the image vector f' has length 2^{16}; for $N_a = 100$ images, the matrix V has size $2^{16} \times 100$; the matrix K would thus be of size $2^{16} \times 2^{16}$ and computation of the eigenvectors would be unfeasible. Instead, we use either iterative methods to compute the eigenvectors [517] or we use a result from singular value decomposition. We compute the eigenvalues λ_i and eigenvectors v'_j of the so-called implicit matrix

$$K' = V^T\ V \tag{3.76}$$

which is much smaller than K. In our example, the size would be 100×100. We note that

$$K' v'_j = V^T \left(V v'_j\right) = \lambda_j v'_j \quad . \tag{3.77}$$

We multiply Equation 3.77 from left by $\boldsymbol{V}$ and get

$$\boldsymbol{V}\left(\boldsymbol{V}^T\boldsymbol{V}\right)\boldsymbol{v}_j' = \left(\boldsymbol{V}\boldsymbol{V}^T\right)\boldsymbol{V}\boldsymbol{v}_j' = \lambda_j\left(\boldsymbol{V}\boldsymbol{v}_j'\right) \quad . \tag{3.78}$$

which shows that the eigenvalues of σ' are also eigenvalues of σ and that the eigenvectors are related by $\boldsymbol{V}$. We use these results to compute the eigenvectors for $\boldsymbol{K}$.

3.4.4 Probabilistic Object Recognition

Probabilistic methods as presented in Section 3.2 are used in several object recognition systems. A Bayesian framework for 3D object recognition requires that the appearance of objects in the image plane is characterized using probability density functions. These densities have to incorporate prior knowledge on objects, rotation and translation, self-occlusion, projection to the image space, the assignments of image and model features as well as the statistical modeling of errors and inaccuracies caused by varying illumination, sensor noise or segmentation errors [335]. We call these densities *model densities*.. The structure of these models can vary: It can be a single multivariate Gaussian density, a hidden Markov model or some other type of density (see Section 3.2).

In [337] we assume N_{K} possible object classes and a set of observations $\boldsymbol{O}$ consisting of feature vectors o_k in a segmentation object $\boldsymbol{O} = \{o_k \in \mathbb{R}^2 | 1 \leq k \leq N_{\mathrm{m}}\}$ where the number N_{m} of observed features like corners or vertices varies for different images. Appearance and position of features in the image show a probabilistic behavior. The statistical description of an object belonging to class Ω_κ consists of a model density $p(\boldsymbol{O}|\boldsymbol{B}_\kappa, \boldsymbol{R}, \boldsymbol{t})$ (see Section 3.2) combined with discrete priors $p(\Omega_\kappa)$ $1 \leq \kappa \leq N_{\mathrm{K}}$ for the probability of an object of class Ω_κ to appear in the scene. The priors are estimated by relative frequencies of objects in the training samples. The set $\boldsymbol{B}_\kappa$ contains the model-specific parameters for the behavior of features as well as the parameters for the assignment of image and model features.

For the explicit definition of $p(\boldsymbol{O}|\boldsymbol{B}_\kappa, \boldsymbol{R}, \boldsymbol{t})$ we use the observed feature set $\boldsymbol{O}$ and the corresponding features $C_\kappa = \{c_{\kappa,1}, c_{\kappa,2}, \ldots, c_{\kappa,n_\kappa}\}$ in the model where in general $n_\kappa \neq N_{\mathrm{m}}$ due to segmentation errors and occlusion. Let the parametric density of the model feature c_{κ,l_k} corresponding to o_k be given by $p(c_{\kappa,l_k}|\boldsymbol{a}_{\kappa,l_k})$, where $\boldsymbol{a}_{\kappa,l}$ $(l = 1, \ldots, n_\kappa)$ characterize model features. For a normally distributed 3D point, for instance, $\boldsymbol{a}_{\kappa,l_k}$ denotes the mean vector and the covariance matrix. A standard density transform results in the density $p(o_k|\boldsymbol{a}_{\kappa,l_k}, \boldsymbol{R}, \boldsymbol{t})$, which characterizes the statistical behavior of the feature o_k in the image plane dependent on the object's pose parameters.

As described in Section 3.2, the probabilistic modeling of the assignment from image to model features is based on discrete random vectors $\zeta_\kappa = (\zeta_\kappa(o_1), \zeta_\kappa(o_2), \ldots, \zeta_\kappa(o_{N_{\mathrm{m}}}))^T$, where ζ_κ defines a discrete mapping from an observed feature o_k to the index $l_k \in \{1, 2, \ldots, n_\kappa\}$ of the corresponding model feature c_{κ,l_k}, i.e., $\zeta_\kappa(o_k) = l_k$. The non-observable assignment is eliminated by the marginalization as shown in Equation 3.51. The probabilistic modeling of the assignment from image to model features is based on discrete random vectors. An assignment function ζ_κ defines a discrete mapping from an observed feature o_k to the index $l_k \in \{1, 2, \ldots, n_\kappa\}$ of the corresponding model feature c_{κ,l_k}, i.e., $\zeta_\kappa(o_k) = l_k$. A set of observed features can thus be associated with the assignment random vector $\zeta_\kappa = (\zeta_\kappa(o_1), \zeta_\kappa(o_2), \ldots, \zeta_\kappa(o_{N_m}))^T$ which is related to the discrete probability $p(\zeta_\kappa)$, i.e., the matching problem is also modelled statistically. The discrete probability of $p(\zeta_\kappa)$ extents the probability density function for observing the set of features $\boldsymbol{O}$. Due to the statistical interpretation of ζ_κ, the non-observable assignment can be eliminated by the following marginalization:

$$p(\boldsymbol{O}|\boldsymbol{B}_\kappa, \boldsymbol{R}, \boldsymbol{t}) = \sum_{\zeta_\kappa} p(\zeta_\kappa) \prod_{k=1}^{N_m} p(o_k | \boldsymbol{a}_{\kappa,\zeta_\kappa(o_k)}, \boldsymbol{R}, \boldsymbol{t}) \quad . \tag{3.79}$$

If the structure of the model density (i.e., the number of model features and the dependency structure of single assignments) is known, algorithms for the estimation of the parameter set $\boldsymbol{B}_\kappa$ exist [335]. The computation of $\boldsymbol{B}_\kappa$ for each object class Ω_κ, $\kappa = 1, 2, \ldots, N_K$ requires $p(\zeta_\kappa)$ and $\{\boldsymbol{a}_{\kappa,l}\}$. Due to the projection of the 3D world to the 2D image plane, the range information is lost. Furthermore, the assignment of image and model features is not a component of the observations. The calculation of $\boldsymbol{B}_\kappa$ thus corresponds to an incomplete data estimation problem which can be solved using the Expectation Maximization algorithm (EM algorithm, cmp. Section 3.2).

The framework introduced so far requires a minor modification of the standard Bayesian decision rule, since a segmentation object $\boldsymbol{O}$ is given instead of a single vector, and the unknown pose parameters are part of the probability density. The modified Bayesian decision rule for the statistical classification of objects is:

$$\lambda = \text{argmax}_\kappa\, p(\Omega_\kappa|\boldsymbol{O}) = \text{argmax}_\kappa\, p(\Omega_\kappa) p(\boldsymbol{O}|\boldsymbol{B}_\kappa, \boldsymbol{R}, \boldsymbol{t}) \quad . \tag{3.80}$$

The a posteriori probabilities $p(\Omega_\kappa|\boldsymbol{O})$ cannot be evaluated explicitly. The pose estimation stage has to compute the best orientation and position $\boldsymbol{R}, \boldsymbol{t}$ before the class decision is possible. This corresponds to the maximization problem

$$\{\widehat{\boldsymbol{R}}, \widehat{\boldsymbol{t}}\} = \text{argmax}_{\boldsymbol{R},\boldsymbol{t}} p(\boldsymbol{O}|\boldsymbol{B}_\kappa, \boldsymbol{R}, \boldsymbol{t}) \quad , \tag{3.81}$$

which requires a global optimization of a multimodal likelihood function.

This framework was used for the recognition of 3D objects based on 2D images [335]: we assume that each input image (e.g. Figure 3.16) is transformed into a segmentation object of 2D feature vectors $\boldsymbol{O} = \{o_k \in \mathbb{R}^2 | 1 \leq k \leq N_m\}$. The

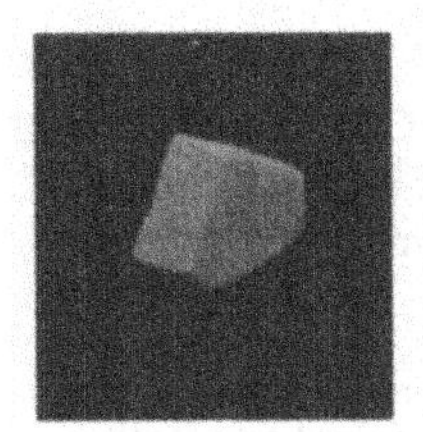   

Figure 3.16. Simple polyhedral 3D objects ($\Omega_1, \Omega_2, \Omega_3, \Omega_4$), used in the experiments in [335].

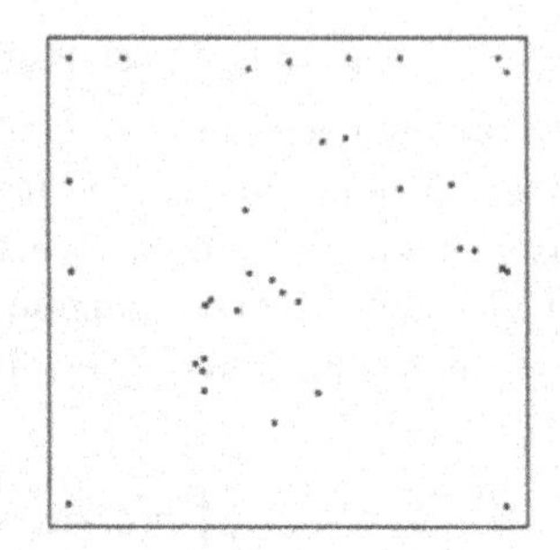
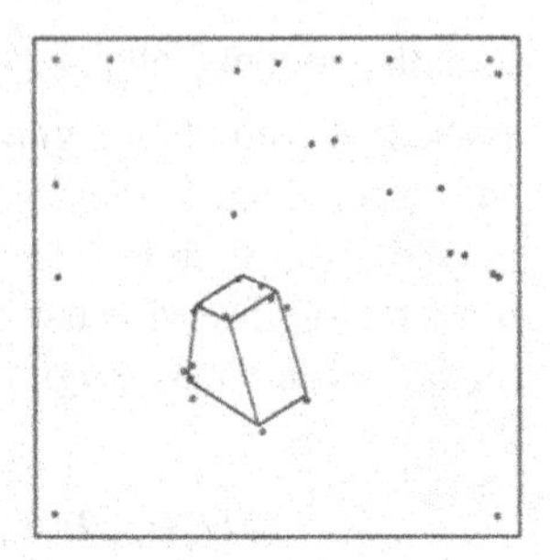

Figure 3.17. Experiment for object recognition with heterogeneous background.

elements o_k may be points (e.g. corners or vertices) or lines, which can be detected by several combinations of segmentation operators which all result in the uniform segmentation object (see Section 3.1).

For segmented 2D point features, $\boldsymbol{B}_\kappa$ provides the parameters characterizing the assignments as well as the accuracy and stability of the object points. Closed form iteration formulas can be found for normally distributed point features, which allow the estimation of mean vectors from projections without knowing corresponding features of different views [335].

This flexible formalism of model densities can also be extended to use multiple views for pose estimation or classification [335] which remarkably improves recognition rates.

A combination of appearance based methods with probabilistic approaches is also feasible. In [554], wavelet features in scale space are used for object representation, requireing no image segmentation. These features are modeled statistically and localization as well as classification is again reduced to a Bayesian decision and statistical parameter estimation problem.

3.4.5 Features and Invariants for Object Recognition

In Section 3.4.2, segmented features are matched to model features. The match is constrained by knowledge on the features, such as, e.g., parallelism. Such information can be provided by segmentation algorithms and can be represented in the relation-slot of a segmentation object. Although appearance based methods use the image directly and do not compute segmented features, they do require that the object figure is separated from the ground. This is usually achieve by image segmentation as well.

If we can find features of an object that are invariant with respect to a certain class of transformations, e.g., illumination changes, changes in size, viewpoint changes, etc., we can use these features to index into a set of models. Clearly, these features must have at the same time invariant properties as well as the discriminative power to separate object classes.

Several intuitive invariant features can be defined on 2D lines and contours. To give an example, the area of a region divided by its contour length is invariant to rotation, scaling, and translation. More complex invariant features can be found which yield, e.g., affine invariant descriptors.

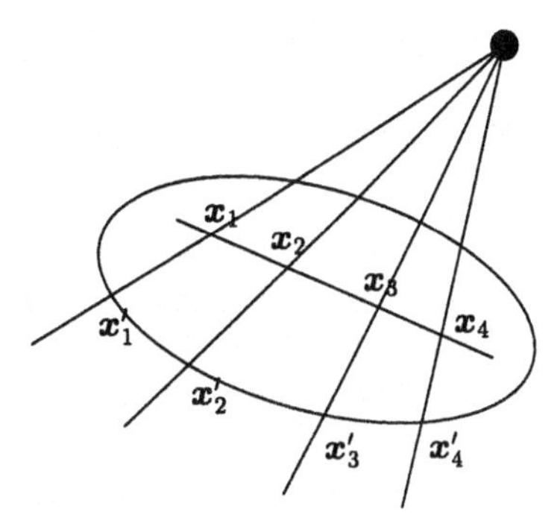

Figure 3.18. Circle skewed to an ellipse under projective transformation and its invariant properties using the cross ratio.

A well-known projective invariant which is of practial interest is the cross-ratio of four 2D points $\boldsymbol{x}_1$, $\boldsymbol{x}_2$, $\boldsymbol{x}_3$, and $\boldsymbol{x}_4$ on a straight line

$$\frac{||\boldsymbol{x}_1 - \boldsymbol{x}_3||}{||\boldsymbol{x}_1 - \boldsymbol{x}_4||} \cdot \frac{||\boldsymbol{x}_2 - \boldsymbol{x}_4||}{||\boldsymbol{x}_2 - \boldsymbol{x}_3||} \quad . \tag{3.82}$$

In fact, the cross ratio is invariant under any collineation, i.e., under every transformation which transforms straight lines into straigth lines. Straight lines can be found with reasonable reliability by image segmentation (see Section 3.1). The four points on such lines can be identified, e.g., when line intersections are searched. The geometric relations for the cross ration are outlined in Figure 3.18.

Acquiring an object model using this approach is simple, as we represent an object class by a set of invariant features. We use one image of each object class to compute feature vector for each object class and store it. For recognition we detect the features in the given image and compare to the stored feature vectors.

Color in many cases provides very useful cues to identify objects. It is relatively insensitive to view point changes and can be regarded as a non-geometric feature, which is invariant with respect to minor changes in the position parameter of an object. Changes which are caused by changes in the lighting conditions have be eliminated to provide such invariance properties.

In order to identify color objects in a scene, color histograms are used in [679]. A color image $[\boldsymbol{f}_{ij}]_{1 \le i \le N_y, 1 \le j \le N_x}$ is searched for an object which is characterized by its histogram in some quantization. The color space is divided into cubes and the number of vectors is determined which fall into the cubes. This number is divided by the total number of color vectors. The histogram is denoted by $\boldsymbol{T} = [T_l]_{l=1 \ldots N_{\mathrm{L}}}$ where N_{L} denotes the number of cubes. A function ζ maps a color pixel to the index in the histogram, and permits to use arbitrary quantizations; e.g. for an RGB histogram with $4 \times 4 \times 4$ bins ($L = 64$) and for color components in the range from 0 to 255 we might choose

$$\zeta : \begin{cases} \mathbb{R}^3 & \rightarrow \{1, \ldots, L\} \\ \boldsymbol{f}_{ij} = (r_{ij}, g_{ij}, b_{ij})^T & \rightarrow \lceil r_{ij}/64 \rceil \cdot 16 + \lceil g_{ij}/64 \rceil \cdot 4 + \lceil b_{ij}/64 \rceil \end{cases} . \tag{3.83}$$

The elements of the histogram $\boldsymbol{T}$ are defined as

$$T_l = \frac{1}{N_x N_y} \left|\{(i,j)|\zeta(\boldsymbol{f}_{ij}) = l, i = 1,\ldots,N_y, j = 1,\ldots,N_x\}\right| \quad . \tag{3.84}$$

A close-up view of the object is recorded and the histogram $\boldsymbol{T}$ is computed from this image. Then, an image $[\boldsymbol{f}'_{ij}]_{1\leq i\leq N'_y, 1\leq j\leq N'_x}$ of the scene is recorded and a color histogram $\boldsymbol{S} = [S_l]_{l=1\ldots L}$ of this image is computed with the same quantization. The sizes of the images may however be different. The elements of the histogram $\boldsymbol{S}$ are defined as

$$S_l = \frac{1}{N'_x N'_y} \left|\{(i,j)|\zeta(\boldsymbol{f}_{ij}) = l, i = 1,\ldots,N'_y, j = 1,\ldots,N'_x\}\right| . \tag{3.85}$$

The ratio histogram $\boldsymbol{Q} = [Q_l]_{1\leq l\leq L}$ is computed from the object histogram $\boldsymbol{T}$ and the histogram $\boldsymbol{S} = [S_l]_{1\leq l\leq L}$ of the image $\boldsymbol{f}$ by

$$Q_l = \begin{cases} 0 & \text{if } S_l = 0 \\ \min\left\{1, \frac{T_l}{S_l}\right\} & \text{otherwise} \end{cases} \quad . \tag{3.86}$$

If for a bin with index l is empty, i.e., $S_l = 0$, then the corresponding color is not present in the scene, i.e., no pixel can be found that is mapped to this bin. Thus, this case never occurs in backprojection. In addition, the approximate size of the object in the image is needed for the algorithm; this size is represented by a mask $\boldsymbol{D} = [\boldsymbol{D}_{ij}]$, $\boldsymbol{D}_{ij} \in \{0,1\}$ covering the object.

The values of the ratio histogram are thus in the range $[0,1] \subset \mathbb{R}$. An intermediate image $\boldsymbol{g} = [g_{ij}](1 \leq i \leq N_y, 1 \leq j \leq N_x)$ of the same dimension as the input image is computed as $g_{ij} = Q_{\zeta(\boldsymbol{f}_{ij})}$ where the function ζ is used to find the appropriate bin for the color vector $\boldsymbol{f}_i j$ at position $(i,j)^T$. The convolution of $\boldsymbol{g}$ with the object mask $\boldsymbol{D}$ yields the output image $\boldsymbol{h}$. Local maxima in the output image indicate possible positions of the object.

A disadvantage of the color backprojection method is its sensitivity to illumination changes. This can be helped by preprocessing with a color constancy algorithm.

If an object is supposed to be at position $(i,j)^T$ in the image and the size is estimated to be $a \times b$, then it is reasonable to create a local histogram at position $(i,j)^T$ for a sub-image of the estimated size and to compare this histogram to the histogram given for the close-up view of the object. The size of this mask has to be known before using this method.

A simple distance measurement is the sum of distance squares (sum of squared differences, SSD):

$$SSD(\boldsymbol{S}, \boldsymbol{T}) = \sum_{l=1}^{L} (T_l - S_l)^2 . \tag{3.87}$$

The following quadratic form can be used as weighted version of L_2-norm:

$$d_A(\boldsymbol{S}, \boldsymbol{T}) = \sqrt{(\boldsymbol{S} - \boldsymbol{L})^T \boldsymbol{A} (\boldsymbol{T} - \boldsymbol{L})} \tag{3.88}$$

where $\boldsymbol{A}$ weighs the color distances in the color channels. Statistical methods can be used to compare the histograms, for example, the χ^2-test:

$$\chi^2_{S,T} = \sum_{l=1}^{L} \frac{(S_l - T_l)^2}{T_l + S_l} \tag{3.89}$$

The number of degrees of freedom is $L - 1 - |\{l | T_l = S_l = 0\}|$. If the histograms are given in absolute frequencies and their sum is equal, then this degree is reduced by one. More information on object recognition using color histograms can also be found in [610].

3.4.6 Active Object Recognition

For real scenes and computer vision problems of practical interest, the examples described so far are too much academic. Real-world scenes rarely contain isolated objects and homogeneous backgrounds, the scenes are rather cluttered, the objects are partially occluded, relatively small, etc.

The strategy of active vision which will be introduced in Section 3.6 can be successfully applied to combine recognition modules, 3D estimation, and model-based camera actions in order to solve the object recognition task in real-world indoor scenes [6]. This system actively moves the camera and changes the parameters of the lens to create high resolution detail views. Camera actions as well as top-level knowledge are explicitly represented in a semantic network; the control algorithm of the semantic network used to guide the search is independent of the camera control; this semantic network will be introduced in Section 3.6. Objects are hypothesized using color information (see Section 3.4.5) and verifyed using geometric matching (see Section 3.4.2) based on segmented color regions (see Section 3.1). The active component is necessary, since the objects are too small to be detected in the wide-angle view robustly, using any of the above mentioned methods.

An example scene containing office objects, object hypotheses by color backprojection, and close-up views of the objects is shown in Figure 3.19 (top). In a sequence of images which is recorded when the camera is moved lateeally in front of the scene, interesting points are selected by one the methods proposed in Section 3.1.4. Tracking of these points and analysis of their trajectory is used to recover the distance of these points to the camera. Figure 3.19 (bottom) shows two projections of these points. This sparse 3D information permits to estimate the size of the object mask $\boldsymbol{D}$ used in histogram backprojection Figure 3.19 (bottom right).

3.4.7 Conclusion

Most systems on object recognition reported in the literature use models which are represented explicitely. The choices for object models are geometric descriptions, appearance-based models or statistical models. None of the approaches guarantees satisfactory results for general problems, if it is used in an isolated module. Instead, a combination of modules is required.

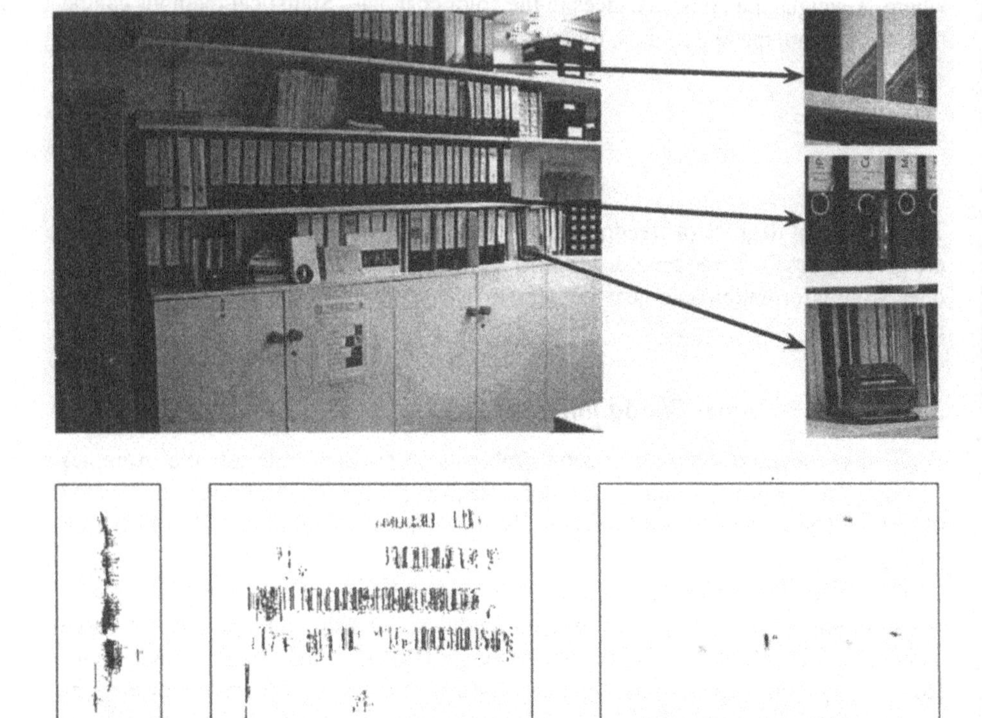

Figure 3.19. Office scene (top left), close up views used for hypotheses (top right), two projections of 3D points (bottom left), color backprojection for red punch (bottom right). The red punch searched for is shown in the close-up views on the top right.

Traditionally, geometric approaches were dominant e.g., in [333]. Recently, statistical methods are gaining more and more interest, e.g., in [262, 335, 406]. Support vector machines as proposed in [723, 708] promise a further step into this direction.

Although the object recognition problem is a central task in almost any computer vision system, it is still not completely solved and there is still room for improvement.

3.5 IMAGE UNDERSTANDING

H. NIEMANN

The term model-based image understanding is used in different meanings, of which we mention in particular:

- The interpretation of images or image sequences using a semantic model in the sense of Section 3.2.

  Examples of this approach are [473, 509, 562, 410]; a main source of knowledge are the geometric and task-specific properties of objects and events; its usage depends on suitable control algorithms which may employ (meta) rules, graph search, or combinatorial optimization.

- The recursive estimation of the static and dynamic parameters, summarized in a state vector, of a spatio-temporal model of relevant objects in the sense of Section 3.2.2.

  Examples of this approach are [178, 496, 377, 752]; a main source of knowledge is the dynamics of motion of objects derived from physical laws, the formation of a 2D-image of a 3D-object by a camera, and the geometrical properties of the involved objects; its usage mainly depends on analysis-by-synthesis employing recursive state estimation with *Kalman* filters or the extended *Kalman* filters.

  The *Kalman* filter, like the EM-algorithm (cf. Section 3.3.2), can recover 'missing information', e.g. estimate 3D-parameters from 2D-images.

- The configuration of an image processing or interpretation system using existing program libraries and/or knowledge about the processes of imaging, and so on.

  Examples of this approach are [131, 439].

Our presentation will emphasize the usage of semantic models, give a short account on the usage of spatio-temporal models, and leave the configuration problem to the references.

3.5.1 General Approach

Based on statistical decision theory (cf. Equation 3.36), the classification of 2D- and 3D-objects and shapes (as well as spoken words and utterances) traditionally is treated as an optimization problem. Based on logic inferences, knowledge-based interpretation often is still related to rule-based and heuristic approaches. It is shown here that also interpretation can be formulated and solved as an optimization problem.

In any case, given an image the task is to compute a symbolic description which is maximally compatible both with the observed images *and* the explicitly represented task-specific knowledge and in addition contains the information requested by the subsequent user.

The process of image understanding usually starts with a phase of mainly data-driven processing followed by a phase of mainly model-driven processing. In a flexible system these two phases are not strictly sequential in the sense that the first phase must be finished before the second may start. This flexibility is achieved by a control algorithm which determines the timing of the phases, the data used by them, and the order of switching between phases. During data-driven processing no explicitly represented task-specific knowledge is used (e.g. no knowledge about cars or diseases of an organ), whereas during model-driven processing this knowledge is the essential part.

By preprocessing (e.g. filtering, morphological operation, or normalization) the image $\boldsymbol{f}$ is transformed into another image $\boldsymbol{h}$. We will not treat this step here. Preprocessing is followed by the computation of an *initial segmentation* $\mathcal{A}$ of the image. It is represented by a network of segmentation objects O, for example, lines, vertices, and regions, together with their attributes, for example, length, contrast, velocity, and coordinates in space or in the image. Initial segmentation will not be treated here.

Model-driven processing requires an explicit *model* $\mathcal{M}$ of the task domain. Let us assume that this model is represented by a network of concepts C

$$\mathcal{M} = \langle C \rangle \tag{3.90}$$

as defined in Equation 3.26. The model has to contain all the information relevant to solve a given task of image understanding and to compute from an image the information requested by a user; it need not contain anything else.

The *interface* between segmentation and understanding is provided by the segmentation objects O and the concepts C. There must be so called *primitive concepts* which directly can be linked to results of segmentation. For example, assume a model $\mathcal{M}$ for the diagnostic interpretation of MR images of the knee contains, among others, concepts defining the cartilage by curved contour lines. In this case segmentation should extract appropriate curved contour lines which may be assigned to the primitive concepts of the cartilage.

It is up to the system designer to define the border between parts of the system implemented in a semantic network and parts of a system implemented, for example, by a set of procedures. Preprocessing and initial segmentation are well handled by procedures, knowledge-based inferences by a semantic network, and object recognition and localization either by a semantic network (if the search complexity is low) or by algorithms tailored to object recognition, for example, statistical approaches or neural networks.

The result of interpretation is a *symbolic description* $\mathcal{B}$ which is a network of instances of concepts defined in the model. Depending on the application the symbolic description may contain a complete description of all objects and their relations, a list of changes occurring between two successive recordings of the same scene, or only a symbolic name (e.g. 'normal chest X-ray'). Due to the hierarchical organization of a semantic network it contains both condensed information in the form of instantiated goal concepts as well as details in the form of instances of parts and concretes.

Primitive concepts are instantiated by segmentation results, and every object, motion, event, situation, and inference is defined by a concept in the model $\mathcal{M}$. Also the final goal of interpretation is defined in a concept, the goal concept C_g. Intermediate results of processing give rise to, possibly competing, instances $I_i(C)$ of a concept C. As pointed out in Section 3.2.1, every concept has an associated judgment G, which allows the computation of the score, value, or alternatively of the cost, of an instance $I(C)$. Therefore, also the goal concept has an associated judgment.

The best possible interpretation of an image is given by the best scoring instance $I^*(C_g)$ of the goal concept, that is

$$I^*(C_g) = \arg \max_{\{I(C_g)\}} \{G(I(C_g) \mid \mathcal{M}, \mathcal{A}\} \,. \tag{3.91}$$

In this sense image interpretation corresponds to an optimization problem. The actual computation of interpretations is done by a *control module*. Its task is to compute an efficient strategy for the interpretation of an image, given a model. This requires in particular to determine which available processing algorithm should be applied to which subset of intermediate results. One approach to do this is by graph search as described below. If the semantic network is converted to a function-centered representation, an alternative approach to solving the control problem is by combinatorial optimization which is also described below.

A somewhat different approach is adopted if knowledge is represented by rules, cf. Section 3.2.3. However, the basic idea of attaching a judgment to all results and then searching for the best scoring result is the same.

3.5.2 *Interpretation by Graph Search*

We now describe how the goal concept of a semantic network can be instantiated in an optimal (i.e. best scoring) manner by handling competing instances by a well-known graph search algorithm. Since our model only consists of concepts with identical syntactic definition, the control algorithm can be implemented task-independently. Only the judgment functions G defining the quality of instances are task-dependent.

In graph search every state of analysis, characterized by the current intermediate results, is associated with a node v in a search tree. If a node is not a goal node v_g of the search process, new nodes are generated by applying transformations. In our context the transformations are given by the task-independent rules mentioned in Section 3.2.1. The search tree is an implicitly defined graph which is given by a start node and by transformations to generate new nodes from a given node. One is only interested to generate as few nodes of the search tree as are necessary to find an optimal instance $I^*(C_g)$ of the goal concept. Therefore, an algorithm for graph search is needed finding efficiently the optimal path to the goal node, cf. Equation 3.93.

We denote the start node of the search tree by v_0, the current node by v_i, and the goal node by v_g. Furthermore, let $\varphi(v_i)$ be the cost of an *optimal path* from v_0 to v_g via v_i, $\psi(v_i)$ the cost of an optimal path from v_0 to v_i, and $\chi(v_i)$ the cost of an optimal path from v_i to v_g. In order to obtain an efficient search algorithm three conditions are imposed. The first condition is that the additive combination

$$\varphi(v_i) = \psi(v_i) + \chi(v_i) \tag{3.92}$$

of costs must hold. Since one usually does not know the true costs, they are replaced by their estimates $\widehat{\varphi}(v_i), \widehat{\psi}(v_i), \widehat{\chi}(v_i)$. The second condition is that the estimate $\widehat{\chi}(v_i)$ must be *optimistic*, that is,

$$\widehat{\chi}(v_i) \leq \chi(v_i) \qquad \text{for } all \ v_i \quad . \tag{3.93}$$

An optimistic estimate is $\widehat{\chi}(v_i) = 0$. More efficient search can be expected if a 'more informed' estimate $0 < \widehat{\chi}(v_i) \leq \chi(v_i)$ is known. The third condition is that the estimate is *monotone* in the sense

$$\widehat{\chi}(v_j) - \widehat{\chi}(v_k) \leq r(v_j, v_k) \quad , \tag{3.94}$$

where $r(v_j, v_k)$ is the true cost of an optimal path from v_j to a successor v_k. A graph search algorithm meeting these constraints is called an *A^*-algorithm*, [508, 511].

For initialization of the algorithm one provides a nonempty set V_s of start nodes, a nonempty set V_g of goal nodes, a judgment function $\widehat{\varphi}(v)$ for a node v, a set T

of transformations to generate successor nodes of a node v, a list *OPEN* initially containing $V_s \cup v_0$, and a list *CLOSED* initially containing v_0. In each interation step one removes from *OPEN* the best scoring node v_k and puts it on *CLOSED*. If this v_k is in V_g, then the algorithm STOPs with *'success'*, that is the best scoring path to a goal node was found. Otherwise one expands v_k to generate the set V_k of its successor nodes. If no successors can be generated, then the algorithm STOPs with *'failure'*, that is, no gaol node can be reached. If successors can be generated, one computes the scores of the nodes in V_k, and adds to the list *OPEN* those successor nodes which are not yet on *OPEN* and not on *CLOSED*. One checks whether a better path to a node in *OPEN* was found and in this case adjusts the scores on *OPEN* accordingly. Then one continues with removing from *OPEN* the best scoring node, and so on.

The theoretical properties of this algorithm are proven, for example, in [511]. It is only mentioned here that in the worst case the A*-algorithm has exponential complexity; however, for 'good' scoring functions its complexity may only be linear. Hence, the design of a proper judgment is essential for the feasibility of this approach.

The A*-algorithm may be adapted to the problem of finding an optimal instance of a goal concept in a semantic network. An elementary control algorithm starts with the top-down expansion of the goal C_g until primitive concepts are reached. As mentioned above, primitive concepts do not have parts and/or concretes and can be instantiated by results of initial segmentation. This expansion is independent of the input image and can be done once for the knowledge base. The result of expansion is the so-called *instantiation path* which is a list of those concepts which must be instantiated prior to the instantiation of the goal C_g. After expansion the concepts on the instantiation path are instantiated bottom-up until the goal is reached, and this step depends on the input image. Due to alternative definitions of a concept, due to competing instances, and due to a combinatorial multiplicity of assigning results of initial segmentation to primitive concepts, it is necessary to concentrate the instantiation on the most promising alternatives which is precisely what the A*-algorithm does. This approach to knowledge-based processing and control has already been applied successfully to the diagnostic interpretation of scintigraphic image sequences of the heart, [509].

The importance of a proper judgment function was mentioned above. One has to distinguish between the judgment $G(I(C))$ of an instance $I(C)$ of a concept C and the judgment $\phi(v)$ of a node v in the search tree. It is common to base the score of an instance always on the quality of the match between data and model or on the quality of fulfilment of fuzzy inference rules. The score of a search tree node is always based on the current estimate of the quality of an instance of the associated *goal concept*. This is requested for graph search by Equation 3.92, and is also used for combinatorial optimization as defined by Equation 3.96. This insuere that instantiation is always directed towards *optimal* instantiation of a *goal concept* C_g as requested in Equation 3.93. The judgment of an instance of a concept is a combination of the judgments of its parts, concretes, attributes, and relations.

The problem that we always need a judgment of an instance of a goal concept, but in the early stages of processing do not yet have such an instance is solved by replacing the actual costs by an optimistic estimate as is requested by Equation 3.93. As usual, we denote the goal concept by C_g, some other concept by C, its set of parts and concretes by $D = \{D_i, i = 1, \ldots, n\}$, a measure of cost normalized to $0 \leq G \leq G_{\max}$ by G, and its estimate by $\widehat{G}$. Then an estimate of the cost of an instance of the goal C_g associated with a node v in the search tree and denoted by $\widehat{G}(I_v(C_g))$ can be computed recursively at every stage of analysis by using the judgment $G(I_v(C))$ of a instance $I_v(C)$ of a concept C associated with search tree node v. If C is already instantiated, then the actual value of $G(I_v(C))$ is available and $\widehat{G}(I_v(C)) = G(I_v(C))$. If C is primitive and still uninstantiated, our optimistic cost estimate is $\widehat{G}(I_v(C)) = 0$. If C is not primitive and still uninstantiated having the set D of parts and concretes, then we use the cost estimate $\widehat{G}(\widehat{G}(I_v(D_1)), \ldots, \widehat{G}(I_v(D_n)) \mid C)$. The cost estimates of the instances $I_v(D_i)$ are obtained from the same recursion. Finally, we obtain the cost of the search tree node by

$$\phi(v) = \widehat{G}(I_v(C_g)) \, , \tag{3.95}$$

which is an optimistic estimate of the cost of an instance of the goal concept in the current search tree node v.

3.5.3 Interpretation by Combinatorial Optimization

Optimization problems usually can be solved in various ways and this is also true for the control problem. If the semantic network is converted to a network of functions as outlined above in Section 3.2.1, one may apply combinatorial optimization techniques to compute an optimal instance of the goal concept. This is an interesting alternative to the design of a *control module*. In this setting a particular assignment of segmentation results to primitive concepts and a particular assignment of concept definitions (i.e. sets of modality) is considered as a *state* in a combinatorial optimization problem and this assignment is optimized with respect to a predefined judgment function. The optimization can be done iteratively by algorithms like simulated annealing, great deluge, or genetic programming. Details are given, for example, in [224].

The function-centered representation of a semantic network provides an approximate interpretation at every iteration and therefore supports the fast computation of suboptimal interpretations, which may be used by other processing modules if less processing time is available. This is the so called *any time property*. A control algorithm at first initiates a bottom-up instantiation and the computation of values and judgments for each attribute, structural relation, link, or concept of the function-centered network. This will lead to many competing interpretations having low values of judgment. In the second stage assignments are modified iteratively by applying a combinatorial optimization procedure. The goal is to obtain a unique interpretation having a high value of judgment.

The process of bottom-up instantiation starts with the computation of attributes that provide an interface to the initial segmentation, and proceeds upwards to the goal concepts. The result of bottom-up instantiation are instances for the goal concepts together with a judgment $G(I(C_g))$. If there are n goal concepts, we create a vector from the instances having the best judgment

$$g = (G(I^*(C_{g_1}), \ldots, G(I^*(C_{g_n})) \tag{3.96}$$

and this represents the final result of one iteration step.

The computation of instances and their judgment is completely determined by assigning to each interface (attribute) node A_i a (possibly empty) subset $\{O_j^{(i)}\}$ of segmentation objects and selecting for each concept node C_l a unique set of modality $H_l^{(k)}$. This allows us to characterize the current *state of analysis* by a vector

$$q_c = \left[(A_i, \{O_j^{(i)}\}) \mid i = 1, \ldots, m \; ; \; (C_k, H_l^{(k)}) \mid k = 1, \ldots, m'\right], \tag{3.97}$$

where m is the number of interface nodes and m' the number of concepts having more than one member in its set of modality. This shows that Eqaution 3.96 is a function of the state $g = g(q_c)$, Furthermore, the computation of an optimal instance according to Equation 3.93 now can be solved as a combinatorial optimization problem whose optimum is defined by a suitable cost function ϕ given below. This approach was used, for example, to find street markers in image sequences of traffic scenes [224].

A *cost function* ϕ suited for combinatorial optimization is based on the fact that an error-free segmentation would support a single interpretation resulting in an ideal judgment $G(I^*(C_{g_i})) = 1.0$ of the instance of the correct goal concept. On the other hand, at the same time this segmentation would give no evidence to any other goal concept such that $G(I^*(C_{g_j})) = 0.0$, $j \neq i$. Therefore, the desired judgment is the i-th unit vector δ_i as an ideal result of instantiation, if the i-th goal concept provides the correct symbolic description. An approximation of this behaviour is obtained by using

$$\begin{aligned} \phi(q) &= \min\nolimits_{1 \leq i \leq n} \{(\delta_i - \widehat{g}(q))^2\}, \\ \widehat{g}(q) &= \frac{g(q)}{|g(q)|} \end{aligned} \tag{3.98}$$

as a cost function; it computes the minimum distance from $g(q)$ to the unit vectors δ_i. A useful difference to the cost function in Equation 3.95 employed for graph search is that now we have a *problem independent* measure of costs, since no assumptions on the contents of the knowledge base are made. Nevertheless, there is a correspondence between the cost functions since both evaluate the expected quality of goal achievement.

From among the different algorithms *stochastic relaxation* was found to have very good convergence properties. Bottom-up instantiation provides an initial state q_0 which initially also is the current state q_c. A new state is generated by either using a different assignment of interface nodes to segmentation objects or using a different set of modality in Equation 3.97. This gives a new state q_n. Stochastic relaxation accepts the new state according to the rule

$$\phi(q_n) - \phi(q_c) \begin{cases} \leq 0 & : \text{ replace } q_c \text{ by } q_n \\ > 0 & : \text{ no change in } q_c \end{cases} . \tag{3.99}$$

3.5.4 Interpretation by Spatio-Temporal Models

If a vehicle (e.g. a car or a plane) is moving in the 3D-world and its motion is to be controlled by a camera, it has become a standard approach to carry out a recursive estimation of the relevant state parameters by evaluating a TV image sequence. State parameters may be, for example, the position and speed of a car, the width and curvature of a road, and the size of vehicles. Interpretation then amounts mainly to estimation of the task-specific state parameters. Instead of evaluating each image of the image sequence more or less independently of the others and then combining results, during recursive estimation only the relevant state parameters are updated from observations of image features. These features depend on the geometry of the involved objects and on the camera forming an image. This approach is capable of real time processing (at video rate), has no need for storing and processing past images, implements a signal-to-symbol transition, allows the prediction of type and position of features in the next image, and has been proven to yield very good results in experimental studies. The relevant equations for the dyamical system model and the observation were introduced in Section 3.2.2.

State Estimation. The state of the dynamical system has to be estimated from observations of object features. Estimation equations may be derived from the requirements that a *recursive linear* estimate of the state vector is computed, the estimate is *unbiased*, and the estimate is *optimal* in the sense of a quadratic cost function. The solution to these requirements is the *Kalman* filter. It gives the optimal estimate for normally distributed noise, that is, no nonlinear estimate can be better; and it is the optimal linear estimate in case of non-normally distributed noise, that is, in this case there may be a better nonlinear estimate.

The state estimation is separated into the two phases of *prediction* without using the observation and *observation update* (or measurement update) including the new observation. We denote the estimate of the state vector $\boldsymbol{q}$ by $\widehat{\boldsymbol{q}}$, and similar for other estimates. We denote the estimate obtained in the prediction phase by $\widehat{\boldsymbol{q}}^{(-)}$ (without observation), and the estimate obtained in the update phase by $\widehat{\boldsymbol{q}}^{(+)}$ (using the observation). If no observation is given, the state estimate is obtained from Equation 3.29.

$$\widehat{\boldsymbol{q}}^{(-)}[k] = \Phi[k-1]\widehat{\boldsymbol{q}}^{(+)}[k-1] + \boldsymbol{B}[k-1]\boldsymbol{u}[k-1] \,. \tag{3.100}$$

Due to noise and imperfections of image processing and modelling there will be an error between the estimate and the true value

$$\varepsilon^{(-)}[k] = \boldsymbol{q}[k] - \widehat{\boldsymbol{q}}^{(-)}[k] \tag{3.101}$$

which should be reduced during the measurement update. The update due to the *Kalman* filter is given by

$$\begin{aligned} \widehat{\boldsymbol{q}}^{(+)}[k] &= \widehat{\boldsymbol{q}}^{(-)}[k] + \boldsymbol{K}[k]\left(\boldsymbol{o}[k] - \boldsymbol{H}[k]\widehat{\boldsymbol{q}}^{(-)}[k]\right) \\ &= \widehat{\boldsymbol{q}}^{(-)}[k] + \boldsymbol{K}[k]\big(\boldsymbol{o}[k] - \widehat{\boldsymbol{o}}[k]\big) \,. \end{aligned} \tag{3.102}$$

The above equation updates the estimate $\widehat{q}^{(-)}[k]$ depending on the *difference* between the actual observation $\boldsymbol{o}[k]$ and the predicted observation $\widehat{\boldsymbol{o}}[k]$. If the state vector and the model are correct, the difference should be zero. If the difference is not zero, the state vector is updated in order to reduce this difference.

The main problem is the determination of the *gain matrix* $\boldsymbol{K}[k]$ above. One approach is to minimize a criterion derived from the estimation error

$$\varepsilon^{(+)}[k] = \boldsymbol{q}[k] - \widehat{\boldsymbol{q}}^{(+)}[k] \,. \tag{3.103}$$

An appropriate error criterion is the trace of the error covariance matrix

$$\boldsymbol{P}_k^{(+)} = E\left\{\left(q_k - \widehat{q}_k^{(+)}\right)\left(q_k - \widehat{q}_k^{(+)}\right)^T\right\} \,. \tag{3.104}$$

If the noise vectors $\boldsymbol{w}[k]$ in Equation 3.29 and $\boldsymbol{r}[k]$ in Equation 3.31 are uncorrelated white noise with covariance matrices $\boldsymbol{Q}[k]$ and $\boldsymbol{R}[k]$, respectively, the gain matrix and the error covariance matrix are given by

$$\boldsymbol{P}^{(-)}[k] = \Phi[k-1]\boldsymbol{P}^{(+)}[k-1]\Phi^T[k-1] + \boldsymbol{Q}[k-1] \,, \tag{3.105}$$

$$\boldsymbol{P}^{(+)}[k] = \boldsymbol{P}^{(-)}[k] - \boldsymbol{K}[k]\boldsymbol{H}[k]\boldsymbol{P}^{(-)}[k] \,, \tag{3.106}$$

$$\boldsymbol{K}[k] = \boldsymbol{P}^{(-)}[k]\boldsymbol{H}^T[k]\left(\boldsymbol{H}[k]\boldsymbol{P}^{(-)}[k]\boldsymbol{H}^T[k] + \boldsymbol{R}[k]\right)^{-1} \,. \tag{3.107}$$

Enhancing Robustness. Image processing to extract object features is subject to various imperfections, visible (or invisible) features may become invisible (or visible) due to motion, the number of visible features is not constant, and everything has to be worked out in real time. This requires some steps in addition to the basic state estimation approach outlined above.

A proper choice of the coordinate systems may facilitate the mathematical formulation of the model equations 3.29 and 3.31.

To avoid or reduce possible problems of numerical instability with Equation 3.106 the form

$$\begin{aligned}\boldsymbol{P}^{(+)}[k] &= \left(\boldsymbol{I} - \boldsymbol{K}[k]\boldsymbol{H}[k]\right)\boldsymbol{P}^{(-)}[k]\left(\boldsymbol{I} - \boldsymbol{K}[k]\boldsymbol{H}[k]\right)^T \\ &\quad + \boldsymbol{K}[k]\boldsymbol{R}[k]\boldsymbol{K}^T[k] \,.\end{aligned} \tag{3.108}$$

can be used (which, however, is computationally more expensive), one may use double precision, or the U-D factorization [64] of the error covariance matrix $\boldsymbol{P}$.

To avoid the uptdate of estimates by wrongly detected image features, the difference between the actual observation and the predicted observation is a suitable measure. The avoidance of unreliably detected features and the above mentioned occurrence of new as well as the disappearance of old image features results in an observation matrix $\boldsymbol{H}$ of variable size.

3.6 ACTIVE VISION

J. DENZLER

In the early 80's the so called 'Marr paradigm' [468] strongly influenced computer vision work all over the world. The idea that Marr describes in his book is to completely build up a symbolic representation of the whole image from the image matrix ("from pixels to predicates"). He states that with such a complete reconstruction, every vision task can be solved.

The idea itself looks quite promising, although the consequence is an enormous computational effort, especially when processing image sequences. Due to increasing computational power of computer hardware in the late 80's it becomes possible to solve some low-level computer vision tasks in real time. Additionally, new and cheaper special imaging hardware (for example, pan/tilt cameras) allows an active influence on the image acquisition process instead of taking images without considering the particular task that should be solved.

During this period, also motivated by the need of a fast response time of image processing algorithms many researchers were wondering, weather it is useful in any case to completely reconstruct the whole scene. The next question was, whether a good representation can be found without knowing the task to solve? The answer is no. A good example to show the effect of a certain representation is the old-fashioned non-digital telephone book: it is very easy to find the phone number of a certain person, but impossible to find a person given a certain phone number.

This has been the starting point for a new processing strategy which sometimes has led to a controversial between the advocates of the Marr paradigm—also called reconstructionists—and the active vision people. The main point of active vision is, that the images should be taken actively in a way, such that a given problem can be solved optimally in a certain sense ("People just not see, people look" [31]). The two paradigms are compared in Figure 3.20.

This section is divided into two parts: a more philosophical part discussing different aspects of vision, both the Marr and the active vision framework, and a practical part

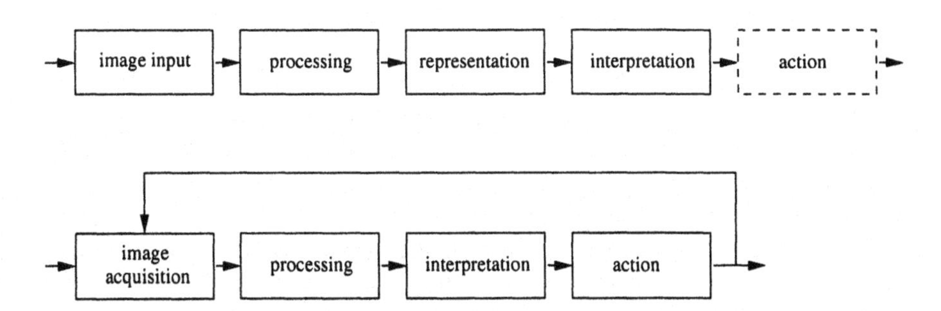

Figure 3.20. Comparison between the processing paths for the Marr paradigm (top) and the active vision paradigm (bottom). An action based on the processing result was not important in the Marr paradigm for a long time. Of course, in principle an action is possible.

summarizing successful stories in implementing visual capabilities for machines using active vision principles. In the paragraphs that follow, active vision will be defined, the main goals and techniques, and some resulting demands in low-level and high-level vision will be presented. Finally some applications will be shown. It is worth noting, that there exist a lot of successful applications, which have been made without using the active vision paradigm. Examples are analysis and classification of medical images, or automatic letter sorting machines. Nevertheless, theoretical work exists, which shows the strength of an active strategy compared with a passive one. One simple example is the complete 3D reconstruction of an object. Without actively changing the position of the sensor in dependence on the shape of the object, it is impossible to get a complete 3D information of the object's shape.

3.6.1 Why Active Vision?

For a better understanding of what active vision means, one has to study the computational framework of vision, created by Marr in his famous book 'Vision' [468]. He states in the introduction of his book that

> "... vision is the process of discovering from images what is present in the world and where it is."

In his opinion this implies that

> "... the brain must somehow be capable of representing this information."

Finally, he defines vision as an information processing task with the duality of representation and processing of information. Step by step the information gathered from the environment must be processed—from pixels to predicates. Then, other modules like planing, decision making, learning, etc., can be applied, because all information is already processed and represented.

Of course, such an approach is quite natural: Build a system for general vision—it is worth noting, that vision must be general to fulfill the demands of arbitrary processing stages, following the vision stage—and build up a complete representation of the environment—again, since the following tasks (planning, learning) are not taken into account, the representation of the environment must be complete.

When looking at this approach two problems can be observed. First, in nature there exists no general vision system. Especially, the human visual system is far away from being general. Examples are a couple of optical illusions fooling human vision. A good illustration of the relationship between general vision, human vision and some other terms like reconstructionist, exploratory, purposive vision has been given by [714] and is shown in Figure 3.21. The second observation is that no animal with the capability to see is doing vision without any task at hand. Vision is always driven by a certain application. Especially for some animals, the visual system is specialized to solve a certain task. One example is the frog, for which scientists have found a specialized module for the detection of bugs [685].

These observations have led to a different approach on constructing machine vision systems. The new framework can be summarized by two main principles: vision should not be seen independently on the application. This results in a purposive, qualitative, and animate processing strategy. And second, each vision system in nature

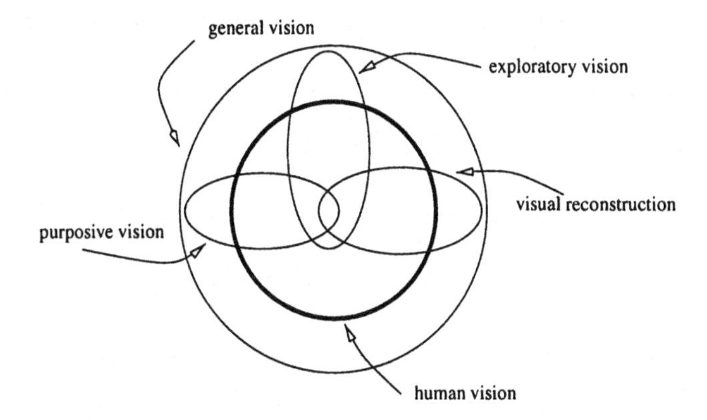

Figure 3.21. Relationship between general vision and vision in the Marr as well as in the active vision paradigm (from [714]).

is active, i.e. in machine vision one main point must be the control of the image acquisition process.

3.6.2 What is Active Vision?

The term active vision has not been used uniquely in the literature. Additionally, several other terms, like 'purposive vision', 'smart sensing', etc. are known. This paragraph is intended to summarize different aspects of the general term active vision. In the next paragraph, we will classify each term into the three levels of processing: image acquisition, low-level, and high-level vision. In [524] an active vision system is defined as follows:

> "An active visual system is a system, which is able to manipulate its visual parameters in a controlled manner in order to extract useful data about the scene in time and space."

Swain [680] focuses that

> "... vision is most readily understood in the context of visual behaviors that the system is engaged in."

so

> "Active vision encompasses attention, selectively sensing in space, resolution and time, whether it is achieved by modifying the physical camera parameters or the way data is processed after leaving the camera."

Aloimonos [13] characterizes active vision in contrast to passive vision:

> "When you work on passive vision, you are given a set of images which you will process with the algorithms you are going to develop. On the other hand, when you work on active vision you do not want prerecorded data. You are given an active observer which has control over the image acquisition process and which acquires images which are relevant to what it intends to do."

In the preface of [70] it is defined what active vision does:

> "It picks out the properties of the images which it needs to perform its assigned task and ignores the rest."

This definition of the selection process corresponds to that of Swain, which was given above. In the same book Brooks [70] divides active vision into:

> "... active operations in the world in order to change the images that are being collected in a way which enhances task achievement. "

and into

> "... active autonomous processes (e.g. snakes) which exploit the coherence of images in a sequence in order to efficiently and reliably track aspects of interest over time."

Krotkov [404] limits active vision to the control of the sensor parameter, so an active sensor is a:

> "... passive sensor employed in an active fashion, with sensing strategies purposefully changing the sensors' state."

Several other terms have been used, like *active and exploratory perception* [31, 32]. There, the term 'active' means that a system

> "... actively seeks for information (*to look around*) and not just rely passively on information falling accidentally on the sensor (*to see*)."

This definition is extended in [32]:

> "Hence the problem of Active Sensing can be stated as a problem of controlling strategies applied to the data acquisition process which depends on the current state of the data interpretation and the goal or the task of the process."

Aloimonos uses the term *active, qualitative and purposive vision* [11]. *Active* means the adaptive change of the camera parameters during analysis, *qualitative* to favor partial instead of complete reconstruction for complexity reduction. Finally, *purposive* means to focus on the task that must be solved to find certain restrictions. These might help to come closer to real-time processing.

In *Animate Vision* [34] the key points are gaze control, attention mechanism, learning, and real-time realizations of algorithms. They propose that if a cooperative processing strategy of different modules is used, the explicit representation can be dropped. In [87] it is distinguished between bottom-up (Marr) and top-down vision (qualitative vision) and it is claimed to use a mixture of Marr paradigm and active vision. The term *Smart Sensing* has been introduced by [100]. Summarized, it

> "... makes carefully use of selective analysis and fast algorithms to achieve efficiency..."

and it is a

> "... selective task oriented gathering of information from the visual world."

There, the main points are hierarchical data structures and coarse to fine search strategies.

3.6.3 Goals and Techniques

The main motivation of active vision is to reduce the limitations of the Marr paradigm in certain applications. Basically, this mean that vision cannot be seen independently on the application, i.e. vision should be purposive which includes selective. And the image acquisition process should be active, which includes not only the movement of a camera but also the change of the camera parameters, like zoom, focus, or aperture. The latter one is quite natural, because the camera parameters should always be adjusted to an optimal setting in a certain situation; sometimes this is done by the camera itself (e.g. autofocus), but it is obvious, that the machine must at least know the changed parameter values. It is of course better to let the image processing system control the image acquisition completely.

Purposive, Selective Vision. The main point of criticism on the active vision paradigm with the purposive, selective framework is, that the main goal is the construction of running systems instead of the development of a deep theory. Indeed, such a demand can be heard from the active vision people. In [123] the development of systems, working outside the laboratory, in [713] the reduction of computation time is important. But this is *not* the main goal. Having systems, which run outside the laboratory—this implies running in real time if interaction with the environment is wanted—opens a completely new area of research for computer vision, with a highly increased complexity of situations. Then it is very important for meeting the real-time constraints to work selectively, in space, time, and resolution [680]. This motivates that the complete scene must only be analyzed periodically (at a rate lower than video frame rate), unless an attention mechanism detects something interesting for fulfilling the task.

Thus, to steer selection, attention mechanisms must be developed to detect unexpected events in images or image sequences. At present this is an unsolved problem in general. One attention mechanism could be based on motion, since we observe in our world that mostly the moving objects, i.e. the changes in the scene are important. Thus, it can be understood, that one main technique for selectivity should be focussing on a moving object by object tracking [70]. This can be done completely driven by data without high-level knowledge and control.

Beside the selectivity on the low level processing layer also a selection on the model level must be done, which makes it necessary to have hierarchical models [713].

Purposive and selective vision can be done not only on the low-level layer of a vision system. Aloimonos [12] proposes so called reduced behavior and action sets following the RISC architecture of processors. Such a RBAS module might be very general therefore efficient, since no special assumption have to be made (like the optical flow based detection of obstacles). He argues that with a purposive combination of such modules—controlled by the high level layer—more complex visual behavior can be constructed. Also such modules could work in parallel, and instead of sequentially processing from low to high level or vice versa, concurrent processing is necessary.

Active Image Acquisition. For active image acquisition there exists theoretical work as well as practical systems (cf. Section 2.4). In [14] the advantage of an active strategy compared to a passive one is shown in the area of shape from X problems (e.g. X = shading, contour, motion, etc.). In concrete systems the active image acquisition process, which includes control of zoom, focus, aperture, and of course the position of the camera in the world is very important [680]. In natural environment the global illumination and the distance to relevant objects cannot be controlled. Thus, the control of the camera and its parameters must be seen as an action itself as well as an *re*action to the environment. To be active implies that such an action must be planned, especially if high level knowledge is necessary to make a decision. Thus, the sensing itself must be planned [86].

The camera parameters can be controlled not only each alone but also in a purposive cooperative way to come up with sensor data fusion, for example, the combination of zoom, vergence, and stereo [404].

Besides the optimization of the optical, sensorical and mechanical parameters of a vision system the algorithmic parameters can also be optimized [524].

Construction of Active Vision Devices. To get such abilities as described above, there is the need to develop visual sensor systems with the same or at least similar degrees of freedom than the human visual system [70]. A very deep discussion of the design of an stereo camera system can be found in [524]. Today, a couple of commercial stereo cameras, and active vision heads exist[1]. They differ mainly in speed and degree of freedom of the optical system and of course in the price. A good choice for beginning with active vision might also be a low-cost steerable multimedia camera. Some public domain camera systems can also be found, which means, that only construction plans are available and the construction must be done on one's own.

3.6.4 Image Acquisition, Low-Level, and High-Level Vision

The techniques to reach the goals in active vision can be classified by the processing level, with which the goals should be achieved: image acquisition level, low-level vision, and high-level vision. In Figure 3.22 the different demands for active vision on the different processing levels are summarized. It is important to realize that the image acquisition layer is added, which is not present in the Marr paradigm. The low- and iconic-levels are merged. Finally the high-level layer is connected to a knowledge base. At the end, as already mentioned, the action or reaction can be found. The arrows between the layers indicate the following: the data flow from image acquisition passes not only sequentially through the low-level over the iconic- to the high-level layer, like in the Marr paradigm. Additionally there is a direct path from image acquisition to the high-level layer, and there is not only top-down or bottom-up processing provided, but processing of low-level and high-level concurrently and in parallel. Finally, the results from the low-level layer as well as from the high-level layer influence back to the image acquisition process.

3.6.5 Selected Applications

In the following two example applications, ranging from low-level active vision to high-level active vision are presented. More examples in the area of focus on attention can be found in [144], for selectivity in [93, 441], for phase based binocular vergence control [694], and cooperative focus and stereo [404, 405]. Also [12] is a good reference for a wide range of applications in active vision, in [143] the results of a huge research program on 'vision as process' are summarized.

[1] Examples are heads from HelpMate Robotics (formerly TRC), Robosoft, or Directed Perception

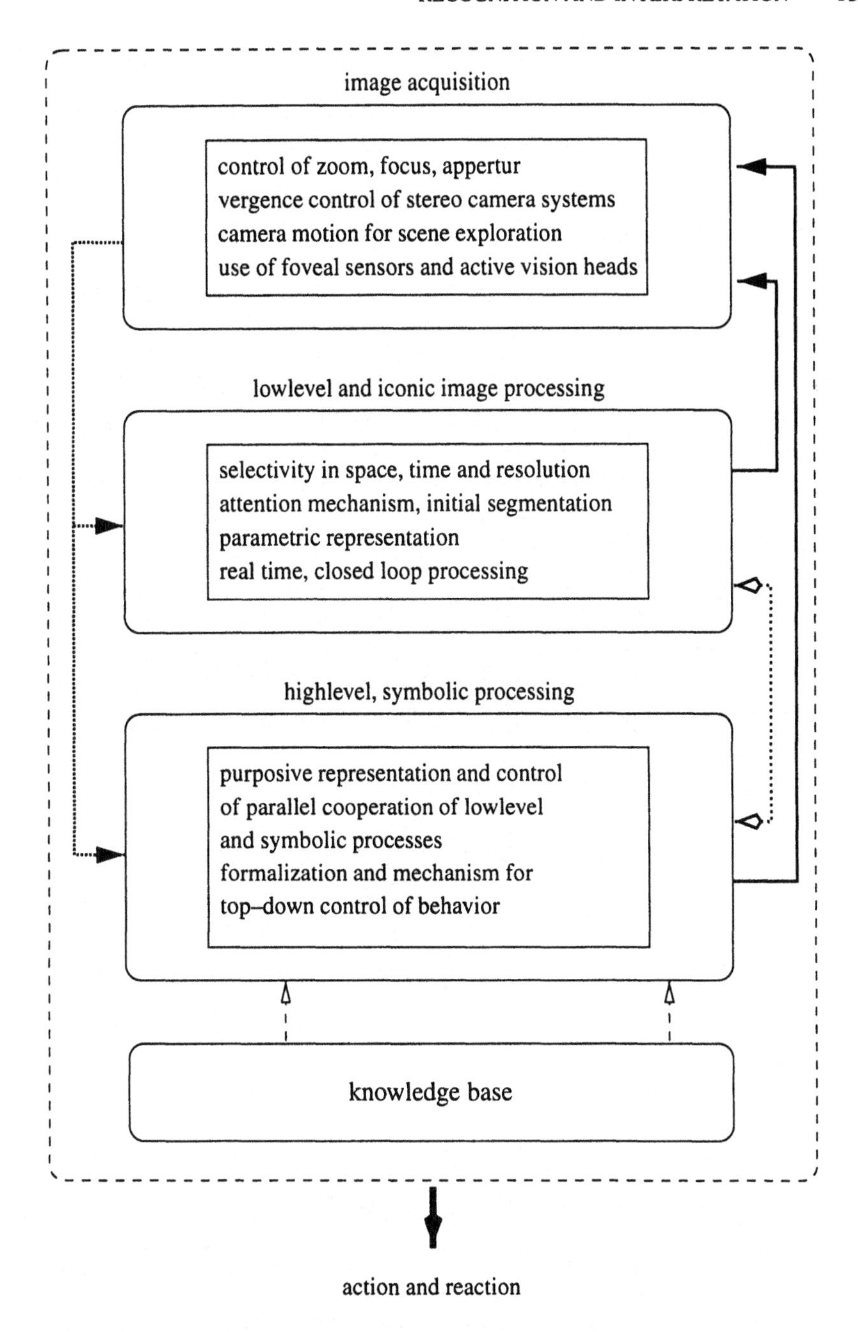

Figure 3.22. Demands for active vision on the different processing levels.

Real-Time Object Tracking. The first example, described in more detail, is a system, using the selectivity principle of active vision mainly at the low-level layer [170]. It has been already presented in Section 2.4 in the context of real-time object tracking. The key points for the success of this system has been

Selectivity in space: the contour tracking by active contours and rays, which are [70]

> "... active autonomous processes which exploit the coherence of images in a sequence in order to efficiently and reliably track aspects of interest over time."

Selectivity in time: the attention module, looking periodically for errors in the extracted contour to detect errors due to rapid changes in the contour.

Selectivity in resolution: motion detection on a low-resolution image and tracking in a high-resolution window of the complete image frame.

It was one of the main demands of the system to track moving objects in real time without specialized hardware. Thus, the selectivity was more because of practical reasons, than a paradigm shift. Nevertheless it shows, how selectivity can help in certain applications. In addition to that, the system has been designed of a couple of modules which work in parallel and concurrently [12]. The communication is done via PVM (parallel virtual machine), which allows to distribute the modules on a heterogeneous cluster of machines. In the current realization, the camera control module runs on an extra machine without synchronization with the tracking process. Of course, this sometimes results in errors due to timing problems and lost camera control messages over the link. But in general it supports the demand for a set of simple modules (cf. Figure 3.23) which in combination solve more complex tasks.

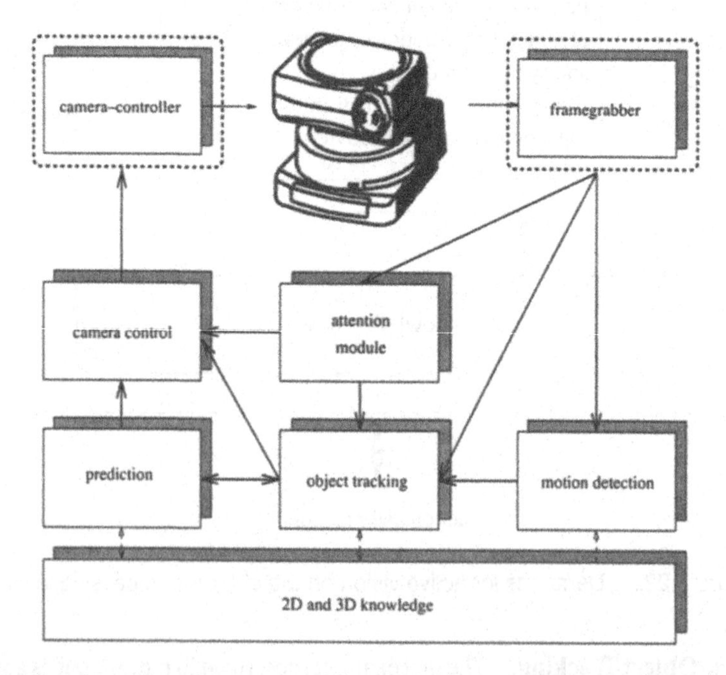

Figure 3.23. Overview of the system COBOLT (COntour Based Object Localization and Tracking), which makes use of several aspects of active vision to achieve real-time performance.

Knowledge Based Scene Exploration. The second example tackles the problem of active vision on the high-level layer of a vision system. Some results of this ongoing research can be found in [527]. The aim is to integrate actions in the formalism of semantic networks for active vision. In Figure 3.24 the semantic network based on the formalism of ERNEST [510] is shown, which uniquely models an office scene as well as actions to analyze a certain scene. The advantage of such an approach is that the control algorithm for applying the knowledge to scene analysis and understanding remains problem independent.

The gray ovals in Figure 3.24 stand for concepts of the knowledge base and are used to represent knowledge about the problem domain, as it is used in most knowledge based systems [510]. The white ovals are concepts for camera actions at different levels of abstraction, like global search strategies (direct vs. indirect search [762]) at the highest level (explOffice) or physical change of the camera parameters (zoom, focus), which are low-level and purely data driven (explOfficeImage). The global search strategies (direct vs. indirect search) are concurrent concepts which should not be instantiated at the same time. For this, modalities are used. During analysis, ambiguities are resolved by the problem independent control algorithm based on the well known A^* search algorithm.

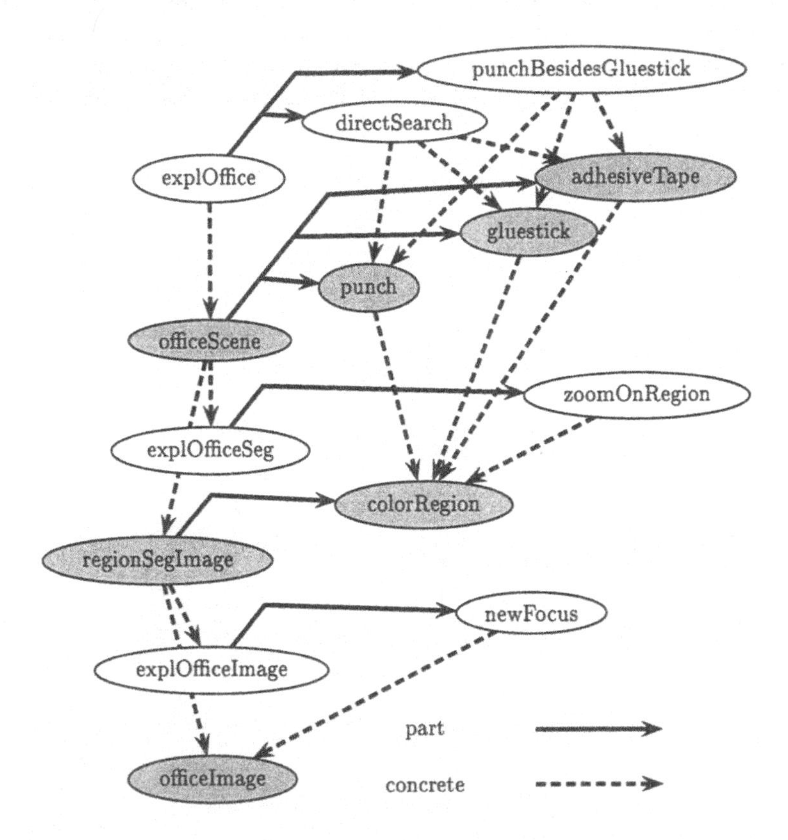

Figure 3.24. A semantic network integrating declarative knowledge about objects and actions.

The approach has been applied to the exploration of an office scene, where different objects (punch, clue stick, adhesive tape) are located. These objects must be found by the system. At the initial focal length of the camera the complete scene can be seen, but the objects looked for are too small to be distinguished at this resolution. Thus, it is necessary to actively change the camera parameters (e.g. zoom, pan, tilt) as well as the search strategies, in the case, that relational knowledge (like "the punch is located mostly besides the adhesive tape") is available. At present, in 85% of all exploration of the office scene the objects are correctly found. The time for completely analyzing an office scene including the camera movements, segmentation, and goal concept instantiation currently lasts about 5 minutes.

3.6.6 The Future of Active Vision

Today, a shift in the vision paradigm towards an active strategy is already in effect. The question arises, if there is still special active vision research necessary. Indeed, many problems remain open until now, like robust object tracking in natural scenes, a general attention mechanism, or structure from motion in a general environment. But the pioneering work of active vision people in the late 80's and 90's has achieved that more and more systems deal with real-time aspects, natural scenes, and processing of image sequences in a closed loop of sensing and reaction. The problems detected in such highly dynamic environments opened the eyes for new theoretical problems that must be solved now, supported by and based on the theoretical results from the Marr paradigm.

4 REPRESENTATION AND PROCESSING OF SURFACE DATA

G. GREINER

INTRODUCTION

In recent years optical 3D sensors have become powerful tools for *reverse engineering*. The shape of a three-dimensional object is sampled for that purpose and turned into a description for computer aided design (CAD). The method enables processing of physical design models on a computer (see Section 8.1). Using computer aided manufacturing (CAM) techniques like numerical controlled (NC) milling or stereolithography, three-dimensional replicas of the digitized objects can be made. In dentistry such methods are used to scan teeth or plaster casts and to automatically produce crowns and inlays from the data.

The raw data delivered by the 3D sensors (*range images*) are not well suited for direct use in CAD systems, as the data are given in the local coordinate system of the sensor. Moreover, the range images do not really describe surfaces, but clouds of point coordinates in 3D space. The amount of data points may be very large (from millions to hundreds of millions). Furthermore, data points are usually distorted by measuring errors like noise, aliasing, outliers, etc. Thus, several problems have to be addressed before a complete surface description can be achieved:

1. The *preparation of the raw data* refers to noise removal, data reduction with small loss of information, detection of outliers, compression or any combination of it. We present problems and solution techniques in Sections 4.1 and 4.3.

2. Detection of features. These can be used for classification purposes or for analysis and interpretation. In addition, features are often used for registration or matching

(see Section 4.2). Here the task is to combine several data sets into best possible alignment. This is necessary, e.g. when an object is scanned from different view points, which often cannot be avoided due to the special geometry of the object.

3. The transformation of the single range images into a common global coordinate system (*registration*, see Section 4.2).
4. Fusion of different views obtain global, topologically correct and geometrically exact representation of the complete model (see Section 4.4).
5. The construction of an analytic surface description (see Section 4.6) in order to further process it in CAD/CAM application, or to do a comprehensive surface analysis (surface interrogation). An example of the latter will be given in Section 8.2.

In practice these items will not be treated separately or necessarily in the specified order. At present, the most frequently used method for the final surface description is a polynomial *tensor product (TP)* approximation to the data points (Bézier or B-spline). This is the de facto standard in automotive industry, aircraft design and many other CAD/CAM-based industries. Up to now these methods require much interactive control. A simpler and more direct way is to generate a polygon mesh, which is sufficient for some applications, and sometimes even desired, e.g. for visualization purposes or computer graphics oriented applications. Tensor product surfaces are necessary for *reverse engineering*, where designers want to modify, or evaluate free form surfaces reconstructed from digitizing real objects.

4.1 POLYGON MESHES

S. KARBACHER, S. CAMPAGNA

Triangle meshes are the simplest type of polygon meshes. Since polygon meshes can be converted into triangle meshes simply by triangulating all n-gons with $n > 3$, this introduction focuses on triangle meshes.

4.1.1 Advantages of Triangle Meshes

At present triangles are the only surface primitive that can be rendered directly by graphics hardware. Hence, all surface descriptions must be approximated by triangle meshes for interactive visualization. Triangle meshes are very flexible. In contrast to tensor product surfaces, for example, they can describe surfaces of arbitrary topology, even with non-manifold elements. Furthermore, the density of vertices can be locally adapted to the surface curvature. Since triangles are a very simple kind of geometric primitive, algorithms for triangle meshes are usually efficient and robust.

4.1.2 Topology and Geometry

The neighborhood structure of a triangle mesh (the triangles and edges) is called the *topology* of the mesh, while the coordinates of the vertices describe its *geometry*. The basics on topology, geometry and graph theory can be found in textbooks on computer graphics [741, 227, 207, 622].

A triangle mesh consists of *vertices* (*knots*, *points*), *edges* and *triangles*. Important topological features are *holes*, *genus* and number of *connected components*. An edge that is bounded by a single triangle is called a *boundary edge*. A closed polygon of boundary edges encloses a *hole*, which usually is an artifact. Physical objects never have such kind of holes, as there must always be an outside and an inside. Physical holes have the shape of 'donuts'. The *genus* of an object is the number of its 'donuts'. A sphere has genus 0, a torus genus 1, etc. The relation between these features is given by the *Euler-Poincaré Equation*

$$V - E + F - H = 2(C - G), \tag{4.1}$$

where V is the number of vertices, E the number of edges, F the number of facets (triangles), H the number of holes, C the number of components, and G the sum of the geni of all components. For meshes with very many triangles ($V, E, F \to \infty$) or for objects which are homomorphic to a sphere ($H = 0, G = 0, C = 1$) this simplifies to the *Euler Equation*

$$V - E + F = 2. \tag{4.2}$$

The fact that every triangle is bound by three edges leads to

$$2E = 3F. \tag{4.3}$$

Inserting this into Equation 4.2 results in

$$F \approx 2V \tag{4.4}$$

and

$$E \approx 3V. \tag{4.5}$$

Thus, a large mesh consists of approximately twice as much triangles and three times as much edges as vertices.

4.1.3 Mesh Representations

A two-dimensional image is usually stored as a 2D array of pixels. The implicit order of that structure enables fast and simple access to adjacent pixels. In general it is not possible to describe triangle meshes, other than those generated from single range images, with such a simple structure.

Explicit Mesh Structure. The simplest structure to describe a mesh with m triangles is a list of $9m$ float values. Each triangle is represented by 3 coordinate triples $(x, y, z)_i$, $0 \leq i < 3$, that define the positions of the vertices. A mesh with n vertices needs approximately $18n$ float numbers. Since a vertex is usually shared by several triangles (6 on average), each vertex is stored several times. Thus, this structure is very inefficient. Beyond that, the topology is not represented explicitly. Shared vertices must be identified by identical coordinates, which is expensive (float compare) and sometimes difficult, as the geometrical positions may vary due to numerical limitations.

Indexed Mesh Structure (Shared Vertex). The above mentioned disadvantages can be avoided by storing two separate lists for geometry and topology. The geometry is stored in an array of n coordinate triples $(x, y, z)_i$, $0 \leq i < n$, for the n vertices (*vertex list*), the topology in an array of m integer indices, $(a, b, c)_j$, $0 \leq j < m$, that address the positions of the three triangle vertices in the coordinate list (*index list*). Since the geometry is stored without any redundancy, this structure needs only $3n + 3m \approx 9n$ numbers ($3n$ floats and $6n$ integers). Although vicinity data are not stored explicitly, adjacent triangles can easily be detected in linear time by identical indices. Since many algorithms repeatedly request this information, usually more elaborate structures that enable extraction of vicinity information in constant time are used [374].

Hierarchical Ring Structure. This data structure [624] enables direct access to the neighbors of each vertex and to the triangles that share a certain vertex. Like in the indexed mesh structure, the triangles are defined by an array of index triples. The vertex list contains the coordinates of each vertex, a list of pointers to all triangles that share that vertex and a list of its direct neighbors. Since the number of joining triangles and adjacent vertices is not constant and bound, these structures are realized by chained lists. Inserting and deleting triangles thus are rather complex operations and may result in fragmented memory. Memory demands depend on the number of neighbors of each vertex. With an average of 6 neighbors, $(3 + 6 + 6)n + 3m \approx (15 + 6)n \approx 21n$ integer and float numbers are required for a mesh with n vertices and m triangles (plus additional overhead for the chained lists).

Winged-Edge Mesh Structure. The most popular mesh structure is the winged-edge representation for arbitrary polygon meshes [46]. The focus of this data structure is the edge. Each edge e contains pointers to its endpoints v_0 and v_1, the two adjacent faces f_0 and f_1 and to the 4 edges e_{0-}, e_{0+}, e_{1-} and e_{1+} (the 'wings' of e) that bound f_0 or f_1 and end in v_0 or v_1, respectively (see Figure 4.1). For each vertex a pointer to an arbitrary one of its joining edges is added to the coordinates in the vertex list. Likewise each face points to an arbitrary one of its bounding edges. This structure enables direct access to all topology information that may be required. A triangle mesh with n vertices, m triangles and l edges needs $8l + m + 4n \approx (24 + 2 + 4)n \approx 30n$ numbers. For storing non-manifold structures, special considerations are required.

Directed Edge Mesh Structure. This is an *half-edge*[1] based data structure for exclusive description of triangle meshes [105, 104]. Each vertex contains a pointer to an half-edge that originates from it (see Figure 4.1). Triangles are not represented explicitly. The half-edges are sequentially stored in an array, in such a manner that the half-edges $3i$, $3i + 1$ and $3i + 2$ define the i-th triangle. Each half-edge contains a pointer to its predecessor e_{prev}, its neighbor e_{neig} and its endpoint v_1. A mesh with n vertices and m triangles needs $(3 \cdot 3)m + 4n \approx 18n + 4n = 22n$ numbers. It is possible to omit e_{prev}, as it can be extracted from the edge array, resulting in a slower

[1] A half-edge is an oriented edge. Each ordinary edge consists of two half-edges with opposite orientation.

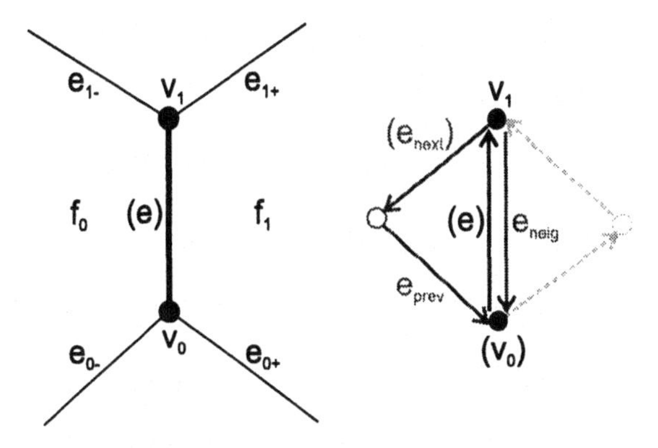

Figure 4.1. Associated pointers of edge e in the winged edge (left) and the directed edge (right) data structure.

performance. In this case only $(2\cdot3)m+4n \approx 12n+4n = 16n$ numbers are required. By default this data structure fails in describing non-manifold objects, too.

Quad-Edge Mesh Structure. The quad-edge structure [268, 521] is extremely general, representing any subdivision of 2-manifolds, permitting distinction between the two sides of a surface, allowing the two endpoints of an edge to be the same vertex, permitting dangling edges, etc. Each edge record contains four circular lists: for the two endpoints, and the two adjacent faces. In contrast to the previous data structures, these pointers do not address positions in the vertex or face list, but the next edge record in the vicinity of the corresponding vertex or edge. Vertices and faces are represented by *rings* (cycles) in Figure 4.2. For example, face A is the ring of edges (a, e, f) and vertex 2 is the ring (a, b, e). The vertex and face lists contain pointers to an arbitrary edge on the corresponding ring, to give access to that ring. The dual of a given graph is simply found by interpreting the vertex rings as faces and vice versa (no computation is necessary). A triangle mesh with n vertices, m triangles and l edges needs $4l + m + 4n \approx (12 + 2 + 4)n \approx 18n$ numbers.

4.1.4 Meshes with Attributes

The previous data structures only consider the geometry of three-dimensional objects. Frequently, additional attributes like color, material, normals, texture coordinates, tension or pressure are required. They can be assigned to triangle meshes in different ways:

- to each vertex of the mesh,
- to each triangle (needs twice the space of the previous structure),
- to each vertex of each triangle (needs six times the space of the first structure).

The last structure requires the most memory but offers high flexibility, as it includes the others as well. On the other hand it is difficult to use this structure for vertex- or

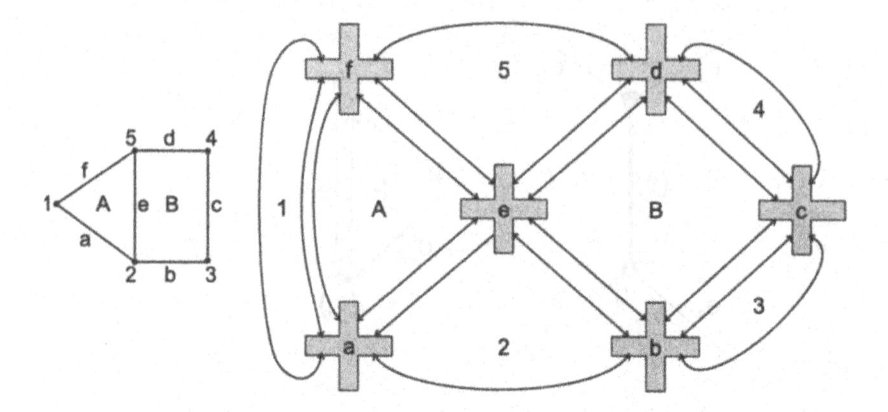

Figure 4.2. A simple plane graph with faces A and B, edges a–f and vertices 1–5 (left) and its quad-edge representation (right). Each edge record (gray crosses) consists of 4 pointers to the next edge that shares the same vertex (2 pointers) or bounds the same face (2 pointers). Thus, vertices and faces are represented by circular lists.

triangle-based standard algorithms which usually cannot handle multiple attributes per vertex or triangle.

4.1.5 Parametric Meshes

Some kinds of meshes can be represented in parametric form which allows to use algorithms that are simpler and more efficient than those for general meshes. Range images, for example, are sometimes named 2.5D surfaces, as each vertex is defined by its height above a parameter plane. Triangle meshes that were generated by tesselating a parametric surface are parametric as well. It is not necessary to store the triangles of parametric meshes explicitly, as they can be reconstructed by Delaunay-triangulating the vertices in parameter space [521]. It is sufficient to save the parameters of each vertex instead. Attributes must be assigned to the vertices then, not to the triangles.

4.1.6 Hierarchical Mesh Representations

Conventional file formats store the data serially. The first p percent of a file contain p percent pixels of an image in full resolution. A hierarchical data stream, in contrast, transmits the whole image at any time, starting with a low resolution. Details are added while the transmission proceeds, until the whole image information is transferred (see Figure 4.3). The regular structure of 2D images enables progressive encoding without any overhead. Simple methods for hierarchical representation of two-dimensional images are interlacing techniques and decomposition by Haar wavelets [674]. 3D models may be represented hierarchically as well (see Figure 4.4). The CPU only reads as many triangles as fit into memory or can be rendered in real-time. Because of the irregular structure of general triangle meshes, it is not possible to find hierarchical representations in a straightforward manner.

Figure 4.3. Sequential and hierarchical representation of an image. Images of the same column need the same storage space.

Discrete Levels of Detail. The most popular method for the hierarchical representation of triangle meshes is to store discrete *levels of detail (LOD)* in a sequence of independent meshes with increasing resolution. Each new level contains the whole image information of its predecessor, resulting in a large data overhead. For that reason only a few levels are usually used and switching between different levels is clearly visible (*popping*). Mesh reduction techniques are usually required to generate different resolution levels (see Section 4.1.7).

Figure 4.4. Sequential and hierarchical representation of a 3D model. Models of the same column need the same storage space.

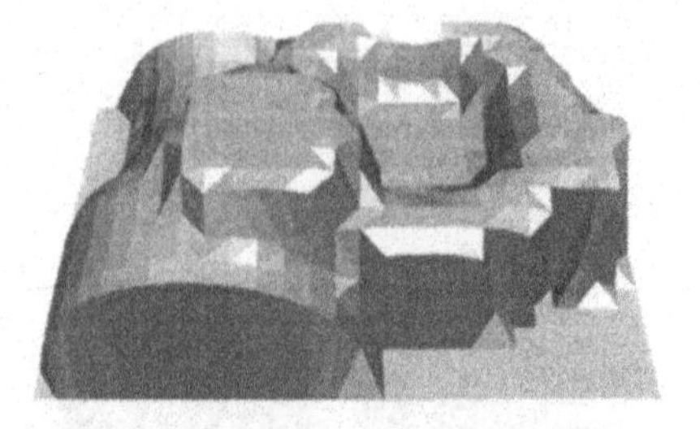

Figure 4.5. Approximation (right) by subsampling a dense regular mesh (left).

Subsampling. Regular triangle meshes like single range images may be stored by interlacing techniques. Similar to two-dimensional images a coarse approximation of a mesh that is defined on a rectangular grid can be found by subsampling every i-th column and every j-th row of the data array. A hierarchy is constructed by reducing i and j iteratively jumping over cells which are already stored in preceding levels (see Figure 4.5). This method solely depends on the topological structure of the data. The geometry is not considered. As a result, details with high frequency are lost in low resolution levels.

Wavelet Decomposition. For regular structures (e.g. images) *multiresolution analysis* based on *wavelets* is possible [674]. The data are decomposed by a series of high and low pass filters. In contrast to the sine and cosine functions of Fourier analysis the wavelet basis functions are spatially and temporally limited. Thus finite signals are easier to process while avoiding any artifacts. The simplest type of wavelets for images

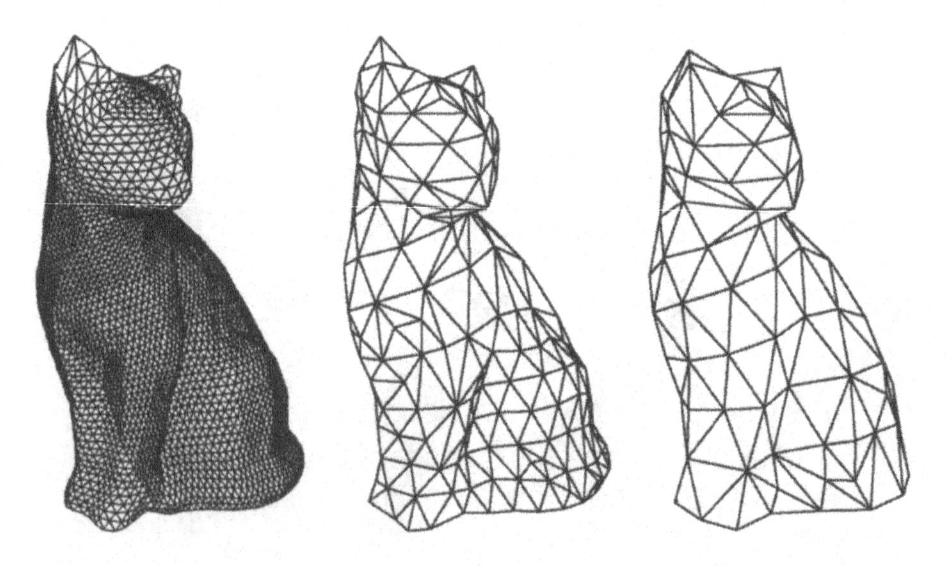

Figure 4.6. A mesh with subdivision connectivity (left) and different decomposition levels using wavelets (middle and right).

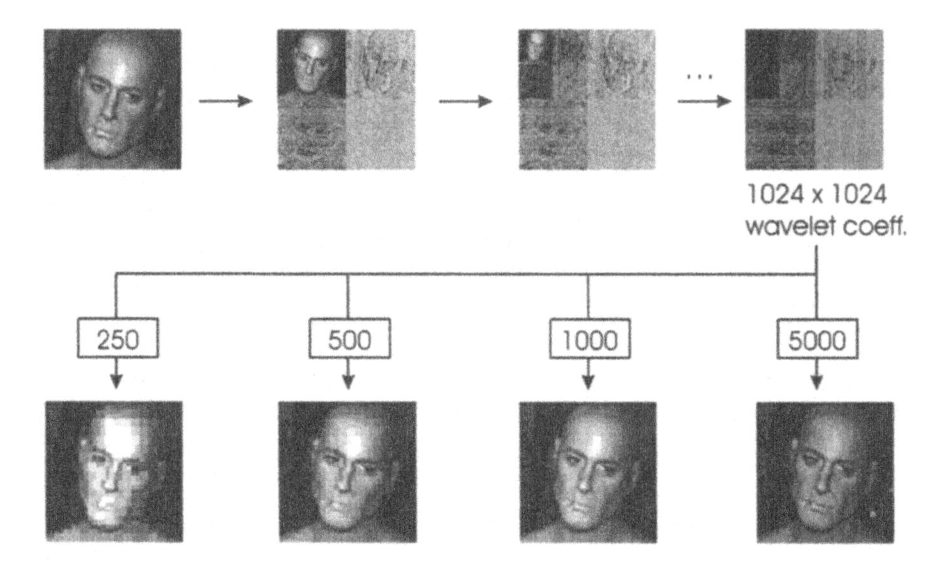

Figure 4.7. Image compression by wavelets (upper: different decomposition levels of a 1024×1024 image resulting in 1024^2 wavelet coefficients, lower: reconstruction solely using the 250, 500, 1000, resp. 5000 most significant coefficients).

are *Haar functions* which simply add (low pass) or subtract (high pass) neighboring pixels. Lounsbery et al. [197, 454] have generalized this approach for meshes with *subdivision connectivity*: all vertices (with singular exceptions) must have the same number of neighbors (see Figure 4.6). The original mesh is approximated by a coarse one that is adequate just to describe the topology of the object. Usually a few hundred triangles are sufficient. Objects which are homomorphic to a sphere may even be approximated by a tetrahedron. A series of correction terms (*wavelet coefficients*) is computed. These are necessary to refine the basic mesh by recursive subdivision until the original mesh is reconstructed. Each subdivision level owns a complete record of wavelet coefficients. It is possible to interpolate continuously between sequent levels. Structure dependent mesh reduction is simply done by eliminating small coefficients (see Figure 4.7). General meshes must be *remeshed* in order to achieve subdivision connectivity [197, 422]. In this case the original mesh can only be reconstructed approximately.

Progressive Meshes. In order to get a coarse approximation of a general dense mesh, details can be eliminated by successively removing vertices, edges or triangles (see Section 4.1.7). It is possible to invert this process by recording all executed operations (see Figure 4.8). Each step of this reconstruction process represents a complete approximation of the original mesh. The most popular implementations of this type of multiresolution hierarchy are *progressive meshes* [324] and its generalization to arbitrary dimensions, the *progressive simplicial complexes* [552]. These representations generate no data overhead and enable exact reconstruction of the original mesh, hierarchies with fine graded levels and sequential access to different levels.

Figure 4.8. A coarse mesh with 256 triangles (left) is successively refined by adding vertices and triangles until the original mesh with 5030 triangles is reconstructed exactly (right).

4.1.7 Mesh Reduction

Mesh reduction techniques are used to reduce the number of triangles of a dense triangle mesh. In recent years researchers have proposed a variety of methods. Surveys are published by Schroeder [620] and Cignoni et al. [130]. Mainly three approaches are used: multiresolution analysis by wavelets, retiling (remeshing, clustering) and iterative algorithms.

Retiling, remeshing and *clustering* methods generate completely new meshes by sampling new vertices. Different resolution levels are independent from each other, so that only discrete levels of detail (LOD) can be created. Turk [715] randomly places new vertices with curvature dependent density into the original mesh and thereafter removes the original ones (*retiling*). Rossignac and Borrel [581] use a coarse three-dimensional grid to merge all vertices within one voxel (*vertex clustering*).

Most frequently *iterative* approaches are used [624, 327, 111, 389, 324, 579, 237, 374, 621, 622, 104]. Topological and geometrical operations are used to remove vertices, edges or triangles from the dense mesh. This process is iterated until a given approximation error is reached or until no further thinning is possible. Alternative algorithms start with a coarse approximation that is refined iteratively by inserting new elements [218, 726]. Both methods enable the creation of progressive LODs.

Geometrical and Topological Operations. Triangle meshes can be modified using geometrical and topological operations. *Geometrical operators* change the geometrical positions of the vertices and leave the topology (connectivity) unchanged, while *topological operators* modify solely the connectivity of the mesh. The following sections introduce the ones most frequently used.

Vertex Change. The geometrical operator *vertex change* modifies the position of a vertex (see Figure 4.9). It may be used for filtering (see Section 4.3.2) or may be a component of more complex operators. Triangles which are influenced by this operator are marked gray in Figure 4.9.

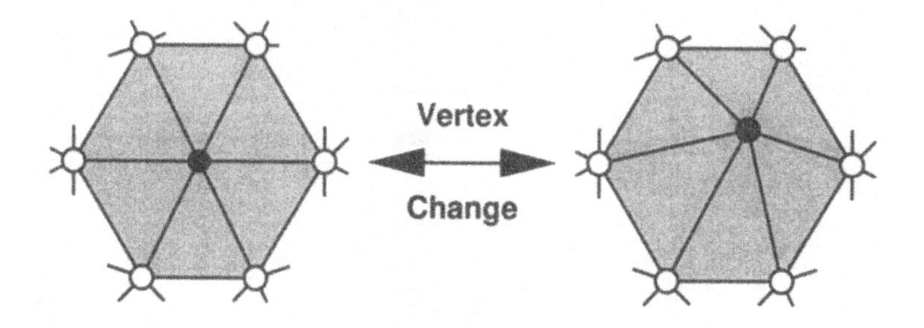

Figure 4.9. The vertex change operator moves the geometrical position of a vertex.

Vertex Removal and Vertex Insertion. The topological operator *vertex removal* eliminates a vertex and re-triangulates the modified region (see Figure 4.10). Its inverse, the *vertex insertion* operator, is not purely topological, as it modifies geometry as well. Vertex removal may be used for iterative mesh reduction, vertex insertion for merging overlapping meshes (e.g. for mesh reconstruction, see Section 4.4, or for refining a coarse mesh by subdivision, see Section 4.3.3).

Edge Collapse and Vertex Split. The *edge collapse* operator removes a vertex by collapsing an edge (see Figure 4.11). Its inverse is called *vertex split*. In general the remaining vertex gets a new position. Hence, the edge collapse operator modifies geometry and topology as well. It is purely topological if one of the end points remains unchanged (sometimes it is named *half-edge collapse* then). In Figure 4.11 *supporting* (influenced) triangles are marked in light gray, triangles which are removed are marked in dark gray. Edge collapse and vertex split are often used for progressive meshes (see Section 4.1.6).

Algorithms for Mesh Reduction. Iterative mesh reduction is usually carried out by repeatedly using edge collapse or vertex removal operations. The quality of the resulting mesh is mainly determined by a proper choice of the candidates to be removed and by the order in which they are eliminated. Simple algorithms choose a certain element (edge or vertex) from the input list (random or sequential), remove it if possible (e.g. if a cost function does not exceed a certain threshold) and proceed to the next

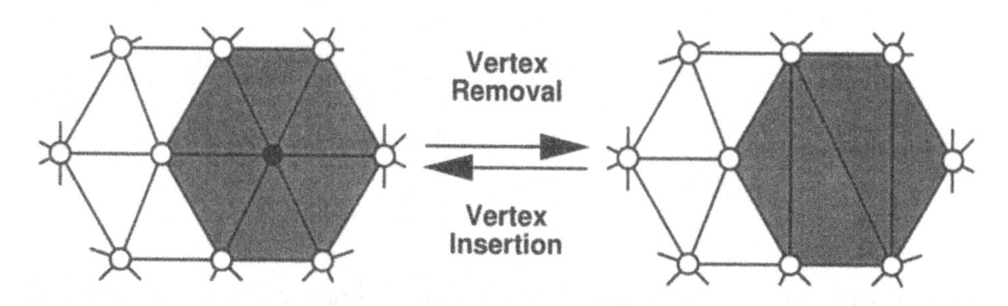

Figure 4.10. Vertex removal and its inverse vertex insertion remove or add two triangles.

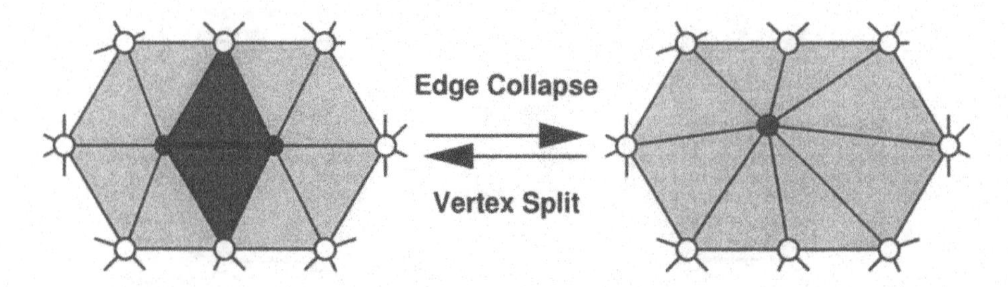

Figure 4.11. Edge collapse and its inverse vertex split remove or add two triangles (dark gray).

element [624, 327, 605, 384, 374, 622]. Best results are achieved if the candidates are sorted according to their costs [395]. The candidate with minimum cost is removed from the mesh and the candidate list is updated. This process is iterated until no candidate with a cost beyond the given threshold is left.

Cost Functions. Error bound mesh reduction techniques compute a cost function for the elimination of each single element (vertex or edge). This function may simply be the maximum distance (*global error*) between elements of the new mesh and the original one [389]. Since it is computationally expensive and impractical to compare the thinned mesh with the original one after each iteration, *local* methods estimate the global error [134, 579, 237, 374] (e.g. see Section 4.3.4) or simply evaluate the difference between two iterations [624, 605, 384, 621].

Merely using distance measures may result in surfaces with small approximation error but poor quality (see Figure 4.12). Hence, some authors use additional quality measures that evaluate the curvature characteristics of the resulting mesh [374, 104] or minimize the energy of a spring model [327].

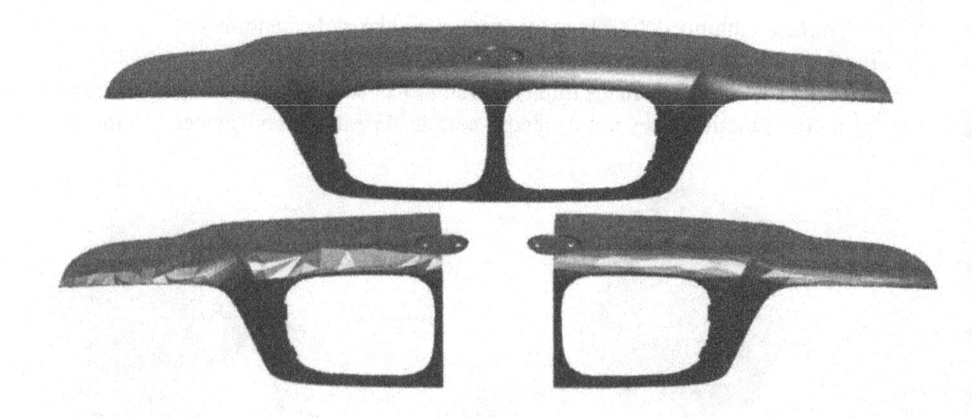

Figure 4.12. A dense mesh (top) is reduced with two different cost functions (bottom left: distance measure only; bottom right: distance measure and curvature dependent cost function).

4.2 FEATURE EXTRACTION AND REGISTRATION

S. SEEGER, X. LABOUREUX

Registration or equivalently *matching* is the process of bringing two data sets into best possible alignment. This is achieved by determining the transformation that transforms *corresponding areas* or *points* into each other. The terms *best possible alignment* and *corresponding areas/points* of two data sets are intuitively quite easy to understand but need a precise mathematical definition for a computational approach (see Section 4.2.4). Since features are one possibility to define corresponding areas/points, we discuss in addition some feature extraction methods (see Section 4.2.3). Features can be defined as modified data formed from a collection of the original data set which might be combined in linear or non-linear ways [65].

In this section only registration methods applicable to surface descriptions are described. In order to illustrate the above definitions, before starting the discussion of registration methods, we present some typical examples of data sets, corresponding areas/points and transformations.

4.2.1 Examples of Data Sets to be Registered

In the following examples we present three kinds of data structures:

- Intensity pictures

  $I : (n_x, n_y) \mapsto I[n_x, n_y] \in \mathbb{R}$, e.g. from photo cameras.

- Surface descriptions (range images)

  $F : (n_x, n_y) \mapsto (x[n_x, n_y], y[n_x, n_y], z[n_x, n_y])^T \in \mathbb{R}^3$, e.g. from a tactil or optical sensor.

- Volume data

  $\rho : (n_x, n_y, n_z) \mapsto \rho[n_x, n_y, n_z] \in \mathbb{R}$, e.g. from medical 3D scanners like CT (Computer Tomography), MR (Magnetic Resonance) or sonography (ultrasound) which show anatomical structure, or PET (Positron Emitting Tomography), SPECT (Single Photon Emission Computed Tomography) or MRS (Magnetic Resonance Spectroscopy) which show functional and metabolic activity.

Intensity images and volume data sets are introduced in addition, since the registration methods used there may also be used for the registration of surfaces:

- Many optical 3D sensors supply in addition to a range image a pixel identical intensity image. In this way the registration of the two intensity pictures is already sufficient to match the underlying range images.

- Since iso-surfaces ($\rho(n_x, n_y, n_z) = \text{const}$) are often extracted to register volume data sets, the registration methods used can be directly applied to the surface matching task.

Examples.

- The data sets to be registered could be two intensity images of an object taken from different viewpoints. Due to varying illumination corresponding points do not have the same intensities which makes the matching process even more difficult. The searched transformation is a 3D rotation and a translation also known as *rigid transformation*. To find this transformation at least seven corresponding points [766] in the images have to be found. The related range image can be calculated from the given intensity images (see Section 2.3).

- The data sets are two or more range images of an object taken from *different* viewpoints. The searched transformation is a rigid one. With at least three corresponding points in the range images the transformation can be found. In this way a complete model of the object can be generated from several views.

- The data sets are two intensity images of *different* objects e.g. two different human faces. Corresponding areas may be manually defined by features like eyes, the mouth and the nose. The searched transformation depends on the application. For example, in a face recognition system it may be useful to find the best rigid transformation between the two given faces. The magnitude of the difference decides whether the same face or different faces are presented in the images. On the other hand for morphing (the smooth transition from one data set into another) a *non* rigid transformation is searched which allows to locally match all parts of the object in the two images to each other.

- The data sets are two range images of an object taken from the same viewpoint but at *different* times so that the object may have changed its shape in the meantime. For example, the object might be a human face before and after a dental operation [132]. Corresponding areas result from features that can be detected in both range images and have not changed in the meantime, e.g. the eyes, the nose and/or the forehead. The searched transformation depends again on the application. For example, if the detection of the post operational swelling is to be visualized, it might be useful to find the rigid transformation that transforms the unchanged parts of the object into each other. With the help of this rigid transformation the difference volume of the two range images can be visualized. In other applications it might be useful to parameterize the time variations of the object by a non rigid transformation which allows to match all parts of the object in the two range images [220, 678]. Such a non rigid transformation can be a (global or local) affine, projective or curved transformation [464], depending on the difference between the data sets.

- The data sets are two volume data sets of the same object measured by the same device but at *different* times. With a rigid transformation variations of the object (e.g. a skull or a brain) can be determined (e.g. to study the evolution of a disease).

- The data sets are two volume data sets of *different* objects measured by the same device (e.g. the heads of different patients). Corresponding areas seem obvious for a human being. The searched transformation is a non rigid one. The registration of different patients' images could allow to contrast a healthy and a sick person [678].

- The data sets are two volume data sets of the same object measured by *different* devices, e.g. a CT and a PET scan of the human head. It may be difficult even for a non expert human being to define corresponding areas. The searched transformation is a rigid one (e.g. to improve the diagnosis by using multimodality data).
- The data sets to be registered can be of different dimensions, e.g. it is possible to match an intensity image $I : (n_x, n_y) \mapsto I[n_x, n_y] \in \mathbb{R}$ with a range image $F : (n_x, n_y) \mapsto (x[n_x, n_y], y[n_x, n_y], z[n_x, n_y])^T \in \mathbb{R}^3$ of the same object [731]. The searched transformation is a rigid one.

The above examples can be classified according to two basic criteria:

1. Nature and domain of transformation
   - rigid (local, global)
   - non rigid (local, global)
2. Modalities involved
   - monomodal
   - multimodal

Now we come to the presentation of several registration methods.

4.2.2 Overview of Registration Methods Applicable to Geometry Data

The task of determining the best spatial transformation for the registration of data sets can be divided into four major components [88]:

- feature space,
- search space,
- similarity metric.
- search strategy.

The choice of feature space determines what is matched. Since features can be independently found in each data set in a preprocessing step, the amount of data to be matched can thus be reduced. Some examples are:

- raw data (intensities in intensity images, 3D points in range images, density values in volume data sets),
- attributes defined for all points: curvatures, principal frames, point signatures,
- special collections of points: edges, surfaces, crest lines,
- salient point features: corners, line intersections, points of high curvature, extremal points,

- statistical features: moment invariants, centroids, principal axes; (they refer to measures over a region that may be the outcome from a segmentation preprocessing step),
- higher-level structural and syntactic descriptions.

The search space is the class of transformations from which we want to find the optimal transformation to align the data sets (global/local, rigid/nonrigid). The similarity metric determines how matches are rated (e.g. sum of squared euclidian distances, normalized cross-correlation, mutual information). The search strategy describes how to find this optimal transformation and depends on the search space (e.g. ICP, Hough method (clustering), correlation, relaxation, prediction-verification, indexing schemes, tree + graph matching).

We begin by presenting the extraction of more sophisticated features. In the whole discussion of registration methods we restrict ourselves to the search space of global rigid transformations (3D rotations + translations). We present some widely used similarity metrics. At the end we describe some search strategies.

4.2.3 Extracting Features from Surface Descriptions

Principal Curvatures. Minimal and maximal curvatures (also called principal curvatures) are shift and rotation invariant local features of an object surface [107]. Given a parametrical suface description $\boldsymbol{p}(u,v) = (x(u,v), y(u,v), z(u,v))^T$, the principal curvatures κ_i $(i = 1,2)$ at $\boldsymbol{p}(u_0,v_0)$ can be calculated as the eigenvalues of the Weingarten map (also called shape operator),

$$\begin{pmatrix} E & F \\ F & G \end{pmatrix}^{-1} \begin{pmatrix} L & M \\ M & N \end{pmatrix} \begin{pmatrix} \alpha_i \\ \beta_i \end{pmatrix} = \kappa_i \begin{pmatrix} \alpha_i \\ \beta_i \end{pmatrix} \tag{4.6}$$

where E, F, G, L, M and N depend on first and second order partial derivatives to u and v at $\boldsymbol{p} = \boldsymbol{p}(u_0, v_0)$:

$$\begin{array}{llll} E = \boldsymbol{p}_u \cdot \boldsymbol{p}_u & , \quad F = \boldsymbol{p}_u \cdot \boldsymbol{p}_v & , \quad G = \boldsymbol{p}_v \cdot \boldsymbol{p}_v, \\ L = \boldsymbol{p}_{uu} \cdot \boldsymbol{n} & , \quad M = \boldsymbol{p}_{uv} \cdot \boldsymbol{n} & , \quad N = \boldsymbol{p}_{vv} \cdot \boldsymbol{n}, \end{array} \tag{4.7}$$

where $\boldsymbol{n}$ is the normal at $\boldsymbol{p}(u_0, v_0)$,

$$\boldsymbol{n} = \frac{\boldsymbol{p}_u \times \boldsymbol{p}_v}{\|\boldsymbol{p}_u \times \boldsymbol{p}_v\|}. \tag{4.8}$$

It is straightforward to determine the curvatures κ_i from Equation 4.6 as the roots of the characteristic polynomial,

$$\kappa^2 - \frac{NE - 2MF + LG}{EG - F^2}\kappa + \frac{LN - M^2}{EG - F^2} = 0. \tag{4.9}$$

Principal Frames. With the help of the components of the eigenvectors α_i and β_i the directions of minimal and maximal curvatures can be determined as:

$$\boldsymbol{e}_i = \frac{\alpha_i \boldsymbol{p}_u + \beta_i \boldsymbol{p}_v}{\|\alpha_i \boldsymbol{p}_u + \beta_i \boldsymbol{p}_v\|}. \tag{4.10}$$

It can be shown that the unit vectors e_1 and e_2 are perpendicular ($e_1 \cdot e_2 = 0$). Since both e_1 and e_2 lie in the tangential plane (defined by p_u, p_v), they are perpendicular to the normal n at $p(u_0, v_0)$. Therefore e_1, e_2 and n define a local orthogonal frame at $p(u_0, v_0)$ also called the principal frame or trihedron. Note that the principal frame is not uniquely determined, since there is no way to choose between the frames (e_1, e_2, n) and $(-e_1, -e_2, n)$ [220].

Crest Lines. One of the principal curvatures is maximal in *absolute* value: it is called in [695] the *largest curvature* $\kappa_{\max}$, in order not to be mistaken with the maximal curvature. The associated principal direction is $e_{\max}$. Crest lines are the loci of the surface where the largest curvature $\kappa_{\max}$ is locally maximal (in absolute value) in the associated direction $e_{\max}$. In [695] crest lines are extracted from volume data. Then a crest line is the intersection of an iso-surface $\rho(n_x, n_y, n_z) = \text{const}$ with the implicit surface $\nabla\kappa_{max} \cdot e_{\max} = 0$ (i.e. change of $\kappa_{\max}$ in direction $e_{\max}$ is zero, which implies that $\kappa_{\max}$ is extremal in direction $e_{\max}$).

Extremal Points. For the definition of crest lines only the largest curvature with its associated principal direction is used. In the same way there are also extremal lines associated with the curvature with minimal absolute value. Extremal points are defined as the points of intersection of such extremal lines with crest lines. In [695] extremal points are extracted from volume data. Then an extremal point is the intersection of an iso-surface $\rho(n_x, n_y, n_z) = \text{const}$ with the implicit surfaces $\nabla\kappa_1 \cdot e_1 = 0$ and $\nabla\kappa_2 \cdot e_2 = 0$. Note that the extremal points are generally not the points of the extremal lines whose curvature is locally maximal. It is only stated here that there are 16 different types of extremal points that can be distinguished. In addition there are several geometric invariants associated with extremal points: the geometric invariants of the surface (principal curvatures), the geometric invariants of the extremal lines (curvature, torsion) and the geometric invariants corresponding to the relative position of the extremal lines with respect to the underlying surface [695].

In order to extract all these features, partial derivatives have to be calculated on the given data. If no parametrical surface description but only point clouds are given, in order to calculate partial derivatives, usually a polynomial surface is locally approximated at each data point. In this way the partial derivatives are always proportional to one of the polynomial coefficients [412].

Point Signatures. Similar to principal curvatures, this kind of rotation and translation invariant feature can be defined for each surface point but with the great advantage that no derivatives have to be calculated [126]: For a given surface point p a sphere of radius r, centered at p, is placed. The intersection of the sphere with the object surface is a 3D space curve C, whose orientation can be defined by an orthonormal frame formed by a 'normal' vector n_1, a 'reference' vector n_2, and the cross-product of n_1 and n_2. n_1 is defined as the unit normal vector of a plane P fitted through the space curve C. In the limit r tends to zero, n_1 approximates the surface normal at the point p. A new plane P' can be defined by translating the fitted plane P to the point p in a direction parallel to n_1. As well, if r tends to zero, P' approximates the tangential plane. The perpendicular projection of C to $\boldsymbol{P}$ forms a new planar curve C'. The distances of the points of C to the corresponding projected points of C' form a signed distance profile that is called the signature of the point $\boldsymbol{p}$ in [126]. The reference direction $\boldsymbol{n}_2$ is defined as the unit vector from $\boldsymbol{p}$ to the projected point on C' which gives the largest positive distance. Note that $\boldsymbol{n}_2$ is orthogonal to $\boldsymbol{n}_1$ since it is located on $\boldsymbol{P}'$.

Extended Gaussian Image (EGI). The EGI is another way to represent the data of a surface data set. In this approach [330] the normal vector at each point of the data set is computed and mapped into a unit sphere where its tail is at the center of the sphere and its head lies on the surface. In addition, each point on the surface of the sphere (i.e. head of a normal vector) is weighted by the Gaussian curvature of the corresponding surface point. In this way the EGI can be considered as the weighted orientation histogram of the data set.

This representation of the data sets has two interesting properties for the pose estimation problem:

- it is translation invariant,
- the EGI rotates in the same way as the corresponding data set.

However such an EGI approach assumes that the mapping between a point in the data set and a point on the sphere is uniquely defined. It can be shown that this is only the case if the object is convex.

4.2.4 Typical Similarity Metrics

The problem of aligning two data sets can be generally defined in the following way: Given two data sets $u(\boldsymbol{x}')$ and $v(\boldsymbol{x})$ describing parts of the same object at $\boldsymbol{x}'$ in the 'u-frame' respectively $\boldsymbol{x}$ in the 'v-frame', i.e.

$$v(\boldsymbol{x}) = F\left(u(\boldsymbol{x}')\right) \tag{4.11}$$

where F is the transfer function from u to v, we want to find the pose transformation T from $\boldsymbol{x}$ to $\boldsymbol{x}'$, i.e.

$$\boldsymbol{x}' = T(\boldsymbol{x}). \tag{4.12}$$

Combining Equations 4.11 and 4.12 this is equivalent to resolve

$$v(\boldsymbol{x}) = F\left(u\left(T(\boldsymbol{x})\right)\right) \qquad \forall \boldsymbol{x}. \tag{4.13}$$

In general it is difficult to determine the transfer function F.

Correlation. However, if the effects of F can be neglected, e.g. u, v are intensity images from different views by negligible illumination variations between u and v, we get

$$v(\boldsymbol{x}) = u\left(T(\boldsymbol{x})\right) \qquad \forall \boldsymbol{x}. \tag{4.14}$$

Due to noise, different occlusions and partial overlapping, a solution T that is valid for all $\boldsymbol{x}$ cannot be found. Therefore it is a common way to search for the transformation T that minimizes

$$E(T) = \sum_{\boldsymbol{x}} [v(\boldsymbol{x}) - u(T(\boldsymbol{x}))]^2 \tag{4.15}$$

$$= \sum_{\boldsymbol{x}} [v(\boldsymbol{x})]^2 - \sum_{\boldsymbol{x}} 2v(\boldsymbol{x})u(T(\boldsymbol{x})) + \sum_{\boldsymbol{x}} [u(T(\boldsymbol{x}))]^2. \tag{4.16}$$

Such a function that determines the ideal model parameters as the arguments that maximizes or minimizes the function is also often called a cost function or objective function. Since the first term in Equation 4.16 is independent of T, minimizing $E(T)$ is equivalent to maximizing

$$C(T) = \frac{\sum_{\boldsymbol{x}} v(\boldsymbol{x})u(T(\boldsymbol{x}))}{\sum_{\boldsymbol{x}} [u(T(\boldsymbol{x}))]^2}, \tag{4.17}$$

called the normalized cross-correlation function [88]. A related measure, which is advantageous if an absolute measure is needed, is the correlation coefficient

$$\tilde{C}(T) = \frac{\sum_{\boldsymbol{x}} [v(\boldsymbol{x}) - \mu_v][u(T(\boldsymbol{x})) - \mu_u]}{\sqrt{\sum_{\boldsymbol{x}} [v(\boldsymbol{x}) - \mu_v]^2 \sum_{\boldsymbol{x}} [u(T(\boldsymbol{x})) - \mu_u]^2}} \tag{4.18}$$

where μ_u and μ_v are the mean values of u and v. The denominator is given by the product of the standard deviations of u and v.

Mutual Information. In the case where the effects of F must be taken into account, e.g. if u represents the normals at each point of a 3D scan and v the corresponding intensity image, more sophisticated methods have to be applied. u and v can be interpreted as random variables with probability distributions P_u and P_v. Intuitively, if u and v are well aligned the randomness of u given knowledge of v is maximally reduced. In statistics this intuition can be formalized as follows.

Firstly the randomness of a random variable X is measured by its entropy, defined by

$$H(X) \equiv -E_X \{\log(P(X))\}. \tag{4.19}$$

Thereby $E_Z\{Z\}$ is the expected value of random variable Z and $P(X)$ the probability distribution of X. Secondly the randomness of random variable Y given knowledge of random variable X is measured by the conditional entropy

$$H(Y|X) \equiv -E_X \{E_Y \{\log P(Y|X)\}\}. \tag{4.20}$$

Then the intuition stated above can be formulated as maximizing

$$I(u(\boldsymbol{x}), v(T(\boldsymbol{x}))) = H(u(\boldsymbol{x})) - H(u(\boldsymbol{x})|v(T(\boldsymbol{x}))), \tag{4.21}$$

called the mutual information of u and v [731].

Least-Squares Sum of Corresponding Points. In the special case where the data sets are two point clouds $\{\boldsymbol{p}_i\}$ and $\{\boldsymbol{p}'_i\}$ of an object measured by a 3D sensor from two different viewpoints we describe the standard similarity metric below. For every pair of corresponding points $\boldsymbol{p}_i$ and $\boldsymbol{p}'_i$ we want to find the rotation $\boldsymbol{R}$ and translation $\boldsymbol{t}$ so that

$$\boldsymbol{p}_i = \boldsymbol{R}\boldsymbol{p}'_i + \boldsymbol{t} \tag{4.22}$$

with $\boldsymbol{p} = (x, y, z)^T$. However, due to noise in the measurements the transformation $(\boldsymbol{R}, \boldsymbol{t})$ calculated from three arbitrary point correspondences is not the best one. To find the best transformation usually the least-squares solution to the overdetermined system of Equations 4.22 is searched,

$$\sum_i \|\boldsymbol{p}_i - \boldsymbol{R}\boldsymbol{p}'_i - \boldsymbol{t}\|^2 \quad \longrightarrow \quad \text{minimum.} \tag{4.23}$$

If using Euler angles for the representation of $\boldsymbol{R}$ the rotation will be given by the following matrix product,

$$\boldsymbol{R} = \begin{pmatrix} \cos\alpha & -\sin\alpha & 0 \\ \sin\alpha & \cos\alpha & 0 \\ 0 & 0 & 1 \end{pmatrix} \begin{pmatrix} \cos\beta & 0 & \sin\beta \\ 0 & 1 & 0 \\ -\sin\beta & 0 & \cos\beta \end{pmatrix} \begin{pmatrix} \cos\gamma & -\sin\gamma & 0 \\ \sin\gamma & \cos\gamma & 0 \\ 0 & 0 & 1 \end{pmatrix}. \tag{4.24}$$

Therefore the parameters α, β, γ are not quadratic in Equation 4.23 and the solution to the minimization problem cannot be reduced to a simple system of linear equations (by calculating the partial derivatives with respect to the parameters and setting them to zero i.e. solving the so called system of normal equations). Such least-squares problems are also called nonlinear least-squares problems. Of course, the problem can be solved by standard optimization techniques like gradient descent, conjugate gradients, Newton's method or the Levenberg-Marquardt algorithm (that has become the standard technique for nonlinear least-squares problems [560]). Amazingly, there are however closed form solutions to this problem that are significantly faster (2 to 5 times dependent on the number of point correspondences) than iterative approaches [21]. The known solutions are based on

- singular value decomposition[2] (SVD) [21],
- quaternions [331],
- dual number quaternions [774],
- orthonormal matrices/polar value decomposition [332].

[2]The SVD method used here should not be confused with the standard SVD method which is used in favor of solving the system of normal equations when we have a linear least-squares problem. Here a completely different matrix is decomposed.

4.2.5 Search Strategies

Hough Method. Before presenting the application of the Hough method to the registration of 3D point sets, we describe its basic idea in a general framework.

Let $\{\boldsymbol{x}_i\}_{i=1...N}$, $\boldsymbol{x}_i \in \mathbb{R}^l$ be a data set and $\{q_j\}_{j=1...m}$, $q_j \in \mathbb{R}$ a parameter set related by a function $f : \mathbb{R}^m \times \mathbb{R}^l \to \mathbb{R}^w$

$$f(q_1, \ldots, q_m, \boldsymbol{x}_i) = 0 \quad \forall i. \tag{4.25}$$

In addition, suppose that all parameters q_i are uniquely determined by a subset of k samples of $\{\boldsymbol{x}_i\}_{i=1...N}$

$$\{\boldsymbol{x}_i\}_{i=1...k} \quad \Rightarrow \quad q_1, \ldots, q_m. \tag{4.26}$$

We search for the parameters $\{q_j\}_{j=1...m}$ which realize Equation 4.11 'as best as possible' for all $\boldsymbol{x}_i$.

Now for the Hough method the following steps have to be performed: For each subset of k samples of $\{\boldsymbol{x}_i\}_{i=1...N}$ (there are $\frac{N!}{(N-k)!k!}$ possibilities) the parameters $q_1, \ldots, q_m$ are calculated and at the corresponding position $(q_1, \ldots, q_m)$ in an m-dimensional accumulation table (Hough table) a counter is incremented by one. In this way every subset of k samples of $\{\boldsymbol{x}_i\}_{i=1...N}$ resulting in the same parameter set $q_1, \ldots, q_m$ contributes to the same position counter in the table. Therefore the position whose counter has the highest score corresponds to the parameter set that is in best accordance with the given data set $\{\boldsymbol{x}_i\}_{i=1...N}$.

Let us apply the Hough method to the registration of two 3D point sets $\{\boldsymbol{p}_i\}$ and $\{\boldsymbol{p}'_i\}$ of extracted point features. In this case$\{\boldsymbol{x}_i\}_{i=1...N}$ is the set of tupels $(\boldsymbol{p}, \boldsymbol{p}')$ of all combinations of points from the first set with points from the second set. The searched parameters are the $m = 6$ parameters of the rigid transformation $(\boldsymbol{R}, \boldsymbol{t})$ between both point sets, so that f from Equation 4.11 is given by

$$\boldsymbol{R}\boldsymbol{p}' + \boldsymbol{t} - \boldsymbol{p} = \boldsymbol{0}. \tag{4.27}$$

Since three non-collinear points uniquely define the 6 parameters of $\boldsymbol{R}$ and $\boldsymbol{t}$, $k = 3$ in the general description. When we now apply the Hough method all transformations calculated from *correct* point correspondences result in the same transformation while all other transformations are distributed more or less randomly in the parameter space [574]. Therefore, the points of accumulation give us the correct set of parameters. Note that in the case where the Hough method is used to find the parameters of a transformation the method is also often called *clustering*.

Correlation. In the previous subsection we have already introduced the similarity metric *correlation* that has to be maximized with respect to the transformation T. One way to find the maximum correlation is to compute $C(T)$ from Equation 4.17 for all possible transformations T. Since the number of possibilities may be very large, the complete search in the parameter space may not be feasible.

Relaxation. Relaxation is a technique to resolve ambiguities between match candidates of two data sets. These candidates are the outcomes of a feature extraction process. There are ambiguities between the match candidates since in general a given feature attribute does not uniquely determine a candidate: Even after using a correlation technique a feature point in the first data set may be paired to several feature points in the second data set.

To overcome these ambiguities for a given feature the relaxation technique makes in addition use of relations (to features in the neighborhood) which are 'more or less' preserved under the considered transformation. For point features these relations are typically the distances between two features in the same data set [766]. A further possibility could be the angles formed by three features. Since distances between features in intensity images (taken from 3D objects) are distorted by perspective projections, these relations are only invariant for point features close to each other.

Indexing Schemes. In general, indexing schemes precompute invariant feature values (e.g. principal curvatures, point signatures) in a data set and hash them into a look-up table (called hash table) with references to the corresponding feature positions [126]. In order to match two data sets $\{\boldsymbol{x}_i\}$ and $\{\boldsymbol{x}_i'\}$ the following steps are performed:

- Firstly, the extracted invariant feature values in $\{\boldsymbol{x}_i\}$ are hashed in a table.
- Secondly, for each point feature in $\{\boldsymbol{x}_i'\}$ with feature values $(v_1, \ldots, v_n)$ we find the possible corresponding positions in $\{\boldsymbol{x}_i\}$ (match candidates) by taking the points at the position $(v_1, \ldots, v_n)$ in the hash table.
- Thirdly, to resolve ambiguities for the match candidates a relaxation, Hough or prediction-verification method is applied.

The advantage of the hash table is the access to possible match candidates in constant time.

A more sophisticated indexing scheme is *geometric indexing* [416]. In this case the indexing scheme is based on the geometrical relationships between extracted features.

Prediction-Verification. The principle of a prediction-verification scheme is quite similar to the correlation approach: Just calculate a given similarity metric for a certain transformation. However there are some differences in practice:

- In the case of correlation all possible transformations are tested; in prediction-verification only the transformations resulting from a preprocessing step (e.g. from an indexing scheme or a feature based approach) are verified.
- In the case of correlation the transformation reaching an extremal value of the similarity metric (e.g. the minimal value of Equation 4.15 or maximal of Equation 4.17) is accepted as the correct one; in prediction-verification the first transformation resulting in a value of the similarity metric better than a given threshold is considered as the right one.
- In the case of correlation in order to reduce the complexity the calculation is usually only applied to extracted features; in prediction-verification the transformation of the whole data set provides the most reliable verification.

Some examples for prediction-verification schemes can be found in [695, 126, 220].

Tree and Graph Matching. After a feature extraction preprocessing step a data set can be described by a tree (also called graph), where the nodes are defined by the features and the links by their geometrical relations. The matching of two data sets is then reduced to the mapping of two graphs [119]. This search process is often called *subgraph isomorphism.*

Standard Optimization Techniques. The previously described search strategies are based on extracted features whenever they are used in practical applications. Otherwise their complexity would be too high. By contrast, standard optimization techniques try to find an extremum of a given similarity metric taking into account the whole data sets. An extremum can be either global or local. Although there is now no guarantee to find the global extremum in practice, a few approaches deal with this problem: for example mean field theory [672], genetic algorithms [94] and simulated annealing [617, 560]. However, if a 'good estimation' of the transformation between two data sets is known there are several standard approaches to find the global extremum. A 'good estimation' means in this context that the local extremum to which the method converges is in fact the global extremum. These techniques usually base on gradient information of the cost function. Typical examples are gradient descent (also called steepest descent), conjugate gradients or the Levenberg-Marquardt algorithm [560].

ICP Algorithm. In the Subsection "Least-Squares Sum of Corresponding Points" we have mentioned that it is straightforward to find the best rigid transformation between two point clouds by minimizing (compare Equation 4.23)

$$\sum_i \|\boldsymbol{p}_i - T(\boldsymbol{p}_i')\|^2 \tag{4.28}$$

where the transformation T is defined by

$$T(\boldsymbol{z}) = \boldsymbol{R}\boldsymbol{z} + \boldsymbol{t} \tag{4.29}$$

and $\boldsymbol{p}_i$ and $\boldsymbol{p}_i'$ are corresponding points. However the point correspondences are not known in advance. In the case where the given data sets are already well aligned to each other (this should be possible with one of the feature based methods described above), the following heuristic assumption may be reasonable: *Corresponding points are the closest points between two given data sets.* In this way, Equation 4.28 can be directly derived from the correlation expression in Equation 4.15 by setting

$$v(\boldsymbol{x}) = \min_{\boldsymbol{p}_k' \in \{\boldsymbol{p}_i'\}} (\|\boldsymbol{p}_k' - \boldsymbol{x}\|), \tag{4.30}$$

$$u(T(\boldsymbol{x})) = \min_{\boldsymbol{p}_j \in \{\boldsymbol{p}_i\}} (\|\boldsymbol{p}_j - T(\boldsymbol{x})\|) \tag{4.31}$$

and by restricting the sum over all $\boldsymbol{x}$ in Equation 4.15 to the data set $\{\boldsymbol{p}_i'\}$. Thus we get $v(\boldsymbol{x}) = 0$ and Equation 4.15 becomes

$$\sum_{\boldsymbol{x} \in \{\boldsymbol{p}_i'\}} \left(\min_{\boldsymbol{p}_j \in \{\boldsymbol{p}_i\}} \|\boldsymbol{p}_j - T(\boldsymbol{x})\| \right)^2. \tag{4.32}$$

By defining the index i so that

$$p_i = \arg\left[\min_{p_j \in \{p_i\}} \|p_j - T(p_i')\|\right] \qquad (4.33)$$

we get the expression in Equation 4.28. Since the transformation T is not known in advance but assumed to be small (the data sets are supposed to be well aligned), T is taken as the identity transformation. In this way corresponding points are defined as:

$$p_i = \arg\left[\min_{p_j \in \{p_i\}} \|p_j - p_i'\|\right]. \qquad (4.34)$$

After calculating the corresponding points and the resulting transformation, it can be expected that—after applying the transformation—the data sets become closer to each other. Thus it seems reasonable to iterate this procedure (therefore the term ICP: Iterative Closest Points [60]) until convergence of the computed transformation:

1. find closest points,

2. calculate rotation $\boldsymbol{R}$ and translation $\boldsymbol{t}$ that minimize the least-squares sum of corresponding points (Equation 4.28),

3. apply the transformation to all points in the first data set.

However, if the estimation of corresponding points during the initialization step is 'too bad' the algorithm will not converge to the right transformation but will get stuck in a local minimum. Nevertheless this algorithm has become the standard for the precise registration of two data sets, described in details in [60, 774] and used as the basis of more sophisticated algorithms in [716, 55, 220, 472, 695, 678].

Since the process of finding closest points is quite time consuming, there are several methods which can considerably speed it up, such as bucketing techniques (in 3D or in 2D by projection), kD trees (abbreviation for k-dimensional binary search tree; here $k = 3$) [774] or octree-splines [681]. The closest point search can be further accelerated by exploiting a coarse to fine strategy during the iterations of the ICP: during the first iterations closest points are only determined for some coarsely sampled points. Then a fine matching using more and more points follows [774, 716].

4.2.6 Robust Registration

In order to improve the registration process with the ICP-algorithm it is recommendable to use weight factors in the least-squares sum (Equation 4.23). For example Turk and Levoy [716] use the dot product of calculated normals and a vector pointing to the light source as confidence values. The least-squares sum used in the ICP-algorithm is based on the assumption that the data are corrupted by Gaussian noise. However this assumption is not valid in practice due to the following arguments:

- Usually there are many points in one data set that should not have a correspondence in the other data set due to different object occlusions in data sets from different views and/or since there is only a partial overlap between the data sets. However point correspondences (i.e., closest points) are always found in the ICP-algorithm.

- Usually there are outliers in the data sets frequently due to some unknown reasons. Perhaps there was a percussion during the measurement process or the camera was overdriven due to light reflections.

Statisticians have developed various sorts of robust methods that can reduce the influence of outliers:

M-Estimators. Instead of minimizing

$$\sum_i \|p_i - T(p_i')\|^2 = \sum_i d_i^2, \tag{4.35}$$

we replace the squared distance (L_2-estimator) by a function ρ of d_i

$$\sum_i \rho(d_i) \tag{4.36}$$

where ρ is a symmetric, positive definite function with a unique minimum at zero and is less increasing than square. It can be shown that minimizing Equation 4.36 is equivalent to minimizing the following iterated reweighted least-squares expression [766],

$$\sum_i \omega\left(d_i^{(k-1)}\right) d_i^2 \tag{4.37}$$

where $\omega(x) = \frac{1}{x}\frac{d\rho}{dx}$, the superscript $^{(k)}$ indicates the iteration number, and $\omega\left(d_i^{(k-1)}\right)$ has to be recalculated after each iteration in order to be used in the next iteration. Note that M-estimators can provide bad results since one far-away outlier can make all other outliers have small distances d_i, so that 'correct' data points are less weighted than outliers. Examples of several weight functions ω can be found in [766, 582, 560].

Least Median of Squares. A really robust approach that overcomes the bad influence of outliers is the Least Median of Squares method (LMedS). This method is not affected by outliers up to a rate of 50% [472, 582]. In this approach the transformation parameters are estimated by minimizing

$$\text{median}_i \left(d_i^2\right). \tag{4.38}$$

Extended Kalman Filtering. Another robust technique coming from the signal processing theory is Extended Kalman Filtering (EKF). This method makes use of a priori knowledge and provides a recursive solution to the least-squares problem. Since it would be quite time consuming to explain this approach, we refer the interested reader for a basic introduction to [742] and to Section 2.4.3, and for an application on 3D registration to [537].

4.2.7 Registration of Multiple Point Sets

So far only the registration between two data sets has been considered. However, for practical applications, such as virtual reality, CAD-processing or NC-manufacturing, it is of interest to match several range views together to reconstruct a 3D-model of the original object. Further details can be found in Section 8.1.3 and some typical approaches are presented in [472, 716, 55].

4.3 DISCRETE MODELING OF POINT CLOUDS

S. KARBACHER

After scanning three-dimensional objects (see Section 4.4) or triangulating CAD surfaces one usually gets *point clouds* or *triangular meshes* as a result. Frequently it is necessary to process this data. Calibration and registration errors, for example, appear after merging multiple range images from different views. These data are not structured and therefore it is not possible to use conventional modeling or signal processing methods like tensor product surfaces or linear filters.

Unfortunately no basic theory exists for handling unstructured data in order to estimate surface normals and curvature, interpolate curved surfaces, subdivide or smooth triangle meshes. Some limited approaches exist that work for special cases. Multiresolution analysis based on wavelets can only be used for triangle meshes with subdivision connectivity [453, 197] (see Section 4.1.6). An approach for general meshes was proposed by Taubin [686, 687]. He generalized the discrete Fourier transformation by interpreting frequencies as eigenvectors of a discrete Laplacian. Defining such a Laplacian for irregular meshes allows to use linear signal processing tools like high and low pass filters, data compression and multiresolution hierarchies. The Laplacian may simply be a weighted sum of all difference vectors to all direct neighbors of a vertex [396, 622]. Laplacian smoothing reduces surface curvature and tends to flatten the surface, so that it must be used very carefully. Since the vertices are isotropically moved, even if the surface is flat, the geometry may be seriously damaged. Guskov et al. [269] introduce a more complex Laplacian that leaves flat surfaces invariant. They apply local parameterizations in order to compute second order derivatives.

However, the translation of concepts of linear signal theory is not the optimal choice for modeling geometry data. Surfaces of three-dimensional objects usually consist of segments with low bandwidth and transients with high frequency between them. They have no 'reasonable' shape, as it is presupposed for linear filters. 'Optimal' filters like *Wiener* or *matched* filters usually minimize the RMS error. Oscillations of the signal are allowed, if they are small. For visualization or milling of surfaces, curvature variations are much more disturbing than small deviations from the ideal shape. A smoothing filter for geometric data should therefore minimize curvature variations and try to reinduce an error that is smaller than the original distortion of the data. This section introduces a method for modeling irregular meshes that was specially designed to achieve these aims [374, 375]. The Laplacian approach is generalized for invariance of surfaces with constant curvature. The method works best with dense meshes like those reconstructed from range images. Since it requires topological information, it is necessary to triangulate pure point clouds in advance (see Section 4.4).

4.3.1 Modeling of Dense Triangle Meshes

It is assumed that the triangle mesh approximates a smooth surface with the vertices as sampled surface points. The sampling density shall be high enough to neglect the variations of surface curvature between adjacent vertices. If this is true, the underlying surface can be approximated more accurately by a mesh of circular arcs. As an example we show how easily curvature can be estimated when using this simplified model.

Figure 4.13 shows a cross section s through a constantly curved object surface between two adjacent vertices v_i and v_j. The curvature c_{ij} of the curve s is ($c_{ij} > 0$ for concave and $c_{ij} < 0$ for convex surfaces)

$$c_{ij} = \pm\frac{1}{r} \approx \pm\frac{\alpha_{ij}}{d_{ij}} \approx \pm\arccos(\boldsymbol{n}_i \cdot \boldsymbol{n}_j), \tag{4.39}$$

which can easily be computed if the surface normals $\boldsymbol{n}_i$ and $\boldsymbol{n}_j$ are known (they can be computed from the normals of the triangles meeting at v_i). The principal curvatures $\kappa_1(i)$ and $\kappa_2(i)$ of v_i are the extreme values of c_{ij} with regard to all its neighbors v_j:

$$\kappa_1(i) \approx \min_j(c_{ij}) \quad \text{and} \quad \kappa_2(i) \approx \max_j(c_{ij}). \tag{4.40}$$

An example for mesh segmentation using this method can be found in Section 3.1.5.

The surface normals are computed separately, hence it is possible to eliminate noise by smoothing the normals without any interference of the data points. Therefore this method is much less sensitive to noise than the usual method for curvature estimation from sampled data which is based on differential geometry [59, 412] (see Section 4.2).

It can be shown that approximation of a mesh of circular arcs requires a sampling density which is at least four times higher than the smallest object detail to be modeled (see Section 4.3.2). This means that the minimum sampling rate must be twice the theoretical minimum given by the Nyquist frequency.

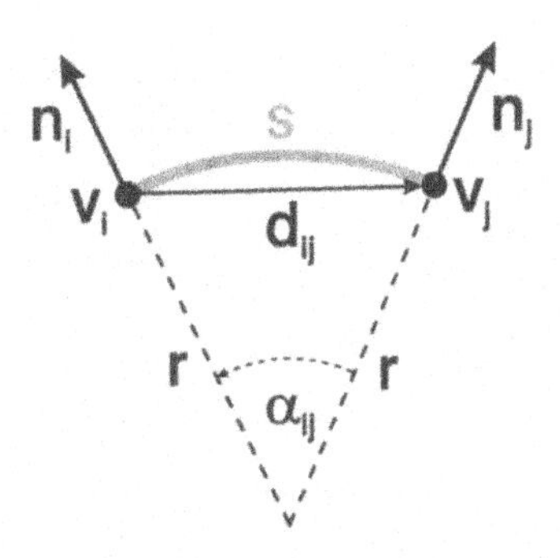

Figure 4.13. Cross section s through a constantly curved surface.

4.3.2 Smoothing Triangle Meshes

This approach can be used for smoothing of measuring errors with minimum interference on real object features like edges. If curvature variations of the sampled surface are actually negligible, while the measured data vary from the approximation of circular arcs, this deviation must be caused by measuring errors. Therefore it is possible to smooth these errors by minimizing the curvature variations.

For this purpose a measure δ is defined to quantify the variation of a vertex from the approximation model. Figure 4.14 shows a constellation similar to Figure 4.13. Now the vertex $\boldsymbol{v}$ is measured at a wrong position. The correct position would be $\boldsymbol{v}'_i$ if $\boldsymbol{v}_i$ and the surface normals $\boldsymbol{n}$ and $\boldsymbol{n}_i$ perfectly match the simplified model (there exist different ideal positions $\boldsymbol{v}'_i$ for each neighbor $\boldsymbol{v}_i$). The deviation of $\boldsymbol{v}$ with regard to $\boldsymbol{v}_i$, given by

$$\delta_i \approx d_i \frac{\cos(\beta_i - \frac{\alpha_i}{2})}{\cos(\frac{\alpha_i}{2})}, \tag{4.41}$$

can be eliminated by translating $\boldsymbol{v}$ into $\boldsymbol{v}'_i$. For regular meshes ($d_i \approx \text{const} \; \forall \; i$) the sum over δ_i defines a cost function for minimizing the variations of $\boldsymbol{v}$ from the approximation model. Minimizing all cost functions of all vertices simultaneously leads to a mesh with minimum curvature variations for fixed vertex normals. Alternatively it is possible to define a recursive filter for general meshes by repeatedly moving each vertex $\boldsymbol{v}$ along its assigned normal $\boldsymbol{n}$ by

$$\Delta n = \frac{1}{2}\overline{\delta_i} := \frac{1}{2N}\sum_i^N w_i \delta_i, \tag{4.42}$$

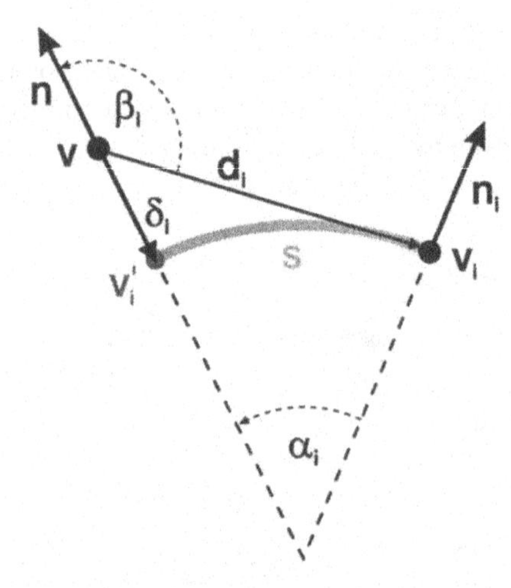

Figure 4.14. Cross section $\boldsymbol{s}$ through a constantly curved surface. The position of vertex $\boldsymbol{v}$ is not measured correctly.

where N is the number of neighbors of $\boldsymbol{v}$ and w_i are weights that attenuate the influences of the $\boldsymbol{v}_i$ according to their distances d_i:

$$w_i = \frac{d_0}{d_i}. \tag{4.43}$$

d_0 is the minimum distance between all vertices of the mesh (usually the sampling distance). Weighting is necessary, since otherwise faraway vertices would have more importance than contiguous ones. After a few iterations all Δn converge towards zero and the overall deviation of all vertices from the circular arc approximation reaches a minimum. Restricting the translation of each vertex to only a single degree of freedom (direction of the surface normal) enables smoothing without seriously affecting small object details (compare with Turk and Levoy [716]). Conventional approximation methods, in contrast, cause isotropic daubing of delicate structures.

Smoothing of Calibration and Registration Errors. This procedure can be used for surface smoothing if the surface normals describe the sampled surfaces more accurately than the data points. In case of calibration and registration errors the previous assumption is realistic. This class of errors usually causes local displacements of the overlapping parts of the surfaces from different views, while any torsions are locally negligible. Oscillating distortions of the merged surface with nearly parallel normal vectors at the sample points are the result. Figure 4.15 shows an idealization of such an error if the sampled surface is flat. The jagged line of vertices derives from different images that were not perfectly registered. It is clear that this error can be qualitatively eliminated, if the vertices are forced to match the defaults of the surface

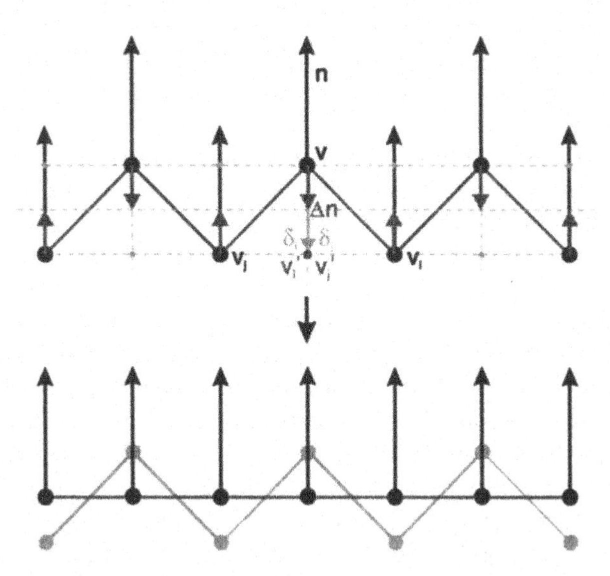

Figure 4.15. Smoothing of an idealized registration error on a flat surface. After moving each vertex by $\frac{1}{2}\delta_i$ along its normal vector, the data points are placed in the fitting plane of the original data set.

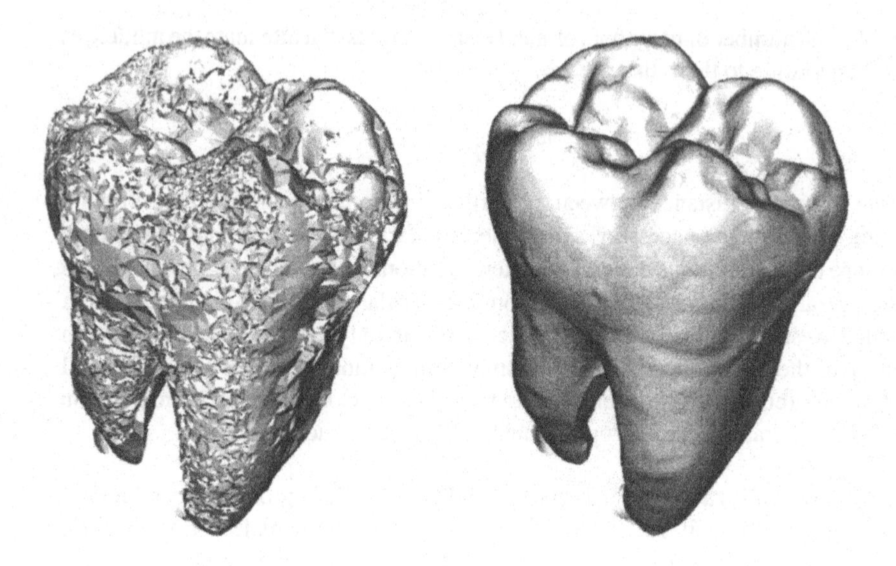

Figure 4.16. Distorted mesh of a human tooth, reconstructed from 7 badly matched range images (left), and the result of smoothing (right).

normals by translating each vertex by Equation 4.42 along its normal. If the sampled surface is flat, the resulting vertex positions are placed in the fitting plane of the original data points. In case of curved surfaces, the regression surfaces are curved too, with minimum curvature variations for fixed surface normals. This method generalizes standard methods for smoothing calibration and registration errors by locally fitting planes to the neighborhood of each vertex [326, 716]. Instead of fitting piecewise linear approximation surfaces, in our case surface patches with piecewise constant curvature are used.

From Figure 4.15 the minimum sampling density for conserving small structures can be deduced. If the oscillations of the pictured data points are not caused by measuring errors but instead by the real structure of the sampled surface, this structure is lost when the normals are computed. This happens, because the surface normals average over ± 1 sampling interval. Therefore the sampling rate ν_n of the normals is only half the sampling rate ν_v of the vertices. In the depicted example ν_n is identical with the maximum space frequency of the sampled surface, which violates the sampling theorem of Shannon [636]. As a result, approximation of a mesh of circular arcs requires a sampling density which is at least four times larger than the smallest object details to be modeled. This means that the minimum sampling rate must be twice the theoretical minimum given by the Nyquist frequency.

Figure 4.16 demonstrates that registration and calibration errors in fact can be smoothed without seriously affecting any object details. The mesh on the left side of Figure 4.16 was reconstructed from 7 badly matched range images of a human tooth. The mean registration error is 0.14 mm, the maximum is 1.5 mm (19 times the sampling distance of 0.08 mm). The mean displacement of a single vertex by smoothing was 0.06 mm, the maximum was 0.8 mm. The displacement of the barycenter was

0.002 mm. This indicates, that the smoothed surface is perfectly placed at the center of the difference volume between all range images.

Smoothing of Noise and Moiré. In case of noise or aliasing errors (Moiré) the surface normals are distorted as well, but can simply be smoothed by weighted averaging. Thus, filtering is done by first smoothing the normals (*normal filter*) and then using the described *geometry filter* to adapt the positions of the data points to these defaults.

Figure 4.17 demonstrates smoothing of measuring errors of a single range image in comparison to a conventional median filter (the simplest and most popular type of edge preserving filters). Although the errors (variations of the smoothed surface from the original data) of the median filter are slightly larger in this example, the described smoothing method shows much more noise reduction. Beyond that, the median filter produces new distortions at the boundaries of the surface. The smoothing filter reduces the noise by a factor of 0.07, whereas the median filter actually increases the noise because of the produced artifacts. The only disadvantage of smoothing is a nearly invisible softening of small details.

Experiments showed that the errors introduced by smoothing are always less than the errors of the original data. In particular no global shrinkage, expansion or displacement takes place, a fact that is not self-evident when using real 3D filters.

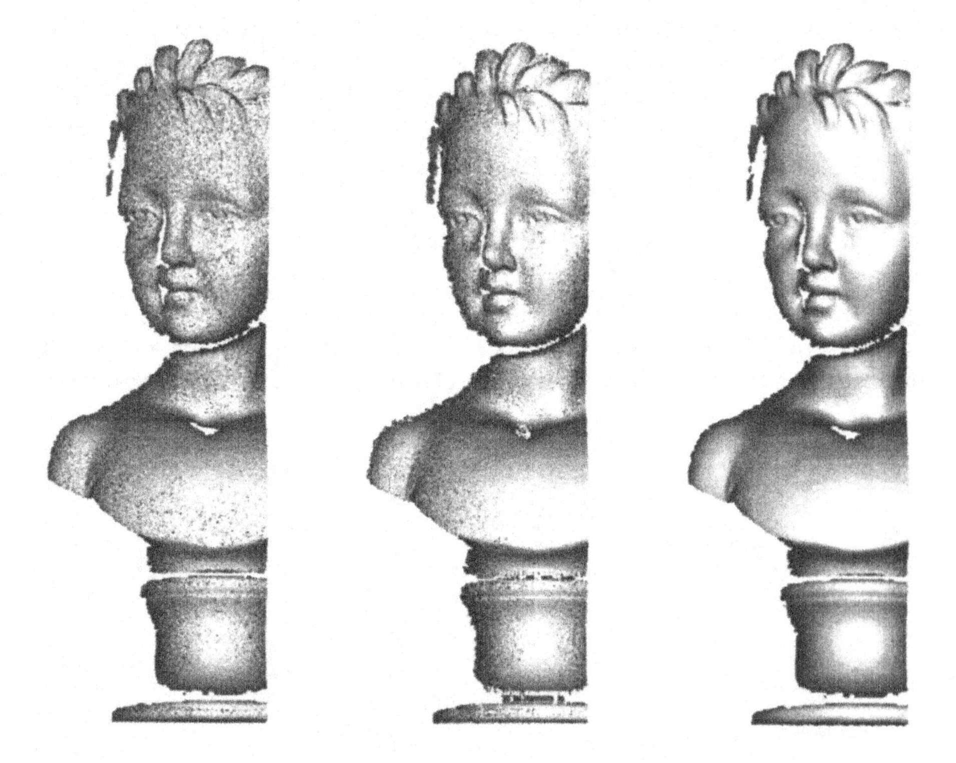

Figure 4.17. Noisy range image of a ceramic bust (left), smoothed by a 7×7 median filter (center), and by the described smoothing method (right).

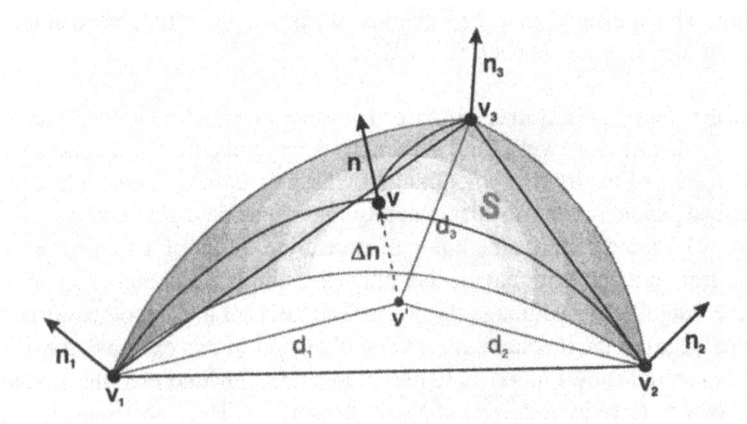

Figure 4.18. Interpolation of a new vertex v over triangle $\Delta(v_1, v_2, v_3)$ $(n = 3)$. The surface s of the curved triangle is the set of all interpolated vertices.

4.3.3 Interpolation of Curves and Curved Surfaces

Interpolation of curves and curved surfaces (e.g. curved triangles) over n-gons can simply be done by blending circular arcs (see Figures 4.18 and 4.19). Solely the vertex positions and the assigned vertex normals are required for that purpose. In order to interpolate a new vertex v, the projection v' of v onto the polygon is constructed from the barycentric coordinates $b_i, 1 \leq i \leq n$ with $\sum_i^n b_i = 1$. A new surface normal n is computed for that position by linear interpolation between all surrounding normals n_i. Then the projection v' is moved by

$$\Delta n = \frac{\sum_i^n b_i^m \delta_i}{\sum_i^n b_i^m} \tag{4.44}$$

along n, with $m \in \{1, 2\}$ (order of interpolation). The deviations δ_i of the resulting vertex v' from the arcs of its neighbors v_i are computed by Equation 4.41. This interpolation scheme can be used to define a subdivision rule for surface interpolation.

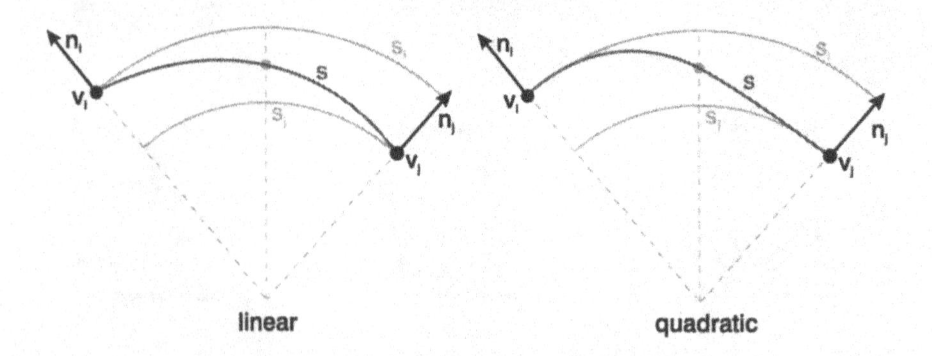

Figure 4.19. Linear $(m = 1)$ and quadratic $(m = 2)$ interpolation of a curve s.

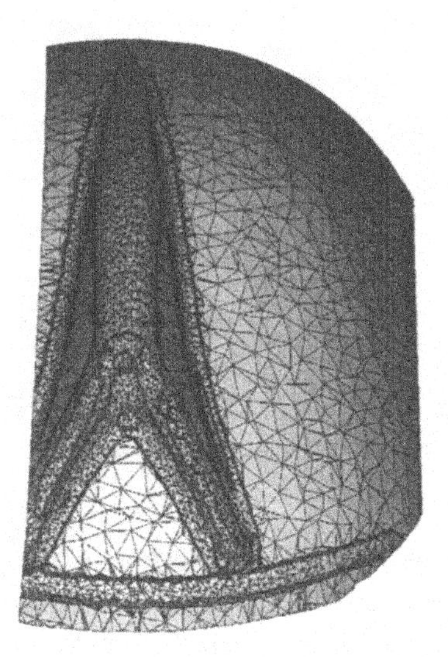
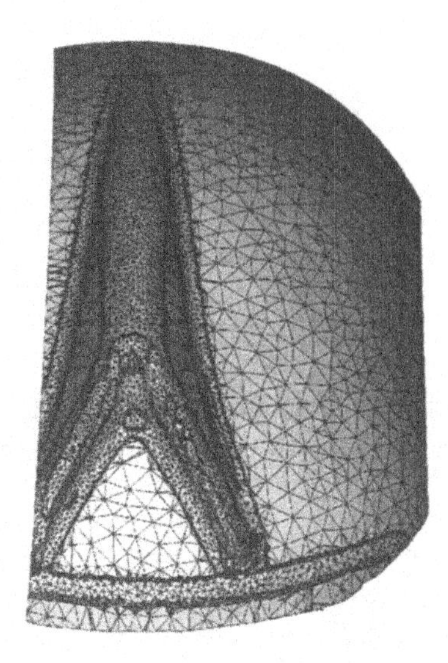

Figure 4.20. Mesh of a fire fighter's helmet after mesh thinning (left) and after geometrical optimization (right).

Order m determines the continuity characteristics of the interpolated curve or surface. Figure 4.19 demonstrates this for curve interpolation ($n = 2$). For $m = 1$ the curve s has no reversal point, whereas for $m = 2$ the curve is forced to interpolate the vertex normals exactly, which enables tangent continuous interpolation (G^1-continuity). The vertices are interpolated by definition, so G^0-continuity is always satisfied. These statements also hold for surface interpolation ($n \geq 2$). However, tangent continuity is only proved at the vertices of the polygons (and for $m = 2$, of course). The transition between adjacent surface patches is nearly tangent continuous. The smoothing method introduced in the previous section minimizes curvature variations between them and therefore keeps deviations from continuity small. Surfaces with constant curvature, which are typical for artificial objects, are by definition interpolated exactly. In contrast, conventional interpolants like splines or Bézier patches require special efforts in order to generate constantly curved surfaces. This simple interpolation scheme may be used for refining an initially coarse mesh by subdivision. For that purpose the patches are repeatedly split into smaller ones e.g. by vertex split or vertex insertion operations (see Section 4.1.7). Then Equation 4.44 may be used to interpolate new geometrical positions for the new vertices. Another application is geometrical mesh optimization (see Figure 4.20). Mesh thinning using *edge collapse* operations (see next section) usually causes awkward triangulations with elongated triangles (left). In order to produce better balanced triangulations, each vertex is replaced by a new one that is interpolated over the polygon of its neighbors (right).

4.3.4 Error Estimation for Mesh Reduction

Error bound mesh thinning techniques compute a cost function for the elimination of each single element (triangle, vertex, or edge). Usually this is the maximum distance

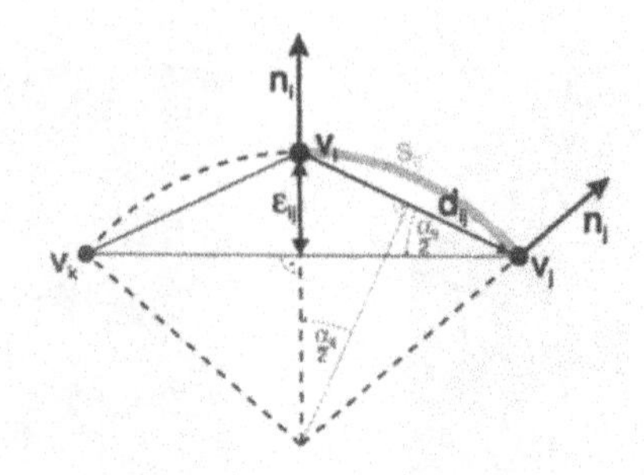

Figure 4.21. Two-dimensional mesh thinning. The arc s_{ij} is approximated by polyline (v_i, v_j, v_k). ϵ_{ij} is the upper limit for the error caused by deleting v_i.

between elements of the new mesh and the original one (see Section 4.1.3). This approach is computationally expensive and impractical as one has to store the original dense mesh for successive generation of meshes with increasing approximation error. The following method allows to estimate that error without comparison with the original data. Instead, the thinned surface is compared with the subdivision surface that is defined in the previous section.

For simplification we first consider the two-dimensional case, shown in Figure 4.21. An arc s_{ij} is approximated by a polyline through vertices v_k, v_i and v_j. The error ϵ_{ijk} caused by deleting v_i can be found by orthogonal projection of v_i onto edge (v_k, v_j). It is maximum if v_i is located on the perpendicular bisector of (v_k, v_j). The maximum value ϵ_{ij} is easier to compute than ϵ_{ijk}, as it only depends on v_i and v_j and their vertex normals:

$$\epsilon_{ij} = d_{ij} \sin(\frac{\alpha_{ij}}{2}) = d_{ij} \sqrt{\frac{1 - n_i \cdot n_j}{2}}, \tag{4.45}$$

where d_{ij} is the Euklidian distance between v_i and v_j and α_{ij} is the angle between the vertex normals n_i and n_j. In the three-dimensional case (see Figure 4.22) s_{ij} is a cross section through a constantly curved surface and (v_k, v_j) is the cross section through the new triangle below v_i. The total error ϵ_i for the elimination of v_i is the maximum of ϵ_{ij} for all its neighbors v_j:

$$\epsilon_i = \max_j(\epsilon_{ij}). \tag{4.46}$$

Mesh thinning is carried out by computing ϵ_i for all vertices v_i of the mesh. The vertices are pushed into a processing queue according to the magnitude of ϵ_i. After eliminating the vertex with minimum error, the queue has to be updated. This procedure is repeated until all vertices with ϵ_i smaller than a given threshold are deleted. Figure 4.20 shows a thinned triangle mesh that was reconstructed from multiple range images. The threshold for the approximation error was 40% of the sampling distance. The collapse of edges that connect patches with different curvature characteristics was prohibited. The final mesh consists of only 7.9% of the original vertices.

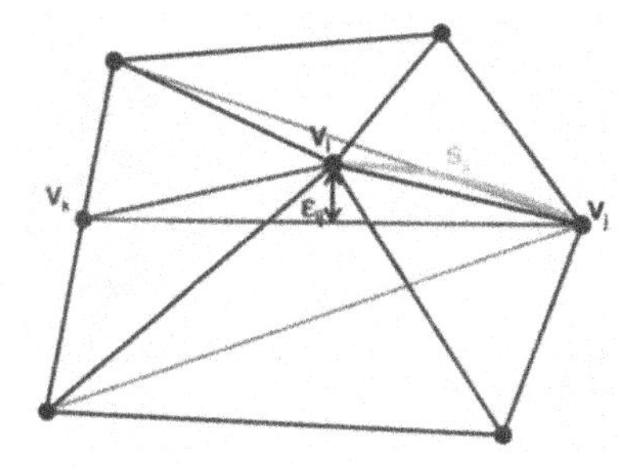

Figure 4.22. Three-dimensional mesh thinning. Now s_{ij} is a cross section through a constantly curved surface. The new triangle below v_i is marked grey.

4.4 FUSION OF DISCRETE MODELS

S. KARBACHER

This section provides a survey on methods to reconstruct the *topology* of a surface from multiple geometric data sets like range images or point clouds. The topology of a sampled object is defined by the neighborhood structure of the data points (see Section 4.1.2). For a single range image this is simply given by the order of the CCD pixel matrix. Topology reconstruction is trivial in this case. Using multiple data sets, the points of the different viewpoints must be merged and the topology has to be recomputed.

4.4.1 Mesh Reconstruction from Unstructured Point Clouds

Most publications discuss topology reconstruction from *unstructured* point clouds, which are unsorted lists of point coordinates [76, 326, 199, 726, 145, 502]. The neighborhood structure of the original pixel arrays is not considered. These methods are very flexible, as they can be used for point probe data and for range images as well. This is particularly important if mechanical probes are used, which usually do not generate structured data sets. However, these methods are costly and error-prone. The results may disagree with the user's expectations. Figure 4.23 demonstrates this by means of a two-dimensional example. The contours of a point set were reconstructed by two different methods with completely different results [522]. None of them agrees

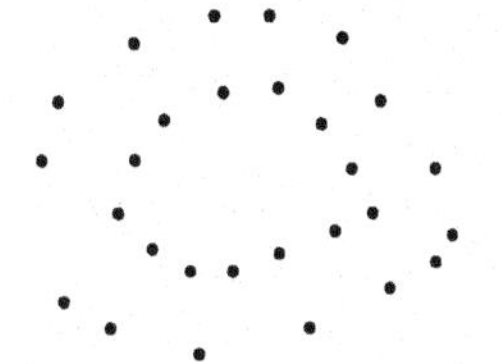

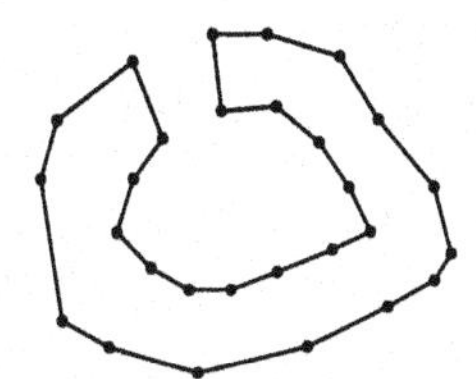

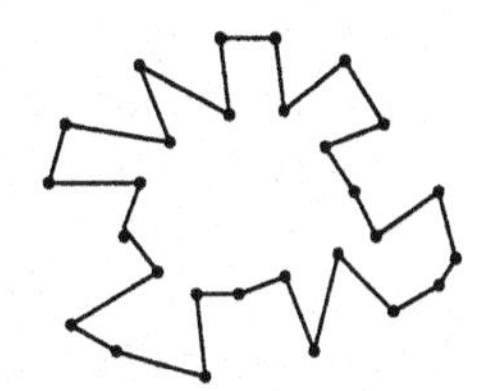

Figure 4.23. In general the topology reconstruction problem cannot uniquely be solved [522] (courtesy of Joseph O'Rourke, Smith College, Northampton, Massachusetts).

Figure 4.24. The results of different methods for topology reconstruction from point clouds [726]; left: synthetic data; right: laser scanner data (courtesy of Remco C. Veltkamp, Centre for Mathematics and Computer Science, Amsterdam, The Netherlands).

with the solution a human would probably find: two concentric rings (the shape of a 'donut'). Figure 4.24 compares three different three-dimensional mesh reconstruction methods [726]. Some of the reconstructed surfaces are even defective.

The reconstruction of the object topology from a point cloud can be solved by means of *graph theory*. Veltkamp [726] gives an introduction to this approach. One usually starts with a *geometric graph* that can be computed efficiently. A geometric graph consists of a set of vertices in $\mathbb{R}^k$ and a set of edges connecting the vertices. A *neighborhood graph* only links adjacent vertices. Thus, the topology of a surface can be described by a geometric neighborhood graph. For surface reconstruction, *hypergraphs* are important. They connect the vertices by high-dimensional *hyper-edges* (triangles, tetrahedrons, etc.).

Some approaches start with the three-dimensional *Delaunay triangulation* which can be computed in $O(n\log(n))$ time (e.g. [76, 199, 726]). A *2D triangulation* links vertices in $\mathbb{R}^2$ with triangles, a *3D triangulation* connects them in $\mathbb{R}^3$ with tetrahedrons (*tetrahedralization*). The Delaunay triangulation features empty circumscribed circles (spheres) that contain no other vertex around each triangle (tetrahedron). The Delaunay triangulation is often used, as it only connects geometrically adjacent vertices.

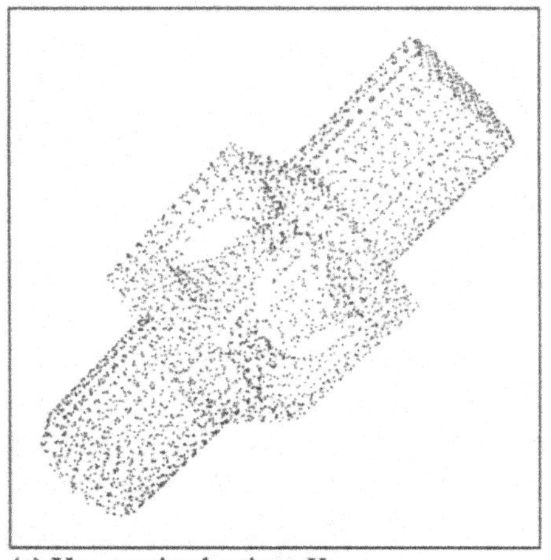

(a) Unorganized points X

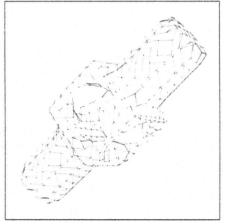

(b) Result of phase 1: initial dense mesh

(c) Result of phase 2: optimized mesh

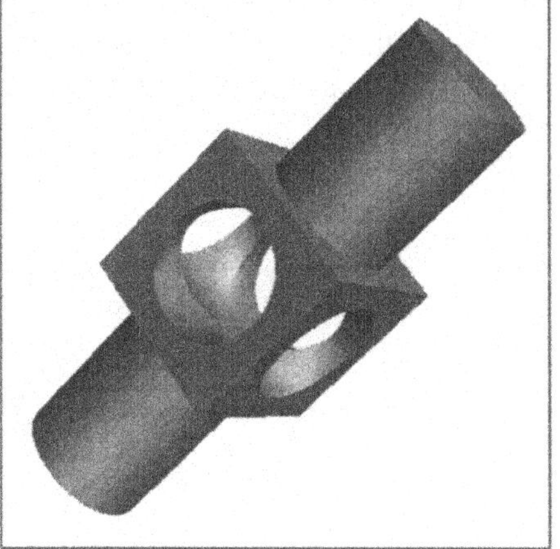

(d) Result of phase 3: piecewise smooth surface

Figure 4.25. Results of the surface reconstruction method of Hoppe et al. [326, 325]. The result of phase 1 (b) demonstrates the problems caused by the loss of any initial topological information. Phase 2 (c) uses special optimizations to solve these problems. Phase 3 (d) generates smooth surfaces by subdivision (courtesy of Hugues Hoppe, Microsoft Research, Redmond, Washington).

The 2D Delaunay triangulation generates triangles that are as equilateral as possible. After tetrahedralization, the interior of the *convex hull* of the point cloud is filled with tetrahedrons, thus with triangles as well. Those triangles that do not cover the sampled surface must be eliminated. This procedure sometimes gets stuck before all interior and exterior triangles are removed or generates illegal surface topologies (e.g. dangling triangles). *Parameterized graphs* were introduced to avoid this problem. Veltkamp [726] uses the γ-graph. Depending on the parameter γ this graph varies from the convex hull to the complete graph, which enables subtle controlling of the final topology. Edelsbrunner's *α-complex* [199] varies from the empty graph to the Delaunay triangulation. The 'cored' hull of the α-complex is named *α-shape*. This is exactly the shape that results from scanning the point cloud with a spherical probe of radius α.

At present these methods have only little importance, as it is difficult to handle the amount of data, which are provided by modern optical 3D-sensors. Furthermore, it is possible only to exactly interpolate the measured data points, but not to smooth errors. For a few years mainly volumetric approaches have been used (e.g. [326, 325, 145, 320, 502]). These are based on well established algorithms of computer tomography and therefore are easy to implement. Usually a three-dimensional density function is computed from the point cloud. A surface of constant density (*iso-surface*) that approximates the object surface is extracted by the *marching cubes* [452] or the *marching triangles* algorithm [319]. Because of surface approximation, error smoothing is carried out automatically. The method of Hoppe et al. [326, 325] is able to detect and model sharp object features. It generates thinned, CAD-incompatible meshes of curved triangles (*subdivision surfaces*), which approximate the original point cloud with high accuracy (see Figure 4.25). Since no initial topological information can be utilized, the result of pure topology reconstruction is rather poor (see Figure 4.25b). Since costly optimizations are necessary in order to improve the quality of the mesh (Figure 4.25c), only a few 10,000 points can be processed, even on fast workstations.

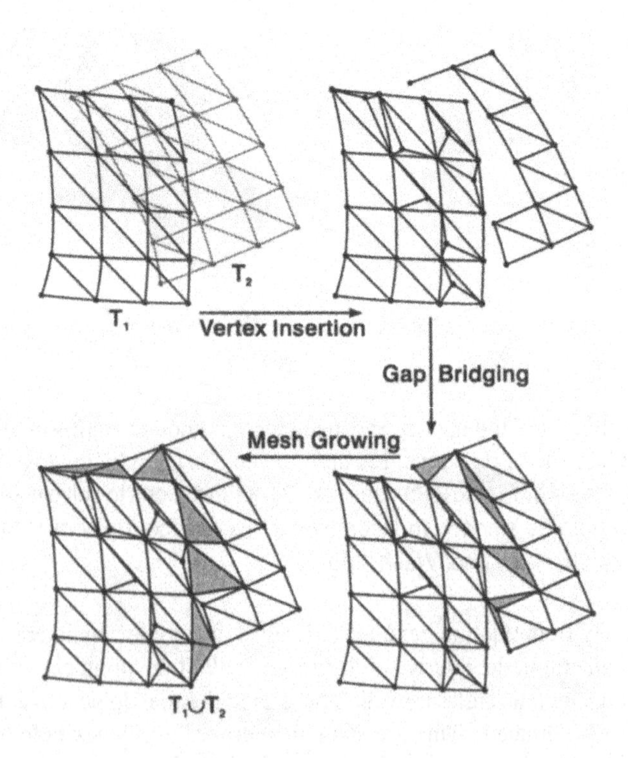

Figure 4.26. Merging of two meshes using vertex insertion, gap bridging and mesh growing operations.

4.4.2 Mesh Reconstruction from Structured Range Images

The usage of topology information provided by the range images enables faster algorithms and more accurate results. For that reason, researchers have proposed several methods for merging multiple range images into a single triangular mesh [716, 586, 299, 374]. Because of the matrix-like structure of the range images, it is easy to turn them into separate triangle meshes with the data points as vertices. These meshes are merged by geometrical and topological operations (see Section 4.1.7). Merging methods usually work incrementally. Furthermore, pure topology reconstruction without any interference of the data points is possible. On the other hand, special efforts for error smoothing are necessary. The method of Rutishauser et al. adds the non overlapping regions of the remaining meshes to the initial master mesh by *mesh growing* (see Figure 4.26). Turk and Levoy [716] merge the meshes by *mesh zippering*.

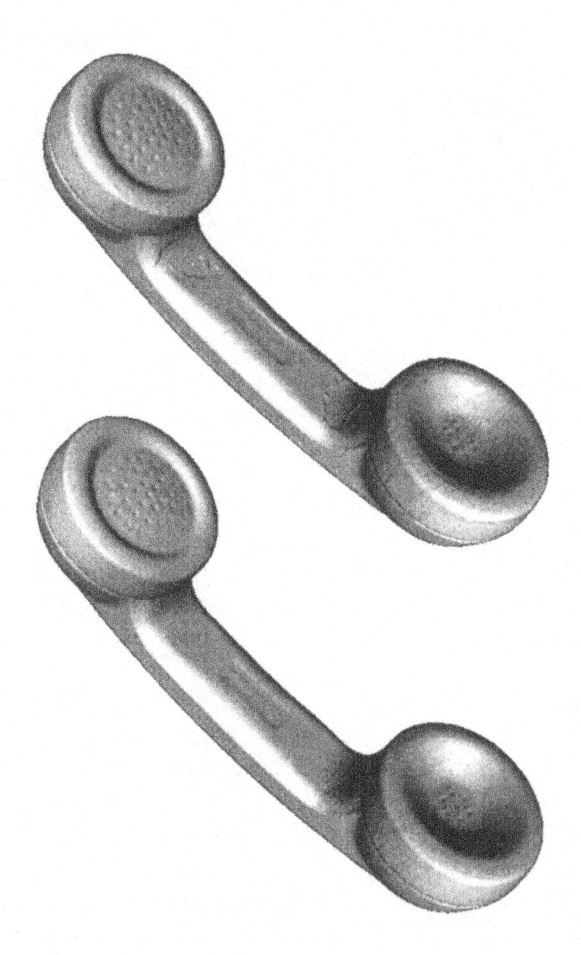

Figure 4.27. Result of the mesh zippering method; top: surface after mesh zippering (160,000 triangles); bottom: mesh after smoothing (courtesy of Marc Levoy, Stanford University, Palo Alto, California).

Figure 4.28. Result of the surface reconstruction method of Curless and Levoy [145]. 71 range images with 12 million data points were processed. The final mesh consists of 1.8 million triangles (courtesy of Marc Levoy, Stanford University, Palo Alto, California).

The overlapping regions are discarded and the boundaries of the remaining parts are linked together (see Figure 4.27). For smoothing errors, the overlapping data points are averaged. Thus, filtering only affects regions measured multiple times. Therefore, it is possible to smooth calibration and registration errors, other measuring errors like noise and aliasing are smoothed incompletely. This method generates dense meshes of flat triangles.

The method introduced in Section 8.1 [299, 374] uses *vertex insertion*, *gap bridging* and *mesh growing* operations for merging two meshes (see Figure 4.26). It includes an effective smoothing filter (see Section 4.3.2). In contrast to other surface reconstruction methods it is able to smooth single images without significant loss of details. The method produces meshes of flat or curved (but CAD incompatible) triangles with curvature dependent density.

Curless and Levoy [145] have proposed a volumetric method that requires topological information provided by the matrix-like structure of range images. As a benefit it achieves good results without any post optimizations (see Figure 4.28). Unlike Hoppe's method it is able to handle millions of data points. As neither curved triangles nor mesh thinning techniques are used, dense meshes containing a huge amount of small triangles are usually produced.

4.5 SPLINES

G. GREINER

In CAD design, the term *free form surface* is used to describe any kind of geometry that can not be expressed by regular geometry, (lines, circles in 2D and planes, spheres, cylinders, cones or quadrics in 3D). The most common way to describe free form curves or surfaces is a parametric representation $C : u \mapsto C(u)$ and $F : (u,v) \mapsto F(u,v)$. The parameter domains are intervals and rectangles or triangles respectively. In practical application only special types of functions are used for the mappings C and F respectively: either polynomials or rational functions. Depending whether one has a local (piecewise) or a global description by polynomials or rationals, different names are commonly used.

	polynomial	rational
globally	Bézier	rational Bézier
piecewise	B-spline	NURBS

Polynomial functions are simpler to deal with. But rationals give a greater variety of shapes. For example, conics (spheres, ellipsoids, hyperboloids) can be represented exactly as NURBS surfaces. This makes these surfaces interesting as a universal format for describing regular geometry and free form geometry in a uniform way.

There is a standard approach to extend a curve scheme to a surface scheme: the *tensor product* approach. It also allows that most of the algorithms for curves can be adapted to algorithms for surfaces. For this reason and the fact that for curves the notations, and fundamental ideas are easier to understand (the formulas are much simpler) we first consider free form curves and then describe the tensor product approach to surfaces.

4.5.1 Bézier Curves

Polynomial representations have been studied extensively throughout the literature [57, 1, 341]. Numerical analysis, e.g., has traditionally used orthogonal polynomials as well as the monomial, Lagrange, or Hermite basis. Perhaps the most important representation within CAD is the Bernstein Bézier representation

$$C(u) = \sum_{i=0}^{n} B_i^{\Delta,n}(u)\mathbf{b}_i, \quad \mathbf{b}_i \in \mathbb{R}^3$$

where

$$B_i^{\Delta,n}(u) = \binom{n}{i} \left(\frac{u-a}{b-a}\right)^i \left(\frac{b-u}{b-a}\right)^{n-1}$$

are the *Bernstein polynomials* w.r.t. the interval $\Delta = [a,b]$. We remark that the Bernstein polynomials are positive on Δ, form a partition ofunity (sum up to 1), and satisfy the recursion formula:

$$B_i^{\Delta,n}(u) = \left(\frac{u-a}{b-a}\right) B_{i-1}^{\Delta,n-1}(u) + \left(\frac{b-u}{b-a}\right) B_i^{\Delta,n-1}(u).$$

The points $\mathbf{b}_i \in \mathbb{R}^3$ are the Bézier *control points*. They form the *control polygon*.

The importance of Bézier curves stems from the following two facts: First, the shape of the curve closely mimics the shape of the control polygon, i.e., the coefficients have geometric significance. Secondly, the Bernstein basis is extremely stable: As recently shown by Farouki and Rajan [216], the condition number of the simple roots of polynomial is smaller in the Bernstein basis than in almost any other basis. Bézier curves satisfy the following shape properties:

Convex Hull Property: A Bézier curve is contained in the convex hull of its Bézier polygon.

Endpoint-Interpolation and -Tangency: A Bézier curve interpolates the endpoints of its control polygon and is tangent to the control polygon there.

Variation Diminishing Property: The number of intersection points of a Bézier curve with an affine hyperplane *H* is bounded by the number of intersection points of *H* with the control polygon. Intuitively, this means that a Bézier curve wiggles not more than its control polygon does.

Affine Invariance: The relationship between a Bézier curve and its control polygon is invariant under affine transformations.

An algorithm for the stable evaluation of Bézier curves is the recursive

Algorithm of de Casteljau. *Consider the recurrence*

$$\mathbf{b}_i^0(u) = \mathbf{b}_i \qquad \text{for } i = 0,\ldots,n,$$

$$\mathbf{b}_i^k(u) = \frac{b-u}{b-a}\mathbf{b}_i^{k-1}(u) + \frac{u-a}{b-a}\mathbf{b}_{i+1}^{k-1}(u) \qquad \text{for } i = 0,\ldots,n-k \text{ and } k = 1,\ldots,n.$$

Then $\mathbf{b}_0^n$ *is the point* $C(u)$ *on the curve.*

Note that the de Casteljau Algorithm uses convex combinations throughout and hence is very stable. The de Casteljau Algorithm produces much more than just evaluation of the curve at specified parameters. in particular, it allows to *subdivide* a given Bézier curve at the specified parameter u. According to the formula above, the Bézier points of the left curve segment (with respect to the interval $[a, u]$) are given by

$$\mathbf{b}_i^{left} = \mathbf{b}_0^i(u)$$

while the Bézier points of the right curve segment (with respect to the interval $[u, b]$) are given by

$$\mathbf{b}_i^{right} = \mathbf{b}_i^{n-i}(u)$$

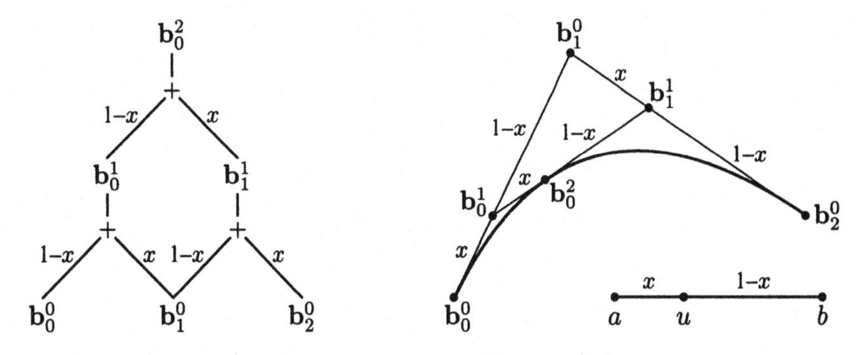

Figure 4.29. The de Casteljau algorithm at u with $x = \frac{u-a}{b-a}$ and $1 - x = \frac{b-u}{b-a}$.

Note that these points appear along the sides of the de Casteljau triangle (see Figure 4.29). Bézier subdivision exhibits quadratic convergence [148] and thus can be used to approximate the curve by a polygon, e.g. for drawing it. Another side effect of the Casteljau Algorithm is the computation of the tangent vector $t_C(u)$ to the curve at $C(u)$, which is given by $\mathbf{b}_1^{n-1} - \mathbf{b}_0^{n-1}$.

4.5.2 B-Spline Curves

Using the results of the preceding section, it is possible (and has actually been done in practice), to build a CAD package based on Bézier curves (see [63]). However, in order to model complex shape, either one has to use curves with very high polynomial degree, or one has to compose the desired shape using several Bézier segments (of low degree). The first approach is numerically demanding and unstable, in the second method, keeping track of the continuity constraints at the joints causes additional problems.

A better solution is to eliminate the continuity constraints once and for all, and to choose a basis where the necessary constraints are already built in. This idea leads to the concept of *B-splines*. In the following we briefly sketch the fundamendal ideas and the basic properties. Details (including proofs) and many more information can be found, e.g., in the following books [77, 627, 44, 215, 341].

A B-spline curve of degree n over a non-decreasing knot sequence $T = (t_i)_{i \in \mathbb{Z}}$ (with $t_i \leq t_{i+1}$ and $t_i < t_{i+n+1}$) is defined as

$$C(u) = \sum_i N_i^{T,n}(u)\mathbf{d}_i, \quad \mathbf{d}_i \in \mathbb{R}^3,$$

where $N_i^{T,n}(u)$ are the *normalized B-splines basis functions* over the *knot vector* T, defined recursively by

$$N_i^{T,0}(u) = \begin{cases} 1, & u \in [t_i, t_{i+1}) \\ 0, & \text{otherwise} \end{cases}$$

and

$$N_i^{T,k}(u) = \frac{u - t_i}{t_{i+k} - t_i} N_i^{T,k-1}(u) + \frac{t_{i+k+1} - u}{t_{i+k+1} - t_{i+1}} N_{i+1}^{T,k-1}(u).$$

The points $\mathbf{d}_i$ are the *B-spline control points*, sometimes also called *de Boor points.* The normalized B-spline basis functions $N_i^{T,n}$, $i \in \mathbb{Z}$ are piecewiese polynomials of degree n, non-negative everywhere and form a partition of unity. In addition, they have local support, i.e., $N_i^{T,n}(u) = 0$ for $u \notin [t_i, t_{t+n+1})$, and they are $C^{n-\mu}$-continuous at a knot of multiplicity μ (i.e., $t_{i-1} < t_i = t_{i+1} = \ldots = t_{i+\mu-1} < t_{i+\mu}$).

Hence, in the case of single knots, B-splines are C^{n-1}-continuous everywhere. B-spline curves have similar shape properties as Bézier curves and some additional features:

Convex Hull Property: A B-spline curve is not only contained in the convex hull of its de Boor points. In addition, the following local convex hull property holds true: For $u \in [t_j, t_{j+1})$ we have $C(u) \in \text{conv}\{\mathbf{d}_{j-n}, \ldots, \mathbf{d}_j\}$.

Multiple Control Points: If n control points $\mathbf{d}_{j-n+1} = \ldots = \mathbf{d}_j = \mathbf{d}$ coincide, then $C(t_{j+n}) = \mathbf{d}$, i.e., the curve interpolates this point and is tangent to the control polygon there.

Collinear Control Points: If $n+1$ control points, $\mathbf{d}_{j-n}, \ldots, \mathbf{d}_j$, lie on a line L, then $C([t_j, t_{j+1})) \subset L$, i.e., the curve contains a line segment.

Multiple Knots: If n knots $t_{j+1} = \ldots = t_{j+n} = t$ coincide, then $C(t) = \mathbf{d}_j$, i.e., the curve interpolates this control point and is tangent to the control polygon there.

Variation Diminishing Property: The number of intersection points of a B-spline curve with an affine hyperplane H is bounded by the number of intersection points of H with the control polygon. Intuitively this means that a B-spline curve wiggles not more than its control polygon does.

Local Control: Moving the control point $\mathbf{d}_j$ only changes the B-spline curve locally. More precisely, the curve points $C(u)$ with $u \notin [t_j, t_{j+n+1})$ are not influenced by $\mathbf{d}_j$.

Affine Invariance: The relationship between a B-spline curve and its control polygon is invariant under affine transformations. Thus, e.g., the translation of a B-spline curve having control points $\mathbf{d}_i$ by a vector b, is the B-spline curve with control points $\mathbf{d}_i + b$.

Since the Curry-Schoenberg Theorem (see [146, 77]) states that every piecewise polynomial can be represented as a linear combination of B-splines basis functions over the corresponding knot vector, this approach offers the full variety of all piecewise polynomial curves.

We also remark that Bézier curves are a special case of B-splines: For the knot vector $T = (\ldots t_0 < t_1 = t_2 = \ldots = t_n < t_{n+1} = t_{n+2} = \ldots = t_{2n} < t_{2n+1} \ldots)$, the B-spline basis function $N_i^{T,n}(u)$ coincides with the Bernstein polynomial $B_i^{\Delta,n}(u)$ for $i = 0, \ldots, n$ and $u \in [t_1, t_{n+1}]$. Therefore, piecewise Bézier curves may be considered as B-spline curves with n-fold knots. B-splines can be evaluated recursively by the

Algorithm of de Boor. *For $u \in [t_j, t_{j+1})$ recursively compute*

$$\mathbf{d}_i^0 = \mathbf{d}_i \qquad \text{for } i = j-n, \ldots, n,$$

$$\mathbf{d}_i^k = \frac{t_{i+n+1} - u}{t_{i+n+1} - t_{i+k}} \mathbf{d}_i^{k-1} + \frac{u - t_{i+k}}{t_{i+n+1} - t_{i+k}} \mathbf{d}_{i+1}^{k-1} \qquad \begin{array}{r} \text{for } i = j-n, \ldots, j-k \\ \text{and } k = 1, \ldots, n. \end{array}$$

Then $\mathbf{d}_{j-n}^n$ is the point $C(u)$ on the curve.

Even more important than the evaluation is *knot insertion*. In order to have more control points available, e.g. for modeling more detail, one can insert additional knots in specified regions of the parameter domain. The problem then is as follows:

Given a B-spline curve $C(u) = \sum_i N_i^{T,n}(u)\mathbf{d}_i$ over a knot vector $T = (t_i)_{i \in \mathbb{Z}}$ plus a new knot $\tilde{t}$, say $t_j \leq \tilde{t} < t_{j+1}$, compute the coefficients $\mathbf{d}_i^*$ of the B-spline representation of C over the refined knot sequence $\tilde{T} = (\ldots \leq t_{j-1} \leq t_j \leq \tilde{t} < t_{j+1} \leq t_{j+2} \leq \ldots)$ that is obtained inserting the new knot $\tilde{t}$ in T. This can be done by the

Algorithm of Boehm. *The new control points $\mathbf{d}_i^*$ are given by*

$$\mathbf{d}_i^* = \alpha_i \mathbf{d}_i + (1 - \alpha_i)\mathbf{d}_{i-1}$$

with

$$\alpha_i = \begin{cases} 1, & \text{if } \quad i \leq j-n \\ \frac{t - t_i}{t_{i+n} - t_i}, & \text{if } j-n+1 \leq i \leq j \\ 0, & \text{if } \quad j+1 \leq i \end{cases} .$$

4.5.3 Rational Curves: NURBS

Many CAD systems use conic sections (ellipses, parabolas and hyperbolas) as basic components for the construction of more complex objects. Parabolas can be easily represented as B-splines, e.g. as quadratic Bézier curves. However, it is impossible to represent an ellipse or a hyperbola exactly by a B-spline. One only can approximate ellipses and hyperbolas with arbitrary prescribed tolerance. Rational curves, which we are going to describe in this part, will allow an exact representation of a conic section. NURBS, the rational extension of B-splines, have all the nice features of B-splines, and include conic sections.

A rational function is the quotient of two polynomials. Here, the term 'polynomial' may be replaced by 'piecewise polynomial'. Thus a *rational curve* has the form $R(u) = \frac{C(u)}{w(u)}$ where C is a (piecewise) polynomial curve and w is a (piecewise) scalar polynomial. After possible knot insertion, we can assume w.l.o.g. that C and w are defined over the same knot vector.

More intuitive than the *analytic description* given above is the following *geometric description*. First, we consider planar curves (see Figure 4.30). Embed the plane $\mathbb{R}^2$ in the three-dimensional Euclidean space $\mathbb{R}^3$ and consider a point $O \in \mathbb{R}^3$ *not* contained in this plane. Then a polynomial (or piecewise polynomial) curve in $\mathbb{R}^3$ can be projected into the plane with O as center of projection. The resulting curve in the plane is a rational curve.

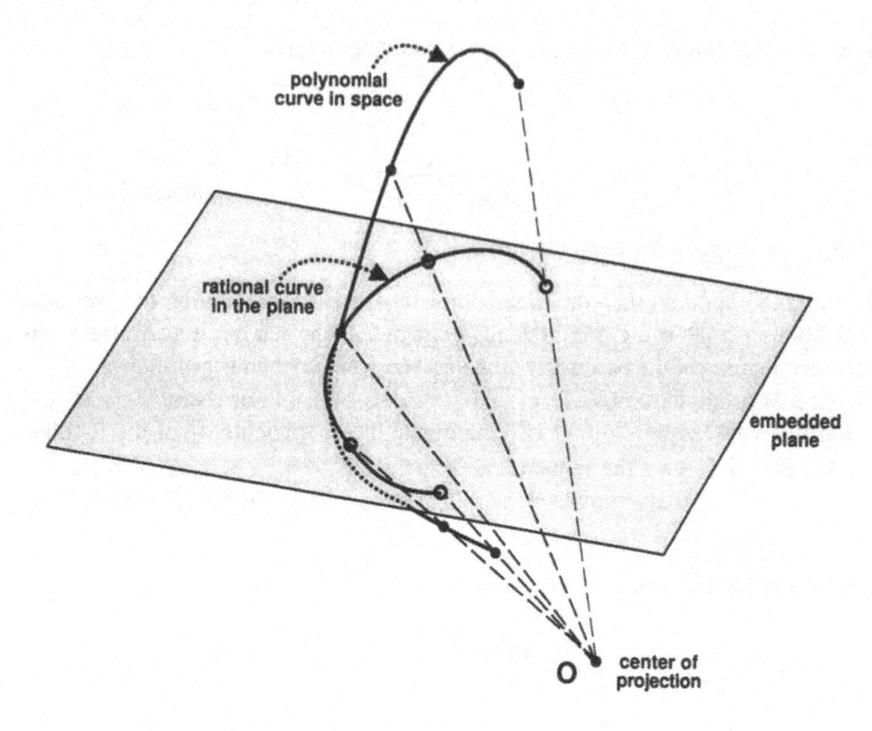

Figure 4.30. Geometric description of a rational curve.

The *standard embedding* of the plane $\mathbb{R}^2$ is obtained by identifying it with the hyperplane $\{z = 1\}$ in $\mathbb{R}^3$. Thus $(x, y) \in \mathbb{R}^2$ will be identified with $(x, y, 1) \in \mathbb{R}^3$. Choosing the center of projection O to be $(0, 0, 0)$ then the projection of the space curve $\hat{R}(u) = (x(u), y(u), w(u))$ into the plane will be $R(u) = \left(\frac{x(u)}{w(u)}, \frac{y(u)}{w(u)}, 1\right)$.

The same construction works for rational curves in $\mathbb{R}^3$: project a (piecewise) polynomial curve in $\mathbb{R}^4$ to a hyperplane of codimension 1 and the result will be a rational curve in $\mathbb{R}^3$.

Rational Bézier Curves. A *rational Bézier curve of degree* n in $\mathbb{R}^d$, $d = 2, 3$, with *control points* $\mathbf{b}_i \in \mathbb{R}^d$ and *weights* w_i has the representation

$$R(u) = \frac{\sum_{i=0}^{n} w_i \mathbf{b}_i B_i^n(u)}{\sum_{i=0}^{n} w_i B_i^n(u)}, \quad u \in [0, 1], \tag{4.47}$$

where B_i^n, $i = 0, \ldots, n$ are the Bernstein polynomials of degree n (say for simplicity over the interval $[0, 1]$).

We will assume throughout this section that the weights are positive, i.e., $w_i > 0$, and that $w_0 = w_n = 1$. Among other things, this ensures the denominater in Equation 4.47 never to vanish and also implies that a rational Bézier curve satisfies the convex hull property. If all the weights are equal to 1 we obtain a Bézier curve as considered in the previous part. This follows from the fact that the Bernstein polynomials form a partition of unity.

A more geometric interpretation of control points and weights is as follows. Let the rational curve R in $\mathbb{R}^d$, $d = 2, 3$, be obtained by a projection of the (piecewise) polynomial curve $\hat{R} = \sum_i \hat{\mathbf{b}}_i B_i^n$ in $\mathbb{R}^{d+1}$. Here, $\hat{\mathbf{b}}_i$ are the control points of $\hat{R}$ and B_i^n are the Bernstein polynomials. Then we have the following properties:

- The control points $\mathbf{b}_i$ of the rational curve R are the projections of the control points $\hat{\mathbf{b}}_i$ of the curve $\hat{R}$.
- The weight w_i of a control point $\mathbf{b}_i$ is the distance of $\hat{\mathbf{b}}_i$ to the hyperplane through O which is parallel to the embedded hyperplane.

Using the standard embedding, the analytic description of control points and weights is as follows.

Planar curve: Assuming that $\hat{\mathbf{b}}_i = (x_i, y_i, w_i)$, then $\mathbf{b}_i = (\frac{x_i}{w_i}, \frac{y_i}{w_i})$ is the control point of the rational curve R and w_i is the weight.

Space curve: Assuming that $\hat{\mathbf{b}}_i = (x_i, y_i, z_i, w_i)$, then $\mathbf{b}_i = (\frac{x_i}{w_i}, \frac{y_i}{w_i}, \frac{z_i}{w_i})$ is the control point of the rational curve R and w_i is the weight.

If we look at this statement from the other way around and are given the control points $\mathbf{b}_i = (x_i, y_i)$ and corresponding weights w_i of a plane rational curve R, then the associated polynomial curve $\hat{R}$ in $\mathbb{R}^3$ has control points $\hat{\mathbf{b}}_i = (w_i x_i, w_i y_i, w_i)$.

Control points and weights of a rational Bézier curve uniquely determine the shape of the curve. In case of all weights being positive, the influence of the control points on the shape of the curve is basically the same as for polynomial curves. In particular, we have:

Convex Hull Property: A rational Bézier curve lies in the convex hull of its control polygon.

Endpoint-Interpolation and -Tangency: A rational Bézier curve interpolates the endpoints of its control polygon and is tangent to the control polygon there.

Variation Diminishing Property: The number of intersection points of a rational Bézier curve with an affine hyperplane H is bounded by the number of intersection points of H with the control polygon. Intuitively, this means that a rational Bézier curve wiggles no more than its control polygon.

Affine Invariance: The relationship between a rational Bézier curve and its control polygon is invariant under affine transformations.

Algorithms: The de Casteljau algorithm can be applied to $\hat{R}(u)$ for evaluation and subdivision.

The influence of the weights can be described as follows (see Figure 4.31). Increasing a single weight has the effect of the curve being pulled towards the corresponding control point. Decreasing a single weight, has the opposite effect: the curve will be pushed away from the corresponding control point. One also can observe from Figure 4.31 that the convex hull property fails for negative weights.

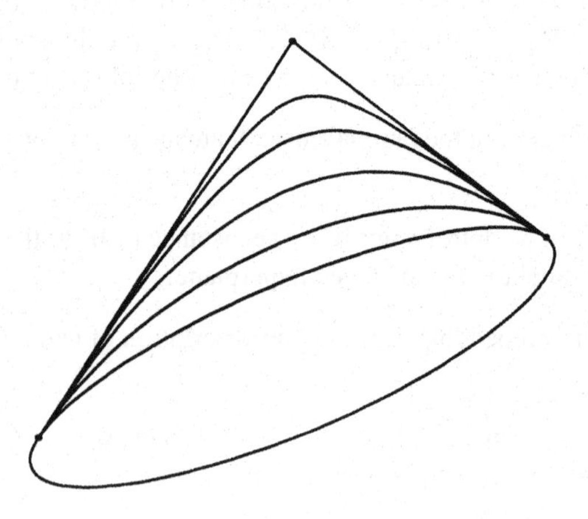

Figure 4.31. Effect of the weight w_1 on the shape of a rational quadratic Bézier curve. The interior weight w_1 has been changed. Top to bottom: $w_1 = 4, 2, 1, \frac{1}{2}, \frac{1}{4}, -\frac{1}{4}$.

The weights carry some redundance. One easily observes that only the ratios of the weights are important. In fact, multiplying all weights by the same non-zero factor does not change the rational curve at all.

A more sophisticated analysis shows that only the so-called *cross ratio's* of four subsequent weights are important. The cross ratios are given by $\frac{w_i}{w_{i+1}} : \frac{w_{i+2}}{w_{i+3}}$ for $i = 0, \ldots, n-3$. Thus replacing the weights w_i by $\tilde{w}_i = c \cdot \alpha^i \cdot w_i$ with fixed scalars $c > 0$, $\alpha > 0$ does not change the shape of the curve. To be more precise, it yields another parametrization of the same curve.

Choosing $c = \frac{1}{w_0}$ and $\alpha = \sqrt[n]{\frac{w_0}{w_n}}$, one obtains another set of weights $\tilde{w}_i$ satisfying $\tilde{w}_0 = \tilde{w}_n = 1$. These weights in combination with the original control points describe the same curve. The conclusion of this discussion: There is no loss of generality when one restricts to rational curves whose weights satisfy $w_0 = w_n = 1$.

NURBS. The most important class of rational spline curves is the set of **N**on-**U**niform **R**ational **B**-**S**plines, briefly NURBS. In this section we want to give a description of this class. Let's begin with an explanation of the notation.

NU Non-uniform refers to the fact that the knot vector used to construct such a curve is not necessarily equidistant. Also, multiple knots are allowed. A typical example of a knot vector would be $(0, 0, 0, 0.5, 1, 2, 2, 3.5, 3.5)$.

R Rational curve as described above: Analytically, this is the quotient of B-splines, or, geometrically, it is the central projection of a usual B-spline curve.

BS B-spline as described in Section 4.5.2.

In order to describe a NURBS curve of degree n in $\mathbb{R}^d$, $d = 2, 3$, one has to specify

- a *knot vector* $T = (t_i)_i$ (see Section 4.5.2),
- *control points* $\mathbf{d}_i \in \mathbb{R}^d$ and
- positive *weights* w_i, associated to the control points $\mathbf{d}_i$.

The analytic representation of the corresponding NURBS curve R is given by

$$R(u) = \frac{\sum_i w_i \mathbf{d}_i N_i^{T,n}(u)}{\sum_i w_i N_i^{T,n}(u)}, \tag{4.48}$$

where $N_i^{T,n}$ are the normalized B-spline basis functions of degree n corresponding to the knot vector T. Since the weights are assumed to be positive, the denominator in Equation 4.48 does not vanish. If all weights coincide, the curve is a B-spline curve.

As described in the introduction to this subsection, the NURBS curve in Equation 4.48 can be obtained from the B-spline curve $\hat{R}$ in $\mathbb{R}^{d+1}$ having the same knot vector and control points $\hat{\mathbf{d}}_i = (w_i \mathbf{d}_i, w_i)$. As a consequence, the following properties generalize from B-spline curves to NURBS curves.

Convex hull property: For $u \in [t_j, t_{j+1})$ we have $R(u) \in \text{conv}\{\mathbf{d}_{j-n}, \ldots, \mathbf{d}_i\}$.

Interpolation in the Interior: If a knot has multiplicity n, e.g. $t_{j+1} = \ldots = t_{j+n}$, then the curve is tangent to the control polygon in $\mathbf{d}_j$ (and has a cusp there).

Variation Diminishing Property: No line (plane) has more intersections with the curve than with the control polygon.

Local control: Both, $\mathbf{d}_j$ and w_j only have influence on the the part of the curve corresponding to parameter values $u \in [t_j, t_{j+n+1})$.

Affine Invariance: The relationship between a NURBS curve and its control points is invariant under affine transformations.

Algorithms: De Boor's algorithm can be applied to $\hat{R}$ for evaluation and Boehm's algorithms for knot insertion.

We conclude the discussion with a simple example: A NURBS representation of the unit circle. It is a NURBS curve of degree 2 having 4 rational segments. The idea is to find parabolas in $\mathbb{R}^3$ that will be projected on a circular arc. For the construction see Figure 4.32. The resulting analytic description is as follows:

Knot Vector:	$(0,0,0,1,1,2,2,3,3,4,4,4)$								
Control Points:	$(1,0)$	$(1,1)$	$(0,1)$	$(-1,1)$	$(-1,0)$	$(-1,-1)$	$(0,-1)$	$(1,-1)$	$(1,0)$
Weights:	1	$\frac{\sqrt{2}}{2}$	1	$\frac{\sqrt{2}}{2}$	1	$\frac{\sqrt{2}}{2}$	1	$\frac{\sqrt{2}}{2}$	1

From the NURBS representation of the unit circle and the affine invariance one can deduce easily a NURBS representation of an arbitrary ellipse.

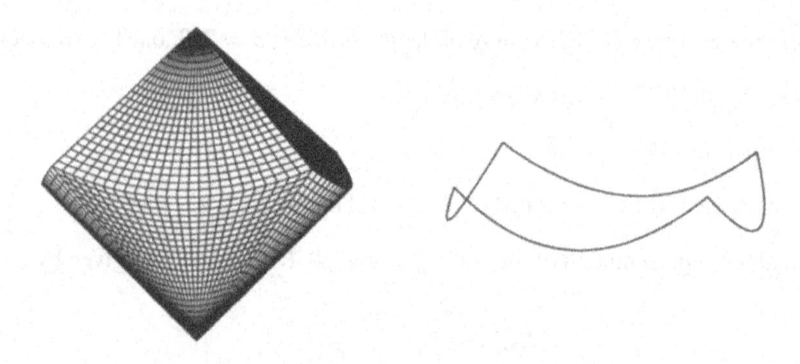

Figure 4.32. Left: Intersection of an ice-cream cone (lower) with a pyramidal cone (upper). Right: The curve obtained by intersecting the two cones. It is composed by four parabolic segments and its central projection is a circle.

4.5.4 Tensor Product Surfaces

The idea of tensor product surfaces is the following: Given a certain class of 3D-curves (e.g. Bézier, B-splines, NURBS) one wants to consider those surfaces for which all the iso-parameter lines belong to the specified class of curves. For example, a tensor product Bézier surface of degree (n, m) defined over the square $[0, 1]^2$ is a surface $F : (u, v) \mapsto F(u, v)$ such that the curves $F_{\overline{u}} : v \mapsto F(\overline{u}, v)$ and $F_{\overline{v}} : u \mapsto F(u, \overline{v})$ are Bézier curves (of degree m and n, resp., defined over the interval $[0, 1]$) for any choice of $\overline{u} \in [0, 1]$ and $\overline{v} \in [0, 1]$.

Thus we can conclude, that F can be written as $F(u, v) = \sum_{j=0}^{m} \mathbf{e}_j(u) B_j^m(v) = \sum_{i=0}^{n} \mathbf{d}_i(v) B_i^n(u)$ and it follows that $\mathbf{e}_j(u) = \sum_i \mathbf{e}_{ji} B_i^n(u)$, $\mathbf{d}_i(v) = \sum_j \mathbf{d}_{ij} B_j^m(v)$ and finally $\mathbf{d}_{ij} = \mathbf{e}_{ji}$. Therefore, F has the representation

$$F(u, v) = \sum_{i=0}^{n} \sum_{j=0}^{m} \mathbf{b}_{ij} B_i^n(u) B_j^m(v).$$

The array $(\mathbf{b}_{ij})$ of 3D-points are called the set of control points. These points form the *control mesh*, a quadmesh of 3D-points.

From the construction it is clear that properties of the curves scheme will cause similar properties for the corresponding tensor product surface scheme. We list a few for tensor product Bézier surfaces:

Convex Hull Property: A Bézier surface is contained in the convex hull of its control mesh.

Endpoint-Interpolation and -Tangency: A Bézier surface interpolates the four corners of the control mesh $\mathbf{b}_{00}, \mathbf{b}_{n0}, \mathbf{b}_{nm}, \mathbf{b}_{0m}$. At these points the tangent plane to the surface is spanned by the two outgoing edges of the control mesh. E.g., at $\mathbf{b}_{00}$ the tangent plane is spanned by $\mathbf{b}_{10} - \mathbf{b}_{00}$ and $\mathbf{b}_{01} - \mathbf{b}_{00}$.

Affine Invariance: The relationship between a Bézier surface and its control mesh is invariant under affine transformations.

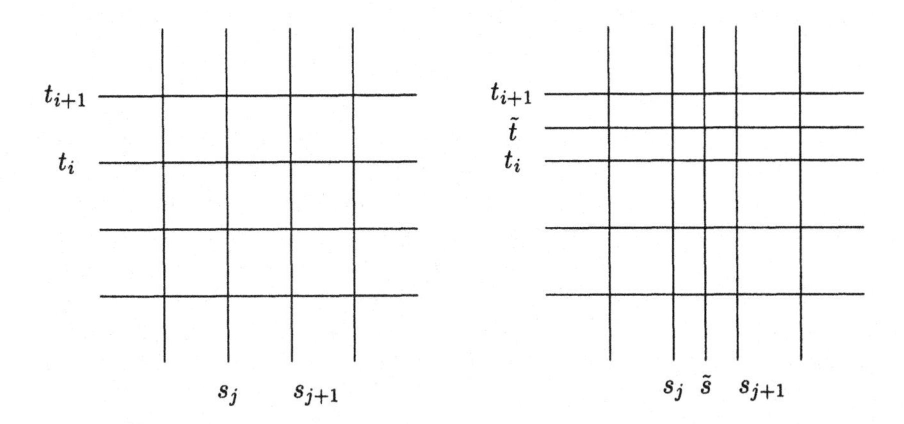

Figure 4.33. Knot grid of a tensor product B-spline surface before (left) and after (right) inserting a knot $\tilde{t}$ in T and a knot $\tilde{s}$ in S.

Algorithms: The de Casteljau algorithm can easily be extended for the evaluation of a tensor product Bézier surface. To evaluate $F(\overline{u}, \overline{v})$, perform $m + 1$ times the de Casteljau algorithm in order to evaluate $b_j(u) = \sum_{i=0}^{n} \mathbf{b}_{ij} B_i^n(u)$, $j = 0, \ldots, m$ at $u = \overline{u}$. Then perform one additional de Casteljau algorithm to evaluate $\sum_{j-0}^{m} b_j(\overline{u}) B_j^m(v)$ at $v = \overline{v}$. The resulting value is $F(\overline{u}, \overline{v})$.

Similarly for a tensor product B-spline surface $F(u, v) = \sum_{i,j} \mathbf{d}_{ij} N_i^{T,n}(u) N_j^{S,m}(v)$ with knot vectors T in u- and S in v-direction and a control mesh formed by the array of control points $(\mathbf{d}_{ij})$ the following shape conditions are valid:

Local Convex Hull Property: The local convex hull property guarantees, that for $(u, v) \in [t_i, t_{i+1}) \times [s_j, s_{j+1})$ we have $F(u, v) \in \operatorname{conv}\{\mathbf{d}_{kl} : i-n \leq k \leq i, j-m \leq l \leq j\}$.

Local Control: Moving the control point $\mathbf{d}_{ij}$ changes the B-spline surface only locally.

Affine Invariance: The relationship between a B-spline surface and its control polygon is invariant under affine transformations.

Algorithms: De Boor's algorithm for evaluation and Boehm's algorithm for knot insertion can be extended to the tensor product setting.

Of course knot insertion can be done only row-, or columnwise, adding either to T or S an additional point. This shows one weakness of the tensor product B-splines. In order to add detail locally one has to add knots in regions where it is not needed (see Figure 4.33). One way to overcome this problem is the idea of hierarchical B-splines introduced by Forsey/Bartels [229]. The idea of this approach will be briefly described in the next section, where this type of surfaces is used for surface fitting.

4.6 FITTING FREE FORM SURFACES

K. HORMANN

The problem of reconstructing smooth surfaces from discrete scattered data arises in many fields of science and engineering and has now been studied thoroughly for nearly 40 years. The data sources include measured values (meteorology, oceanography, optics, geodetics, geology, laser range scanning, etc.) as well as experimental results (from physical, chemical or engineering experiments) and computational values (evaluation of mathematical functions, finite element solutions of partial differential equations or results of other numerical simulations). Due to the vast variety of data sources many different methods have been developed, each of them more or less suited to a specific problem.

In the field of geology, meteorology, cartography, a.o., the problem can typically be stated as follows: given data points $(x_i, y_i, z_i) \in \mathbb{R}^3$, find a scalar function $F : \mathbb{R}^2 \to \mathbb{R}$ that approximates or interpolates the value z_i at (x_i, y_i), i.e. $F(x_i, y_i) \approx z_i$. This problem is generally known as *Scattered Data Interpolation* (cf. Figure 4.34) and there exist many solutions to that problem which include Shepard's methods [641], radial basis functions [285] and finite element methods. Good surveys of these methods and further references can be found in [42, 232, 506, 626].

In contrast to this *scalar problem* there is the *parametric problem*, where the task is to find a parametrized surface $F : \mathbb{R}^2 \to \mathbb{R}^3$ that approximates or interpolates the data points. This is usually done by specifying additional parameter values $(u_i, v_i) \in \mathbb{R}^2$ and by determining F such that $F(u_i, v_i) \approx (x_i, y_i, z_i)$ (cf. Figure 4.35). While the theory of parametric surfaces was well understood in differential geometry, dating back to the time of Gauß [184], their potential for the representation of surfaces in the field of engineering remained unknown for a long time. "The exploration of the use of parametric curves and surfaces can be viewed as the origin of *Computer Aided Geometric Design (CAGD)*" ([215], p. xv).

The fundamental ideas in CAGD have been developed at the French and US-American car industry in the 1960ies. While de Casteljau at Citroën and Bézier at Renault independently developed the theory of Bézier curves and surfaces (see

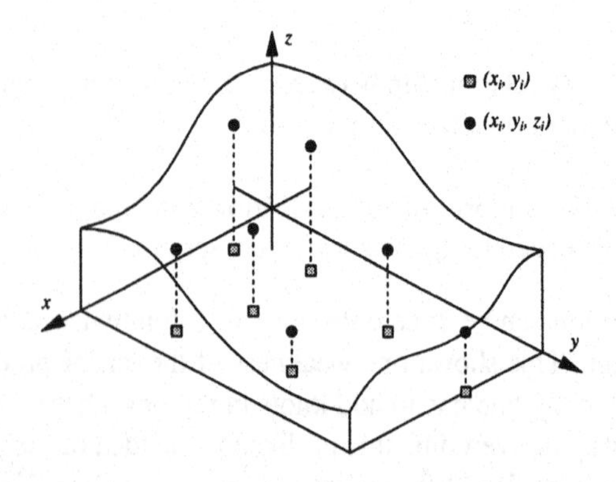

Figure 4.34. Scattered Data Interpolation.

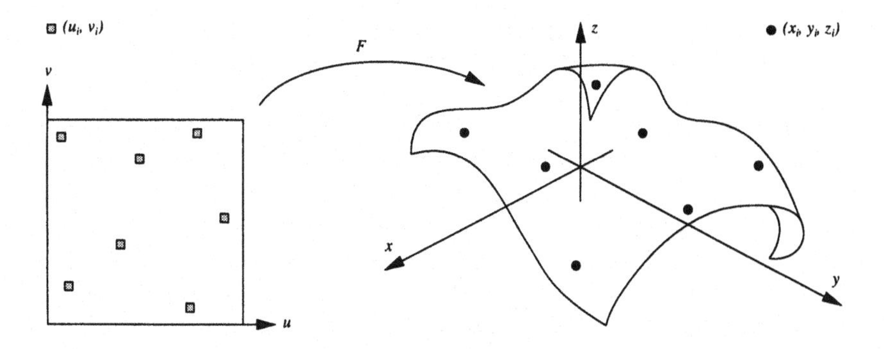

Figure 4.35. The parametric surface fitting problem.

Section 4.5), Coons at Ford and Gordon at General Motors worked on *transfinite interpolation methods*, which can be used to 'fill in' a network of curves. At the same time *mathematical splines*, introduced by Schoenberg in the 1940ies, were used by Ferguson at Boeing and Sabin at the British Aircraft Corporation.

Combining the ideas of Bézier curves and splines naturally leads to an efficient and numerically stable representation of B-splines and further on to non-uniform rational B-splines (NURBS), which is the de-facto standard of today's CAD systems. Detailed information about the theoretical and practical aspects of B-splines and NURBS can be found in Section 4.5 and [77, 180, 215, 630, 675].

4.6.1 Parametrization of Scattered 3D Data

The problem of parametrizing 3D data points is fundamental to many applications in CAGD, especially for the surface fitting problem. In general, a set of data points $P_i \in \mathbb{R}^3$ and a parameter domain $\Omega \subset \mathbb{R}^2$, over which the points are to be parametrized, are given. In addition, the topology of the point set has to be specified as different topologies lead to different approximation (or interpolation) problems (cf. Figure 4.36). This topological information is usually given in terms of a triangulation of the data points, i.e. there exists a list of triangles $T_j = \Delta(P_{j_0}, P_{j_1}, P_{j_2})$.

The task now is to find parameter values $p_i \in \Omega$, one for each data point P_i, so that the topology of the point set is being preserved, i.e. the triangles in the parameter domain $t_j = \Delta(p_{j_0}, p_{j_1}, p_{j_2})$ must not overlap. Note, that the parametrization is

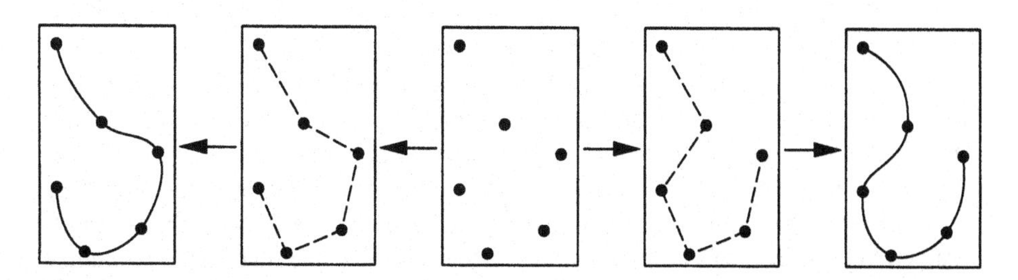

Figure 4.36. A point set (middle) with two different topologies (left and right).

implicitly given in the case of the *scalar problem*: for each point $P_i = (x_i, y_i, z_i)$, the first two components define the parameter value $p_i = (x_i, y_i)$.

While it is quite clear how to solve the *local* problem, i.e. parametrizing a set of points surrounding a reference point R, which can be done e.g. by an exponential mapping or by projection into an adequate tangent plane at R, the *global* problem is more complicated and has been discussed in several papers before.

Bennis et al. [52] propose a method that is based on differential geometry: they map isoparametric curves of the surface onto curves in the parameter domain such that the geodesic curvature at each point is preserved. The parametrization is then extended to both sides of that initial curve until some distortion threshold is reached. But this method as well as the one presented in [463] by Maillot et al. need the surface to be split into several independent regions and can therefore not be seen as solutions to the *global* problem.

Ma/Kruth [460] project the data points P_i onto a parametric base surface $S : \Omega \to \mathbb{R}^3$ and the parameter values of the projected points are taken as p_i. But since this method works only if the shape of the base surface is close to that of the triangulated data, it is not suitable for arbitrary data sets.

The approaches in [197, 226, 259, 329] have the following strategy in common:

1. find a parametrization for the *boundary points*, and

2. minimize an edge-based energy function

$$E = \frac{1}{2} \sum_{\{i,j\} \in \text{Edges}} c_{ij} \|p_i - p_j\|^2 \tag{4.49}$$

to determine the parametrization for the *inner points*.

The edge coefficients c_{ij} can be chosen in different ways. While Floater chooses them so that the geometric shape of the surface is preserved [226], Greiner/Hormann set $c_{ij} = \frac{1}{\|P_i - P_j\|^r}$ for some $r \geq 0$, as they want to minimize the energy of a network of springs [259, 329]. Both methods are generalizations of well-known results for the parametrization of curves, namely the *chord length* and *centripetal* parametrization [215, 228, 423].

A different method is introduced by Pinkall/Polthier in [547] and Eck et al. in [197]. They consider the discrete harmonic piecewise linear function between the surface triangles T_j and the corresponding parameter triangles t_j which leads to minimizing Equation 4.49 with $c_{ij} = \frac{1}{4}(\cot \alpha + \cot \beta)$, where α and β are the angles opposite to the edge $\overline{P_i P_j}$ in the two adjacent surface triangles.

In all cases, minimizing Equation 4.49 is equivalent to solving a non-singular sparse matrix system, that is (apart from Floater's method) even symmetric positive definite. Though this is a comparatively fast way to find a parametrization, it suffers from the fact that it is not clear how to choose the initial parametrization of the *boundary points*. Floater maps them to the boundary of the unit square using chord length parametrization, Greiner/Hormann project them into the plane that fits all *boundary points* best in the least square sense, and Eck et al. use parameter values lying on a circle.

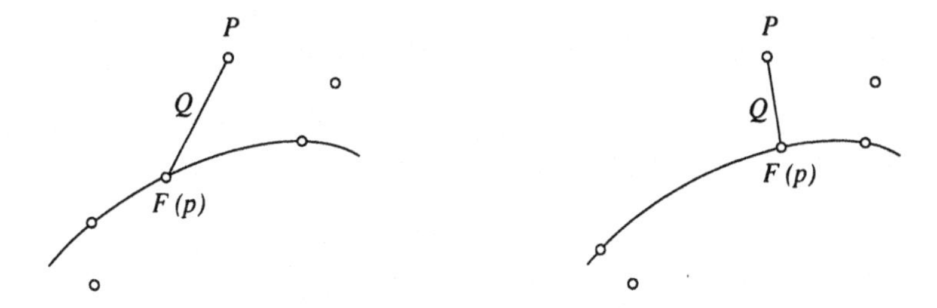

Figure 4.37. p should be corrected.

Figure 4.38. p is well chosen.

4.6.2 Parameter Correction

Once the parameter values are determined and a surface $F : \Omega \to \mathbb{R}^3$ approximating the data points is found, one can try to improve the parametrization, thereby exploiting the additional information given by the approximating surface itself [339, 596]. The standard method, introduced by Hoschek in [339], is derived from the observation that the error vector $Q = P - F(p)$ will not be orthogonal with respect to the approximating surface in general and will thus not represent the minimal distance between the surface and the data point (cf. Figures 4.37 and 4.38). If that happens, the parameter value p should be corrected.

This is done by linearly approximating the surface at $F(p)$ and projecting the data point P orthographically onto this tangent plane (L in Figure 4.39). The correction term $\Delta p = (\Delta u, \Delta v)$ can be determined by solving $L = F(p) + \Delta u F_u + \Delta v F_v$, which can be rewritten as

$$\Delta p = \begin{pmatrix} F_u^2 & F_u F_v \\ F_u F_v & F_v^2 \end{pmatrix}^{-1} \begin{pmatrix} \langle Q | F_u \rangle \\ \langle Q | F_v \rangle \end{pmatrix}.$$

The approximation process will then be repeated with the improved parameter values $\tilde{p} = p + \Delta p$. The underlying idea of this method is to split the non-linear

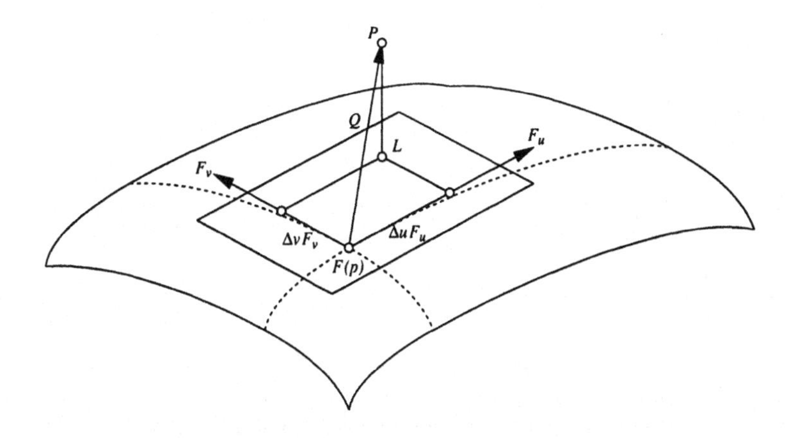

Figure 4.39. Parameter correction.

approximation problem, in which the surface parameters and the parameter values are unknowns, into two steps:

1. an *approximation step*, which finds the optimal surface parameters for given parameter values, and

2. a *parameter correction step*, which determines the optimal parameter values while the surface parameters are fixed.

Solving these problems alternately leads to good results after a few iterations only [329].

4.6.3 Interpolation and Approximation

The process of reconstructing smooth surfaces from discrete data offers two possibilities: interpolation and approximation. Since the solution of the interpolation problem coincides with the data points, one might assume it to be the more accurate reconstruction. But if the samples carry some noise (which is e.g. due to measurement errors) an approximating surface will be more adequate. Therefore, the choice of the appropriate method will depend on the specific structure of the problem (cf. Figure 4.40).

The interpolation problem can be stated as follows: given a set of n data points $P_i = (x_i, y_i, z_i) \in \mathbb{R}^3$ with corresponding parameter values $p_i = (u_i, v_i) \in \mathbb{R}^2$ and some class $\mathcal{S}$ of parametrized surfaces $F : \mathbb{R}^2 \to \mathbb{R}^3$, find $F \in \mathcal{S}$ such that $F(p_i) = (P_i)$ for all i. $\mathcal{S}$ may consist of bivariate polynomials, tensor product B-splines or NURBS, piecewise Bézier patches or triangular splines. If $\mathcal{S}$ is spanned by basis functions $B_1, \ldots, B_k$, the interpolation function $F = \sum_{j=1}^{k} c_j B_j$ satisfies $F(p_i) = \sum_j c_j B_j(p_i) = P_i$ for all i, hence the unknowns $c_1, \ldots, c_k$ can be found by solving the linear system

$$\boldsymbol{Bc} = \boldsymbol{p}$$

with

$$\boldsymbol{B} = \begin{pmatrix} B_1(p_1) & \cdots & B_k(p_1) \\ \vdots & \ddots & \vdots \\ B_1(p_n) & \cdots & B_k(p_n) \end{pmatrix}, \quad \boldsymbol{c} = \begin{pmatrix} c_1 \\ \vdots \\ c_k \end{pmatrix}, \quad \boldsymbol{p} = \begin{pmatrix} P_1 \\ \vdots \\ P_n \end{pmatrix}. \tag{4.50}$$

Note, that this problem will not be solvable in general if $k < n$ and that the existence of a unique solution requires $k = n$.

Figure 4.40. The configuration to the left is suitable for interpolation while approximation should be used for the noisy data points to the right.

In case of curve interpolation with polynomials, $\boldsymbol{B}$ will be a *Vandermonde-Matrix*, that is nonsingular for mutually different parameter values p_i. An interpolating B-spline curve can be found if and only if the *Schoenberg-Whitney* conditions are fulfilled [77, 180, 215]. These results can be extended to surface interpolation with polynomials or tensor product B-splines if the parameter values p_i are distributed on a grid. The problem then decouples into a sequence of curve interpolation problems [230, 460].

However, if the parameter values are not gridded, the problem becomes more complicated. E.g., a biquadratic polynomial ($k = 6$) that interpolates $n = 6$ given data points cannot be found if the corresponding 6 parameter values lie on an algebraic curve of degree 2 (e.g. circles, ellipses and hyperbolas), although $k = n$. Another problem is that the interpolating surfaces might have unwanted oscillations.

To overcome these difficulties, a variational approach can be used for solving the interpolation problem. The idea is to start with a class of surfaces having more degrees of freedom which are strictly necessary to fulfill the interpolation conditions and to use the remaining degrees of freedom to smooth the surface [257, 258]. This is achieved by minimizing a functional that somehow measures the smoothness (see Section 4.6.4).

If $\mathcal{J} : \mathcal{S} \to \mathbb{R}$ is such a fairness functional, the task is now to find $F \in \mathcal{S}$ such that

1. $F(p_i) = (P_i)$ for all i, and
2. $\mathcal{J}(F) \leq \mathcal{J}(G)$ for any $G \in \mathcal{S}$ satisfying 1.

Many reasonable measures of smoothness can be expressed by a quadratic and positive semidefinite functional (see Section 4.6.4), i.e. $\mathcal{J}(F) = \langle F|F\rangle_{\mathcal{J}}$ for a suitable positive semidefinite, symmetric, bilinear form $\langle\cdot|\cdot\rangle_{\mathcal{J}}$. By introducing *Lagrangian multipliers* λ_i, the smooth interpolation function $F = \sum_{i=1}^{k} c_i B_i$ can be found by solving the linear system

$$\begin{pmatrix} \boldsymbol{A} & \boldsymbol{B}^t \\ \boldsymbol{B} & 0 \end{pmatrix} \begin{pmatrix} \boldsymbol{c} \\ \lambda \end{pmatrix} = \begin{pmatrix} 0 \\ \boldsymbol{p} \end{pmatrix} \tag{4.51}$$

with

$$\boldsymbol{A} = \begin{pmatrix} \langle B_1|B_1\rangle_{\mathcal{J}} & \cdots & \langle B_1|B_k\rangle_{\mathcal{J}} \\ \vdots & \ddots & \vdots \\ \langle B_k|B_1\rangle_{\mathcal{J}} & \cdots & \langle B_k|B_k\rangle_{\mathcal{J}} \end{pmatrix}, \quad \lambda = \begin{pmatrix} \lambda_1 \\ \vdots \\ \lambda_n \end{pmatrix},$$

and $\boldsymbol{B}, \boldsymbol{c}, \boldsymbol{p}$ as defined in Equation 4.50.

However, interpolation is not always the best method for surface fitting. If the data points are not exact it is advisable to give up interpolation and better look for an approximating surface. The most common approach to this problem is the classical method of *least-squares*, whereby the approximation error

$$\mathcal{E}(F) = \sum_{i=1}^{n} (F(p_i) - P_i)^2 \tag{4.52}$$

is to be minimized. With the definitions from above this approximation error can be rewritten as

$$\mathcal{E}(F) = (\boldsymbol{Bc} - \boldsymbol{p})^t(\boldsymbol{Bc} - \boldsymbol{p}) = \boldsymbol{c}^t\boldsymbol{B}^t\boldsymbol{Bc} - 2\boldsymbol{c}^t\boldsymbol{B}^t\boldsymbol{p} + \boldsymbol{p}^t\boldsymbol{p}$$

and the minimum can be found by solving

$$B^t B c = B^t p. \tag{4.53}$$

Equation 4.53 is called the *normal equation*. Note, that the matrix $M = B^t B$ is symmetric, positive semidefinite. In case of approximation with B-spline curves, M is also sparse and has band structure. Furthermore, if the *Schoenberg-Whitney* conditions are fullfilled, the matrix is actually invertible and the normal equation can be solved by the method of Cholesky, QR decomposition or the conjugate gradient method [249]. However, in the bivariate case such simple conditions do not exist unless the given data is gridded and the uniqueness of the solution cannot be guaranteed.

Although straightforward least-squares fitting is often appropriate, the same problem as with interpolating surfaces may occur: it produces a surface that is not sufficiently smooth. In such cases it may be better to minimize a combination of the error functional $\mathcal{E}$ and the fairing functional $\mathcal{J}$:

$$\mathcal{K}_\omega(F) = \omega\mathcal{J}(F) + \mathcal{E}(F)$$

with a *smoothing parameter* $\omega \geq 0$ that controls the tradeoff between the smoothness and the approximation quality of the surface. Using the definitions from above, the minimum of this combined functional $\mathcal{K}_\omega$ can be found by solving

$$(\omega A + B^t B)c = B^t p. \tag{4.54}$$

Assuming $\mathcal{J}$ to be a symmetric positive definite smoothing functional, Equation 4.54 always has a unique solution, call it F_ω, for which some interesting properties hold [329]:

1. the function $\omega \to \mathcal{E}(F_\omega)$ is monotone increasing,
2. the function $\omega \to \mathcal{J}(F_\omega)$ is monotone decreasing,
3. $F_0 := \lim_{\omega \to 0} F_\omega$ exists,
4. F_0 minimizes $\mathcal{E}$ and is among all minima the one that additionally minimizes $\mathcal{J}$.

The first and second property support the statements on the effect of the smoothing parameter ω: the smaller ω is chosen, the smaller the approximation error is, while the smoothness of the surface increases by increasing ω. An immediate consequence of the last property is, that if there exist interpolating surfaces in $\mathcal{S}$, then F_0 is the interpolant with optimal fairness. Thus, it will be the solution to Equation 4.51, too. A detailed discussion of penalized least square methods can be found in [248].

4.6.4 Fairness Functionals

The construction of 'fair' or 'visually pleasing' surfaces is of vital importance in many areas of geometric modeling, especially in industrial design and styling [595]. While the human eye can easily rate the quality of a surface, the translation of this rating strategy to mathematical formulas is a crucial step. A lot of work has been dedicated to this problem and lots of different approaches how to measure the quality of a surface have been proposed by different authors.

In principal there are two approaches: one can either construct a functional by physical analogy (e.g. minimizing the energy of thin plates) or by geometric reasoning (e.g. minimizing area, curvature or the variation of curvature). But since every reasonable physical quantity has to be a generic property of the surface, it does not depend on the special parametrization that is used to describe the surface. This is, by definition, a geometric property, and therefore every physical quantity is a geometric one.

When selecting a fairness functional one has to take the following points into consideration.

1. Does the functional yield surfaces with a pleasant shape?

2. Can the functional be minimized in a reasonable amount of time?

It turns out that the functionals involving surface curvature yield surfaces of high quality, but an enormous amount of time is necessary to compute these solutions. The *thin plate energy functional*

$$\int a(\kappa_1^2 + \kappa_2^2) + 2(1 - b)\kappa_1\kappa_2 \, d\omega$$

that describes the energy of a thin plate [140] is of that type. Here, κ_1 and κ_2 are the principle curvatures of the surface and a and b are constants describing properties of the material of the thin plate (resistance to bending and sheering). In [272] a special case ($a = b = 1$) of this functional,

$$\int \kappa_1^2 + \kappa_2^2 \, d\omega, \tag{4.55}$$

is used to determine smooth tensor product B-spline surfaces. To overcome the difficulty that this functional is highly nonlinear and thus the numerical solution is very time consuming, the integrand of Equation 4.55 is evaluated only at the corners of the rectangular control grid. In [491] the functional

$$\int \left(\frac{\partial \kappa_1}{\partial \epsilon_1}\right)^2 + \left(\frac{\partial \kappa_2}{\partial \epsilon_2}\right)^2 d\omega$$

with ϵ_1 and ϵ_2 denoting the directions of principle curvature is considered. Although this MVC functional that minimizes the variation of the curvature has proven to yield surfaces of perfect shape, the complexity and hence the numerical treatment is even worse than for Equation 4.55.

In contrast to these highly nonlinar functionals there are simpler, quadratic functionals that can be minimized by solving a linear system and are thus suitable for interactive use. The most commonly used is the *simplified* version of the thin plate energy functional in Equation 4.55:

$$\int F_{uu}^2 + 2F_{uv}^2 + F_{vv}^2 \, dudv,$$

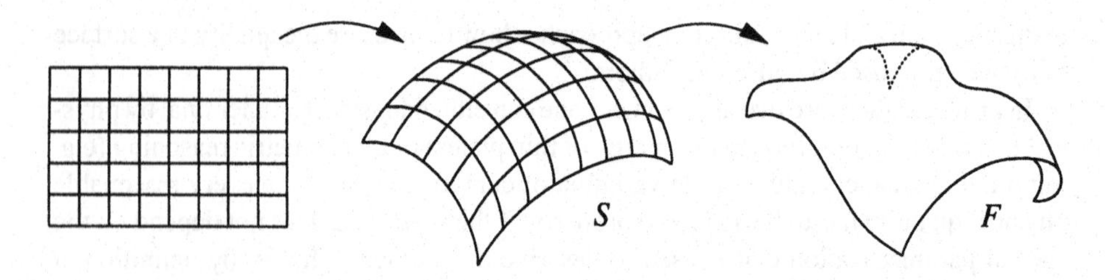

Figure 4.41. Parameter space, reference surface S and surface F.

where F_{uu}, F_{uv} and F_{vv} denote the second order partial derivatives of the surface F. As this functional is quadratic and much easier to minimize than the exact version, it is widely used [217, 260, 279, 541, 747]. However, it is a good approximation to Equation 4.55 only in the functional case, whereas the surfaces obtained in the parametric case may fail to have a pleasant shape. The more general quadratic, second order functional

$$\int \alpha_{11}F_u^2 + \alpha_{12}F_uF_v + \alpha_{22}F_v^2 + \beta_{11}F_{uu}^2 + \beta_{12}F_{uv}^2 + \beta_{22}F_{vv}^2 \, dudv$$

has been used in [110], where the coefficients α_{ij}, β_{ij} are chosen by physical reasoning, and in [743], where the choice is based on geometric properties.

In contrast to the functionals based on the surface curvature, these simplified versions depend on the parametrization of the surface, i.e. different parametrizations of the same surface will lead to different values of the functional. This statement also holds for energy functionals based on the third order derivatives, like

$$\int F_{uuu}^2 + F_{vvv}^2 \, dudv,$$

as proposed in [74] and

$$\int (F_{uuu} + F_{uvv})^2 + (F_{uuv} + F_{vvv})^2 \, dudv,$$

introduced in [258]. A good compromise between the simplicity of the quadratic functionals and the quality of the exact ones are data dependent functionals, introduced in [257, 258, 260].

The basic idea is the following: since the simple quadratic functionals approximate the exact energies well if the shape of the surface is nearly planar, i.e. being similar to the shape of the parameter space, one can take the inverse approach and adapt the parameter space in such a way that its shape is close to that of the surface (cf. Figure 4.41). The concepts needed to do so are well-known in differential geometry [107, 390]. Considering a *reference surface* $S : \mathbb{R}^2 \to \mathbb{R}^3$ as parameter space and *gradient* grad_S, *divergence* div_S and *Laplacian* Δ_S with respect to that reference surface, we can introduce the data dependent energy functionals

$$\begin{aligned} &\int \mathrm{grad}_S(F)^2\, d\omega_S \\ &\int \Delta_S(F)^2\, d\omega_S \\ &\int \mathrm{grad}_S(\mathrm{div}_S(\mathrm{grad}_S(F))^2\, d\omega_S \end{aligned} \tag{4.56}$$

with $d\omega_S$ denoting the surface element $||S_u \times S_v||\, dudv$. These functionals are still quadratic and can be minimized by solving a linear system.

If the reference surface S is close to F, the data dependent functionals will be good approximations to the exact energies. Note that the functionals in Equation 4.56 do not depend on the specific parametrization of the reference surface S.

4.6.5 Hierarchical B-Splines

A major problem that arises by using the surface fitting methods discussed in Section 4.6.3 is the size of the linear systems that have to be solved. One way to overcome this drawback is the use of hierarchical surface models like the *Hierarchical B-Splines* introduced by Forsey/Bartels in [229]. Although these surfaces were originally designed for the purpose of modeling, the ideas have successfully been transfered to the surface fitting problem [230, 259, 424].

The basic idea is to subdivide the *global* approximation problem adaptively into several *local* problems where only a comparatively small part of the data has to be taken into account. These local problems will lead to linear systems of small size which can be solved efficiently. The first step of this method is to start with a class of surfaces $\mathcal{S}_0$ defined on a coarse control lattice, thus having only a small number of control points, and determine the surface $F_0 \in \mathcal{S}_0$ that solves the problem

$$\min_{F \in \mathcal{S}_0} \mathcal{K}_\omega(F),$$

where $\mathcal{K}_\omega$ is the combined fairness/error functional defined in Section 4.6.3.

If the quality of F_0 suits our needs, i.e. the approximation error is within a specified tolerance and the surface is sufficiently 'fair', nothing has to be done. Otherwise we will either have to adapt the smoothing parameter ω or increase the number of degrees of freedom. Halving the grid size of the control lattice yields a new class of surfaces $\mathcal{S}_1 \supset \mathcal{S}_0$ whose dimension is roughly four times the dimension of $\mathcal{S}_0$. However, this *global refinement* step increases the size of the linear system that has to be solved by factor 16. The strategy of *local refinement* offers a better alternative: only at those regions where the approximation error exceeds the tolerance, local overlay patches $\mathcal{T}_k$ with finer control lattices are added to the surface (cf. Figure 4.42). Now, instead of solving the approximation problem in the globally refined space $\mathcal{S}_1$ we only look for the optimal surface contained in $\tilde{\mathcal{S}}_1 = \mathcal{S}_0 + \mathcal{T} \subset \mathcal{S}_1$ with $\mathcal{T} = \sum_k \mathcal{T}_k$.

If the overlay patches are chosen such that they are orthogonal, i.e. having disjunct support, this problem decomposes into several local problems. Indeed, if F_1 solves

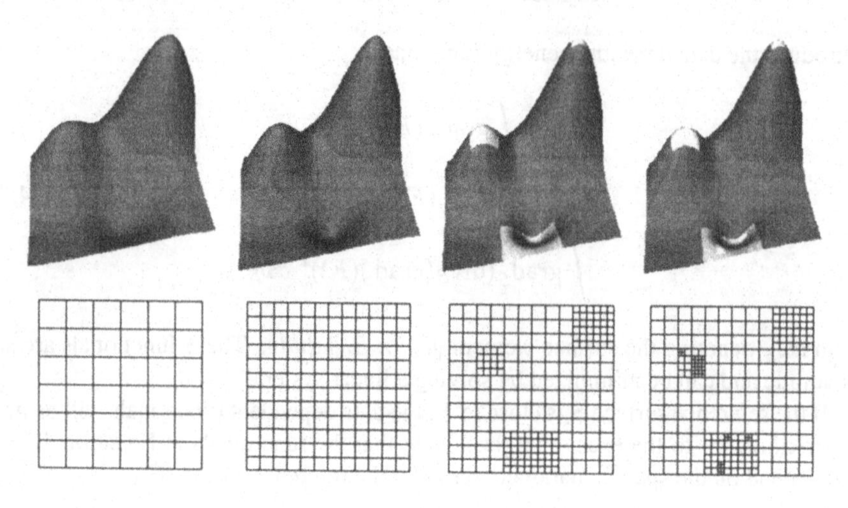

Figure 4.42. Approximating surface with one global and two local refinement steps.

the optimization problem

$$\min_{F \in \bar{\mathcal{S}}_1} \mathcal{K}_\omega(F) = \min_{G \in \mathcal{T}} \mathcal{K}_\omega(F_0 + G),$$

then F_1 can be written as $F_1 = F_0 + \sum_k G_k$ with overlay patches $G_k \in \mathcal{T}_k$ determined by solving the local problems

$$\min_{G \in \mathcal{T}_k} \mathcal{K}_\omega(F_0 + G),$$

which can be solved efficiently due to their small size. Obviously, this strategy can be applied recursively.

5 REALISTIC RENDERING

H.-P. SEIDEL

INTRODUCTION

Realistic, computer-generated images have found their way into a broad range of applications in everyday life. Highly sophisticated image synthesis techniques have become essential parts of TV and movie productions as well as video games. Architectural projects are visually validated under various environmental conditions using illumination simulation. Augmented reality techniques are used in transportation and medical science. Interactive virtual worlds are moving rapidly towards photorealism. These and many more examples prove the exciting development in concurrent rendering techniques. This chapter presents the most important foundations for realistic rendering, along with a number of recent approaches that deserve attention.

Photo-realistic images basically capture the light that has been reflected off a number of objects in a scene. Section 5.1 reviews the most important physical and geometrical aspects of light and reflection, dealing with the *local* interaction of light and matter at some object surface. Section 5.2 outlines the area of *global* illumination, describing the exchange of light between different scene objects, light sources, and the virtual camera. Another important question is how to add detail to the scene without rising the rendering cost too much. Section 5.3 presents various techniques to this end which are based on the principle of texture mapping.

While the traditional approaches to realistic rendering rely heavily on geometric surface models (see also Chapter 4), recently a lot of attention has been drawn towards *image based* rendering. Section 5.4 sketches some prominent examples of that genre, in which image data is used for generating novel views of captured objects.

5.1 MODELS FOR ILLUMINATION AND REFLECTION

W. HEIDRICH, P. EISERT, M. STAMMINGER, A. SCHEEL

Since much of the work in realistic rendering is based on physics, we will in the following review the basic principles of light transport and interaction of light and matter. A more detailed discussion of the underlying foundations can be found in [79] and [535], while the specific applications for image synthesis are discussed in [280] and [246].

5.1.1 Radiometry

Light is a form of electromagnetic radiation. As such, it can be interpreted both as an electromagnetic wave, and as a flux of particles, called *photons*. These different interpretations lead to wave optics and particle optics, respectively.

In the case of wave optics, the energy is carried by oscillating electrical and magnetic fields. The oscillation directions for the electrical and the magnetic field are perpendicular to each other and to the propagation direction of the light. Light that only consists of waves whose electrical fields (and thus all magnetic fields) are aligned, is called *linearly polarized*, or simply *polarized*.

In particle optics, the energy is carried in the form of photons moving at the speed of light. Each photon has a certain amount of energy, depending on its frequency.

Wave and particle optics are largely complementary in that one can explain physical phenomena the other cannot explain easily. In computer graphics, it is possible to abstract from both wave and particle optics, and describe phenomena purely based on geometrical considerations (*geometrical* or *ray optics*) most of the time. In order to explain the laws of geometrical optics, however, results from wave and particle optics are necessary. For a derivation of these laws and a detailed discussion of optics in general [79, 535].

Radiometry is the science that deals with the measurement of light and other electromagnetic radiation. The most important quantities in this field are also required for understanding the principles of digital image synthesis.

Radiant Energy is the energy transported by light. It is denoted by Q, and measured in *Joules* $[\mathrm{J} = \mathrm{Ws} = \mathrm{kg\,m^2/s^2}]$. Radiant energy is a function of the number of photons and their frequencies.

Radiant Flux or **Radiant Power,** is the power (energy per time) of the radiation and denoted as Φ. It is measured in *Watts* [W].

Irradiance and **Radiant Exitance** are two forms of *flux density*. The irradiance $E = d\Phi/dA$ represents the flux $d\Phi$ per unit area dA *arriving* at a surface, while the radiant exitance B, which is often also called *radiosity* in computer graphics, describes the flux per unit area *leaving* a surface. Both quantities are measured in $[\mathrm{W/m^2}]$.

In wave optics, flux density is defined as the product of the electrical and the magnetic fields [535], and is therefore proportional to the product of their amplitudes. Since the two fields induce each other, their amplitudes are linearly dependent. As a consequence, flux density is proportional to the square of either amplitude.

Radiance is the flux per projected unit area and solid angle arriving at or leaving a point on a surface: $L(x, \omega) = d^2\Phi/(\cos\theta \, d\omega \, dA)$, where θ is the angle between the direction ω and the surface normal. Thus, radiance is measured in [W/m^2 sr], where sr stands for *steradian*, the unit for solid angles.

The relationship between irradiance and incoming radiance is

$$E(x) = \int_{\Omega(n)} L_i(x, \omega') \cos\theta' \, d\omega',$$

where $\Omega(n)$ represents the hemisphere of directions around the surface normal n. L_i, the *incoming radiance*, is the radiance arriving at the surface point x from direction ω'. A similar equation holds for the relationship between radiosity and the radiance L_o leaving the surface (*outgoing radiance*).

Radiance is a particularly important quantity in computer graphics, since it is constant along a ray in empty space. Thus it is the quantity implicitly used by almost all rendering systems including ray-tracers and interactive graphics systems.

Intensity. *Point light sources*, which assume that all the radiant energy is emitted from a single point in 3D-space, are a common model for light sources in computer graphics. Unfortunately, radiance is not an appropriate quantity to specify the brightness of such a light source, since it has a singularity at the position of the point light.

The *intensity* I is a quantity that does not have this singularity, and can therefore be used for characterizing point lights. Intensity is defined as flux per solid angle ($I = d\Phi/d\omega$). Thus, an isotropic point light (a light source that emits the same amount of light in each direction) has intensity $I = \Phi/4\pi$sr.

All of the above quantities can, and, in general will, additionally vary with the wavelength of light. For example, the *spectral radiant energy* Q_λ is given as $dQ/d\lambda$, and its units are consequently $[J/m]$, while *spectral radiance* $L_\lambda := dL/d\lambda$ is measured in [W/m^3 sr].

Although wavelength dependency can cause impressive visual effects, most rendering systems do not deal with spectral quantities due to the high computational and storage costs this would impose. Therefore the discussion in this chapter will largely ignore spectral dependency.

5.1.2 Bidirectional Reflection Distribution Functions

In order to compute the illumination in a scene, it is necessary to specify the optical properties of a material. This is usually done in the form of a *bidirectional reflection distribution function* (BRDF). It is defined as follows:

$$f_r(x, \omega' \to \omega) := \frac{L_o(x, \omega)}{L_i(x, \omega') \cos\theta' \, d\omega'} = \frac{dL_o(x, \omega)}{dE(x, \omega')}. \tag{5.1}$$

The BRDF is the radiance L_o leaving a point x in direction ω divided by the irradiance arriving from direction ω'. Its unit is [1/sr]. The BRDF describes the *reflection* of light at a surface. Similarly, the *bidirectional transmission distribution function* (BTDF) can be defined for refraction and transmission. The combination of BRDF and BTDF is usually called *bidirectional scattering distribution function* (BSDF).

Although BSDFs are used very often in computer graphics, it is important to note that they cannot model all physical effects of light interacting with surfaces. The simplifying assumptions of BRDFs (and BTDFs) are:

- the reflected light has the same frequency as the incoming light. *Fluorescence* is not handled.
- light is reflected *instantaneously*. The energy is not stored and re-emitted later (*phosphorescence*).
- there are no participating media. That is, light travels in empty space, and if it hits a surface, it is reflected at the same point without being scattered within the object. This is the most restrictive assumption, since it means that atmospheric effects as well as certain materials such as skin cannot be treated adequately.

In general, the BRDF is a six-dimensional function, because it depends on two surface parameters (x) and two directions with two degrees of freedom each (ω' and ω). Often however, it is assumed that a surface has no *texture*, that is, the BRDF is constant across an object, thus reducing the dimensionality to four.

The dimensionality can be further reduced by one through the assumption of an *isotropic* material. These are materials which are invariant under a rotation around the normal vector. Let $\omega' = (\theta', \phi')$ and $\omega = (\theta, \phi)$, where θ and θ' describe the angle between the normal and the respective ray, and ϕ and ϕ' describe the rotation around the normal. Then the following equation holds for isotropic materials:

$$f_r(x, (\theta', \phi' + \Delta\phi) \to (\theta, \phi + \Delta\phi)) = f_r(x, (\theta', \phi') \to (\theta, \phi)).$$

All other materials are called *anisotropic*.

Reflectance and Transmittance. While the BRDF is an accurate and useful description of surface properties, it is sometimes inconvenient to use because it may have singularities. For example, consider the BRDF of a perfect mirror. According to Equation 5.1, $f_r(x, \omega' \to \omega)$ is infinity if ω is exactly the reflection of ω', and zero otherwise.

Another quantity to describe the reflection properties of materials is the reflectance ρ. It is defined as the ratio of reflected flux to incoming flux:

$$\rho := \frac{d\Phi_o}{d\Phi_i}. \tag{5.2}$$

From this definition it is obvious that ρ is unitless and bounded between 0 and 1. Unfortunately, the reflectance in general depends on the directional distribution of the incoming light, so that a conversion between BRDF and reflectance is not easily possible. In the important special case of a purely diffuse (Lambertian) reflection, however, f_r is a constant, and $\rho = \pi \mathrm{sr} \cdot f_r$.

In analogy to the reflectance, the *transmittance* τ can be defined as the ratio of transmitted to received flux. The fraction of flux that is neither reflected nor transmitted, but absorbed, is called *absorptance* α. The sum of reflectance, transmittance, and absorptance is always one.

Physical Reflection and Transmission Properties of Materials. Independent of the actual dimensionality of the BRDF, it has to obey certain physical laws. First of all, this is the conservation of energy. No more energy must be reflected than is received by the surface. This is guaranteed if the following equation holds [436]:

$$\int_{\Omega(\boldsymbol{n})} f_r(\boldsymbol{x}, \omega' \to \omega) \cos\theta \, d\omega \leq 1 \quad \forall \omega' \in \Omega(\boldsymbol{n}). \tag{5.3}$$

The second physical law that a BRDF should obey is known as *Helmholtz reciprocity*. It states that, if a photon follows a certain path, another photon can follow the same path in the opposite direction. In the case of reflections, this means that

$$f_r(\boldsymbol{x}, \omega \to \omega') = f_r(\boldsymbol{x}, \omega' \to \omega). \tag{5.4}$$

For refraction this relation does not hold, since the differential diameter of a refracted ray is different from that of the original ray.

In addition to being physically valid, models for surface materials should also be *plausible* in the sense that they model the reflection characteristics of real surfaces. The basic principles underlying plausible BRDFs are the reflection at a planar surface, Snell's law, and the Fresnel formulae. These principles will be reviewed in the following.

Consider a ray of light arriving from direction $\boldsymbol{l}$ at a perfectly smooth, planar surface between two materials with optical densities n_1 and n_2. This situation is depicted in Figure 5.1. Since the surface is perfectly smooth, this ray will be split into exactly two new rays, one for the reflected, and one for the refracted part. The reflected ray is given via the relation

$$\theta_r = \theta_l, \tag{5.5}$$

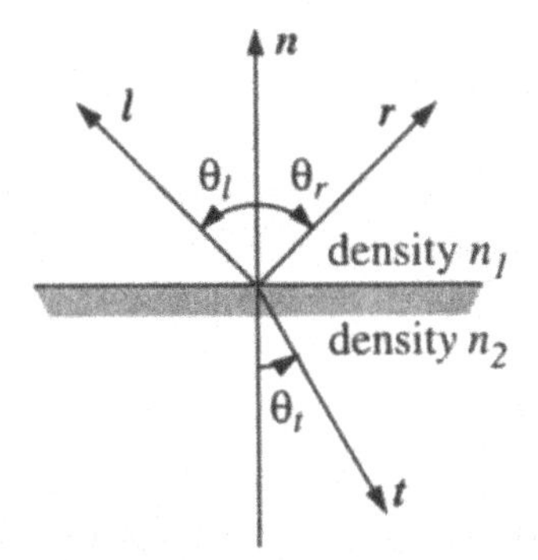

Figure 5.1. Reflection and refraction on a planar surface.

while the refracted ray direction is given by Snell's law [79]:

$$n_1 \sin \theta_l = n_2 \sin \theta_t. \tag{5.6}$$

But how is the energy of the incoming ray split between the reflected and the refracted part, i.e. what are the values for the reflectance and the transmittance? This depends on the polarization of the light. Let $r^\perp$ be the ratio of the reflected to the incoming amplitude of the electrical field perpendicular to the plane formed by the surface normal and the incoming light ray. Let $r^\parallel$ be the same ratio for an electrical field parallel to this plane, and let $t^\perp$ and $t^\parallel$ be the corresponding ratios for the transmitted amplitudes.

The Fresnel formulae specify these ratios in terms of the angles θ_l and θ_t for non-magnetic materials (permeability ≈ 1) without absorption ($\alpha = 0$):

$$r^\perp = \frac{n_1 \cos\theta_l - n_2 \cos\theta_t}{n_1 \cos\theta_l + n_2 \cos\theta_t}, \quad r^\parallel = \frac{n_2 \cos\theta_l - n_1 \cos\theta_t}{n_2 \cos\theta_l + n_1 \cos\theta_t} \tag{5.7}$$

$$t^\perp = \frac{2n_1 \cos\theta_l}{n_1 \cos\theta_l + n_2 \cos\theta_t}, \quad t^\parallel = \frac{2n_1 \cos\theta_l}{n_2 \cos\theta_l + n_1 \cos\theta_t}. \tag{5.8}$$

Since $\rho = d\Phi_o/d\Phi_i = dB/dE$, and because the flux density is proportional to the square of the amplitude of the electrical field (see Section 5.1.1), we need to square these ratios in order to get the reflectance and transmittance. For unpolarized light, which has random orientations of the electrical field, the reflectance for the perpendicular and the parallel components need to be averaged (see [79]):

$$\begin{aligned} \rho &= \frac{(r^\perp)^2 + (r^\parallel)^2}{2}, \\ \tau &= \frac{n_2 \cos\theta_l}{n_1 \cos\theta_t} \cdot \frac{(t^\perp)^2 + (t^\parallel)^2}{2}. \end{aligned} \tag{5.9}$$

Note that $\rho + \tau = 1$ due to the assumption of a non-absorbing material. Figure 5.2 shows the reflectance for unpolarized light for the example of a surface between air ($n_1 \approx 1$) and glass ($n_2 \approx 1.5$).

As mentioned above, these formulae describe the reflection at a perfectly smooth, planar surface. In reality, however, surfaces are rough, and thus light is reflected and refracted in all directions, not only $\mathbf{r}$ and $\mathbf{t}$. In this case, the surface is assumed to consist of small planar regions, or *facets*, for which the above formulae can be applied. The reflection on such a rough surface then depends on the statistics of the orientations for these facets. This approach to describe the reflection on rough surfaces is called *micro facet theory*. In Section 5.1.3 we review some of the related models that have gained importance in the context of computer graphics.

5.1.3 Reflection Models

While it is in principle possible to use measured BRDF data for image synthesis, often this is not feasible in practice due to the large amount of data required to faithfully represent this high-dimensional function. *Reflection models* or *lighting models* attempt to describe classes of BRDFs in terms of simple formulae using only a few parameters

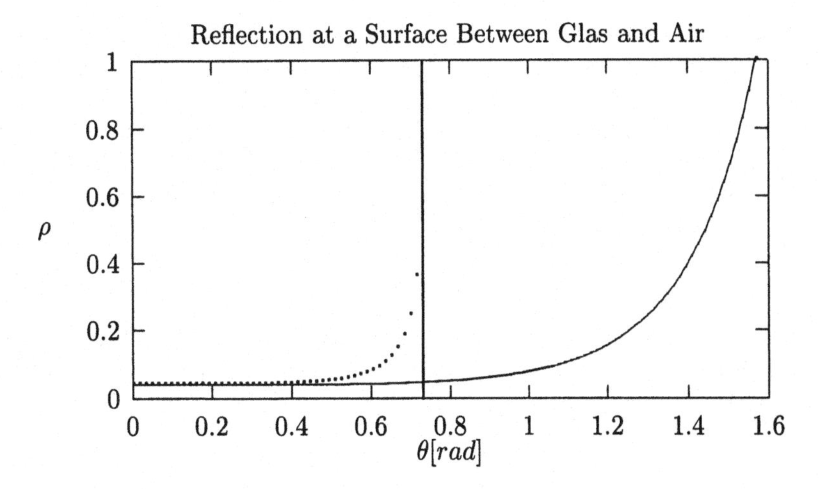

Figure 5.2. Reflectance ρ according to Fresnel for the interaction of unpolarized light with a surface between glass ($n = 1.5$) and air ($n \approx 1$). The solid curve is for light rays arriving from the air, while the dotted curve is for rays arriving from the glass side. Note the total reflection at an angle $\sin\theta = 1/1.5$ in the latter case.

to customize the model. These parameters can either be adjusted manually by a user or programmer, or they can be chosen automatically to best fit measured BRDF data.

The choice of a specific reflection model is a trade-off between performance and physical accuracy. Since the lighting model is numerically evaluated very often during the synthesis of an image, its computational cost has a strong impact on the total rendering time. Nonetheless, for realistic lighting effects, a model should follow basic physical principles, as laid out in Section 5.1.2.

For the following discussion we use the vectors depicted in Figure 5.3. $\boldsymbol{n}$ is the surface normal, $\boldsymbol{v}$ the viewing direction, $\boldsymbol{l}$ the light direction, $\boldsymbol{h} := (\boldsymbol{v}+\boldsymbol{l})/|\boldsymbol{v}+\boldsymbol{l}|$ the halfway vector between viewing and light direction, and $\boldsymbol{r_l} := 2 < \boldsymbol{l}, \boldsymbol{n} > \boldsymbol{n} - \boldsymbol{l}$ the reflection of the light vector on the surface normal. $\boldsymbol{t}$ is the tangent direction for any microscopic features that may cause anisotropy. All these vectors are assumed to be unit length.

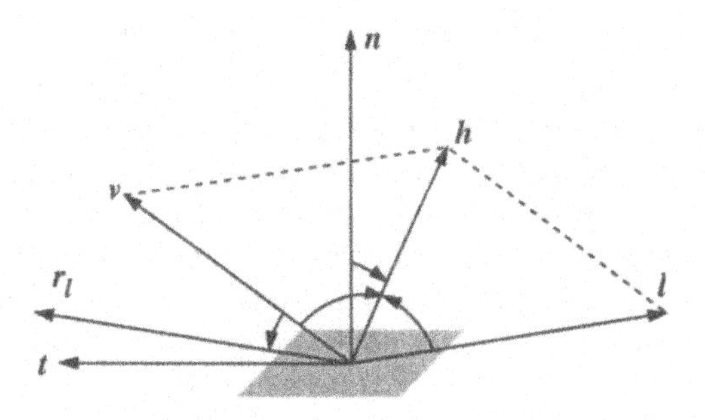

Figure 5.3. Vectors used for describing the reflection models in this section.

Ambient and Diffuse Reflection. The most simplistic reflection model is caused by *ambient illumination*. It assumes that light is arriving uniformly from all directions. The ambient model is not physically valid by any means, but it can be used to approximate the indirect illumination in a scene. Due to the uniform distribution of the incoming light, the ambient light reflected off a surface is simply a constant k_a times the ambient illumination, which is described by L_a, the radiance in the scene averaged over all points and directions:

$$L_o(\boldsymbol{x}, \boldsymbol{v}) = k_a \cdot L_a. \tag{5.10}$$

Many materials, especially in architectural scenes, appear roughly equally bright from all viewing directions. This is known as *diffuse* or *Lambertian* reflection. The BRDF of a Lambertian surface is a constant

$$f_r(\boldsymbol{x}, \boldsymbol{l} \to \boldsymbol{v}) = k_d. \tag{5.11}$$

Note that, according to Equation 5.3, the reflection coefficient k_d should be less than $1/\pi$ for reasons of energy conservation.

Models by Phong and Blinn-Phong. The Phong lighting model [98] was one of the first models in computer graphics to account for specular reflections. For efficient computations, but without a physical justification, the model uses powers of the cosine of the angle between reflected light vector and viewing vector for the scattering of light on the surface:

$$L_o(\boldsymbol{x}, \boldsymbol{v}) = k_s \cdot \cos(\boldsymbol{r_l}, \boldsymbol{v})^n \cdot L_i = k_s \cdot < \boldsymbol{r_l}, \boldsymbol{v} >^n \cdot L_i. \tag{5.12}$$

In this equation and in the following, all cosine values are implicitly clamped to the range [0..1]. This is not shown explicitly in order to simplify the formulae. The corresponding BRDF is

$$f_r(\boldsymbol{x}, \boldsymbol{l} \to \boldsymbol{v}) = k_s \frac{\cos(\boldsymbol{r_l}, \boldsymbol{v})^n}{\cos(\boldsymbol{l}, \boldsymbol{n})} = k_s \frac{< \boldsymbol{r_l}, \boldsymbol{v} >^n}{< \boldsymbol{l}, \boldsymbol{n} >}. \tag{5.13}$$

As pointed out by Lewis [436], this BRDF does not conserve energy due to the denominator, which may become arbitrarily small. The Helmholtz reciprocity is also violated, since the equation is not symmetric in $\boldsymbol{l}$ and $\boldsymbol{v}$. [436] therefore proposed to use the term $k_s < \boldsymbol{r_l}, \boldsymbol{v} >^n$ directly as the BRDF, which fixes both problems if an additional scaling factor of $(n+2)/2\pi$ is introduced for energy conservation (Equation 5.3). This factor can either be merged with the reflection coefficient k_s, or, more conveniently, be added as a separate term, which then allows for $k_s \in [0..1]$. This modified Phong model has come to be known as the *Cosine Lobe Model.*

A slightly modified version of the Phong model, known as the Blinn-Phong model has been discussed in [73]. Instead of the powered cosine between reflected light and viewing direction, it uses powers of the cosine between the halfway vector $\boldsymbol{h}$ and the surface normal:

$$f_r(\boldsymbol{x}, \boldsymbol{l} \to \boldsymbol{v}) = k_s \frac{< \boldsymbol{h}, \boldsymbol{n} >^n}{< \boldsymbol{l}, \boldsymbol{n} >}. \tag{5.14}$$

This models a surface consisting of many small, randomly distributed micro facets that are perfect mirrors. The halfway vector is the normal of the facets contributing to the reflection for given viewing and light directions. The cosine power is a simple distribution function, describing how likely a certain micro facet orientation is.

Like the Phong model, the Blinn-Phong model itself is not physically valid, but with similar modifications as described above, Helmholtz reciprocity and energy conservation can be guaranteed.

Both the Phong and the Blinn-Phong model are very popular in interactive computer graphics due to their low computational cost and simplicity. They are usually combined with a diffuse and an ambient term as described before in this section.

Generalized Cosine Lobe Model. As the name suggests, the generalized cosine lobe model [414] is a generalization of Lewis' modifications to the Phong model. The term $k_s < \boldsymbol{r}_l, \boldsymbol{v} >^n$ can be written in matrix notation as $k_s[\boldsymbol{l}^T 2 \cdot (\boldsymbol{n} \cdot \boldsymbol{n}^T - \boldsymbol{I}) \cdot \boldsymbol{v}]^n$. The generalized cosine lobe model now allows for an arbitrary symmetric matrix $\boldsymbol{M}$ to be used instead of the Householder Matrix $2\boldsymbol{n} \cdot \boldsymbol{n}^T - \boldsymbol{I}$:

$$f_r(\boldsymbol{x}, \boldsymbol{l} \to \boldsymbol{v}) = k_s[\boldsymbol{l}^T \cdot \boldsymbol{M} \cdot \boldsymbol{v}]^n. \tag{5.15}$$

Let $\boldsymbol{Q}^T \boldsymbol{D} \boldsymbol{Q}$ be the singular value decomposition of M. Then, $\boldsymbol{Q}$ can be interpreted as a new coordinate system, into which the vectors $\boldsymbol{l}$ and $\boldsymbol{v}$ are transformed. In this new coordinate system, Equation 5.15 reduces to a weighted dot product:

$$f_r(\boldsymbol{x}, \boldsymbol{l} \to \boldsymbol{v}) = k_s(D_x r_{l,x} v_x + D_y r_{l,y} v_y + D_z r_{l,z} v_z)^n. \tag{5.16}$$

The advantage of this model is that it is well suited to fit real BRDF data obtained through measurements [414]. In contrast to the original cosine lobe model, the generalized form is also capable of describing anisotropic BRDFs by choosing $D_x \neq D_y$.

Torrance-Sparrow Model. The illumination model by Torrance and Sparrow [702, 701] is one of the most important physically-based model for the interreflection of light at rough surfaces (the variation by Cook and Torrance [139] is also quite widespread for spectral renderings). It is given as

$$f_r(\boldsymbol{x}, \boldsymbol{l} \to \boldsymbol{v}) = \frac{F \cdot G \cdot D}{\pi \cdot < \boldsymbol{n}, \boldsymbol{l} > \cdot < \boldsymbol{n}, \boldsymbol{v} >}, \tag{5.17}$$

where F, the Fresnel term, is the reflectance ρ from Equation 5.9. It is usually given in the form

$$F = \frac{(g-c)^2}{2(g+c)^2}\left[1 + \frac{(c(g+c)-1)^2}{(c(g-c)+1)^2}\right], \tag{5.18}$$

with $c =< \boldsymbol{h}, \boldsymbol{v} >$ and $g^2 = n^2 + c^2 - 1$. The equivalence to Equation 5.9 can be shown using trigonometric identities.

The term D in Equation 5.17 is the distribution function for the micro-facets. Multiple choices have been proposed for this term by several authors, including [48].

The function given in [701] assumes a Gaussian distribution of the angle between normal and halfway vector:

$$D = \exp(\ (k \cdot \sphericalangle(\boldsymbol{n}, \boldsymbol{h}))^2\ ), \tag{5.19}$$

where k is the standard deviation of the surface angles, and is a surface property. Many other researchers have since used a Gaussian distribution of the *surface heights*, which also results in a Gaussian distribution of *surface slopes* [654, 307]. Let σ be the RMS (root mean square) deviation of the surface height. Then the RMS deviation of surface slopes is proportional to σ/τ, where τ, a parameter of the model, is a measure for the distance of two surface peaks [307].

Finally, the term G describes the geometrical attenuation caused by the self-shadowing and masking of the micro-facets. Under the assumption of symmetric, v-shaped groves, it is given as

$$G = \min\left\{1, \frac{2 < \boldsymbol{n}, \boldsymbol{h} >< \boldsymbol{n}, \boldsymbol{v} >}{< \boldsymbol{h}, \boldsymbol{v} >}, \frac{2 < \boldsymbol{n}, \boldsymbol{h} >< \boldsymbol{n}, \boldsymbol{l} >}{< \boldsymbol{h}, \boldsymbol{v} >}\right\}. \tag{5.20}$$

This model for shadowing and masking has later been improved in [654] for the specific case of a Gaussian height distribution function D (erfc denotes the error function complement):

$$\begin{aligned} G &= S(< \boldsymbol{n}, \boldsymbol{v} >) \cdot S(< \boldsymbol{n}, \boldsymbol{l} >) \\ \text{where} \quad S(x) &= \frac{1 - \frac{1}{2}\mathrm{erfc}(\frac{\tau \cot x}{2\sigma})}{\Lambda(\cot x) + 1}, \\ \text{and} \quad \Lambda(\cot x) &= \frac{1}{2}\left(\frac{2}{\sqrt{\pi}} \cdot \frac{\sigma}{\tau \cot x} - \mathrm{erfc}(\frac{\tau \cot x}{2\sigma})\right). \end{aligned} \tag{5.21}$$

Several variations of this model have been proposed, which will not be discussed in detail here. In particular, [613] has proposed a model in which all terms are approximated by much simpler formulae. Moreover, it is possible to account for anisotropic reflections by changing the micro facet distribution function D [368, 557, 613]. An overview of the variations to the Torrance-Sparrow model can be found in [277].

[307] introduced an even more comprehensive model, which is also capable of simulating polarization effects. This model is extremely complex, and a description would go beyond the scope of this section.

Anisotropic Model by Banks. A very simple anisotropic model has been presented by Banks [38]. The model assumes that the anisotropy is caused by thin, long features, such as scratches or fibers. Seen from relatively far away, these features cannot be resolved individually, but their orientation changes the directional distribution of the reflected light.

This suggests that anisotropic surfaces following this model can be illuminated using a model for the illumination of lines in 3D. The fundamental difficulty in the illumination of 1D manifolds in 3D is that every point on a curve has an infinite number

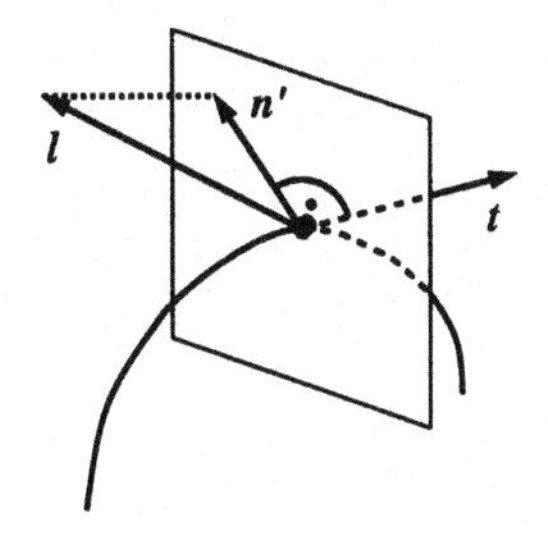

Figure 5.4. In order to find the shading normal n', the light vector l is projected into the normal plane.

of normal vectors. Thus, every vector that is perpendicular to the tangent vector of the curve is a potential candidate for use in the illumination calculation.

For reasons described in [38], and [663], the vector n' selected from this multitude of potential normal vectors is the projection of the light vector l into the normal plane, as depicted in Figure 5.4.

Applied to anisotropic materials, this means that the projection of the light vector l into the normal plane of the tangent vector t is used as the shading normal n'. At this point, any of the isotropic reflection models described above can be used, but usually the Phong model is chosen.

5.2 GLOBAL ILLUMINATION

A. SCHEEL, K. DAUBERT, M. STAMMINGER

So far in this chapter, we have mainly been concerned with the *local* interaction of light and matter at a specific point on an object's surface. However, in order to simulate the complete illumination in virtual environments, it is necessary to take into account the *global* exchange of light energy between the different virtual light sources, objects, and the virtual camera.

A virtual scene or environment is defined by a set of surfaces and their light reflection and emission properties. Such descriptions are created for instance using a CAD (computer aided design) modeler. Since these models are purely virtual, the lighting within the scene has to be simulated in order to yield realistic renderings of the virtual world. Here lies the difference to image-based rendering (treated in Section 5.4), which generally does not need a geometric description of the objects it renders, but instead requires several reference views (i.e., photographs) of the scene.

The typical application for lighting computations of geometric models, as treated in this section is design. Imagine an interior designer, who wants to present his ideas to his clients. This can best be achieved by images of his models, with realistic lighting including shadows, highlights, and so on. Different simulations could be computed to demonstrate how the room looks, lit by daylight through the windows, or by artificial light from lamps.

For a realistic lighting simulation it is not sufficient to compute the direct illumination due to the light sources only, because indirect illumination by reflective objects also significantly contributes to the illumination in the rest of the scene. The simulation of this global illumination is a complex task and in the last twenty years a lot of research has been committed to finding efficient algorithms for its computation.

In 1986 Kajiya [369] published the *Rendering Equation* 5.22, which describes the equilibrium of light exchange in closed environments and serves as a basis for all global illumination algorithms

$$L(x,\omega) = L^e(x,\omega) + \int_\Omega L(h(x,\omega'),-\omega') \cdot f_r(\omega',x,\omega)\cos\theta' d\omega'. \quad (5.22)$$

The term $L(x,\omega)$ is the radiance of surface point x into direction ω. This equation states how the radiance at a point x in direction ω can be expressed by the self-emittance at x in direction ω (defining light sources), plus the light arriving at x from any direction ω' and reflected into direction ω. The reflected light is obtained by integrating the incident light, weighted by the BRDF f_r (see Section 2.3). The incident light from direction ω', in turn, is described by shooting a ray (x,ω') and evaluating the exitant radiance of the hit point $h(x,\omega')$.

After an approximation for the radiance function L has been computed, arbitrary views of the scene can be generated. In order to do this, the perceived brightness of some surface point x is computed by evaluating L at x into the observer's direction.

There are also algorithms that use the adjoint equation to the rendering equation, the so called potential equation, which is not based on radiance but on the *potential* $W(x,\omega')$:

$$\begin{aligned} W(x,\omega') &= W^e(x,\omega') \\ &+ \int_\Omega W(h(x,\omega'),-\omega) \cdot f_r(\omega',h(x,\omega'),\omega)\cos\theta d\omega. \end{aligned} \quad (5.23)$$

The direct potential $W^e(x,\omega')$ describes, how the radiance of surface point x in direction ω' directly affects the light intensity measured by some light measuring device, for instance a single pixel of a CCD in a video camera. For a pinhole camera, W^e is only non-zero, if the ray (x,ω') hits the current film pixel directly through the pinhole.

The potential $W(x,\omega')$ is W^e plus the influence via reflection, that is the influence of light emitted at x that does not directly hit the measuring device, but is first reflected by other scene objects and then contributes to the measured intensity indirectly. Informally, this potential propagates like radiance, with the difference however, that the direction is exactly the opposite, namely from the camera into the scene. Therefore, the camera can be viewed as a potential source.

$W(x,\omega')$ describes the direct or indirect influence of light emission at x in direction ω' on the measuring device. So if we know the potential distribution in the scene for a particular pixel, integrating the light source distribution L^e weighted by the potential directly gives us the measured intensity.

Both the rendering and the potential equation are integral equations. In order to solve them, they are expanded into an infinite Neumann series, as we will demonstrate in the following paragraph.

For the expansion it is more convenient to use another formulation of the rendering equation which involves the definition of the integral operator $\mathcal{T}$ describing light-surface interactions:

$$(\mathcal{T}L)(x,\omega) = \int_{\Omega} L(h(x,\omega'), -\omega') \cdot f_r(\omega', x, \omega) \cos\theta' d\omega'. \tag{5.24}$$

Now the rendering equation is reformulated as:

$$L = L^e + \mathcal{T}L. \tag{5.25}$$

The expansion step is accomplished by substituting the L on the right side by $L^e + \mathcal{T}L$:

$$L = L^e + \mathcal{T}L = L^e + \mathcal{T}(L^e + \mathcal{T}L) = L^e + \mathcal{T}L^e + \mathcal{T}^2L = \ldots \tag{5.26}$$

After having done that n times we obtain:

$$L = \sum_{i=0}^{n} \mathcal{T}^i L^e + \mathcal{T}^{n+1} L = \sum_{i=0}^{\infty} \mathcal{T}^i L^e, \tag{5.27}$$

because $\mathcal{T}$ is a contraction and therefore $\lim_{n\to\infty} \mathcal{T}^{n+1} L = 0$.

The terms of the infinite Neumann series have an intuitive meaning: $\mathcal{T}^i L^e$ is the light reflected exactly i times, the entire sum is the light after an arbitrary number of reflections. The expansion of the potential equation is carried out analogously.

Two main methods to solve the expanded integral equation have been established: *finite element* methods with the most prominent representative being the radiosity algorithm, and *Monte Carlo* methods including for example the ray tracing technique. These two approaches will be described in more detail in the following two sections. The last section then briefly sketches two hybrid approaches.

5.2.1 Finite Element Methods

Radiosity. The first finite element method for global illumination treated a simpler case of reflectance behavior: all surfaces are considered to reflect light diffusely, that is uniformly in all directions, independent of the viewing direction.

In this case, instead of radiance the quantity radiosity is used, which is defined as the total power leaving a surface per unit area. As it only depends on the position on the surface, but not on the outgoing direction, it is only two-dimensional.

The basic radiosity method was developed in the 1950's to simulate the radiative heat transfer between surfaces. In 1984 Goral et al. [253] applied this principle to the global illumination problem for diffuse surfaces.

The surfaces are subdivided into a set of elements of finite size (patches); the exitant light of each patch is described by its average radiosity value B_i. Radiosity methods simulate the exchange of light between all patches. By discretization the Rendering Equation 5.22 becomes a linear equation in the radiosity values B_i:

$$B_i = E_i + \rho_i \sum_j F_{ij} B_j. \tag{5.28}$$

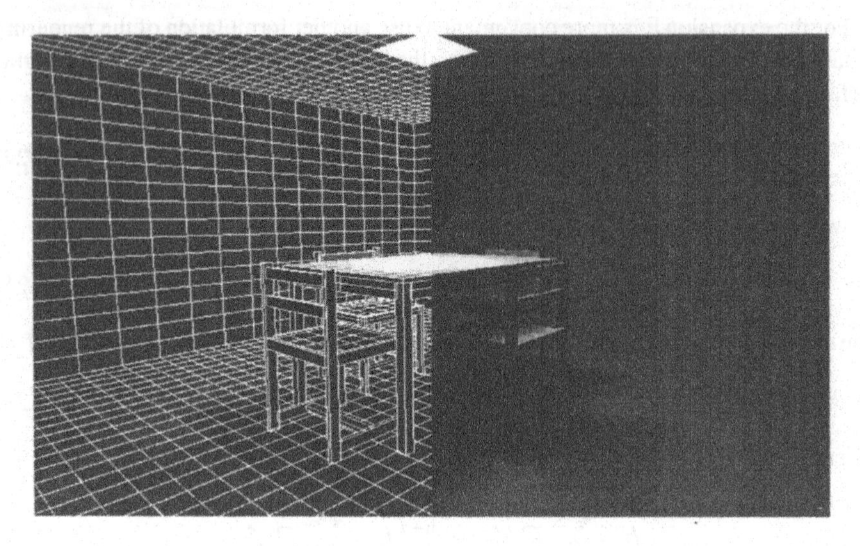

Figure 5.5. Radiosity image. Left part: Piecewise constant shading of the solution, along with the patch boundaries. Right part: reconstructed smooth solution.

Informally, the radiosity B_i of patch i consists of the self-emission E_i and the sum of the radiosity arriving from all other patches and being reflected at this patch. The BRDF for diffuse patches is reduced to a single reflection coefficient ρ_i. The term F_{ij} is called *form factor* and gives the fraction of radiosity emanating from patch j that arrives at patch i. It takes into account the orientation of the patches relative to each other, their distance and their mutual visibility. For more information about form factors and their computation please refer to, e.g. [136].

One big advantage of the radiosity method is its view-independence. Once a solution has been computed, different views of it can quickly be rendered, even achieving interactive frame rates is possible if appropriate hardware is available. A smoothing rendering step is always necessary to cover up the division into patches (see Figure 5.5).

Solving the System. The linear system in Equation 5.28 can be solved iteratively by Jacobi or Gauß-Seidel iteration. These schemes recompute the radiosity B_i for each patch i by gathering all incident light via the form factors F_{ij} and writing the newly computed value to B_i, which is why these methods are called *gathering* schemes.

As an alternative, the *shooting* scheme is introduced by Cohen et al. along with the *Progressive Refinement* method [135]. The goal of this algorithm is to obtain significant images already in a very early stage of the iteration. Radiosity images are usually very dark after the first few iterations, because only a small fraction of the light is distributed so far. With every iteration the solution becomes brighter. To make also the early images brighter, Cohen et al. add an ambient light term to the solution, which is then decreased with every iteration. In addition, they apply a ranking to the set of all patches: for one iteration step, the patch with the highest amount of radiosity is selected and its undistributed light is then shot into the scene. This method is equivalent to the so-called *Southwell relaxation*.

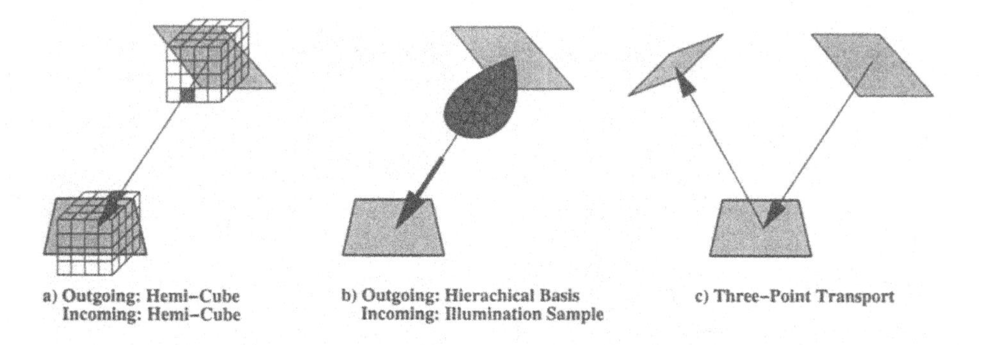

Figure 5.6. Different parametrizations of radiance.

Radiance Methods. The methods introduced so far are only capable of simulating global illumination for diffuse environments. This is not sufficient, however, for realistic scenes, because glossy, i.e. non-diffuse, reflections are an important feature for lively environments.

The problem of the simulation of glossy reflections is that we have to compute the four-dimensional quantity radiance instead of radiosity, which is only two-dimensional. Radiance does not only vary over its position on the patch, but also directionally, such that a discretization of the directional domain becomes necessary also.

Immel et al. [352] were the first to publish a finite element method for surfaces with glossy as well as diffuse reflection behavior. For directional discretization they use pixels on a *global cube*. A global cube is attached to every patch, describing the directional distribution of its exitant light (Figure 5.6, left).

Over the past years, a variety of different radiance representations and consequently also modifications of the algorithm have been presented. Sillion et al. [647] use spherical harmonics as basis functions to represent incoming as well as outgoing radiance. In [664] a hierarchical Haar-Basis is utilized for representing exitant light, whereas incident light is stored in the form of *illumination samples*. An illumination sample can be interpreted as an incident radiance Dirac peak; it consists of the arriving energy and its main direction of incidence. For each energy transport to a receiving patch, such an illumination sample is created, so that the set of all illumination samples describes the entire incoming light (Figure 5.6, center).

A different formulation of the rendering equation called *Three-point Transport* was introduced by [26]. Here, radiance is not parameterized by a position on a patch and a direction on the hemisphere, but as radiance flowing from one point towards another point. Consequently, instead of computing the transport of radiance between two patches and subsequent reflection into arbitrary direction, the light is transported from a sending patch over a reflecting patch towards a receiving patch (Figure 5.6, right).

Hierarchical Methods. The computational effort of the plain radiosity method is immense. In order to obtain visually pleasing images, the scene has to be subdivided into a huge number of patches. If the transport between all pairs of patches has to be computed, the complexity is of order $\mathcal{O}(n^2)$, with n being the number of patches.

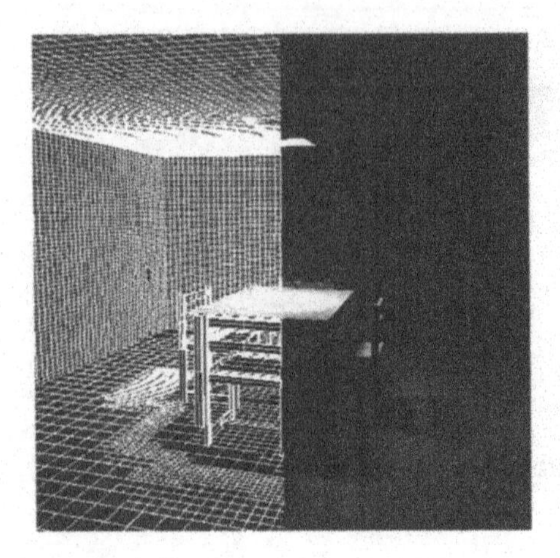

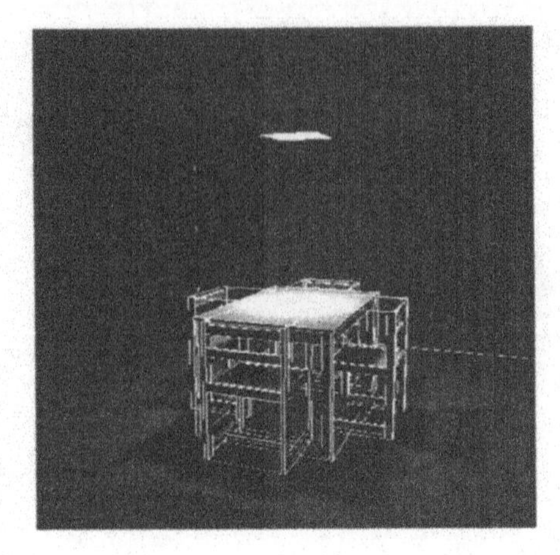

Figure 5.7. Left: Mesh and solution created by hierarchical radiosity. Right: Clustering of the same scene.

Therefore, only scenes of very modest complexity are computable with this method. For the radiance method presented in the previous paragraph [352], the behavior is even worse, because here a four-dimensional function has to be computed instead of a two-dimensional one. Furthermore, memory consumption is a big problem. Consequently, hierarchical methods which adaptively steer the accuracy of the solution are mandatory. Not surprisingly, the basic method was first invented for the simpler case of radiosity before it was transferred to radiance.

The advent of *hierarchical radiosity* (HR) was a large step for radiosity towards handling scenes of higher complexity [282]. The idea is to start with a very rough subdivision of the scene, and to adaptively subdivide only those patches where the radiosity changes rapidly (cf. also Figure 5.7, left).

In order to decide about refinement, the algorithm investigates the transport of energy between all pairs of sending and receiving patches. The variation of the radiosity at the receiver induced by the sender is estimated. If this variation is larger than an error threshold, the transport step is refined. This is done by subdividing either the sender or the receiver (depending on which patch is larger), and considering the interactions between the children recursively.

In 1993 a first hierarchical radiance method was presented by Aupperle et al. [26]. They use the previously described *three-point transport*. In their method, always a triple (P, Q, R) of a sending, a reflecting and a receiving patch is considered and form factors are computed which describe the amount of light emitted from P that is reflected at Q towards R. The transport between the sender and the reflector is refined if the form factor variation is high, whereas refinement of the transport between reflector and receiver is accomplished in case of high variation of the BRDF.

Clustering. Hierarchical radiosity is a large step towards handling more complex scenes. A further important step is *Clustering*, which extends the patch hierarchy 'upwards' [657, 646]. All objects of the scene are packed into clusters at the beginning (see also Figure 5.7, right). The variation of the energy transport between two entire clusters is estimated, and only if it exceeds an error threshold, the clusters are unpacked either into smaller sub-clusters or directly into patches, and the refinement proceeds. The clustering hierarchy is established before the main algorithm. During the algorithm, the patch hierarchy is built like in conventional HR (if necessary).

A cluster radiates light inhomogenously, even if it contains diffuse objects only. Therefore, it is straightforward to integrate clustering into a radiance algorithm for glossy environments. This was done by [647, 242, 124, 664]. In [665] tight bounds on the light transport between clusters are found, with all types of objects (plain polygons as well as curved surfaces) being handled in a uniform manner.

A comparison with classical radiosity shows the superiority of clustering. For example, the scene shown in Figure 5.5 was computed using 4916 patches and took 10.673 seconds, whereas a clustering method (for example [664]) generated a much more precise solution with 33,984 (final) patches in 104 seconds (see Figure 5.7, right).

Importance. Even earlier than the invention of clustering, another method for reducing the complexity of radiosity was presented by Smits et al. [658]. They incorporate the viewing position into the radiosity method and mainly refine those parts of the scene which are important for the final image to be computed.

This *importance* is closely related to the potential in Equation 5.23 and transported through the scene similarly to light, with the difference that it is emitted from the camera [125]. This method can decrease the computational effort drastically. However, this way the solutions' independence from the actual viewpoint is lost.

Aupperle et al. [27] integrated this concept of importance into a radiance algorithm for glossy environments.

5.2.2 Monte Carlo Methods

Monte Carlo methods use random numbers to approximate the answer to a given problem. Their most common application in the context of computer graphics is for the approximation of integrals. The main object of this section is to very briefly introduce the main concepts of Monte Carlo integration and then to present some algorithms based on this technique. As the number of stochastic methods used in computer graphics is quite large, this section can only focus on a very limited number of approaches. Interested readers are referred to [683] for a very comprehensive report on stochastic methods in global illumination.

Monte Carlo Integration. The key to Monte Carlo integration is the fact that the formula for the expected value of a function f of a random variable $\boldsymbol{x}$ with the probability density function p can be approximated by a sum:

$$E\{f(\boldsymbol{x})\} = \int_{\boldsymbol{x}\in S} f(\boldsymbol{x})p(\boldsymbol{x})d\mu_x \approx \frac{1}{N}\sum_{i=1}^{N} f(\boldsymbol{x}_i). \qquad (5.29)$$

S is the continuous space from which $\boldsymbol{x}$ is taken. In graphics this is often an area or a set of directions. μ_x denotes the measure on the probability space. The interesting part is the right side, which tells us that the integral in the middle can be estimated by averaging a number of samples $f(\boldsymbol{x}_i)$. To make things easier to handle, the integrand is substituted by $g = fp$, leaving us with:

$$\int_{\boldsymbol{x}\in S} g(\boldsymbol{x})d\mu \approx \frac{1}{N}\sum_{i=1}^{N}\frac{g(\boldsymbol{x}_i)}{p(\boldsymbol{x}_i)}. \tag{5.30}$$

By intelligently designing the probability density function $p(\boldsymbol{x})$, the quality of the estimate can be greatly improved, as will be explained later on (see *importance sampling* below). In the context of global illumination, Monte Carlo integration can be used for approximating the integral in both the rendering or potential equation.

Gathering Walks—General Idea. Algorithms obtained by applying Monte Carlo integration to the expanded rendering equation are called *gathering* walks. They generally work in the following way: A ray is sent from the eye through a pixel $\boldsymbol{p}$ on the image plane into the scene, and the first intersection $\boldsymbol{x}_1$ with the scene is computed. Then L^e, that is the emittance of this surface in the direction of $\boldsymbol{p}$, is obtained and contributes directly to the pixel intensity.

Next we want to compute the light which reaches the eye via one reflection at $\boldsymbol{x}_1$. In order to do this, the ray is stochastically reflected at $\boldsymbol{x}_1$ into a new direction, and the intersection of this ray with the scene, $\boldsymbol{x}_2$, is found. The emitted radiance at $\boldsymbol{x}_2$ into the direction of $\boldsymbol{x}_1$ is weighted according to the surface BRDF at $\boldsymbol{x}_1$, before adding the contribution to the pixel intensity. Again the ray is reflected by computing a new direction in order to compute light which reaches the eye via two reflections and so on.

This algorithm relates directly to the terms in the expanded rendering equation, as will become clear if we write it in its slightly longer form:

$$L = L^e + \int_\Omega f_r \cos\theta' L^e d\omega' + \int_\Omega f_r \cos\theta' \int_\Omega f_r \cos\theta'' L^e d\omega'' d\omega' + \ldots \tag{5.31}$$

Each part of the sum corresponds to one step in the algorithm. For example, the second summand is approximated by sending one ray from $\boldsymbol{p}$ and reflecting it at $\boldsymbol{x}_1$. In other words, we have approximated the whole integral by choosing exactly one sample direction. In the third term we use this direction for sampling the outer integral and chose a new direction to sample the inner integral and so on. It is easy to imagine that taking only one sample per integral cannot possibly lead to good results. Another major drawback of gathering walk algorithms lies in the low probability for a ray, emitted from the eye and reflected several times, to actually hit a light source. Therefore a vast number of rays would have to be traced in order to obtain an image. The majority of these rays would never hit a light, and thereby not contribute to the pixel intensity at all.

Cook proposed a Monte Carlo algorithm called *distributed ray tracing* [138] that yields far better result by using more than one sample to approximate each integral. In this algorithm several child rays are generated randomly each time a ray hits a diffuse surface, taking into account the surface BRDF.

Of course the algorithm sketched above is only a very rough idea of how a gathering walk ray tracer could work. The important subject of termination will be handled further down after we have taken a closer look at how to generate sample directions.

Importance Sampling. Clearly the quality of the approximation can be improved by using as many samples $f(x_i)$ as possible. As the computation of a sample involves the recursive computation of other samples, it follows that by raising the number of samples, the computation time is also significantly increased. On the other hand the number of samples must be quadrupled in order to halve the error, which can be explained by looking at the standard deviation of the sum in Equation 5.30, which is proportional to $1/\sqrt{N}$.

There is a better method for improving the quality of the estimate though, without vastly raising the number of samples and thereby painfully prolongating computation time. Evidently the quality of the estimation heavily depends on which samples are used. When looking at Equation 5.30 it is obvious that the variance of the sum can be reduced if g and p are functions of similar shape. In other words, we want the probability p for a sample to be taken in a certain region to be large in those regions where g contributes a lot to the sum, and small in others. This strategy is called *importance sampling*.

One way of applying this to the rendering equation is to get $p(x_i)$ to be large in regions where the integrand, that is $\int f_r(\omega', x_i, \omega) \cdot \cos\theta' \cdot L(x_i, \omega')d\omega'$, is large. Evidently we cannot take $L(x_i, \omega')$ into account easily, as we are in the middle of evaluating it (although there are techniques for doing so, see [683] for references). However, we do know values for $f_r(\omega', x_i, \omega)$ and $\cos\theta'$, so a good idea would be to generate samples with a density $p(x_i)$ proportional to $f_r(\omega', x, \omega) \cdot \cos\theta'$. This means that when computing an outgoing direction for a ray, we want to generate a direction into which the incoming ray is likely to be reflected. This technique of importance sampling is called *BRDF sampling*, because the likelihood for the generation of sample directions is based on the surface BRDF (and the cosine angle).

Russian Roulette. An important subject we have not brought up yet is the issue of termination. In our case this means how often do we want to reflect the child rays or, in other words, how do we want to handle the infinite sum in Equation 5.27.

One possibility would be to argue that the contribution of light that has been reflected more than a certain number of times is negligible. This means, though, that we would simply truncate the infinite Neumann series after a certain number of terms, which would obviously introduce bias.

A statistically correct method for handling this problem is *Russian roulette*. The idea is to include higher order terms randomly, with a probability decreasing with the order of term. In order to compensate for the missing terms, the computed terms are multiplied by an appropriate factor. Applied to our problem this means that we decide for every sample (i.e., child ray) whether we want to use it or not, depending on a certain probability. If we have made up our mind not to use it, this child ray is not traced any further. In the other case we weight the contribution of this computed sample in an appropriate way.

There are a variety of methods for deciding which sample to use. A possible approach which works well in combination with BRDF sampling (described in the previous paragraph) is to set the probability of continuing a walk to the so-called *albedo* of the surface at x in the outgoing direction ω:

$$a(x, \omega) = \int f_r(\omega', x, \omega) \cdot \cos\theta' d\omega' \tag{5.32}$$

and setting the probability density to the same value. This way the direction is sampled with a density which integrates to $a(x, \omega)$ instead of 1 and to compensate for this the walk is stopped with the remaining probability of $1 - a(x, \omega)$. By this technique paths of different lengths are generated and the introduction of bias is avoided.

Expansion of the Potential Equation. By expanding the potential equation instead of the rendering equation we obtain a different class of algorithms called *shooting walks*. They begin at a known point x on a light source and simulate the photon reflection for a few times, finally arriving at a pixel whose radiance this walk contributes to. In comparison to the gathering walk these algorithms generate paths in the opposite direction. Of course techniques like importance sampling and Russian roulette can be applied similarly.

An example of a Monte Carlo method based on expansion of the potential equation is an algorithm called *Light tracing* [194]. It performs random walks through the scene, beginning at the light sources. At each intersection point a ray is traced to the eye and the contribution is added to the selected pixel (if the path is not occluded). The next direction is determined by BRDF based importance sampling.

Bidirectional Walks. With gathering walks the major drawback is the low probability for a path to end in a light source. The disadvantage of shooting walks is the low probability of a path, beginning at the light source, to end at the eye. Bidirectional walks overcome these problems by combining gathering and shooting walks. Basically, two walks are initiated at the same time, one beginning at the light source, one at the eye. After a few steps these two walks are connected in some way or other. In [725] for instance one single deterministic ray is cast, joining the part of the walk from the light source with the part to the eye. In [415], all points of the gathering walk are connected to all points of the shooting walk using deterministic rays. If a deterministic ray intersects an object the contribution of the walk is zero.

In the photon map algorithm [364] the two simulation directions are separated to two passes. In the first one, a large number of photons is emitted from the light sources and traced through the scene stochastically. The photon hits (e.g., on a surface) are stored in a spatial data structure, the so called photon map.

The second pass is similar to distributed ray tracing, but is optimized by exploiting the information from the photon map. Either the photon map can be used to directly approximate the illumination at any point in the scene by searching the n closest photons. This can be done in reasonable time if the maps are stored accordingly, e.g. in a balanced binary spatial partition tree. Alternatively, if the illumination at some point is required more accurately, at least a valuable importance map of the incident light can be built from the information of the photon map, guiding the reflection directions of the secondary rays.

5.2.3 Hybrid Methods

Finite element and Monte Carlo methods are suited for contrary types of scenes. The first can handle scenes with predominantly diffuse reflection well whereas the latter is ideal for specular environments. Therefore, combining the two techniques is an obvious approach to obtain a general purpose method. Two such ideas are described in the following, for more detail the reader is referred to [643].

Two-Pass-Methods. Wallace et al. [733] and Sillion et al. [648] propose to use a radiosity method only as a first pass for simulating the diffuse indirect light. The form factor is extended by including the light exchange between diffuse patches via glossy patches. A subsequent ray tracing pass then adds specular reflection towards the eye.

This method was extended by [118] towards a Multi-Pass method. Direct illumination, highlights on specular surfaces, caustics on diffuse surfaces due to specularly reflected light as well as diffuse distribution of light are computed with combinations of appropriate methods.

Monte Carlo Radiosity. Another approach is to integrate stochastic principles into a radiosity algorithm. In order to do this, Neumann et al. [503] modified the progressive refinement (see *solving the system* earlier in this section) to such extent that the patches which shoot radiosity into the scene are selected stochastically by importance sampling. Additionally, the form factor is approximated stochastically by shooting randomly selected rays using an appropriate probability distribution.

5.3 TEXTURE MODELS

W. HEIDRICH

The first two sections of this chapter have layed out the basic principles of how to generate realistic images by simulating the illumination in a virtual scene. As mentioned in the previous section, the objects in a scene are usually described by a number of geometric primitives and their material properties. In order to yield convincing images, it is necessary to provide a considerable amount detail in the scene. However, it is imperative not to represent all this detail by increasing the number of geometric primitives, since this would also result in a drastical increase of computational complexity.

Texturing or *texture mapping* [109] is the process of mapping images (textures) onto geometric primitives. It represents the most efficient way of increasing the visual complexity of a synthetic image without increasing the geometric complexity of the scene. Since very complex effects can be achieved at only moderate computational cost, texturing has become quite popular in recent years. It is used in both high-quality off-line rendering systems and interactive computer graphics.

Textures come in a variety of different flavors describing different kinds of detail information. For example, a texture can simply contain color values to be applied to a surface, but it can also be used to perturb the surface normal (bump map) and the surface point (displacement map), or it can store information about the incoming illumination at a certain point in space (environment map). These different variants will be surveyed in this section. The dimensionality of textures can be anywhere from the more traditional 1D and 2D textures, to 3D volumes, to 4D light fields. It is to be expected that textures of even higher dimensionality will be introduced in the future as soon as hardware capacities allow for it.

An important issue for texture mapping is that of reconstruction filtering and anti-aliasing. In the context of this discussion, a texture is assumed to consist of a regularly spaced array of texels (texture elements), just as each image consists of an array of pixels (picture elements). Furthermore, each texel or pixel is a *point sample* from which the continuous picture can be reconstructed [653]. This is in contrast to many older publications that assume pixels to be small squares (or rectangles), which prescribes the use of a box filter for the reconstruction [653].

5.3.1 Texture Mapping—The Concept

Given a sampled texture of arbitrary dimension, i.e. a regular array of point samples, the question is how this information can be mapped onto a geometric primitive. Most often geometric primitives are 2-manifolds, that is, surfaces, but they can also be points, lines, or curves, or even volumetric objects.

Since a texture is essentially stored in a multi-dimensional lookup table, it is necessary to associate a set of indices into this table (texture coordinates) with each point on the geometric primitive. We will call these texture coordinates u, v, s, and t. If the texture has less than 4 dimensions, only a subset of these is actually used. A 2D texture, for example, only uses the u and v coordinates.

There are multiple ways of associating such a set of texture coordinates with each point on the geometric primitive. The easiest and by far most often used approach assumes that the primitive is a *parametric object*, i.e. defined analytically as a function over a certain parameter domain. Examples for parametric objects are points, lines, polygons, Bézier surfaces and NURBS, as well as spheres, cones, cylinders and so forth. For these surfaces, each point on the object can be associated with its corresponding position in parameter space, and these parameter values can directly be used as texture coordinates. For more flexibility in the positioning of the texture, it is also possible to apply an affine transformation to the parameter values before they are used as texture coordinates. This allows for scaling, rotation, and translation of the texture on the object.

For piecewise linear objects, such as lines and triangles, it is also customary to explicitly specify a set of texture coordinates for each vertex, and then to interpolate these across the primitive. The same effect could also be achieved using an affine transformation, but the interpolation method is more efficient, and easier to use with compound objects such as triangle meshes.

These techniques, however, do not work with objects that do not have a natural parameterization, such as implicit surfaces or fractals. One way of handling this kind of objects is to find *local parameterizations* for parts of the object that can then be used for texturing these parts individually [534]. The more frequently used approach, however, is to project the texture from a parametric surface (which can be textured easily, see above) onto the non-parametric object. For example, the texture can be mapped on a sphere surrounding the object, and the color for each object point can then be found by finding the intersection of the sphere with the ray from the center of the sphere through the object point.

An important special case for this kind of projection is when the intermediate parametric geometry is a rectangle, or, more generally, a polygon. This case is called projective texturing. In this case the texture coordinates for each object point can be found simply through a perspective transformation of the point's location in world space. This allows one to specify the projection in terms of a 4×4 homogeneous matrix $\boldsymbol{P}$. The texture coordinates for a homogeneous point $[x, y, z, 1]^T$ are then u/t, v/t, and s/t for

$$[u, v, s, t]^T := \boldsymbol{P} \cdot [x, y, z, 1]^T . \tag{5.33}$$

Since there are only three degrees of freedom for each point on the object, projective textures only make sense for textures of at most dimension 3.

Once the texture coordinates have been determined for a certain point on the object, the value corresponding to this location needs to be reconstructed from the texture. This is described in more detail in Section 5.3.3. For the moment, we can simply assume that the closest entry in the texture map is determined from the texture coordinates, and that this entry is used for the object point (nearest neighbor sampling). Once the texture value has been determined for an object point, the further steps depend on the specific kind of texture. These are discussed in the following.

5.3.2 Applications of Textures with Various Dimensions

The most well-known variant of textures is simply a 2D image that is being mapped onto a surface without further processing. While this is a commonly used form of textures, there are other forms that differ both in dimensionality and the kind of information stored for each sample. We now give an overview of the most important applications of texture mapping.

One- and Two-dimensional Textures. As mentioned the texture can simply contain a set of color values to be applied directly to the surface at the corresponding locations. Instead of using a 2D texture, it is sometimes also useful to apply a 1D texture as shown in Figure 5.8. Here, the texture coordinate u for each point was set to its world-space z coordinate, and then a 1D texture was used to draw contour lines.

Quite often, the colors reconstructed from the texture are not directly applied to the surface, but are used as a diffuse reflection coefficient in an illumination computation (see Section 5.1.3 for details on reflection models).

Instead of storing a surface color or reflection coefficient, textures can also hold information about the incoming light from all directions at a certain point in space. Since a direction in 3D has two degrees of freedom, a 2D texture suffices to represent all incoming illumination for one point. This kind of texture is called an *environ-*

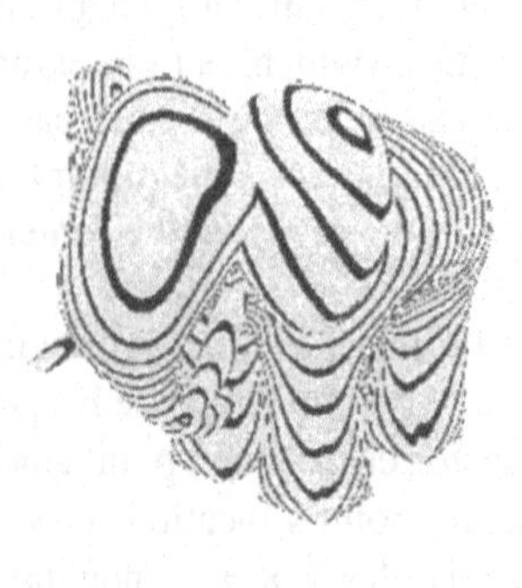

Figure 5.8. Elephant textured with a 1D texture to show iso-lines for world-space z.

Figure 5.9. A reflective torus rendered with an environment map.

ment map [73]. The idea underlying the concept of environment maps is that, if a reflecting object is small compared to its distance from the environment, the incoming illumination on the surface only depends on the direction of the reflected ray. The ray origin, that is the actual position on the object, can be neglected. Therefore, the incoming illumination at the object can be precomputed for each reflected ray direction, and stored in a two-dimensional texture map. Figure 5.9 shows a reflecting torus that has been rendered with this technique. As shown in [156], environment maps can also be used for global illumination purposes with arbitrary material characteristics, not just mirrors.

One of the core issues of environment maps is the choice of an appropriate parameterization. On the one hand, it should sample the sphere of directions as uniformly as possible, and on the other hand it should be easy to compute, especially for hardware implementations. The two most commonly used parameterizations are spherical environment maps [270] and cubical maps [256]. Spherical maps are easy to compute, and can be found on most contemporary graphics hardware, but they exhibit a singularity and areas of extremely poor sampling, so that they can only be used for one specific viewing direction. Cubical maps, on the other hand, can be used for all viewing directions, but are quite difficult to handle. Recently, a new, parabolic parameterization was proposed in [312] which is even more uniform than cubical maps, and also lends itself to efficient hardware implementations.

So far, textures were only used for storing color values, but the concept of texturing also extends to other data types. With *bump mapping* [72], for example, the texture only contains a single height value that is used as a displacement of the underlying surface along its normal. Instead of actually changing the geometry, bump mapping only perturbs the normal vector in each point *as if* the surface was displaced. The resulting normal vector for each surface point is then used to compute the illumination in the point by applying one of the reflection models described in Section 5.1.3. Figure 5.10 shows a single polygon to which such a bump map has been applied.

Figure 5.10. A single, bump-mapped polygon with a Phong reflection model and an environment map for the mirror reflection.

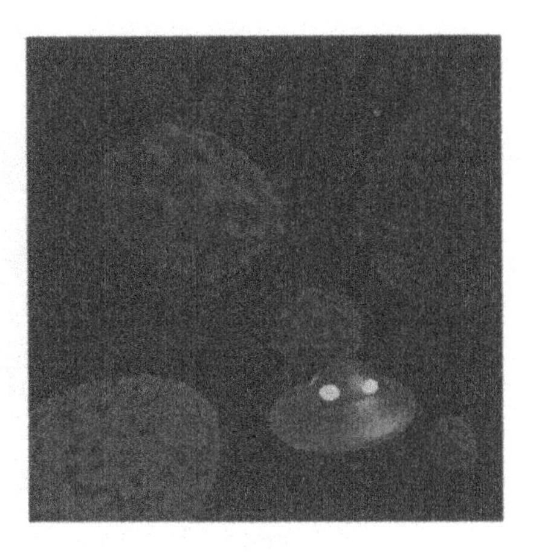

Figure 5.11. A number of surfaces with displacement maps. The underlying geometry for each object is a sphere.

The calculation of the perturbed normal vector requires the computation of partial derivatives and is quite costly. This is the reason why bump mapping hardware has only recently begun to show up. A computationally less expensive alternative is to directly store the normal vectors in the texture (*normal map*). Normal maps also bear the advantage that they can be directly generated through measurements [584] and mesh simplification [133]. On the down side, they require more space and are not independent of the underlying surface geometry.

As mentioned above, bump mapping only uses the displacement information stored in a texture for perturbing the surface normal. If the same information is used to actually change the geometry, this is called *displacement mapping* [137]. This approach results in a further improvement of the visual quality, since the silhouettes of the object change. This can be observed in Figure 5.11. Displacement maps are typically not provided in terms of a sampled texture, but as a procedural description. The rendering of these maps is difficult, and most often achieved through tessellation in many small triangles. A direct method for ray-tracing procedural displacement maps has been proposed in [311].

Another application of 2D textures are *shadow maps* [757]. They are an efficient means for computing shadows, and can also be implemented in hardware [629]. A shadow map contains the depth values of a scene as seen from the light source. To test whether a specific object point is illuminated by the light source, the 3D location of the point is projected into the local coordinate system of the light source, and the depth value is compared with the result from applying the shadow map as a projective texture. If distance of the point is larger than the depth stored in the map, the point lies in shadow, and otherwise it is illuminated.

Shadow maps can be flexibly combined with a projective texture describing the illumination that the light source emits in each direction. This approach is called a

Figure 5.12. An elephant with a solid wood texture.

light map [629]. It allows for implementing more complex light sources such as spot lights and simple slide projectors. However, light maps can still only model light sources of infinitely small size. Light sources with a finite extent can be represented as light fields (see below).

Solid Textures and Volumes. As an alternative to 2D texture mapping, it is also possible to use 3D textures, which are also called *solid textures*, or *volumes*. As mentioned above, solid textures have the advantage that no parameterization is required for the geometry. Instead, the 3D location of an object point is directly used as the texture coordinate. This way, solid textures can be used for the same applications as 2D textures, but with the additional benefit that geometric primitives without a parameterization can be textured. An example for this is depicted in Figure 5.12. Although the elephant is composed of many Bézier surfaces which do have a parameterization, it would be hard to generate the same effect with 2D textures. In particular, the generation of the textures for each of the Bézier surfaces would be difficult.

In addition to this straightforward extension of 2D texturing to solid textures, 3D data sets also have additional applications, in particular in the area of medical visualization. Here, the data acquired from a device such as a MR or a CT is typically given as a 3D array of scalar density values. For the visualization of these data sets, there exist two principal approaches. Firstly, it is possible to extract surfaces from the data set, which can then be rendered with traditional means (*iso-surface extraction*). Secondly, the data sets can be rendered *directly* by tracing a ray through the volume and integrating the illumination along this ray. The latter approach has the advantage that many different features are visible at the same time and that these cannot occlude each other. The details of the different approaches to volume rendering are described in Section 6.4.

Light Fields. Even one dimension more than required by volumes is represented by light fields [435] and Lumigraphs [254]. A light field (as used nowadays in computer graphics and computer vision) is a dense sampling of the five-dimensional plenoptic

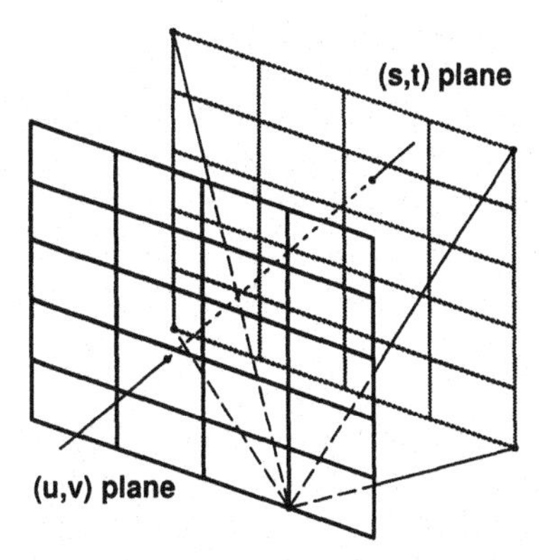

Figure 5.13. A light field is a two-dimensional array of images taken from a regular grid of eye points on the (u, v)-plane through a window on the (s, t)-plane. The two planes are parallel, and the window is the same for all eye points.

function [2], which describes the radiance at every point in space in every direction. Since radiance does not change along a ray in empty space, the dimensionality can be reduced by one, if an appropriate parameterization is found, that reflects this property.

The so-called two-plane parameterization used by the light field representation fulfills this requirement. It represents a ray via its intersection points with two parallel planes, a so-called *light slab*. Since each of these points is characterized by two parameters in the plane, this results in a four-dimensional function that is sampled through a regular grid on each plane (see Figure 5.13).

One useful property of the two-plane parameterization is that all the rays passing through a single point on the (u, v)-plane form a perspective image of the scene, with the (u, v) point being the center of projection. Thus, a light field can be considered a two-dimensional array of images with eye points regularly spaced on the (u, v)-plane.

Moreover, since it is assumed that the sampling is dense, the radiance along an arbitrary ray passing through the two planes can be interpolated from the known radiance values in nearby grid points. Each such ray passes through one of the grid cells on the (u, v)-plane and one on the (s, t)-plane. These are bounded by four grid points on the respective plane, and the radiance from any of the (s, t)-points to any of the (u, v)-points is stored in the data structure. This makes for a total of 16 radiance values, from which the radiance along the ray can be interpolated quadri-linearly. As shown in [254], this interpolation can also be approximately computed in hardware by using texture mapping.

When using the two-plane parameterization, a total of six different slabs can be used to represent the complete light field leaving an object. This information can be used to view the object [435, 254], or to illuminate other objects in the scene [24, 25, 310]. Alternatively, a light field can also be used to represent the light arriving at a surface. In this case, light fields are a straightforward extension of environment maps. In fact, they can be interpreted as a collection of environment maps that are interpolated over a convex hull surrounding the object. More on light fields and their applications can be found in Section 5.4.

5.3.3 Texture Filtering and Anti-Aliasing

Independent of the dimensionality of the texture used, the filtering of the texture for reconstructing a continuous signal is very important. This subject is very complex, and a comprehensive treatment would exceed the scope of this section. Therefore, we will limit our discussion to the methods that are most widespread in practice. A more detailed overview of the topic with additional pointers to other literature can be found in [308] and [246].

Nearest Neighbor. The simplest form of reconstruction is nearest neighbor sampling. For each pixel in the image, the texture lookup proceeds as follows. Let u be the (interpolated) texture coordinate for a given pixel, in the range $[0 \ldots 1]$. For simplicity, let us assume a 1D texture array $h[n_u], n_u \in [0, \ldots, N_u - 1]$ of size N_u, in which a value $h[n_u]$ corresponds to a texture sample at position $n_u/(N_u - 1)$. The reconstructed texture color at u is then given by

$$h(u) := h[\text{round}(u \cdot (N_u - 1))].$$

For higher-dimensional textures, the other texture coordinates v, s, and t are treated separately in a similar manner. Nearest neighbor sampling is a very poor quality reconstruction, leading to severe aliasing artifacts. On the other hand, it is very efficient, since it requires only a single access to the texture map.

Linear Interpolation. A somewhat better reconstruction is the linear interpolation of samples. Here, we define

$$n_u := \lfloor u \cdot (N_u - 1) \rfloor,$$

and

$$\Delta u := \frac{u \cdot (N_u - 1) - n_u}{N_u - 1}.$$

Then the reconstructed color value is given as

$$h(u) := (1 - \Delta u) \cdot h[n_u] + \Delta u \cdot h[n_u + 1].$$

This means that two texture accesses are required for a 1D texture. For a 2D texture and bilinear interpolation, 4 lookups are needed. The values are first interpolated in u direction, and then the resulting two values are interpolated in v direction. In general, for a dimension of d, the number of texture lookups is 2^d, and the number of linear interpolations is $2^d - 1$. Obviously, this approach is much more expensive than nearest neighbor sampling, but the reconstruction quality is significantly improved.

Higher Order Reconstruction. As we move to even higher order interpolations, such as quadratic or cubic interpolation, the cost gets even higher (o^d lookups for an order o reconstruction). Nonetheless, these methods are often mandatory for high-quality reconstructions. Since the human eye is particularly sensitive to discontinuities in the first derivative of color values, at least quadratic interpolation is required to avoid artifacts (called Mach bands).

Cubic interpolation is more often used than quadratic interpolation, both because it is symmetric around the sample point and because it allows for a much better approximation of an ideal low pass filter.

Mip-Maps. Linear and higher order reconstruction are suitable to avoid aliasing artifacts as long as the sampling rate in the final image is larger than the sampling rate of the texture, i.e. as long as the texture is 'magnified' on the screen.

If the texture is minified, that is, if a large texture occupies only a few pixels on the screen, it is necessary to average the color values of many texels for a single pixel in the final image. Since objects can be arbitrarily small on the screen if they are viewed from far away, it is in the extreme case necessary to average *all* texels for a single pixel. This would be too expensive, so that several researchers have thought of ways to use precomputation for these filtering steps (*prefiltering*).

The most important of these techniques is *mip-mapping* ('mip' stands for 'multum in parvo', which is Latin for 'many things in one place') [758]. With mip-mapping, the resolution of the original texture is restricted to powers of two. From this original texture, a *pyramid* of textures is generated, where the resolution in each level of the pyramid is reduced by a factor of 2 in each dimension. The reduction of resolution is achieved by computing the average of 2^d neighboring texture values, where d is the dimension of the texture. Storing a complete pyramidal representation of a 2D texture in this way requires only one third more memory than storing the original texture.

For reconstruction, it first has to be determined which level of the pyramid is to be used for texturing. To this end, the vertices of rectangular region corresponding to a pixel in the final image are mapped back into texture space. This yields an arbitrary quadrilateral polygon in texture space. Finding an appropriate level in the pyramid corresponds to approximating this quadrilateral by a square [308]. Instead of selecting a fixed level for texturing, it is also possible to linearly interpolate between two adjacent levels to avoid seams. The cost of this method is 2^{d+1} lookups.

Anisotropic Filtering. One of the core problems of mip-mapping is that it is an isotropic filtering method, which means that the size of the texture is reduced by the same factor in all parametric directions. However, there are many situations where the the quadrilateral in texture space is a long, thin polygon that cannot be approximated well by a square. This situation requires *anisotropic* texture filtering. Examples for such methods are summed area tables [141] and footprint assembly [611].

5.3.4 Shaders: Procedural Textures

Besides by discretely sampled, tabulated textures, surface properties can also be specified by procedural descriptions, so-called *shaders*. These can be used in any of the applications mentioned in Section 5.3.2. In comparison to fixed resolution textures, procedural shaders have the advantage of being resolution independent and storage efficient. Since a shader can be any procedure, both physical and non-physical behaviors can be programmed.

Typically, a specific programming language, the *shading language* is used for programming procedural shaders. One of the most important of these languages is the RenderMan shading language [281], which was the first to provide all the features a modern shading language has. The PixelFlow graphics hardware includes a limited subset of this shading language (i.e., no derivatives), and supports procedural shaders in hardware [518].

Shading languages usually introduce a set of specific data types and functions exceeding the functionality of general purpose languages and libraries. Examples for this kind of domain specific functions are continuous and discontinuous transitions between two values, like step functions, clamping of a value to an interval, or smooth Hermite interpolation between two values. The more complicated features include splines, pseudo-random noise, and derivatives of expressions.

Every shader is supplied with a set of explicit, shader specific parameters that may be linearly interpolated over the surface, as well as fixed set of implicit parameters (global variables). The implicit parameters include the location of the sample point, the normal and tangents in this point, as well as vectors pointing towards the eye and the light sources. For parametric surfaces, these values are functions of the surface parameters u and v, as well as the size of the sample region in the parameter domain, du and dv.

5.4 IMAGE-BASED RENDERING

M. MAGNOR, W. HEIDRICH

To model real-world or synthetic 3D-objects, two different approaches can be distinguished. Traditionally, a scene is decomposed into its 3D-geometry, surface texture and illumination. Suitably parameterized, these constituents serve as basis for the rendering process. This *geometry-based* rendering approach, which has been discussed in detail in Sections 5.1–5.3, allows to alter the original scene by changing illumination, texture and geometry.

On the down side, the process of rendering 2D-views from geometric scene descriptions is computationally very demanding. Acceleration hardware for various rendering techniques is available in graphics workstations, and enables rendering of complex geometry models at interactive frame rates. Illumination computations, however, can significantly slow down rendering performance. And often, even after time-consuming global illumination calculations, geometry-based rendered views still give the impression of artificial, 'computer-generated' images.

Other problems arise if accurate geometry, texture and illumination model parameters have to be acquired from real-world objects or scenes. 3D-scanning devices need to be designed specifically for certain natural objects. The Cyberware scanning unit can acquire geometry model parameters and texture of the human head [147]. Figure 5.14 depicts a recorded head model. No geometry information was recorded for the hair, making apparent the fundamental limitation of geometry-based rendering: because hair, fur, smoke and fractal structures cannot be described well by surfaces, geometry-based rendering must fail for objects with such surface characteristics.

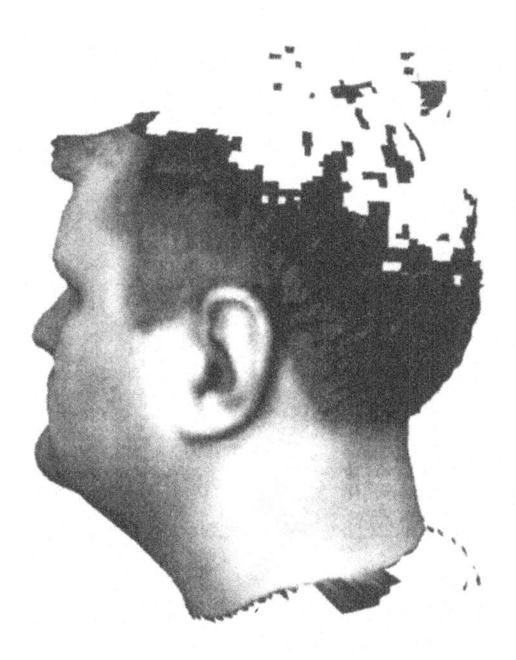

Figure 5.14. Scanned head model.

A different approach to generate 2D-images of natural 3D-scenes bases on the notion of the *light field* as described by Gershun in 1936 [240], which has been renamed the *plenoptic function* by Adelson and Bergen much later, in 1991 [2]. The plenoptic function describes light properties as a function of space, time and propagation direction: an illuminated object fills the surrounding space with light reflected of its surfaces, establishing the object's *light field*. An observer at any point in space (x, y, z) and any moment in time (t) can look in any direction (θ, ϕ) and record the incoming light intensity Φ at all visible wavelengths λ:

$$\Phi = \Phi(x, y, z, t, \theta, \phi, \lambda). \tag{5.34}$$

Φ is the radiation's *spectral intensity*, a scalar value that denotes the incoming light's power per unit area per unit solid angle per unit wavelength interval. It contains all information about the object that can be conveyed by electromagnetic radiation. Correspondingly, if Φ is known, any view of the scene can be reconstructed. This approach to computer-generating 2D-views of natural 3D-scenes has been coined *Image-based Rendering* (IBR), because conventional images can be used to capture the light field.

In principle, IBR can overcome the limitations of geometry-based rendering: any type of object can be imaged and thereby its light field captured, rendered views can be identical to photographs of the original scene, and rendering speed is independent of scene complexity. Because of limited hardware resources, however, the general plenoptic function in Equation 5.34 must be somewhat simplified and its versatility

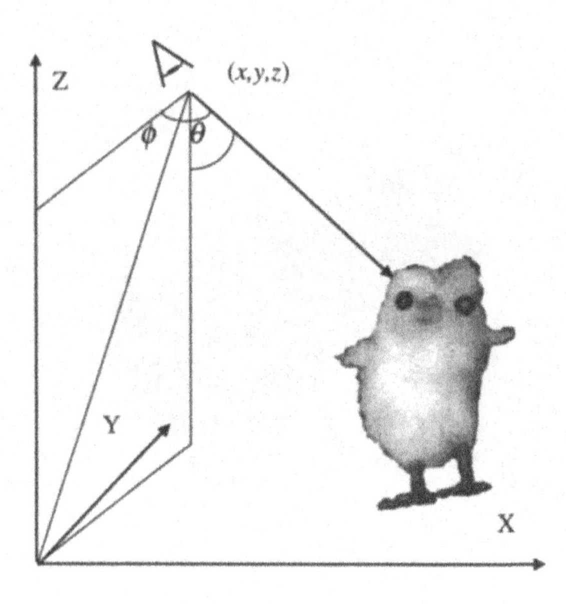

Figure 5.15. The plenoptic function Φ describes light properties depending on viewing position (x, y, z) and viewing direction (θ, ϕ).

restricted. First, the light's spectral information can be described more efficiently by exploiting the human visual system's color perception characteristics, storing color information using some common color space (RGB, YUV, etc.). Further, current IBR-systems do not take temporal changes into account: depicted scenes are static, i.e., neither object geometry nor illumination change. Current efforts to overcome these limitations are described in Section 5.4.9. 5 degrees of freedom remain: position in space (x, y, z) and viewing direction (θ, ϕ) (see Figure 5.15):

$$\Phi_{R,G,B} = \Phi_{R,G,B}(x, y, z, \theta, \phi). \tag{5.35}$$

If free space can be assumed to be transparent, such that light properties do not change during light propagation through unobstructed space, the plenoptic function in Equation 5.35 can be further reduced to 4 dimensions [435].

The main task in IBR consists of discretizing and sampling the continuous, multi-dimensional plenoptic function $\Phi(x, y, z, \theta, \phi)$ to yield a function $\hat{\Phi}[n_x, n_y, n_z, n_\theta, n_\phi]$ that can be stored and processed on digital computer systems. During rendering, the discretized and sampled plenoptic function $\hat{\Phi}$ is used to yield a reconstructed continuous plenoptic function $\Phi'(x, y, z, \theta, \phi)$. Φ' is resampled for a specific location in space as well as viewing direction. Because some information about the continuous function Φ is inevitably lost during sampling, the reconstructed function Φ' is not identical to the original plenoptic function Φ, potentially introducing rendering artifacts.

Even the reduced plenoptic function in Equation 5.35 requires a large amount of image data, too much for most available hardware. Sensible data acquisition [612] in conjunction with efficient data compression [462] are therefore a major concern in most image-based rendering applications.

The line between image-based and geometry-based rendering techniques cannot be clearly drawn: Geometry-based rendering methods have been supplemented by IBR-techniques [652, 512, 313, 305], and most IBR-techniques require additional geometry information as input or derive approximate geometry from the image data (*Image-based Modeling*):

- *View-dependent Texture Mapping* [158] extends geometry-based rendering by variable, image-based texture maps.
- *Hybrid rendering techniques* such as [157] often exploit a-priori knowledge about the depicted object and use an approximate 3D-model to reconstruct the plenoptic function.
- Dense depth maps are needed for *Delta Tree Representations* [149], *Multiple-Center-of-Projection Images* [565], *Layered Depth Images* [634, 113] and *Sprites with Depth* [428, 634].
- In *Plenoptic Modeling* [476], corresponding points between images need to be specified.
- *Environment Maps* [117, 682] allow to reconstruct views from a fixed viewpoint without any geometry information.
- *Light Field Rendering* [435] allows to render arbitrary 3D-objects solely from images. The *Lumigraph* [254] supplements light field rendering by approximate object geometry to yield better rendering results.

5.4.1 View-Dependent Texture Mapping

In geometry-based rendering, object surfaces have fixed attributes and static texture. Conventionally, this surface description is used in conjunction with lighting conditions to calculate the object's appearance. View-dependent texture mapping employs images of the object instead to derive texture depending on viewing direction. Besides a full 3D-description of the object's geometry, multiple images of the object at static illumination conditions are needed. The recording positions and internal camera parameters have to be known to register image pixel positions with object surface points. In the set of images taken from various viewing directions, many surface points are imaged multiple times from different angles. 'View maps' are created for the points on the object surface by collecting their appearance from different viewing angles. In-between angles are interpolated during rendering, and the object's view-dependent texture map is determined by consulting all visible surface points' view maps.

The rendered views appear more realistic than if conventional texture mapping were applied. Rendering-accelerating hardware can be exploited to achieve interactive rendering rates. As with most image-based rendering approaches, however, a lot of memory is required to fit all image data into local RAM. Depending on the number of images available, view-dependent texture mapping returns convincing results for not too specular surfaces. On the other hand, because the images are taken at static illumination, the advantage in geometry-based rendering of being able to alter lighting conditions and object geometry is lost.

5.4.2 Hybrid Geometry- and Image-Based Rendering

In this rendering system, approximative geometry of architectural scenes is derived from photographs (image-based or 'photogrammetric' modeling) before view-dependent texturing is applied to the model. From a (small) number of conventional photographs of a building, basic geometry as well as camera positions are determined. Point correspondences between images are established interactively. The photographs allow to apply view-dependent texturing to the geometry model. The approximate model is used to automatically refine its geometry further ('model-based stereo').

The system allows model acquisition and interactive rendering of real-world architectural scenes from a sparse set of images. No information about the images' recording positions and orientations is needed, but manual guidance during model recovery is required. Because facade appearance is not calculated but rather derived from images, illumination changes (e.g. sunset, overcast skies, fog, snow) cannot be simulated from a single set of images, and the building can be rendered for recorded lighting conditions only.

5.4.3 Delta Tree Representation

The delta tree is a data structure that hierarchically orders images of a 3D-object. Additional depth maps allow to warp images to yield intermediate views. The images are divided into patches, and only a few patches are kept such that each object surface point is depicted just once. The patches are then coded using a block-based discrete cosine transform (DCT) scheme. During rendering, image patches that show visible surface points are warped to match the view. In addition, surface normals are reconstructed and shading effects can be calculated.

Aliasing effects can be suppressed by decoding only low-frequency DCT-coefficients, thereby adjusting the level-of-detail depicted. Because the delta tree representation removes view-dependent object texture information, surface appearance must be calculated from surface orientations and lighting conditions, as in geometry-based rendering. This allows, on the one hand, simulating different illumination conditions. On the other hand, the IBR-advantage of rendering photo-realistic views from any direction is lost.

5.4.4 Multiple-Center-of-Projection Images

In MCOP rendering, a single image is put together using multiple viewpoints. The MCOP image is made up of many one-dimensional image strips, each recorded from a different, but known, viewpoint.

For each pixel in the MCOP-image, additional depth information is required to compute its reprojected location during rendering. An example of an MCOP image is the texture map and the corresponding depth map acquired by the Cyberware scanning unit [147], Figure 5.16.

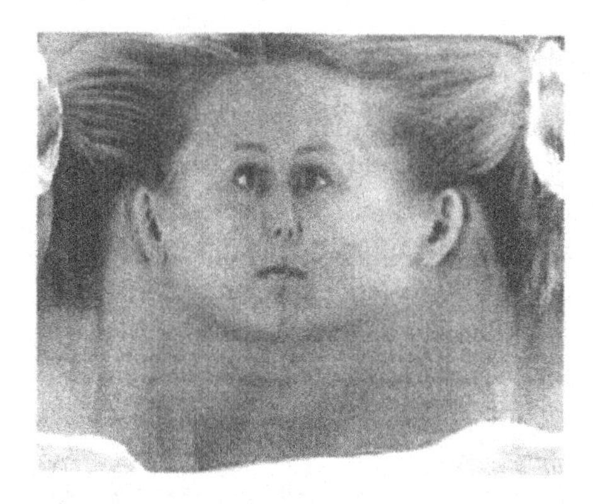

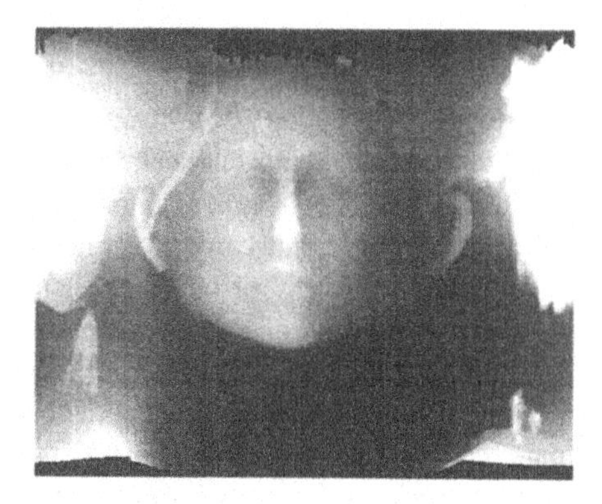

Figure 5.16. An MCOP-image consists of a texture map and corresponding depth information as is recorded by the Cyberware scanning unit [147].

Because any object surface point appears just once in an MCOP image, view-dependent texturing cannot be performed from a single MCOP image. Because of the complex epipolar geometry, MCOP rendering in occlusion-compatible order has proved difficult.

5.4.5 Layered Depth Images

The image-based rendering methods described so far are all object-centered, i.e. an object could be rendered from arbitrary viewpoints. Layered depth images, on the other hand, are viewpoint-centered because the viewpoint location is confined to a small region in space.

Multiple images of a scene with corresponding depth maps and known recording positions are combined to one layered depth image by warping the images to one fixed viewpoint. Each pixel of the layered depth image then actually consists of multiple pixels from different images, ordered in depth. Parallax and occlusions can be faithfully displayed if the viewpoint moves a small distance from the center viewpoint.

Because only partial geometry information is available, positions farther away from the central viewpoint result in distorted rendered images. Also, the different texture layers remain static during movement.

5.4.6 Sprites with Depth

Conventional sprites are textured planar image patches at a fixed scene depth, showing no internal parallax or occlusion effects. Views from a tilted angle reveal the sprites' flatness. Sprite appearance can be improved if the texture is projected onto a surface that (smoothly) changes in depth. After extracting a sprite from an image sequence, its depth map is reconstructed using traditional stereo algorithms.

Depth can be recovered for sprite objects with reasonably smooth surfaces, which guarantees pixel connectivity within the sprite. Sprite extraction and -pursuit in image sequences, however, requires the decomposition of the scene into non-overlapping depth layers which can become an arbitrarily complex task.

5.4.7 Plenoptic Modeling and Environment Maps

In Apple's QuickTimeVR software, 360-degree panoramic images are used to allow an observer to stand inside a scene and look into any horizontal direction from his fixed position. These 'environment maps' or 'panoramas' are cylindrically projected images, taken with a panoramic camera or stitched together from multiple conventional camera recordings. Spherical or cubic projections also allow to render views of the zenith or nadir, at the cost of higher computational complexity during rendering. Because environment maps are view-centered, several maps, recorded at different locations, are needed to move within the scene. The viewpoint then jumps from one environment map recording position to the next.

If, in addition, disparity information can be obtained for all pixels, the environment map can be warped to yield views from nearby the map's recording position. In the 'plenoptic modeling' approach, two environment maps of a scene are used and the optical flow field between the maps is determined.

The optical flow allows to reconstruct both maps' relative recording positions as well as to warp both maps along the line of their respective centers-of-projection. The plenoptic modeling system requires to manually specify corresponding points in both environment maps to reconstruct the optical flow field.

5.4.8 Light Field Rendering and Lumigraphs

The notion behind light field rendering is to reconstruct an object's plenoptic function solely from image data. Because geometry information is neither required nor derived, any surface characteristics can be rendered, making light field rendering the most versatile IBR technique.

A scene's light field is described by a number of light rays intersecting object surfaces. A ray's light properties (like radiance, cf. Section 5.1) are determined at the surface point where it intersects an object in the scene. As free space is assumed to be transparent, the light attributes (if properly chosen) stay constant along the ray. Four parameters are needed to unambiguously describe ray position and orientation in this free space around the desired object.

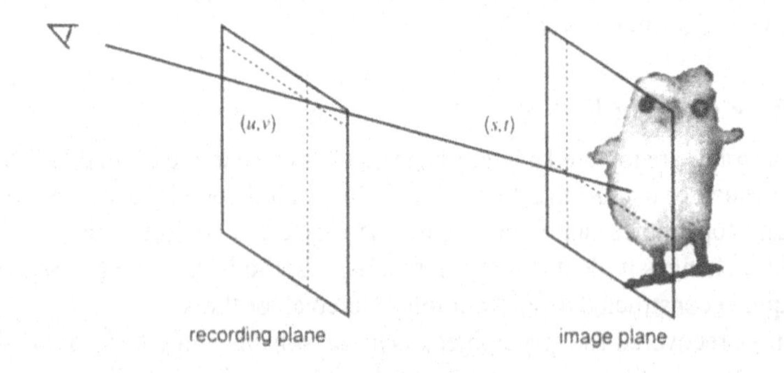

Figure 5.17. Two-plane parameterization of the light field.

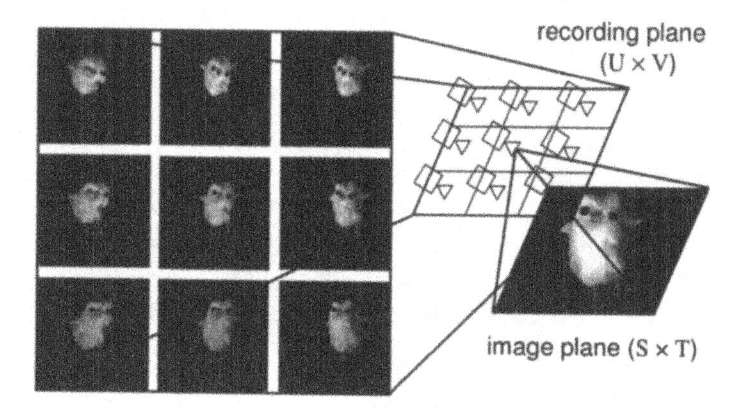

Figure 5.18. Recording an object's light field in the two-plane parameterization.

One convenient parameterization describes ray orientation by the ray's intersection points with two parallel planes, the *recording plane* (UV-plane) and the *image plane* (ST-plane) [435], Figure 5.17. On both planes we select an axis aligned rectangle of finite size. These are regularly subdivided into a lattice with $U \times V$ grid points in the recording plane and $S \times T$ points in the image plane, Figure 5.18. A ray is parameterized by its intersection coordinates (u, v, s, t). The set of all rays that can be parameterized by the two grid planes represents a *light field slab*. Because a light field slab can only capture frontal views of an object, six slabs need to be arranged in a cube around the object to record the entire light field.

Other parameterizations are possible (e.g. [712, 764]), yet the two-plane parameterization exhibits the lowest computational complexity during data acquisition as well as rendering: the light field can be readily recorded in the two-plane parameterization by arranging an array of cameras in the recording plane with their optical axes aligned in parallel. During rendering, rays are traced from the viewpoint, and the intersection coordinates with the UV-plane and the ST-image plane are rapidly calculated. The recording plane intersection coordinates (u, v) determine the image from which the pixel corresponding to the image plane intersection coordinates (s, t) is rendered.

The image plane's sampling resolution $S \times T$ determines the finest renderable details. To achieve aliasing-free rendering at maximum resolution, parallax motion between adjacent light field images must be smaller than one pixel (maximum stereo disparity). Therefore, the recording plane's sampling density $U \times V$ depends on image resolution. The number of images necessary to meet this sampling criterion can be estimated from simple geometry considerations, Figure 5.19:

- Image plane size depends on the object's maximum radius r; to take optimal advantage of image resolution without clipping the object, the image sides must be of length $2\sqrt{2}r$.

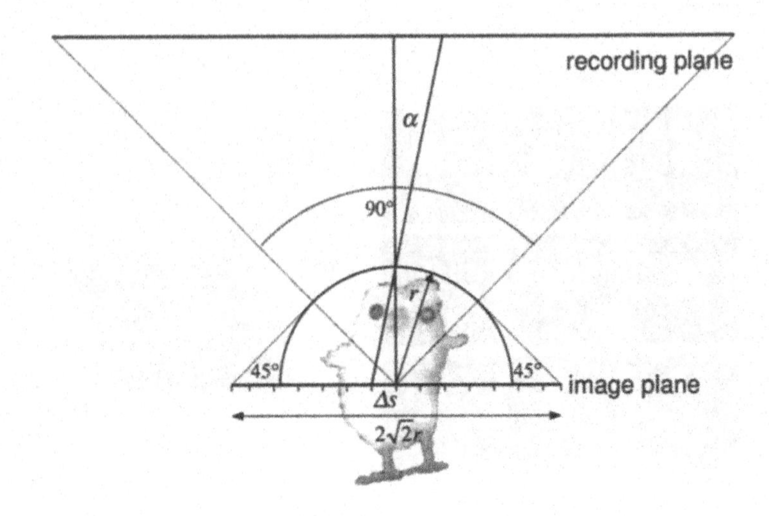

Figure 5.19. Recording geometry: r denotes the object's maximum extend, Δs is the size of a single pixel, and α is the parallax angle for one-pixel disparity motion of the outmost object point.

- The image consists of S pixels along the side, determining pixel size $\Delta s = 2\sqrt{2}r/S$.
- Maximum parallax angle α depends on pixel size Δs and object radius r: $\alpha = \Delta s/r = 2\sqrt{2}/S$.
- A light field slab covers $90° = \pi/2$ radians along one side, resulting in $U = \pi/(2\alpha) = \pi/(4\sqrt{2})\ S$ images.
- For the entire slab plane, $U \times V = \pi^2/32\ S \times T$ images are needed.

Because 6 slabs must be arranged around the object to capture its entire light field,

$$UV \approx 6\frac{\pi^2}{32}ST \tag{5.36}$$

images need to be recorded, stored and processed during light field rendering. For example, at $S \times T = 256 \times 256$ pixel resolution, $U \times V \approx 121000$ images are required.

Because this number of images can neither be acquired nor processed in reasonable time, only subsampled light field representations are available for rendering. The images have to be low-pass filtered prior to rendering in order to avoid aliasing effects. This corresponds to degrading image resolution $S \times T$ to meet the sampling condition in Equation 5.36 for a fixed number of images $U \times V$. Blurred rendered images result.

To estimate missing light field information, the scene's approximate geometry must be used. The *Lumigraph* employs a silhouette-based algorithm to derive approximate object geometry from light field images. Instead of low-pass filtering, geometry-dependent basis functions are derived. The light field images are projected onto the depth-corrected quadrilinear basis functions. If an approximate geometry model can be derived, the Lumigraph yields superior rendered views to light field rendering using the same input imagery, at the cost of higher computational complexity during rendering.

5.4.9 Current Research

Conventional geometry-based rendering methods have been augmented by IBR techniques to yield more convincing rendering results: In [652], an image-based approach to global illumination in virtual scenes is presented. Radiance maps are derived from images for any point within the scene. Another method is described in [512], where images and range data are used to efficiently compute and store global illumination effects in animated (dynamic) scenes. IBR has also been employed to simulate imaging properties of lens systems [313] and complex illumination effects, such as caustics, in virtual scenes [305].

Among IBR-methods, light field rendering has great potential because of its broad range of possible applications. Different light field parameterizations have been investigated [712], and attempts have been made to add variable illumination to light field rendering [764]. Because light field rendering relies on large numbers of images, data compression is a vital aspect; vector quantization [435] and block-based discrete cosine transform (DCT) schemes [481] allow fast data access at low compression ratios. Much higher compression ratios can be achieved if video coding techniques are suitably employed to light field compression [462], at the expense of higher decoding complexity. Approximate object geometry can further enhance coding efficiency as well as rendering performance by eliminating redundant information and estimating missing light field images [461]. Object silhouettes [266] or a volumetric approach [203] can be used to reconstruct the object's approximate geometry. Rendering from only three images is presented in [28], if corresponding points can be found for all pixels.

Future work in image-based rendering will focus on rendering dynamic scenes, simplifying light field acquisition from natural objects and interactive rendering performance from, necessarily compressed, complete light fields. In the not too distant future, IBR techniques might play an important part in the development of three-dimensional, object-oriented television.

6 VOLUME VISUALIZATION

T. ERTL

INTRODUCTION

Scientific visualization is the process of generating a visual representation of the information contained in abstract data fields resulting from computer simulations or sensoric measurements. The standard model of this process comprises a pipeline of three stages as shown in Figure 6.1.

The *filter* stage is a preprocessing step converting the raw input data into visualization data which is usually reduced by operations like sampling, slicing, cropping, etc. The *mapper* stage performs a mapping of the abstract data fields into a visual representation consisting of geometric primitives like points, lines, surfaces or voxels and associated graphical attributes like color, transparency, texture, etc. The *renderer*, finally, uses this scene description to generate images by means of 3D graphics APIs such as OpenGL or OpenInventor, possibly exploiting 3D graphics hardware to achieve interactive frame rates.

Many different mapping algorithms have been developed for various scenarios. A crude classification of these methods distinguishes between the dimensionality of the data set, the underlying data type, such as scalar, vector, multivariate, and the supported grid structure, such as regular, curvilinear or unstructured. Additionally, the continually increasing size of the data sets must be taken into consideration.

Volume visualization is one of the most actively researched topics in scientific visualization. It aims at the generation of comprehensible representations from three-dimensional scalar fields for a variety of application fields, like science, engineering or medicine.

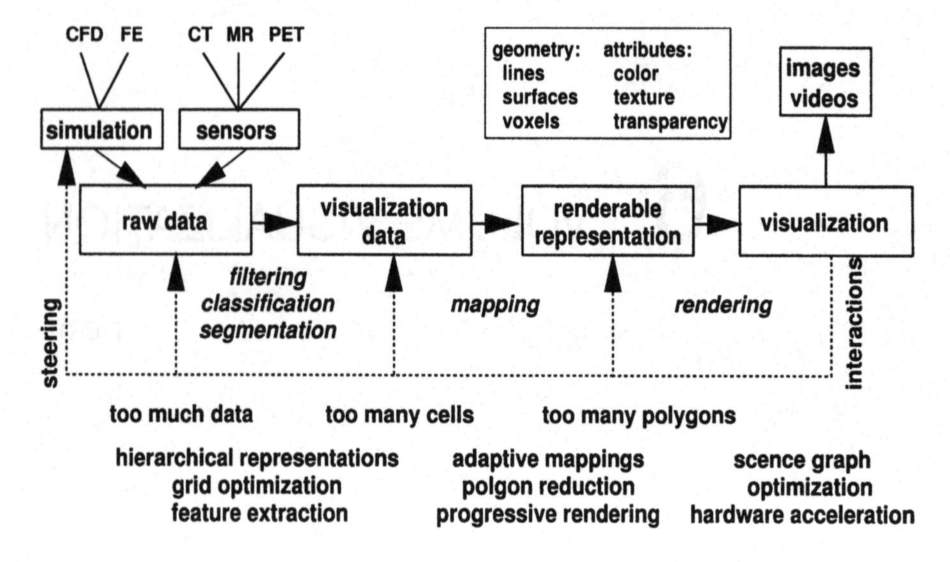

Figure 6.1. The visualization pipeline.

6.1 VOLUME DATA

T. ERTL

In the case of three-dimensional scalar data sets we usually speak of volume data. These data sets are very common in scientific or engineering applications and they result from computer simulations (e.g. pressure distribution in computational fluid dynamics or electric potentials in computational chemistry) and from measurements (e.g. seismic data in geophysics or atmospheric data in meteorology). Another very important class of scalar volume data comes from tomographic medical imaging, e.g. computed tomography (CT), magnetic resonance tomography (MRT), and other modalities. Although there are also some applications dealing with analytically given volume data sets the main input data for volume visualization is data sampled on discrete locations within the underlying domain. The spatial layout of data samples determines the shape of the volumetric object. This kind of information is provided by the *geometric model*.

Even though we are dealing with discrete data most visualization algorithms use the samples to implicitly reconstruct a continuous 3D scalar function which can then be evaluated at any arbitrary position. The *functional model* describes the decomposition of the underlying function spaces. Thus it determines the mathematical equations that have to be evaluated in order to reconstruct the 3D signal from its discrete representation. In some applications the nearest-neighbor model is chosen for algorithmic efficiency reasons, which means that the scalar function is considered to be constant within the volume element (*voxel*) surrounding a sample point. For better accuracy first-order interpolation is used within a volume cell [83]. For tetrahedral cells this means linear interpolation from the four cell vertices, for hexahedral cells trilinear interpolation based on the eight vertices of the cube is used. While the problems of locating a cell containing an arbitrary position or finding a neighboring cell is simple in regular grids, special algorithms have to be devised for these problems for the more complex meshes. Unstructured grids required additional data structures in order to store the neighbor information described by shared faces and vertices. Structured grids are either tetrahedrized resulting in a five-fold increase of the cell count or they are treated with special P-space/C-space algorithms like the stencil walk for cell location [99]. Although quite oftenly there exists a duality between the geometric or spatial decomposition and the decomposition of the underlying function spaces, in general no such unique relationship has to apply.

Finally, the *physical model* describes the physical properties of the underlying medium thereby approximating the visual stimulus of viewing the object. Thus it determines the rendering algorithms appropriate for a certain type of volume data.

6.2 VOLUME MODELS

R. WESTERMANN

Recent advances in the technology of 3D sensors, in the performance of numerical simulations and in the design of volume graphics algorithms result in the generation of volume data showing a broad variety of different characteristics and meanings. In order to allow for a consistent classification of volumetric objects, which is necessary to derive constructive approaches for data analysis and image synthesis, different models have been introduced. As in the real world, where objects are commonly described in terms of appearance and shape, volumetric objects in 3D graphics can be distinguished by means of similar attributes. As described earlier, three different models (*geometric, functional and physical*) exist, which allow for a unique characterization of arbitrary classes of volume data.

In this section we will focus exclusively on the geometric model. The geometric representation consists of a number of grid points spread out across an arbitrary domain in Euclidean space. By assuming that the associated data samples represent a 3D signal that entirely covers the underlying domain, the functional model describes the procedure by which an arbitrary value not necessarily coinciding with one of the discrete data samples has to be reconstructed. For an overview and comparison of different functional models we would like to refer to [36, 80, 211, 264, 154, 515, 501]. The underlying concepts of the physical model are well explained in [22, 309].

6.2.1 The Geometric Model

The shape of the volumetric object is described by the *geometric model*. In particular it accounts for the spatial layout of data samples within the underlying domain. Most significantly the geometric representation of the *lattice* or *grid* on which the samples are located provides the data structure necessary to store and to access the samples efficiently and determines the traversal routines used to visualize the object. In accordance to the nomenclature used to describe polygon meshes the topology of the grid describes the adjacency information while the spatial position of grid points is refered to as the geometry.

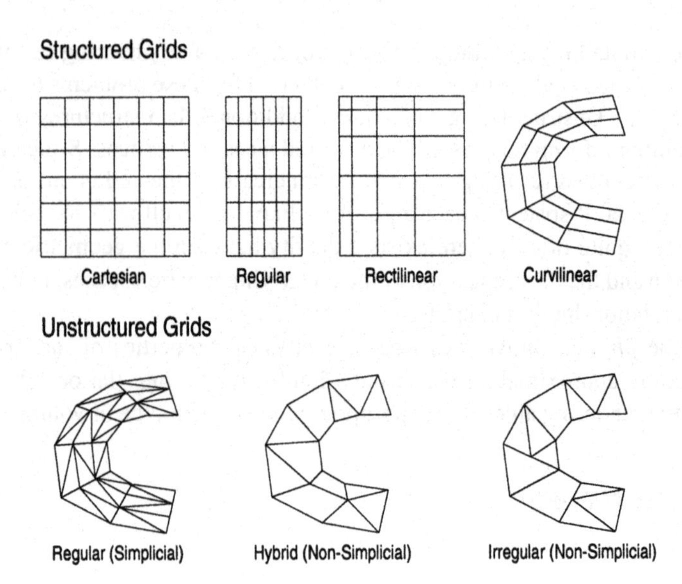

Figure 6.2. Classification of grid types.

Grid Types. There are many types of grids, which differ substantially in the *simplicial elements* or *cells* they are constructed from and in the way the inherent topological information is given. In $\mathbb{R}^n$ we consider a *simplex* to be the convex hull of $n+1$ affinely independent points. A collection C of simplices is called a *simplicial complex* if every face of an element of C is also in C and if the intersection of two elements of C is empty or it is a face of both elements. For instance, a tetrahedral grid forms a simplicial complex composed of simplices of dimension three (*3-simplices*). In the following we will only consider partitions in $\mathbb{R}^3$ which consist of elements that are 3-simplices or that can be constructed of tensor products of simplices of lower dimension, e.g. prisms or hexahedra. These partitions will subsequently be called *triangulations*. Although in some applications non-simplicial complexes may arise they are not discussed here.

In a first approach the geometric structure is usually classified by means of the type of the cells the grid is composed of [379] (see Figure 6.2).

Structured Grids. *Structured* grids are composed of sets of connected hexahedra, with cells being equal or being distorted with respect to possibly non-linear transformations. An aquidistant triangulation of a sub-space of $\mathbb{R}^3$ by means of cubic elements yields a *Cartesian* grid. *Uniform* grids consist of equal cells as well, but the resolution in at least one dimension is different from the resolution in the other dimensions. This kind of grid is most likely to occur in applications where the data is generated by a 3D imaging device providing different sampling rates in either direction. In *rectilinear* grids, on the other hand, the resolution might change arbitrarily within each dimension. As a matter of the geometric structure, all three grid types are necessarily convex. Although rectilinear grids as described exhibit certain useful features, e.g. the visibility ordering of elements with respect to any viewing direction is given implicitly, the rigid layout, in general, prohibits the geometric structure to adapt to local features. In contrast to rectilinear grids, *curvilinear* grids composed of concave elements which grid points have been transformed in physical space reveal a much more flexible alternative to model arbitrarily shaped objects. However, even this flexibility in the design of the geometric shape makes sorting and searching of grid elements a much more complex procedure.

One of the most important benefits structured grids composed of hexahedral elements offer is that the associated data samples can always be stored in a regular 3D array with different size and resolution in either direction. As a consequence arbitrary samples can be directly accessed by indexing a particular entry in the array. In addition, the topological information is implicitly coded in the array structure thus enabling direct access to adjacent elements at random positions.

Unstructured Grids. If no implicit topological information is given the grids are termed unstructured grids. Regular or conforming unstructured grids are composed of different types of elements that are connected in a consistent way. In such a triangulation any two elements are disjoint or they share a common face—a condition that is violated in irregular grids. For instance, a standard solution in finite element analysis to obtain consistent transitions between adjacent grids with different resolution is to perform a conforming split that eliminates *T-vertices* or *hanging nodes* at the element faces [62]. Hexahedral elements are split by inserting an additional grid point located at the cube's center and by decomposing the cube into six pyramidal elements. According to the pattern of T-vertices from finer neighboring cells that have an edge in common with the current cell pyramids are successively split into tetrahedra (see Figure 6.3).

Unstructured tetrahedral representations provide a striking degree of flexibility and they are less restrictive than other types in terms of their ability to accurately model global and local features. Furthermore, since each grid can be converted into

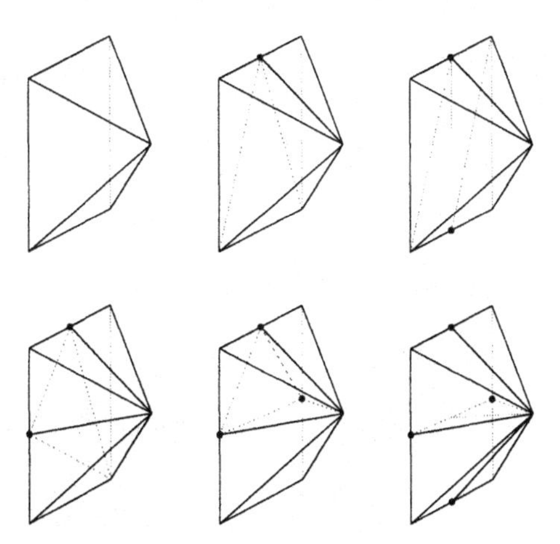

Figure 6.3. Possible configurations for a conforming split yielding an unstructured but simplicial grid.

a tetrahedral grid the algorithms necessary to deal with this type of geometry can be effectively reduced to the efficient handling of one base element. In general, similar data structures as outlined in Section 4.1.3 for triangle meshes can be used to store and manipulate tetrahedral grids. Topological information can be easily derived from the fact that any $(k-1)$-simplex in $\mathbb{R}^k$ can be a face of only two k-simplices. On the other hand, since unstructured grids are typically organized in data structures that consist of indices into the list of grid points rather than regular 3D arrays, direct access to elements according to a given position in space is in general not possible. Access usually involves traversing the grid along the adjacency information across shared faces.

6.2.2 Hierarchical Volume Representations

In contrast to the described geometric representations, which only provide one unique triangulation of the underlying domain, hierarchical or nested grids provide a sequence of triangulations with different resolutions. Throughout all areas of 3D graphics techniques benefiting from the hierarchical nature of multi-resolution representations have gained special attention. These techniques are considerably important to investigate in terms of the representation of objects at various *levels of details* (see Figure 6.4) and with respect to efficient analysis and rendering.

Hierarchical decompositions are constructed recursively by applying finite sets of coarsening or refinement rules to the elements of the previous level. In general, this can be accomplished in two different ways:

- Bottom-up: The coarsening procedure starts with the original grid. Increasingly coarser triangulations are generated by successively removing cells. Holes that occur during the removal procedure have to be closed by re-triangulation of the remaining cells. In order to account for geometric requirements on the resulting triangulation new grid points might have to be inserted.

- Top-down: The refinement procedure starts with a coarse geometric representation. Increasingly fine triangulations are generated by applying local grid refinement operations. New grip points are successively inserted, and those elements including the new points are split and re-triangulated instantaneously.

Figure 6.4. This illustration shows a volumetric object from a bio-medical application that was reconstructed from different resolution levels.

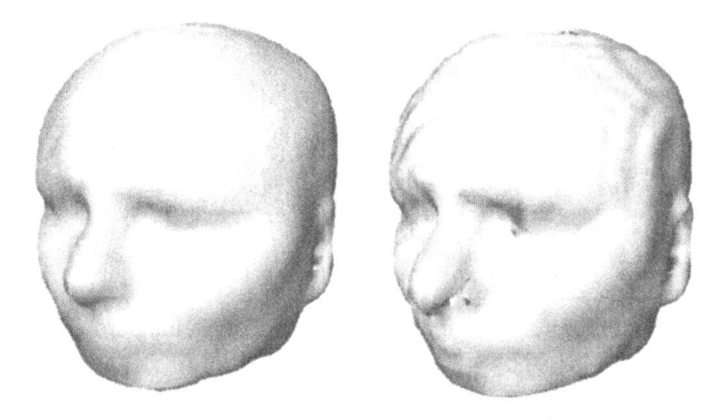

Figure 6.5. Iso-surface reconstruction from averaged (left) and subsampled (right) data.

Octree Representations. The most popular approach usually applied to Cartesian or regularly structured grids is to hierarchically decompose the grid by means of an *octree* representation. The basic idea is to recursively combine eight cubic cells to one coarser cell working bottom-up from the original grid. In practice this means that every other grid point is removed from the geometry thus bisecting the size in each dimension. We should emphasize here that by changing the geometric representation in this way it is in general no longer possible to reconstruct the original data from its discrete counterpart. The functional model, however, specifies the simultaneous decomposition of the underlying function spaces in order to allow for the reconstruction within certain tolerances. In Figure 6.5, for example, an iso-surface was reconstructed from one of the coarser levels of the octree hierarchy. In the left image data samples from cells that were combined in order to build the elements on the next coarser level were successively averaged, while on the right the down-sampling of data values was carried out.

Hierarchical Tetrahedral Grids. In general, the hierarchical decomposition of tetrahedral grids becomes more complex than that of rectilinear grids, since a combination of multiple elements can no longer be easily converted into one single element. Different approaches have been developed for the top-down and bottom-up construction of multi-level hierarchies.

Tetrahedral Grid Refinement. The grid refinement process starts with a coarse triangulation $\mathcal{T}_0$, and generates a sequence $\mathcal{T}_0, \cdots, \mathcal{T}_j$ of increasingly finer triangulations by successive local mesh refinement. Each triangulation in the sequence is required to be *conforming* (i.e. no T-vertices occur) and the sequence has to be *stable* (i.e. only elements with good aspect ratios are generated). In some application it may also be important that a hierarchy of nested spaces is constructed. Therefore, the triangulation sequence has to satisfy the *nestedness* condition, which means that each element in a triangulation is obtained by subdividing an element of the coarser triangulation.

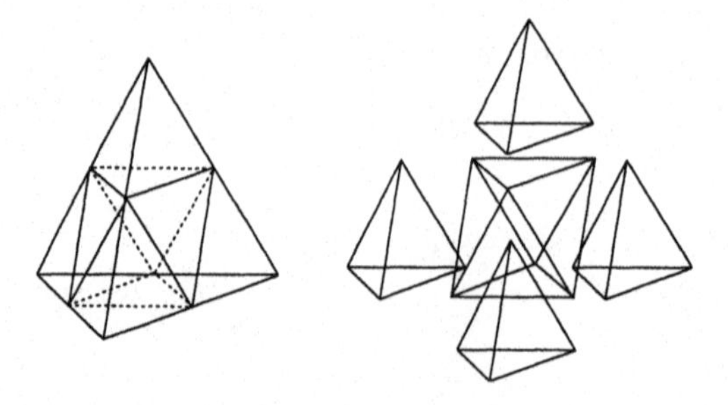

Figure 6.6. Regular or red refinement of a tetrahedron.

In particular, a conforming and stable triangulation of a partiton of $\mathbb{R}^3$ can be obtained by a Delaunay tetrahedralization [198, 365, 127], where the triangulation process starts with an initial complex and new grid points are successively inserted at appropriate positions. Each time a point is added a new triangulation has to be computed in consideration of the Delaunay criterion. Altough this method yields satisfying results it suffers from the well know disadvantages of Delaunay triangulations, i.e. the high numerical complexity.

In [62, 247] the common approach to refine triangle meshes by combining regular (red) and irregular (green) refinement operations [37] was extended to tetrahedral grids. At first, a refinement rule for a single element is defined in such a way that successive refinements generate stable and consistent triangulations. Such a local refinement rule is called *red* or *regular*, but in general it will result in T-vertices. In order to allow for a consistent triangulation a set of *green* or *irregular* refinement rules is defined for elements with hanging nodes. These refinement rules are *local* as well, but they are only carried out to 'repair' the hanging nodes. In order to avoid stability problems green refined tetrahedra must not be refined again. If in a subsequent step a subdivision is required, then the originally green refined tetrahedron must be refined according to the red rule. Finally, these local rules are combined and rearranged into a *global refinement algorithm*, which guarantees stability and conformity. The global refinement algorithm is based upon the following two restrictions: Firstly, irregular elements are never refined. Secondly, for the final hierarchy it holds that if an element remains unrefined passing from level k to $k+1$, then it remains unrefined for all levels $l \geq k$. By the regular refinement rules each element is split into four tetrahedra at the vertices as shown in Figure 6.6. The subdivision of the remaining octahedron into four tetrahedra is ambiguous and depends on the selection of one of the three possible diagonals. The strategy proposed in [62] produces stable regular refinements based upon affine transformations of a reference tetrahedron.

By using a different strategy the octahedra that is generated during the regular refinement can be preserved, but now the grid no longer is a simplicial grid. Thus, an additional regular refinement rule for octahedra is needed. It splits each element into six octahedra by cutting off the vertices and eight tetrahedra that will remain after the cutting off. The tetrahedra are obtained by connecting the triangles at the face center with the barycenter of the octahedron. Only during the closure (green refinement) each octahedron will be split into eight tetrahedra. The closure is then computed on the

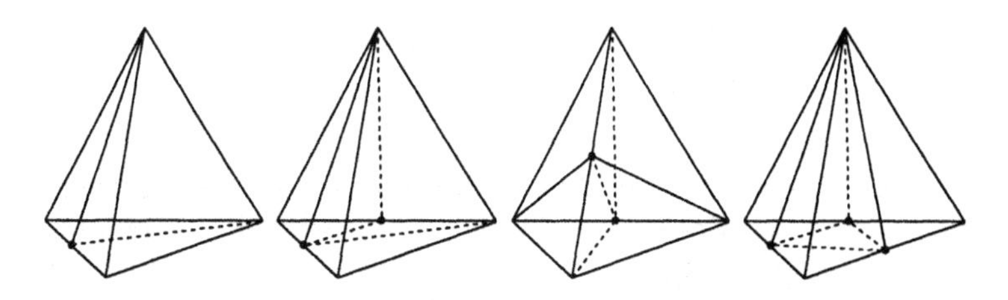

Figure 6.7. Irregular or green refinement of a tetrahedron.

resulting elements. Green refinement is restricted to the four different types shown in Figure 6.7 performing a red refinement on all remaining patterns. The *green closure* can also be computed by a set of refinement rules as described in [264].

Tetrahedral Grid Coarsening. In contrast to the outlined refinement strategies it is also possible to generate the hierarchical decomposition in a bottom-up approach starting with the original grid and building coarser representations until a stopping criterion is reached. Therefore, elements have to be removed and the occuring holes have to be re-triangulated, which usually prohibits a conforming triangulation of the domain.

In general, similar techniques as described in Section 4.1 for the simplification of triangle meshes can be applied to tetrahedral grids [552, 707, 661]. As in the triangular case the removal of a single vertex results in the removal of a number of elements sharing this vertex. However, by the vertex removal a possibly non-convex polyhedra is generated that needs to be re-triangulated. But note that the overall number of elements that are generated by means of the selected triangulation technique might be larger than the original number. An optimal triangulation technique in terms of the number and the shape of the resulting elements needs to be developed. To our best knowledge, no such technique that would be able to handle arbitrary geometries is actually available.

Edge collapses as described for triangle meshes in Section 4.1 can be extended straight forwardly to tetrahedral grids. The simplification procedure can be accomplished by collapsing three edges of the element to be removed into one arbitrary vertex of this element at a time [707] and by collapsing all elements sharing these edges as well. Elements sharing a single vertex of one of the edges that were collapsed might have to be stretched in order to include the remaining vertex. A different alternative is to collapse only one single edge and to remove all tetrahedra sharing this edge [661]. Now a vertex split operation is performed in order to add a new grid point located on the previously collapsed edge. All remaining tetrahedra that share one of the vertices on the collapsed edge have to be re-triangulated with respect to the newly added point.

In both scenarios special attention has to be paid in order to avoid the intersection of tetrahedra generated in this way. Again, the shape of the resulting elements have to be taken into account in order to derive stable triangulations. Furthermore it is most desirable that the boundaries of the original grid remain unchanged by the simplification process since otherwise the volume of the underlying domain might shrink considerably.

6.3 DATA ACQUISITION

P. HASTREITER

Starting with the discovery of x-rays by W. Röntgen in 1895, various technical and technological achievements led to the development of different medical imaging modalities. The respective systems produce two-dimensional *(2D)*, three-dimensional *(3D)* and partly four-dimensional *(4D)* images. They show absorption or emission effects if tissue or an organ was exposed to some kind of energy. The resulting image information describes the distribution and the intensity of the related processes giving insight to structural, anatomical and functional details. The notion 'tomography' is derived from the Greek word 'tomos'. It denotes the representation of a slice image which intersects a 3D object.

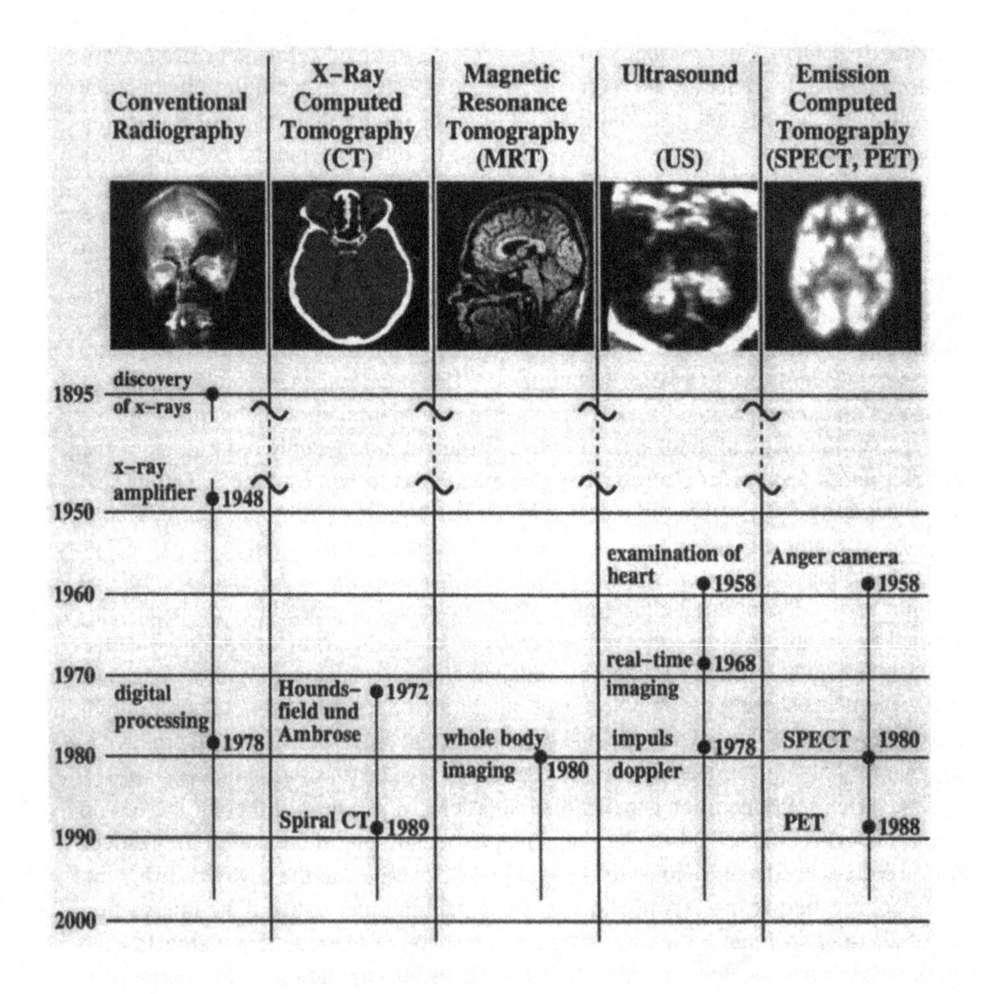

Figure 6.8. Historical overview of medical imaging modalities showing important technical and technological developments.

The historical overview in Figure 6.8 shows the imaging modalities which are mainly used in medicine. Among these modalities conventional radiography represents the origin. Still, it is the most commonly available and most frequently applied technique providing 2D projection images. Contrary to that tomographic imaging modalities are gaining increasing attention since they provide 3D information. This group includes CT (*X-Ray Computed Tomography*), MRT (*Magnetic Resonance Tomography*) and ECT (*Emission Computed Tomography*). Additionally, ultrasound must be mentioned since it is used more frequently due to its ability to produce images non-invasively and in real-time.

Understanding the underlying principles and techniques of every imaging modality contributes considerably to improve the development of modules which are used for the post-processing of the resulting image data. In several circumstances it is even easier to choose a more appropriate imaging technique instead of applying complex algorithmic approaches.

The subsequent overview introduces the above mentioned tomographic imaging modalities which primarily allow to access 3D information. The contributions of Knoll [394], Hutten [350], Cho [121], Morneburg [492] and Lehmann [427] mainly served as reference. In these articles the underlying physics and various technological aspects are explained more comprehensively.

6.3.1 X-Ray Computed Tomography

X-Ray Computed tomography (*CT*) determines the 3D distribution of x-ray densities within a target object. This requires to measure the attenuation of the radiation for many angles of view. On modern CT scanners a rotation of 180 degrees is sufficient while the projections are performed at fine angular increment. Similar to conventional radiography this results in images which represent the line integrals over all local attenuation coefficients along the projection rays. Based on all measurements the spatial distribution of x-ray densities is finally reconstructed by using different algebraic approaches and the laws of the Fourier transform.

CT-scanners mainly differ according to the approach applied to sample the target object. In the beginning of the technological development a simple strategy was used based on parallel projection and a focused x-ray. However, this requires a separate translation of the x-ray source and the detector for every increment of the rotation angle. Moreover, only a very limited portion of the generated energy is exploited. Due to the expensive mechanical procedure this results in time consuming measurements and a long exposure to radiation.

The introduction of fan-beam systems considerably improved CT measurements since the radiation of a single projection includes the whole target object. Thereby, a separate translation for every rotation increment is dispensable. Having used multi-cellular detector systems which rotate the x-ray tube and the detector about the patient, more recent scanners are based on ring detector systems with a fixed detector ring. Finally, the measurement time and the necessary radiation were considerably reduced with spiral CT-scanners. According to the underlying approach the target object is scanned on a spiral by continuously translating the patient table and rotating the x-ray source.

Theory. A particular relation between the target object and the single projection images represents the basis for the following reconstruction process. It is commonly known as Radon transformation. If parallel x-rays, exclusively considered within thin planes, are supposed, a projection with a coordinate s and a unit vector $\boldsymbol{n}_s$ is given with

$$p_\theta(s) = \int_{R^2} \phi(\boldsymbol{x})\delta(\boldsymbol{x}\boldsymbol{n}_s - s)d\boldsymbol{x} \ . \tag{6.1}$$

Thereby, for every projection ray with an angle θ according to the reference coordinate system, the attenuation coefficients $\phi(\boldsymbol{x})$ are integrated. Further on, relating the Fourier transform and the Radon transform, the projection-slice theorem is derived. According to

$$F_{1D}\left\{p_\theta\right\}(\omega) = F_{2D}\left\{\phi(\boldsymbol{x})\right\}(\omega\cos(\theta), \omega\sin(\theta)) \tag{6.2}$$

the 1D Fourier transform F_{1D} of a projection can be obtained by evaluating a line with angle θ within the 2D Fourier transform F_{2D} of the respective plane. Using this relation it is possible to reconstruct the distribution of x-ray densities for slice images if a sufficient number of projections is available. However, an adapted strategy is required since this strategy is very sensitive if discrete data is used.

Depending on the applied scanning technique (e.g. parallel or fan beam), a projection is based on a specific model which requires to use different algorithms for the reconstruction process. However, due to complex relations, it is more efficient to transform the image data to a uniform representation which corresponds to the parallel projection. Thereby, it is possible to apply a general convolution approach [492]. This requires to transform all projections $p_\theta(s)$ using a 1D Fourier transform according to $F_{1D}p_\theta(s)$. Subsequently, the 2D spectrum of the respective plane is reconstructed by applying the Fourier projection-slice theorem and integrating over corresponding locations. However, a substitution is required since efficient implementations of the fast Fourier transform are only available for Cartesian grids. Thereby, the polar coordinate system of the projections is taken into account which causes a change of the differentials. This implies a filtering operation which is expressed by a multiplication with $H(\mu)$ in the spectral domain. Choosing a different order according to linearity this results in

$$\phi(x, y) = \int_0^\pi F_{1D}^{-1}\left\{H(\mu) \cdot F_{1D}\left\{p_\theta(s)\right\}\right\} d\theta \ . \tag{6.3}$$

Since the filtering operation is performed prior to the integration process only a 1D Fourier transform has to be performed. Due to symmetry effects the integration is restricted to an angle of π. According to the Fourier transform the multiplication in the spectral domain equals a convolution with $F_{1D}^{-1}\left\{H(\mu)\right\}$ in the time domain. Therefore, specific filtering functions are applied in order to process the discrete image data in an efficient way. As an alternative, it is also possible to perform the reconstruction on a polar grid. However, the required interpolation operations make the inverse Fourier transform quite problematic resulting in considerable artifacts.

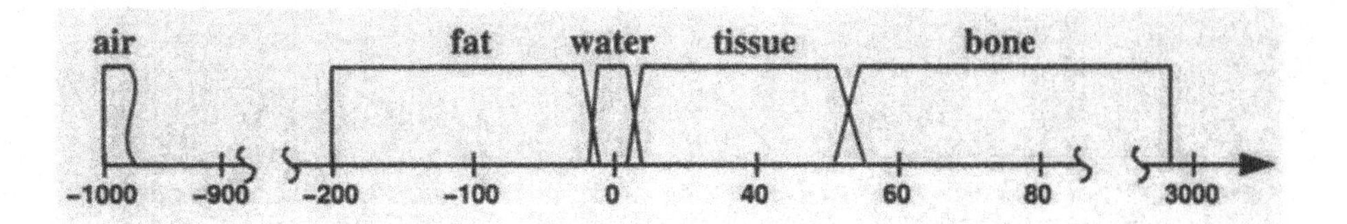

Figure 6.9. Hounsfield scale showing relative attenuation coefficients with water as reference material.

Application. Instead of using the reconstructed attenuation coefficients ϕ_{object} directly, they are normalized to the relative values ϕ_{relative}. Water (H_2O) is used as reference material since it effects x-rays almost the same way as soft tissue or body fluids. Based on the attenuatlon value ϕ_{H_2O} which is determined with a calibration curve [492], the relation

$$\phi_{\text{relative}} = \left(\frac{\phi_{\text{object}} - \phi_{H_2O}}{\phi_{H_2O}} \right) \cdot 1000 \tag{6.4}$$

provides a new scale with the unit HU *(Hounsfield-Units)*. Thereby, the difference of the attenuation ϕ_{object} caused by an object is given with a weighting factor of 1000 relative to the value $\phi_{\text{relative}} = 0$ achieved with water. As can be seen in Figure 6.9, the Hounsfield scale is coarsely subdivided into the areas of air, fat, tissue and bone. Within certain bounds this allows to analyze only a specific part of the data by using simple thresholding operations.

The technological progress related with spiral CT has considerably improved the diagnosis of vascular information based on CT-angiography (CTA) examinations (see Figure 6.10). Since relatively short measurement times are required it is possible to scan vessel trees completely, though it is difficult to delineate the most tiny structures. However, this approach is much less invasive compared to conventional angiography.

As an advantage, CT provides high resolution image data with matrices of usually 512^2 pixels. This allows good localization and analysis of fractures, hemorrhage and other diseases. However, there is a very limited ability to differentiate soft tissue. In comparison to conventional angiography every scan causes higher costs and above all a considerably higher radiation exposure. Moreover, the slice images are restricted to orientations orthogonal to the longitudinal axis of the body. In the future, smaller and mobile scanners will gain importance while the scanning technique will be accelerated with multiple detector arrays.

6.3.2 Magnetic Resonance Tomography

Magnetic resonance tomography *(MRT)*—other notations are magnetic resonance imaging *(MRI)* and nuclear magnetic resonance tomography *(NMRT)*—provides slice images of arbitrary orientation which contain a great variety of information. Chemical elements with a magnetic moment (e.g. H, ^{19}F, ^{31}P, ^{23}Na) represent an important prerequisite. Since hydrogen occurs most frequently in a human body it is mainly used. This allows to show soft tissue with very high level of detail whereas bone structures are almost invisible.

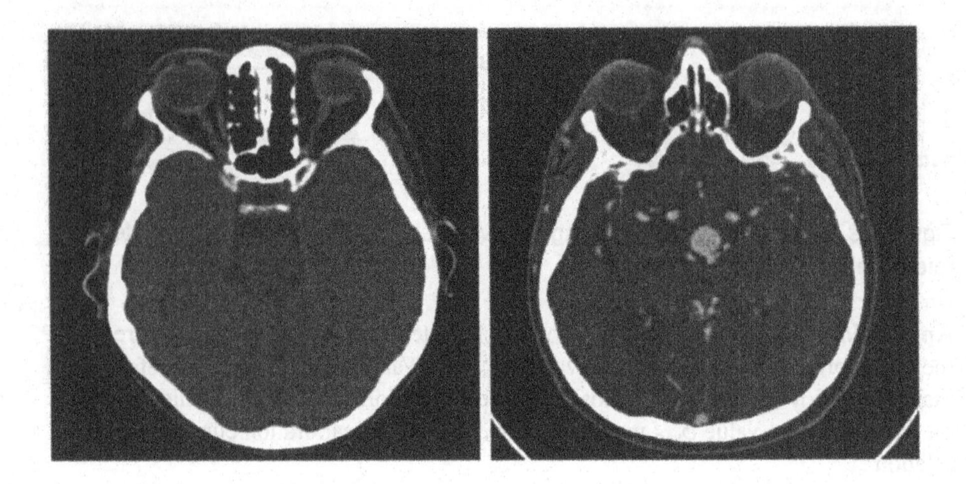

Figure 6.10. Slice image of a conventional CT scan (left) showing the attenuation of x-rays according to the density. Additionally, vessel information is visible if contrast dye is injected during a CTA examination (right).

During the imaging process the electro-magnetic energy is measured which is emitted after excitation of the respective nuclei within the target structure. This process is based on a radio frequency pulse and a locally variable static magnetic field. In order to perform the spatial encoding, the characteristic frequency *(Lamor frequency)* of the selected element is of fundamental importance. The resulting images mainly show the 3D distribution of the nuclear magnetization (see Figure 6.11). In the

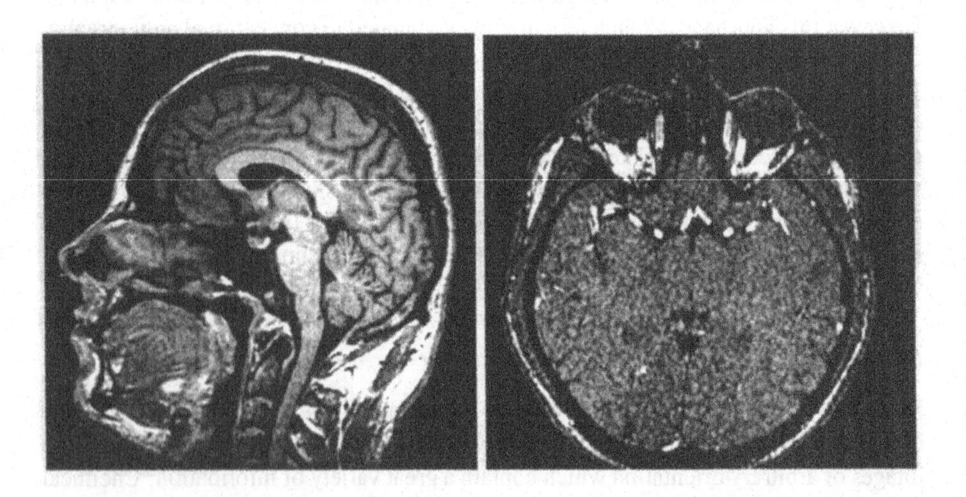

Figure 6.11. Slice images of a T_1 weighted MR scan giving a detailed representation of soft tissue (left). With MRA the vascular information (white) is clearly delineated from the remaining tissue (right).

majority of cases an additional weighting is performed with the specific time constants T_1 and T_2 which describe the main relaxation processes after excitation.

Theory. Protons are charged particles which perform a self-rotation called spin. Therefore, they produce a magnetic field according to the first equation of Maxwell. Taking into account all atomic nuclei with an odd number of protons and neutrons this results in a joint spin and magnetic moment. In consequence, the rotation axes of all protons have the same orientation within a magnetic field of flux density $\boldsymbol{B}\left[\frac{\mathrm{Vs}}{\mathrm{m}^2} = \mathrm{Tesla}\right]$. Considering the gyro-magnetic ratio which is specific for each nuclei, the spins precess with the Lamor frequency $f_L = \left(\frac{\gamma}{2\pi}\right)|\boldsymbol{B}|$ relative to the axis of the external static magnetic field. This situation is comparable to the precession of a gyroscope in a gravitational field.

Using a radio frequency pulse the orientation of the spins is altered. Following the excitation the spins go back to their original state within the static magnetic field. During this process which is called *free induction decay* (FID) the absorbed energy is emitted again. As a result, the intensity of the electro-magnetic signal gives information about the density of the participating nuclei. Additionally, two exponential processes are of importance which describe the induction decay (see Figure 6.12). Directly after the excitation there is interaction among all spins. This relatively short process, which is described with the time constant T_2, is called spin-spin relaxation. However, the absorbed energy is mainly exchanged with the surrounding lattice in order to return to a state of equilibrium. This second process which is described with the time constant T_1, is called spin-lattice relaxation. Based on different pulse sequences the values of these time constants give specific information about soft tissue.

The tomographic imaging procedure requires to encode the FID signal in order to identify specific spatial location according to the measurements. Since the Lamor

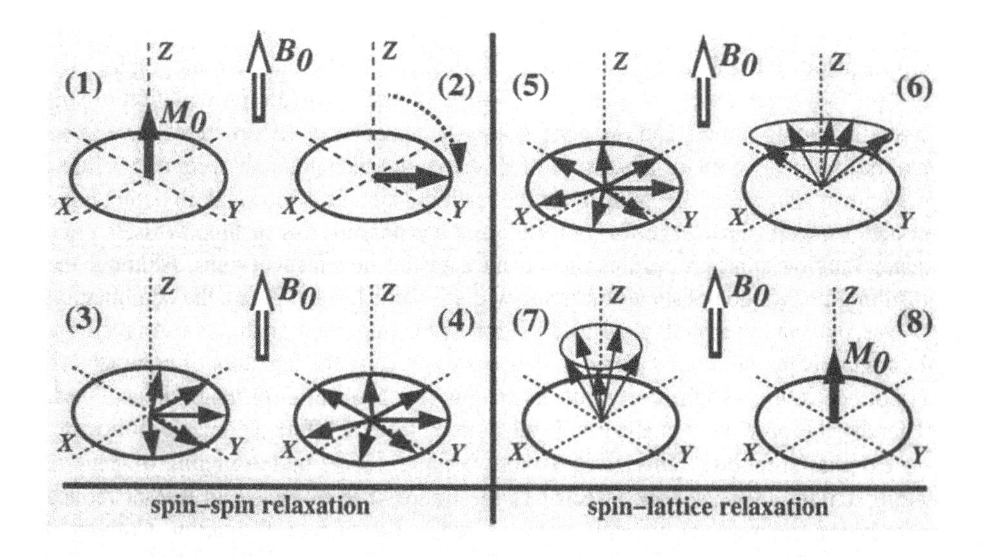

Figure 6.12. Relaxation of spins after excitation with a 90° radio frequency pulse.

frequency depends on the underlying magnetic field, a magnetic field gradient is superimposed onto the main static magnetic field. Performing a Fourier transform of the emitted signal, there is direct correlation between the frequencies of the spectrum and the location of the magnetic gradient. Additionally, the intensity values of the resulting slice image are determined directly, since the amplitude of every spectral component depends on the relative density of the precessing nuclei.

In order to achieve a detailed spatial encoding a magnetic field gradient in z-direction (longitudinal axis of the patient table) selects a specific slice orthogonal to its orientation. Thereby, all nuclei of the selected xy-plane respond within a range of Lamor frequencies of very limit width. Further encoding is performed with a second magnetic field gradient in x-direction. This assigns specific frequencies to columns in y-direction which are used in two different ways.

Using a method based on *projection reconstruction*, a multi-spectral radio frequency pulse is applied for excitation. This causes a response signal whose Fourier spectrum represents a projection of the spin densities onto the x-axis. If the x-gradient is successively rotated with small angular increment, it is possible to reconstruct the spatial distribution of the nuclear magnetization with a great number of these measurements. The reconstruction is performed with a standard CT method which was discussed in Section 6.3.1. However, motion artifacts inhomogeneities of the magnetic field strongly influence the quality of the results.

Using an approach based on *Fourier reconstruction* the x-gradient remains at the original position. In this case the necessary is performed with a third magnetic field gradient in y-direction. It is used to modulate the phase of the nuclei in x-direction allowing to differentiate every column within the Fourier spectrum of a projection. Considering an image matrix consisting of N $\times$ N unknown pixel values, this method requires to perform N measurements, since every projection results in an equation for every column.

Application. The image data obtained with magnetic resonance tomography give answer to a great variety of questions. Above all, the detailed representation of soft tissue including tumors and the good spatial localization of all structures have to be emphasized. As a major advantage of this imaging technique no secondary effects are known. However, the intensity of the resulting signals is too weak to depict bone structures, teeth or cartilage. In contrast to that it is possible to scan blood vessels since special angiographic procedures allow to measure the movement of spins. Although the resulting images do not show soft tissue with the same level of detail, the combination of vascular and morphological information gives important anatomical orientation. In comparison with conventional radiography and CTA the injection of contrast dye is not necessary. Additionally, functional and biochemical information is accessed. Thereby, it is possible to analyze molecular structures and different processes following an external stimulus. This gives a more comprehensive understanding of general brain activity and metabolic effects. In the future, it is envisaged to further reduce acquisition times and to improve ability to access functional information. Combined with technological solutions which provide enhanced local flexibility, MRT will gain increasing importance.

6.3.3 Emission Computed Tomography

The notion emission computed tomography (ECT) comprises all methods of nuclear medical imaging. In comparison to other modalities the related techniques have considerable potential to account for functional information quantitatively. This requires to analyze the decay of radio-isotopes, which concentrate in specific parts of the body depending on the metabolic activity, after administration or injection. Measuring the occurrence and the emitted energy of photons (γ-quantum), which result from the decomposition processes, allows to understand metabolism in-vivo in organs or tissue. The fundamental device for the necessary measurements is the scintillation counter [492]. Using a crystal it converts every hitting γ-quantum to a visible photon which are finally transformed to electrical current with a photomultilpier. The whole system is integrated into a gamma camera which is also called Anger camera.

SPECT. With SPECT *(Single Photon Emission Computed Tomography)* the spatial distribution of radiation is measured which is emitted from instable isotopes. This is achieved with radioactive nuclei which produce a single photon during decay. The measurements is performed with a gamma camera which rotated around the patient. Similar to CT, projections are acquired for a great number angles accounting for the decay events. Typically, a full rotation of the camera is required with an angular increment of 2° between the measurements. The resulting images are mainly used to analyze functional processes like the blood flow and metabolism.

Finally, equidistant slice images are reconstructed which show the spatial distribution of the decay processes. In most cases the approach based on filtered back projection is applied which was explained in Section 6.3.1. This results in image matrices of 128^2 pixels with each pixel having an average size of 3 mm^2. However, there are certain limitations of this approach, since specifically designed filters are required. Above all, this concerns the low signal to noise ratio of the images caused by the limited sensitivity of the collimator within the camera.

PET. The imaging modality PET *(Positron Emission Tomography)*, which was brought to clinical application more recently, uses tracers emitting positrons during decay. Shortly after a positron is created, it combines with an electron. During the subsequent annihilation process two gamma photons are produced which move on anti-parallel directions. Therefore, the photons are measured with detectors arranged on a ring. Thereby, an event is assigned to the line combining the activated detectors. In order to reconstruct the spatial distribution of the occurring activity multiple measurements are required. This results in projection images similar to CT which are used to reconstruct the final slice images (see Figure 6.13). The resolution is restricted to 1–2 mm for head and 2–3 mm for whole-body images since the location of the annihilation process is recorded instead of the positron emission. Moreover, the direction of flight of the photons is not an exact line due to a remaining impulse. However, in comparison to SPECT, the resolution of PET is better by a factor of 2–4 [492].

The radioactive labeling is performed with the positron emitting isotopes ^{11}C, ^{13}N, ^{15}O, and ^{18}F which are applicable for radio-pharmaceutical use due to their chemical characteristics. However, the short half-life period of these elements requires to have a

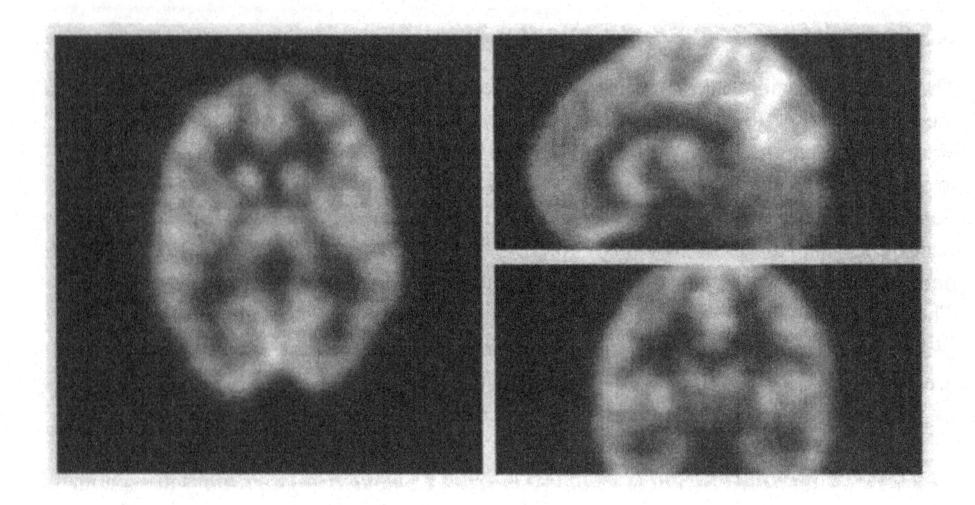

Figure 6.13. Orthogonal slice images taken from a PET volume.

cyclotron as additional imaging equipment. Altogether PET is an important assistance for the analysis of tissue, metabolism and functional processes in cardiology, neurology and oncology.

6.3.4 Tomographic Image Data

Currently, every tomographic data set consists of an arbitrary number of slice images which are scanned at equidistant locations. Therefore, all slice together represent a volume if they are arranged according to their spatial positions. Thereby, every slice forms a matrix of 128^2, 256^2 or 512^2, and in some case even 1024^2 sample points. The spatial resolution, the contrast resolution and the applied form of energy *(ionizing, non-ionizing)* are further parameters which characterize a tomographic data set.

In analogy to the notation of pixels (**pic**ture **el**ement) which is used for the 2D case, the sample points of a tomographic data set are called voxels (**vo**lume **el**ement) since they are spatial objects. In general, the size of a voxels is uniform within a slice image (`xdim` = `ydim`) if the reconstruction process is taken into account. Usually, the size of the remaining direction is different since a separate parameter is used to define the distance between neighboring slices. Therefore, tomographic data sets are described with an anisotropic uniform grid. For further processing, like volume rendering, the respective scaling factor must be considered (see Figure 6.14). The data values at every grid position describe a scalar magnitude like the density or the magnetic resonance within a continuous area. In general, they comprise a maximal range of 12 bits within an unsigned data type of 16 bits.

Due to the discrete nature of tomographic volume data intermediate values have to be calculated in order to avoid artifacts if post-processing is performed. During the interpolation process, this requires to reconstruct the continuous function which is subsequently sampled at discrete positions. Then, according to the sampling theorem, an exact reconstruction is performed if the discrete representation of the data is con-

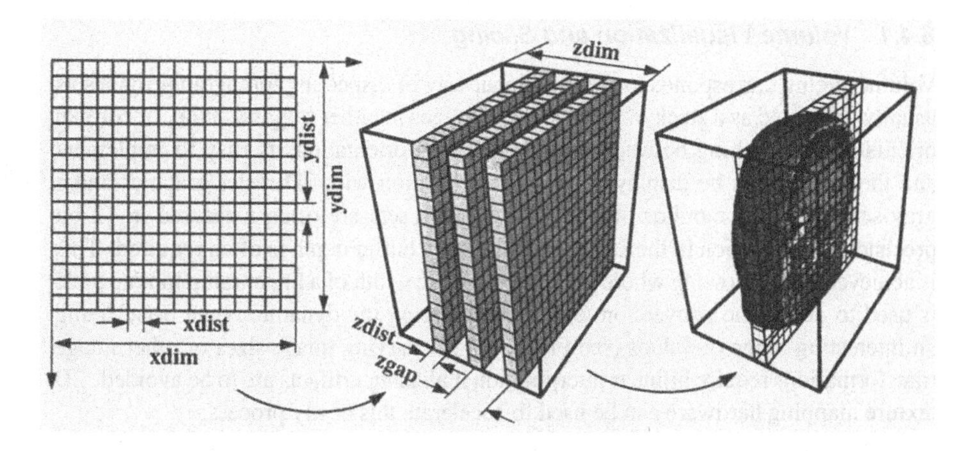

Figure 6.14. Structure of tomographic data sets: The parameters `xdim`, `ydim` and `zdim` represent the number of samples in every dimension with `xdist`, `ydist` and `zdist` describing the size of a voxel. In order to avoid artifacts during volume rendering the distance between the centers of neighboring slices (`zgap`) and the slice thickness should be equal (`zdist = zgap`).

volved with a sinc-function. However, the band-pass filtering which occurs during the tomographic reconstruction restricts this ability. Furthermore, the necessary computational effort is opposed to a fast visualization process. A compromise is achieved if nearest neighbor or linear interpolation is chosen.

6.4 VISUALIZATION OF SCALAR VOLUME DATA

T. ERTL, P. HASTREITER

The problem of computer-assisted imaging of scalar volume data is still one of the main research areas of scientific visualization [580]. Volume visualization techniques deal with the efficient generation of a visual representation of the information contents of volumetric data sets. These data sets result from different applications (e.g. medical imaging, CFD simulation, voxelized geometries, and participating media). For scalar-valued volumes different techniques have been developed to a high degree of sophistication. However, only a few approaches allow parameter modifications and navigation at interactive rates for realistically sized data sets. This fact is due to the huge number of volume cells which have to be processed and to the variety of cell types ranging from regular to completely unstructured grids.

In general, the approaches for visualizing three-dimensional scalar data can be divided into three categories [206]: those that compute two-dimensional slices using color or brightness to indicate the scalar value at each point of the slice, those that extract characteristic surfaces, i.e. isosurfaces, from the volume and render the corresponding polygonal representation appropriately lighted and shaded, and finally those that directly render the volume by assigning different color and opacity values to different objects or value ranges and compositing these values along lines-of-sight.

6.4.1 Volume Visualization and Slicing

Volume slicing corresponds to the traditional way of inspecting tomographic data sets usually presented as a stack of parallel slice images in either saggital, axial, or coronal orientation. Switching between those orthogonal orientations is easy to implement and the images can be displayed on any workstation with 2D raster graphics and a gray-scale or color monitor. Since medical data sets are often measured in 12 bit precision, a conversion to the standard 8 bit frame buffer depth is often required. This is achieved by windowing where the center and the width of a linear ramp lookup table is used to define the conversion, either compressing the dynamic range or selecting an interesting range of values (see Figure 6.15). Varying image sizes or other image transformations require bilinear interpolation if aliasing artifacts are to be avoided. 2D texture mapping hardware can be used to accelerate this costly process.

Extracting arbitrarily oriented slice planes from a volume data set, which is termed multi planar reformatting in the medical community, is even more expensive due to

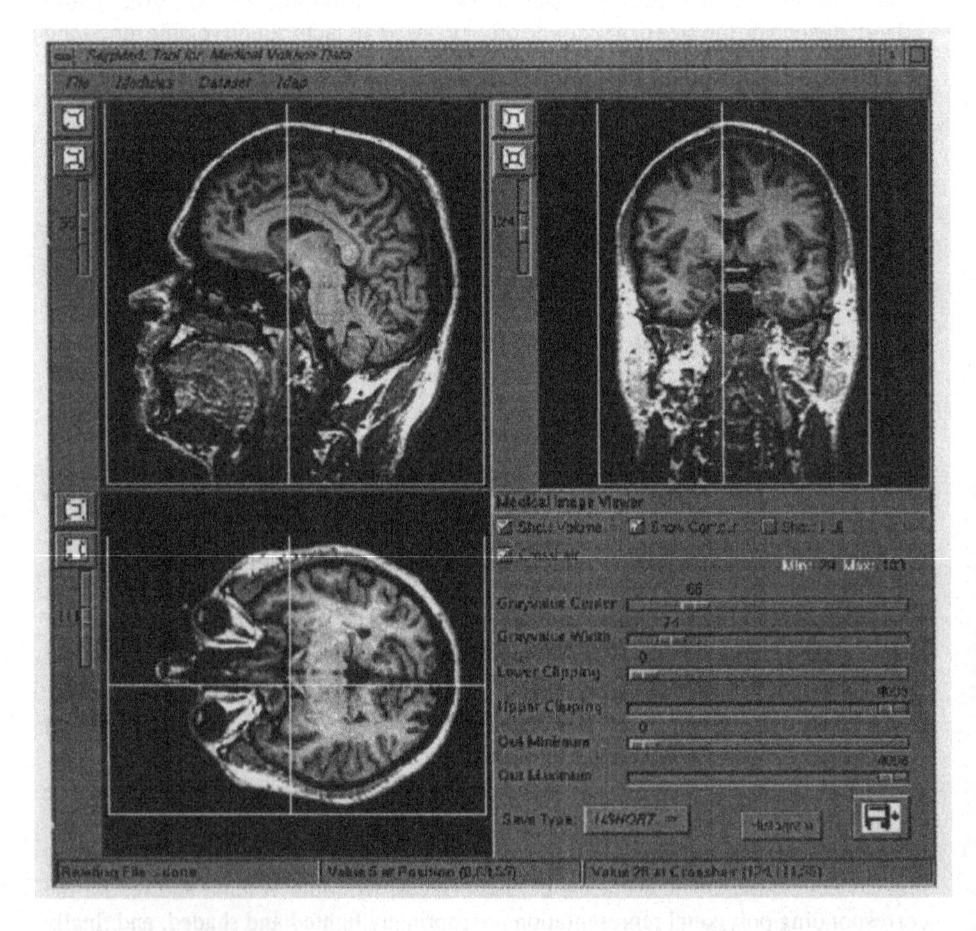

Figure 6.15. Saggital, axial and coronal slices of a MRT data set with gray-level windowing functionality.

the need for trilinear interpolation. Again, exploiting 3D texture hardware enables the user to interactively manipulate the position and orientation of the slice plane and to render the slice with real time frame rates. With 3D texture mapping the entire volume or a certain brick is loaded into texture memory. If a polygon is rendered with texture coordinates of the 3D volume assigned to the vertices, each pixel of the polygon is textured by trilinear interpolation during rasterization [499, 633].

Cutting planes are also a standard technique for a step-by-step visualization of volume data sets from scientific and engineering applications. For data on irregular grids, however, cell search and interpolation are too complex for high rates of interactivity. Besides exploiting graphics hardware [749] common approaches are to limit the size of the slice plane or to use smaller volume probes such as bars or icons.

6.4.2 Isosurface Extraction

The set of all points in a volume with a given scalar value is called level set. If the scalar field is non-constant and sufficiently smooth, this set of points actually does form a surface and is then called level surface or isosurface. Isosurfaces are an indirect way of visualizing a scalar volume. An opaque surface that passes through all cells that include a specific function value only represents one aspect of the data set, since volume features at other values are ignored. This is a good approach for smoothly varying functions and for objects with sharply determined borders like bones in CT, where the illumination of the surface greatly enhances its three-dimensional structure (see Figure 6.16), but it is inadequate for amorphous objects that can hardly be represented by mathematically thin surfaces. Understanding the structure of an unknown data set requires the extraction of surfaces for many different isovalues which can then be viewed on 3D graphics workstations from all directions at interactive frame rates.

The slice-wise representation of medical volume data sets led to early approaches where isosurfaces are constructed by connecting contour lines extracted from neigh-

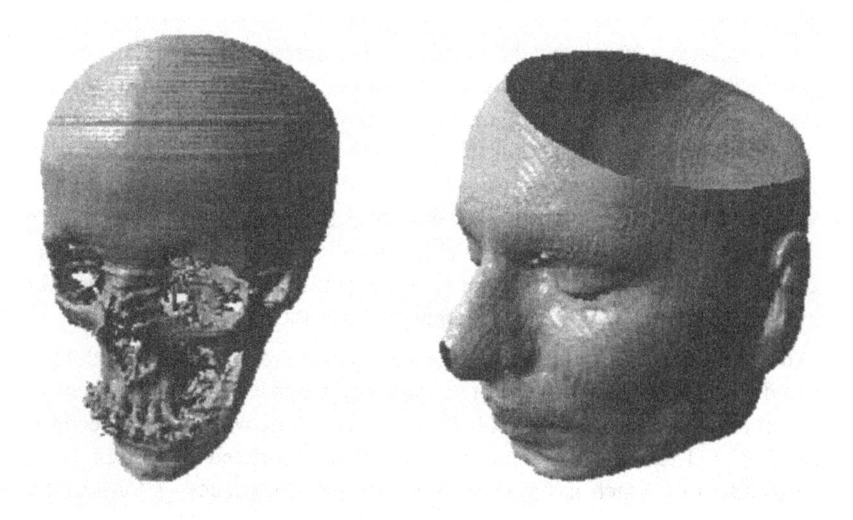

Figure 6.16. Isosurfaces of skull and face extracted from CT and MR data.

boring images. There are sophisticated algorithms for dealing with problems of corresponding vertices, of topological inconsistencies, and of correct triangulation [479, 519], however, the principal problem with surfaces almost tangent to the slice planes remains.

The standard isosurface extraction method used today is the Marching Cubes algorithm [452]. It successively traverses all cells of the volume first deciding whether the cell contains the isosurface by a simple min-max test. Then the cell is classified by an 8 bit index which marks each vertex as either inside or outside. This index points to a table where all possible triangulations of the surface within the cell are stored in 15 equivalence classes. The table lookup specifies the edges of the cube which are involved in the triangulation. The actual position on each of the edges is derived by linear interpolation. Special treatment of ambiguities is required to avoid inconsistencies visible as holes [505].

Before the resulting list of triangles can be rendered shading normals should be determined in order to improve the spatial appearance of the isosurface. For the simplest case of flat shading the normals of the triangles are easily computed from its vertices. For Gouraud shading, however, it is better to derive the normals at the triangle vertices directly from the data set. Since the normal of an isosurface characterizes the direction of maximum variation within the data set, the gradient of the volume can be used as a surface normal. First, the gradient is computed at the position of the volume samples by divided differences, then, it is linearly interpolated at the triangle vertices and appropriately normalized.

While the generated polygonal isosurfaces seem to be an appropriate volume visualization technique when using 3D graphics workstations, it turns out that this is not the case for very large data set sizes. Here, the surface extraction takes on the order of minutes on advanced CPUs and generates up to a million triangles and more, which can hardly be rendered with acceptable frame rates on even high-end graphics hardware. Both aspects severely restrict the interactive explorative visualization of large volume data sets and various suggestions have been made to overcome each of them.

One approach to accelerate the extraction of isosurfaces is to preselect the cells that may contribute triangles to the isosurface. The octree spatial subdivision presented by Wilhelm and Van Gelder [756] annotates each octree node with the span of sample values in its children. Traversal of the octree allows for a fast exclusion of large volume blocks from unsuccessful cell search. Several related techniques have been developed during the last years, like improved partitioning and incremental update techniques [30, 639, 640], as well as efficient cell search with interval data structures [128, 129, 444]. These algorithms generate the same triangles as the standard Marching Cubes does, but they require extra memory to store the interval information.

Another isosurface optimization is the reduction of the amount of triangles generated for rendering. This can be done in a post-processing step with one of the many proposed polygon reduction algorithms [624, 324, 395] or interleaved with the reconstruction [638, 556]. The reduction decisions are always restricted to a fixed iso-surface, the extraction of which is not optimized in these approaches. Closely related are geometry optimizations like compression [161], vertex removal by striping [213, 208] or occlusion culling [443].

6.4.3 Volume Rendering

Direct volume rendering tries to convey a visual impression of an entire volume data set by assigning different color and opacity values to different objects or value ranges within the volume [379, 434]. The resulting image is then computed by taking into account the specified emission and absorption effects as seen by an outside viewer. Volume rendering can be successfully employed for visualization purposes, leading to different visual representations of the same data set depending on the assigned transparency and color. Smooth opacity variations lead to images giving a semi-transparent overview of the total data set, while steep opacity gradients result in enhancing certain volume structures similar to surface rendering (see Figure 6.17). Volume rendering can also be used in photo-realistic settings in order to generate images of clouds, fog, smoke, and other volumetric effects.

The underlying theory of the physics of light transport is simplified to the well-known volume rendering integral in the case of neglecting scattering and frequency effects [407, 309, 588]. Given the emission q and the absorption κ the intensity I along the ray s can be computed from:

$$I(s) = \int_{s_0}^{s} q(s') e^{-\int_{s}^{s'} \kappa(s'') ds''} ds' \quad . \tag{6.5}$$

The discretization of this integral together with the assumption that the mapping from scalars to RGBA values can be described by transfer functions results in the compositing formula for computing the intensity contribution along one ray of sight:

$$I = \sum_{k=1}^{n} C_k \prod_{i=0}^{k-1} (1 - \alpha_i) \quad . \tag{6.6}$$

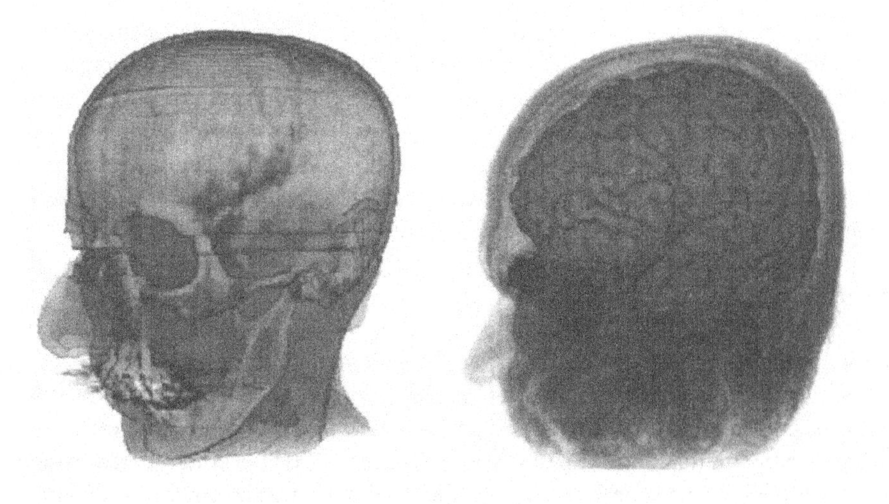

Figure 6.17. Direct volume rendering of a CT and a MR data set of the head.

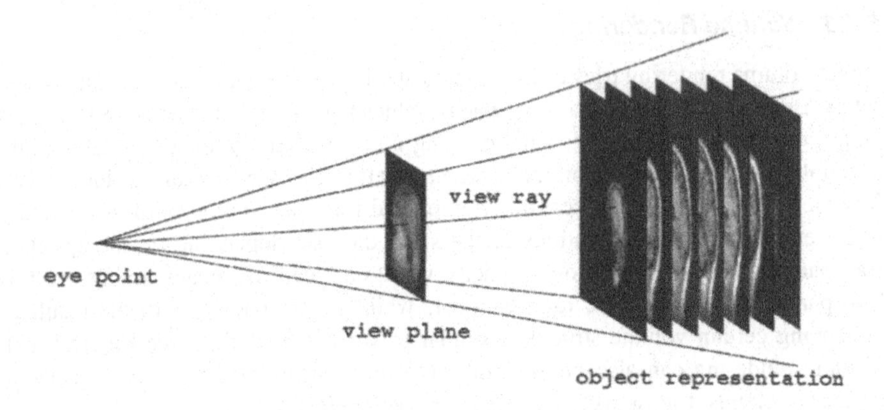

Figure 6.18. Volumetric ray casting.

The emission of the voxel C_k and its opacity α_k are derived from the transfer functions after interpolation of the scalar value from the discrete sample points.

In order to compute these equations three main approaches can be distinguished. Object order techniques loop through all cells of the volume and project them correctly ordered into one ore more pixels on the screen. Image order techniques, in contrast, loop through all pixels on the screen determining for each of them the correct contribution of the volume cells along the line-of-sight. Domain techniques transform the volume from the spatial domain in some other function space, e.g. Fourier domain or wavelet domain, and generate the projection directly in that domain or supported with special information gained from there.

The basic ray casting idea (see Figure 6.18) is to shoot a ray through every pixel into the volume, reconstructing the function value at appropriately chosen sample points along the ray and blending the mapped color and opacity values [433]. Several variants of this technique exist. X-ray like images are generated by neglecting the opacity and just summing up all values. The so-called maximum intensity projection method (MIP) determines the intensity of a pixel to be the maximum function value occurring along the corresponding ray. Even isosurfaces can be rendered without polygonal extraction if an illuminated pixel is displayed only if the difference of the current function value and the isovalue changes sign. The visual perception especially of surface structures can be enhanced by voxel shading. Again the volume gradient is used as normal in Phong-like lighting models. Acceleration of this expensive technique is achieved by early ray termination [434], by adaptive sampling [188, 153, 767] or by fast discrete ray generation.

Projection methods [755] which process cells or voxels independently require sorting with respect to the view point in order for the compositing to work correctly. While this is comparatively simple for regular volumes, visibility ordering of cells from structured or unstructured grids demands for complex algorithms [667, 650] which severely limit interactive performance. Splatting [751] is a crude approximation of the integration within on cell typically applied in object-order algorithms. Here, a Gaussian reconstruction kernel is projected into screen space due to its independence of the density distribution and it is stored in form of footprint tables.

Shear-warp volume rendering [413] is a hybrid approach which factors the viewing transformation into a shear operation to align voxels and pixels for fast compositing and into a final image warp. Massivly exploiting coherence allowed image generation rates of up 1 frame per second for 256^3 volumes on standard graphics workstations.

Frequency domain volume rendering [703], finally, exploits the Fourier projection slice theorem which states, that the absorption-free projection of a 3D data set is equivalent to a 2D slice plane reconstruction appropriately oriented in frequency space. The final image is obtained by 2D inverse Fourier transform. While this approach exhibits an asymptotically lower complexity than the standard methods much care has to be taken in order to avoid aliasing artifacts.

6.4.4 Hierarchical Approaches

Most importantly, the motivation in all these approaches is to develop algorithms that react to changes of mapping parameters (e.g. varying the iso-value or the transfer function) by almost immediately regenerating the image with several frames per second. Only with this type of real-time interaction and navigation it is possible to effectively analyze an unknown data set and to compensate for the information lost during the projection of the 3D scene onto the screen. However, despite all the sophistication incorporated into these methods, does it seem that the data sets are growing faster than algorithmic progress is made. For example data volumes from 3D medical imaging like CT are approaching sizes of 512^3 which amounts to more than 100 million of voxel cells. It is obvious that visualization methods which essentially have to access each cell of a data set in order to derive a visual mapping might not catch up to the goal of interactive processing. Thus, we have to reduce the number of cells which have to be visually mapped, which means that we have to compress the data set in a preprocessing filtering step. Since we strive to reduce the number of cells by at least an order of magnitude, only lossy compression schemes will be employed. However,

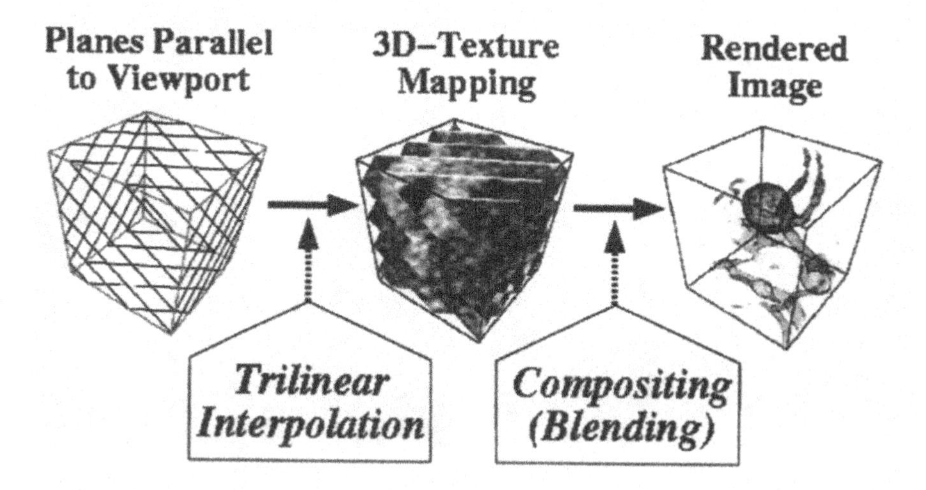

Figure 6.19. Volume rendering with three-dimensional texture mapping.

this does not always have to lead to a significant loss of information. On the contrary, the compression scheme will be chosen in such a way, that only redundant or irrelevant information (e.g. CT voxels containing air) is discarded, while important features like high gradients, edges, etc. are retained or even emphasized.

Nevertheless, if an error is introduced in such a scheme, the user has to be given control over the threshold letting him choose between a fast visualization of a very crude approximation of the data and an almost perfect representation of the data which took perhaps minutes to compute. This requirement can only be met if not only one compressed version of the data, but a complete hierarchy of representations of the data set at different levels of resolution is available or can be generated on the fly.

Various ways to derive such a multiresolution hierarchy have been investigated for isosurface extraction and direct volume rendering ranging from octrees [756, 434, 750] to wavelets [442, 748] and adaptive finite element meshes [515, 263].

6.4.5 Hardware-Based Volume Visualization

Despite the described algorithmic advances single CPU software implementations are usually not capable of volume visualization at interactive frame rates. Many researchers tried to overcome the limitations by parallelizing the algorithms in image [618, 619, 729] and object space [344, 459, 504]. Others designed special purpose volume rendering hardware, some of which [393, 543, 267] were realized with fully functioning hardware prototypes. A competing approach is to exploit standard rasterization hardware in high-end graphics workstations [102].

The basic idea of the three-dimensional texture mapping approach is to use the scalar field as a three-dimensional texture. At the core of the algorithm, multiple equidistant planes (*slices*) parallel to the image plane are clipped against the bounding box of the volume (see Figure 6.19). During rasterization, the hardware is exploited

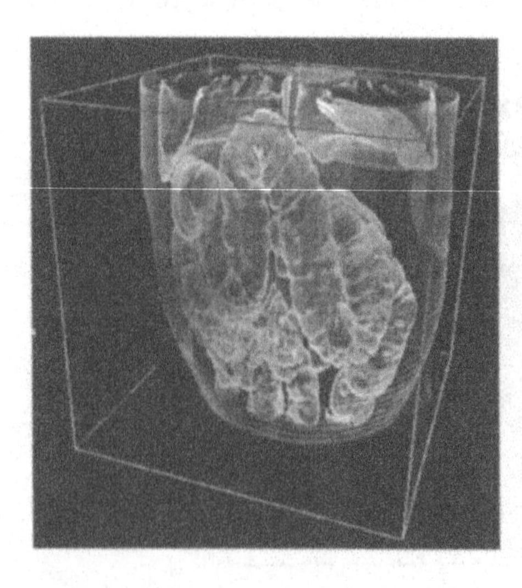

Figure 6.20. Volume rendering of abdominal CT data.

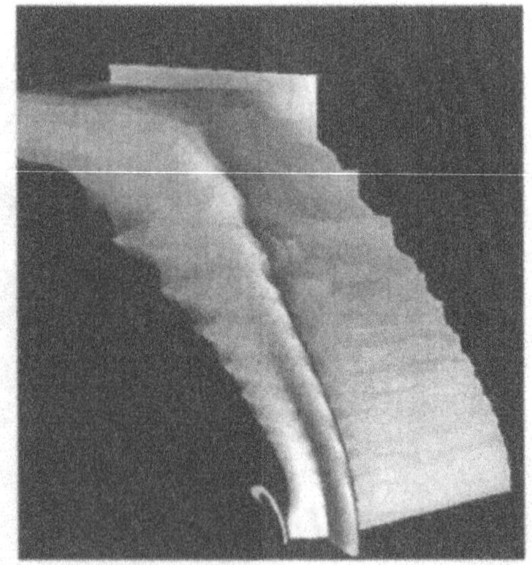

Figure 6.21. Shaded isosurface of blunt-fin CFD data.

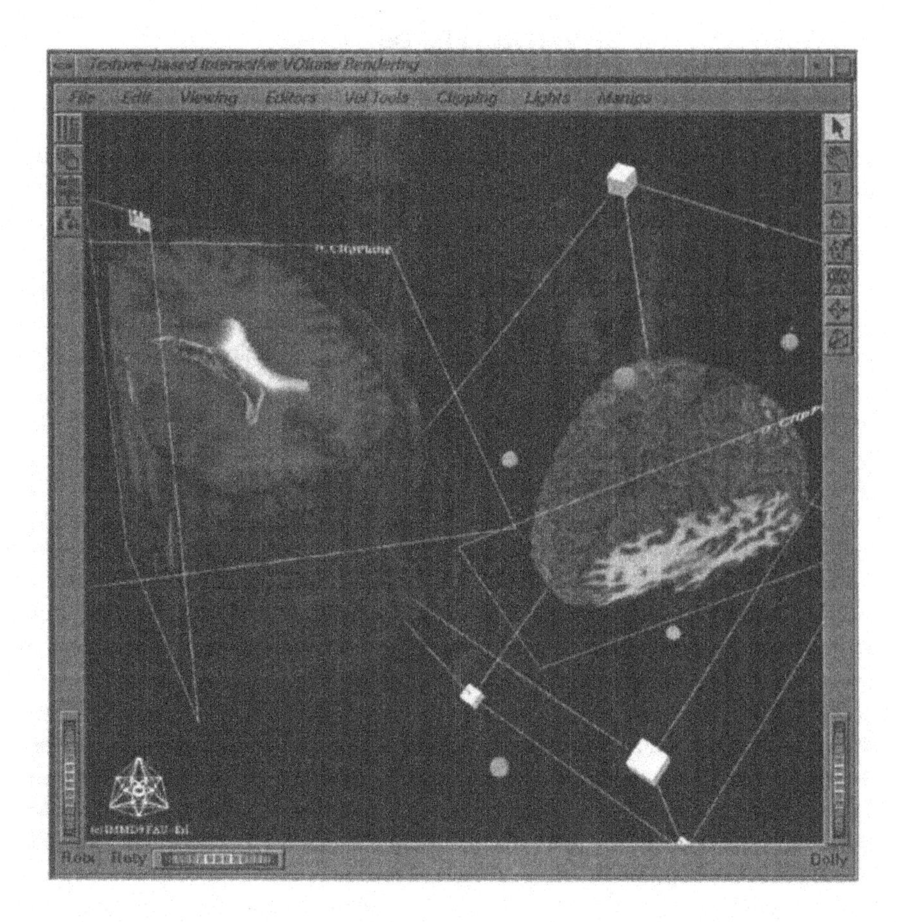

Figure 6.22. Texture-based volume rendering with OpenInventor.

to interpolate three-dimensional texture coordinates at the polygon vertices and to reconstruct the texture samples by trilinearly interpolating within the volume. Finally, the three-dimensional representation is produced by successively blending the textured polygons back-to-front onto the viewing plane. Since this process uses the blending and interpolation capabilities of the underlying hardware, the time consumed for the generation of an image is negligible compared to software based approaches. Interactive frame rates can be achieved even if scalar fields of high resolution have to be subdivided into bricks (see Figure 6.20). Additionally, hardware can be exploited to interactively modify the texture color table. This can be used to enhance or suppress portions of data specified by certain scalar values.

OpenGL-based volume rendering via 3D textures can easily be integrated with scence graph toolkits such as OpenInventor, thus providing a rich set of functionality ranging from 3D manipulators and GUI components to specialized viewers (see Figure 6.22). Clever use of fragment tests, stencil buffers and polygon shading hardware allows to extend these techniques to fast isosurface shading and volume rendering of unstructured grids (see Figure 6.21).

6.5 VISUALIZATION OF VECTOR FIELDS

C. TEITZEL, M. HOPF

In the foregoing part of this chapter techniques for the visualization of scalar field data have been presented. These methods are not suitable, in general, for the exploration of vector field data. Hence, different concepts have been developed in order to reveal the information contained in this kind of data.

Visualizing vector field data is challenging because no existing natural representation can visually convey large amounts of three-dimensional directional information. In fluid flow experiments, external materials such as dye, hydrogen bubbles, or heat energy are injected into the flow. The advection of these external materials creates flow patterns which highlight the inherent structure of the field. Analogues to these experimental techniques have been adopted by scientific visualization researchers.

In the reminder of this chapter we will summarize some of the basic approaches for the visualization of flow fields. We will sketch ideas how to use numerical methods and three-dimensional computer graphics techniques to produce graphical icons such as arrows, motion particles, stream lines, stream ribbons, and stream tubes that act as three-dimensional depth cues. While these techniques are effective in revealing the flow field's local features, the inherent two-dimensional display of the computer screen and its limited spatial resolution restrict the number of graphical icons that can be displayed at one time. In order to overcome these limitations additional techniques for flow field visualization including global imaging techniques have been developed. These techniques can successfully illustrate the global behavior of vector fields; however, it is difficult when using such methods to control the tremendous information density in a way that effectively depicts both the direction structure of the flow and the flow magnitude.

In the following we will deal with three-dimensional flow or vector fields, which associate a vector with each point of the underlying domain. In general, the physical idea of flow can be represented by the mathematical concept of a *flux*. Detailed descriptions of this conception can be found in nearly every textbook about differential topology and in books about ordinary differential equations, e.g. [20, 175, 419, 482].

A more common point of view is to look at the flow field as a family of differentiable curves parameterized over a certain domain: Then, the curve

$$\begin{aligned} \alpha_x : \mathbb{R} &\longrightarrow M \\ t &\longmapsto \Phi_t(x) = \Phi(t, x) \end{aligned}$$

is called *integral curve* or *flow line* of x. The image $\alpha_x(\mathbb{R})$ of the integral curve is called *orbit*, *path*, or *trajectory* of x. If a flux, Φ, is given on a manifold, M, then every point of M belongs to exactly one orbit. In order to get a better geometrical understanding of a given flux or a physical flow respectively, there are two different ways to look at the flux. On the one hand, we can study how the motions Φ_t act on the manifold M. That is, the deformations of lines and volumes by the flux Φ_t are studied. On the other hand, we can look for an overview over the shape of all orbits (see Section 6.5.1). There are three kinds of orbits: An integral curve $\alpha_x : \mathbb{R} \rightarrow M$ of a given flux is either an injective immersion, a periodic immersion, i.e. α_x is immersion and there exists

$p > 0$ with $\alpha_x(t+p) = \alpha_x(t)$ for all $t \in \mathbb{R}$, or α_x is constant, $\alpha_x(t) = x$ for all t. In the latter case x is called *fixed-point* of the flux.

The actual aim in flow visualization is to analyze and to display the properties of flows. The data, which are received from measurements or numerical simulations, consist of velocity information. In the following we will outline a variety of different methods, which can be used successfully to illustrate the local or global behavior of vector fields.

6.5.1 Velocity Field and Flux Visualization

According to the last section, a flow can be analyzed by visualizing its velocity field, orbits, or motion. A number of different techniques have been evolved to make use of these three notations.

Arrow Plots. A traditional standard technique for visualizing a flow is to visualize its velocity field directly by drawing small arrows at discrete grid points. This simple arrow plot algorithm is quite fast but has some disadvantages.

If the arrow plot method is applied to curvilinear or unstructured grids by drawing the arrows at the grid nodes, more arrows are placed in regions where the cells are small than in areas where the cells are large. This variation in arrow density is unrelated to the velocity field itself. A second drawback is that regularity in the grid causes distracting patterns in the output image. Finally, the user has no control over the global arrow density.

A solution to these problems is to re-sample the velocity field in computational space (C-space) or physical space (P-space) and to generate a random distribution of arrows. Thereby, the global arrow density can be chosen by the user. For a detailed description of this approach we refer to [187].

Glyphs. A more advanced method for visualizing the velocity field of a flow is using glyphs instead of arrows. A glyph is a local flow probe that shows some more parameters of the field. Besides the actual velocity, information about the Jacobian of the velocity is revealed. By decomposing the Jacobian into meaningful components and by their mapping using metaphors that are easy to understand, the matrix is presented in an intuitively way. The result is that the probe presents information like velocity, acceleration, curvature, shear, convergence (or divergence), and local rotation, also called torsion by some authors. This iconic visualization method was developed by de Leeuw and van Wijk [426] in 1993.

Glyphs can be used as a tool to study details of a flow field rather than to globally visualize a flow. On the one hand, it is an advantage of this technique that each probe supplies the user with a vast quantity of information. On the other hand, these icons are at first glance difficult to interpret and some users feel overwhelmed by the flood of information.

Topological Methods. Topological methods combine two of the three mentioned approaches for analyzing a given flux by visualizing properties of both the velocity field and the orbits. On the one hand, the singularities of the velocity field are considered, on the other hand, the shape of orbits near fixed-points of the flux. In 1989 Helman and Hesselink [315] introduced techniques for visualizing the topology of vector fields, Since then this method has been of great interest in the visualization community [316]. The concept of Clifford algebra was introduced in order to detect higher order singularities [609] and to handle nonlinear vector fields [608]. Additionally, topological methods can be extended from vector fields to tensor fields [165, 421].

Particle Methods. Particle methods enjoy a good reputation in flow visualization, since they can display orbits or motion of a flux depending on the initial geometry. Both orbits and motions provide a good impression of the fluid flow. It is possible to depict details and to show global behavior as well. Given the velocity field $\boldsymbol{v}(\boldsymbol{x}, t) := \dot{\Phi}(t, \boldsymbol{x})$, the orbit, *path line*, or trajectory of $\boldsymbol{x}_0 \in M$ is given by the solution of an initial value problem for the following ordinary differential equation:

$$\frac{d\boldsymbol{x}(t)}{dt} = \boldsymbol{v}(\boldsymbol{x}(t), t), \qquad \boldsymbol{x}(0) = \boldsymbol{x}_0. \tag{6.7}$$

Physically, such a path line corresponds to a long time exposure photograph of an illuminated fluid particle. *Stream lines* correspond to the solution of the differential equation:

$$\frac{d\boldsymbol{x}(s)}{ds} = \boldsymbol{v}(\boldsymbol{x}(s), t), \qquad \boldsymbol{x}(0) = \boldsymbol{x}_0, \tag{6.8}$$

where the time t is treated as constant and s is the parameter of the resulting curve $\boldsymbol{x}(s)$. That is, we take a snapshot of the vector field $\boldsymbol{v}$ at time t. Stream lines are tangential to the instantaneous velocity at every point, except at points where $\boldsymbol{v} = 0$.

The image at time t of the *streak line* passing through the point $\boldsymbol{x}_1$ is the curve formed by all the particles which happened to pass by $\boldsymbol{x}_1$ during the time $0 < t_1 < t$. Physically, a streak line corresponds to the curve traced out by a non-diffusive tracer injected at the position $\boldsymbol{x}_1$. Note that the particles of the tracer move according to Equation 6.7. If a flow is steady, i.e. it lacks explicit time dependence, then path, stream, and streak lines coincide.

As already mentioned, the orbits of a fluid flow can be visualized applying Equation 6.7. The same equation can now be employed for displaying motions of the flow. Therefore, a set of connected particles is traced instead of a single particle. Lines, surfaces, or volumes can be considered as such sets. That is, we wonder what the evolution of the initial start configuration will be.

If all points of a given line are traced according to Equation 6.7, the result will be a stream ribbon [51, 386], a stream tube [177], or a stream surface [347]. In 1992 Hultquist improved stream surfaces by a new tiling technique and by splitting divergent ribbons [348]. In addition, Stolk and van Wijk introduced the representation of stream surfaces and tubes via surface particles [673]. Moreover, implicit stream surfaces were proposed by van Wijk [754] in order to overcome difficulties with irregular topologies of originating curves and surfaces.

With stream surfaces a good insight into the structure of the flow field can be achieved, because hidden surface elimination and shading can be used to provide cues on depth and orientation. Furthermore, local parameters of the field or other variables can be mapped onto the surface. On the other hand, great efforts have to be undertaken in order to solve problems with convergent, divergent, or shear flows, and with flows around obstacles.

Flow volumes [475] are the volumetric equivalent of stream surfaces. That means, a polygon is given as source object and traced. The resulting flow volume is divided into a collection of tetrahedra, which are rendered by a method of Shirley and Tuchman [644].

Therefore, it is possible to render semi-transparent tetrahedra using hardware texturing and compositing in order to generate images which create the impression as if smoke were released into a gas flow.

Another approach dealing with surfaces as source objects is the method of *time surfaces* [195]. Starting with a surface at a given time, the particles of this surface are traced and further surfaces are rendered after discrete time steps. This time, the different surfaces will not be connected. Bye the way, the particles that lie on the same surface are particles of the same age.

Finally, volumes like balls or hexahedra can be used as start object. Of course, it is sufficient to trace particles of the surface of a closed volume, for instance a sphere is traced instead of a ball. The shape of the volumetric source object is distorted by the flow, which gives an impression of the stretching within the flow field. Floating volumes were presented by Duvenbeck and Schmidt [195] and Stolk and van Wijk [673].

A lot of other techniques for flow visualization based upon particle methods have been presented in the last years, e.g. stream polygons by Schroeder et al. [623] and the stream ball technique by Brill et al. [81].

Line Integral Convolution. As mentioned in the previous paragraphs, vector fields can be visualized in a number of different manners. The approaches presented so far are restricted to a rather coarse spatial resolution. In contrast to them, texture-based methods achieve a much higher resolution. Line integral convolution (LIC) is an effective and versatile technique for visualizing flow fields with small scale structures.

In an early texture-like method, introduced by van Wijk [753] in 1991, oval spots with white noise are distorted along a straight line segment oriented parallel to the local vector direction. LIC itself was introduced by Cabral and Leedom [103] in 1993. In their algorithm convolution takes place along curved stream line segments. In 1995 Stalling and Hege [662] made LIC much faster, more accurate and independent of resolution. Due to these improvements LIC turned out to be very suitable for displaying vector fields on two-dimensional surfaces and became very popular. Hence, a vast quantity of different algorithm and improvements have been developed in the last years. In 1994 Forssell [231] presented an extension that allows to map flat LIC images onto curvilinear surfaces in three dimensions. A problem of this method is the distortion of length during the mapping process. In 1997 Teitzel et al. [689] solved this problem by computing LIC images directly on triangulated surfaces in three-dimensional space without mapping. Another method for creating LIC images on surfaces in three-dimensional space was presented by Mao et al. [467] using solid texturing. Wegenkittl et al. [740] introduced oriented LIC in order to visualize the orientation of the flow and Risquet [573] presented a drastic simplification for accelerating the imaging process. Many other authors have been working on enhancements by color coding or animating LIC, by accelerating the image generation, or by developing especially adapted techniques for applying LIC to unsteady flows. A comparison of LIC and recently enhanced spot noise techniques can be found in [425].

Considering volumetric data, though three-dimensional LIC volumes can be computed in the same manner as two-dimensional LIC images, volume LIC is scarcely used because of difficulties to depict inner structures of the vector field. Modern high-end graphics workstations provide a high number of tri-linear interpolation operations per second and thereby allow to perform direct volume rendering (compare Section 6.4.5) at high image quality and interactive frame rates. Although the ability to interactively manipulate the three-dimensional volume greatly improves the perception of the inner structures, the stream lines inside a three-dimensional LIC texture are too dense and intricate to be visualized as a whole. Semi-transparency and the application of sparse input textures as proposed in [353] can enhance the resulting LIC texture.

Animation is an intuitive way to add information about the absolute value of velocity to static LIC images. Techniques used to animate two-dimensional LIC, which compute LIC textures for each time step, are less applicable for three-dimensional LIC, because of the great computational expense and the immense amount of data. To avoid the performance penalty and the high memory requirements that come with loading and storing large pre-computed three-dimensional textures, the idea was not to animate the three-dimensional LIC itself, but to use an animated clipping object instead. Rezk-Salama et al. provided two different approaches [568], which both use a single three-dimensional LIC texture and a set of clipping objects. To display animated three-dimensional flow at interactive frame rates texture-based volume rendering is performed.

Basics of Line Integral Convolution. The LIC algorithm filters an input volume along path or stream lines of a given vector field and generates a three-dimensional texture as output. In most cases in scientific visualization a texture with white noise is used as input. The Intensity I of an output texture voxel located at $\boldsymbol{x}_0 = \boldsymbol{\sigma}(s_0)$ is

$$I(\boldsymbol{x}_0) = \int_{s_0-L}^{s_0+L} k(s - s_0)T(\boldsymbol{\sigma}(s))ds, \tag{6.9}$$

where $\boldsymbol{\sigma}(s)$ denotes a stream line of the vector field parameterized by arc length, T the intensity of the input texture and k a filter kernel. If we choose a constant filter kernel k and consider that T is constant at each voxel, the convolution integral can be computed by sampling the input texture T at locations $\boldsymbol{x}_i$ along the stream line $\boldsymbol{\sigma}(s)$:

$$I(\boldsymbol{x}_0) = k \sum_{i=-n}^{n} T(\boldsymbol{x}_i), \tag{6.10}$$

where we choose $k = 1/(2n + 1)$ to normalize the intensity. The convolution causes voxel intensities to be highly correlated along individual stream lines but independent in directions perpendicular to them. In the resulting images the stream lines are clearly visible.

6.5.2 Vortex Visualization

Up to now, we have analyzed the flow and its velocity field. What can be done next is to investigate the derivative of the velocity field, the Jacobian field of the velocity. In order to do this, we firstly decompose the Jacobian into a symmetric matrix $\mathbf{\Lambda}$ and a skew-symmetric one $\mathbf{\Omega}$:

$$\begin{aligned} D_{\boldsymbol{x}}\boldsymbol{v} &= \underbrace{\frac{1}{2}\left(D_{\boldsymbol{x}}\boldsymbol{v} + (D_{\boldsymbol{x}}\boldsymbol{v})^T\right)} + \underbrace{\frac{1}{2}\left(D_{\boldsymbol{x}}\boldsymbol{v} - (D_{\boldsymbol{x}}\boldsymbol{v})^T\right)} \\ &= \qquad\qquad \mathbf{\Lambda} \qquad\qquad + \qquad\qquad \mathbf{\Omega}\,. \end{aligned} \tag{6.11}$$

$\mathbf{\Lambda}$ is called deformation or stretching tensor and its proper directions are the maximum directions of stretching. $\mathbf{\Omega}$ represents the local rotation, which is a skew-symmetric matrix, called vorticity or spin matrix. The rotation matrix is of the form

$$\mathbf{\Omega} = \frac{1}{2}\begin{pmatrix} 0 & -\omega_3 & \omega_2 \\ \omega_3 & 0 & -\omega_1 \\ -\omega_2 & \omega_1 & 0 \end{pmatrix} \tag{6.12}$$

with the vector field $\boldsymbol{\omega} = (\omega_1, \omega_2, \omega_3)^T$, which is the curl or rotation of the velocity, i.e. $\boldsymbol{\omega} = \nabla \times \boldsymbol{v}$. The vector field $\boldsymbol{\omega}$ is called *vorticity* and is readily interpreted as twice the local angular velocity in the fluid. Note also that $\mathrm{div}(\boldsymbol{\omega}) = \mathrm{div}(\mathrm{rot}(\boldsymbol{v})) = 0$.

Since vorticity is a vector field, all the methods presented in the previous Subsections can be employed for visualizing the vorticity. Prominent examples are vortex lines [488] and vortex tubes [39], which are the stream lines and tubes of the vorticity. However, the vorticity is only one component of the derivative of the velocity, which is itself the derivative of the flux. That is, it seems rather difficult to gather information about the flow from vortex lines or tubes and even experienced researchers can be surprisingly misled by vortex lines.

6.5.3 Particle Tracing

In order to compute a particle path, Equation 6.7 used to be solved on uniform meshes. In 1989 Buning [99] presented the stencil walk algorithm, which makes it possible to perform particle tracing on curvilinear grids. Sadarjoen et al. [590] compared physical space (P-space) and computational space (C-space) methods for curvilinear grids. The errors introduced in the C-space method by transforming the velocity by means of possibly inaccurate Jacobians favorite the P-space approach. Here however, point location usually computed by the stencil walk algorithm, which also needs the Jacobians, is a slight drawback. Kenwright and Lane [385] suggested a tetrahedral decomposition of the hexahedral cells for speeding up the point location but introducing an additional cell search on the tetrahedral cells. Frühauf [234] and Ueng et al. [718] investigated fast algorithms for computing particle traces in steady flows on unstructured grids.

In the following sections we are going to describe how different properties of a flow field can be visualized by appropriate particle tracing techniques. The visualized flow characteristics are *orbits*, *speed*, and *local rotation*. In addition, extra scalar values like temperature, density, or pressure of the fluid flow can be visualized, for instance by color coding. The geometric primitives used for the particle visualization are *lines*, *tubes*, *balls*, *ribbons*, and *tetrahedra*.

Path. *Lines* are the native approach for visualizing path, stream, or streak lines. The path is numerically computed and then a broken line is drawn to depict the result. Furthermore, a scalar variable, e.g. temperature, density, pressure, or the absolute value of the velocity can by displayed by color coding.

If the velocity field $v(x, t)$ of the flow is given in an analytical form, integration algorithms of high order are preferable like extrapolation methods or high order Runge-Kutta schemes. However, in real applications vector fields arise that are defined on discrete grids, since these velocity fields are given by numerical simulation or by measurement. For such rough vector fields higher order algorithms are useless. As a result of careful analysis of numerical efficiency and accuracy of different integration methods on discrete data, it can be shown that an adaptive RK3(2) scheme is accurate enough in relation to the interpolation error and significantly more efficient than higher order integration algorithms [688]. Furthermore, so-called implicit integration methods can be applied to handle stiff systems of ordinary differential equations. These algorithms are slower than explicit methods but in stiff data sets, where explicit methods fail, they can create proper trajectories.

Tubes are lines with spatial extent. In principle, a line is computed and each point of the line is replaced by a circle lying in the plane perpendicular to the current velocity direction. A benefit of tubes is that they are polygonal objects and therefore support the spatial perception by both hidden surface removal and shading. Moreover, tubes have the advantage that the radius of the tube can be varied in order to visualize an second scalar variable.

Another possibility to create tube-like stream objects is to choose a circle as source element and to trace all its points. Hereby, the convergence and divergence in the flow can be easily recognized. On the other hand, it often happens that the tube is getting very thin or fat, which means that the tube nearly either disappears or occludes all the scene.

Speed. If *balls* are used, the speed of the flow can be perceived in a natural way from the distance between successive balls. Since size and color of balls are utilized as before, balls can visualize an additional scalar value compared to tubes. However, in case of adaptive integration methods, the distance between successive particles does not depict the absolute value of the velocity but the size of the current integration step.

In this case, we can just visualize the same properties of the flow applying either tubes or balls. Nevertheless, balls have the advantage that they give valuable insight into the adaptive integration process.

Local Rotation. In addition to the visualization of the orbits, speed, and some extra scalar variables, we now concentrate on the visualization of the local rotation in the fluid flow (compare Section 6.5.2).

In fact, *bands* or *ribbons* reveal the local rotation in an intuitive way. There are two different ways to generate a band that visualizes the local rotation. The first method

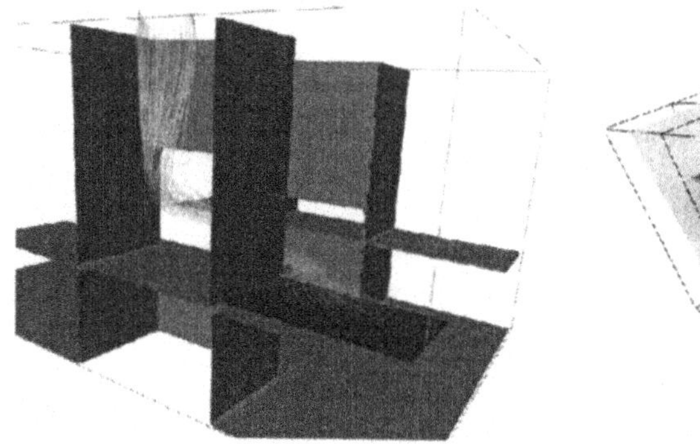

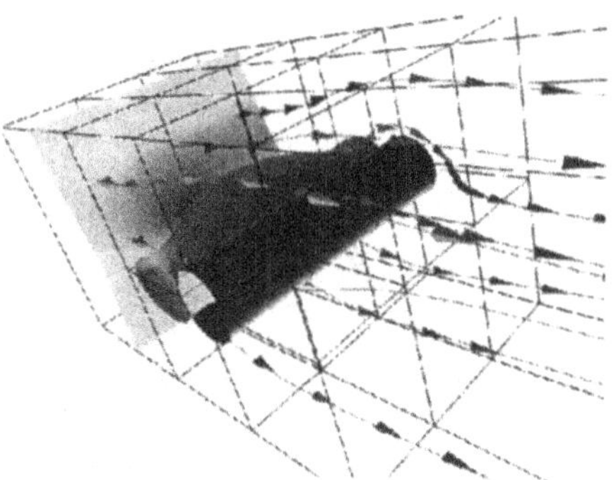

Figure 6.23. On the left hand side, the air flow in a clean air laboratory is investigated. The ventilation is improved step by step by displacing the workbenches, hoods, and gasper fans, starting a new numerical simulation, and visualizing the air flow again. On the right, the time-dependent flow around a cylinder is visualized by means of streak tetrahedra. The curvilinear multi-block data sets consists of 24 blocks. The pressure is visualized by the color of the tetrahedra. In addition, an iso-surface of the pressure has been computed in the right image. This time, the difference between successive tetrahedra does not visualize the speed of the fluid, since an adaptive Runge-Kutta method (RK3(2)) has been used for calculating the streak lines. Hence, the tetrahedra near the cylinder reveal the small integration step sizes that have been used in this region.

is to compute two particle traces that are close together and to fill the space between the lines with a shaded polygon [51]. In order to obtain a ribbon of constant width, the distance between the particle lines has to be normalized after each integration step. Moreover, since this technique is just an approximation to the local rotation, the bands have to start close enough to achieve proper results. Another drawback of this method is the fact that two lines have to be computed. Hence, it is a good idea to calculate the local rotation directly. One can use the vorticity matrix $\boldsymbol{\Omega}$ (see Section 6.5.2) to generate the band. Assume, just for simplicity, that we were using Euler's method for integrating the orbit:

$$\boldsymbol{x}_n = \boldsymbol{x}_{n-1} + \tau \cdot \boldsymbol{v}(\boldsymbol{x}_{n-1}) \quad . \tag{6.13}$$

Furthermore, let the band vector $\boldsymbol{b}$, which determines the direction of the ribbon, be a small vector perpendicular to the tangent vector of the orbit. Then, $\boldsymbol{b}$ is transformed according to the expression

$$\boldsymbol{b}_n = \boldsymbol{b}_{n-1} + \tau \cdot D_{\boldsymbol{x}}\boldsymbol{v}(\boldsymbol{x}_{n-1}) \cdot \boldsymbol{b}_{n-1} + O(||\boldsymbol{b}_{n-1}||^2) \tag{6.14}$$

if Euler's integration scheme is used. Neglecting higher order terms, expansion, and shear, we obtain the following equation:

$$\boldsymbol{b}_n = \boldsymbol{b}_{n-1} + \tau \cdot \boldsymbol{\Omega}(\boldsymbol{x}_{n-1}) \cdot \boldsymbol{b}_{n-1} \quad . \tag{6.15}$$

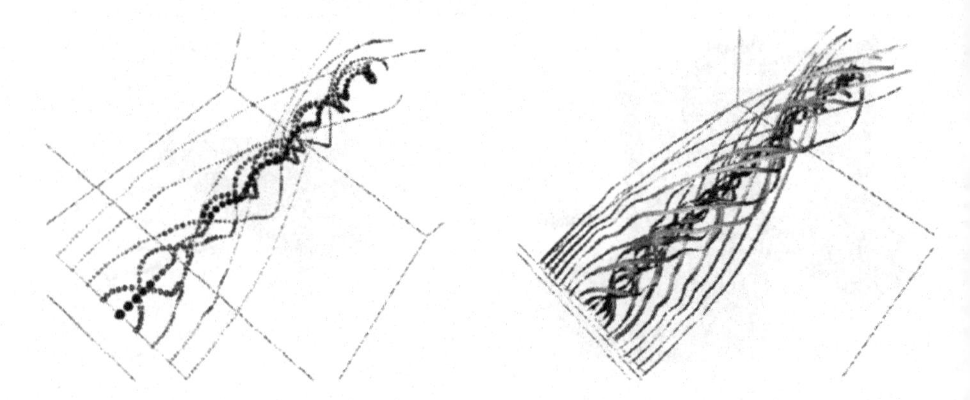

Figure 6.24. The turbulent jet stream in the Pegase data set is visualized by means of streak balls (left) and streak bands (right). The absolute value of the velocity is revealed by color coding and in the left image by the size of balls as well.

Because of approximation and rounding errors due to using finite differences for calculating the Jacobian $D_{\boldsymbol{x}}\boldsymbol{v}$ and due to neglecting terms of higher order, it is safer to project the new band vector $\boldsymbol{b}$ after each integration step onto the plane perpendicular to the tangent vector of the orbit. This results in ribbons that spin longitudinally in a swirling flow.

Another variable that can be visualized by bands is the torsion of the current path, stream, or streak line [386]. However, notice that the local rotation of the flow is in general different from the torsion of the stream line through the current position.

Further geometric primitives for visualization purposes are tetrahedra as shown in Figure 6.23. They have the advantages from both balls and bands (see Figure 6.24).

7 ACOUSTIC IMAGING, RENDERING, AND LOCALIZATION

R. RABENSTEIN

INTRODUCTION

The preceding chapters covered the acquisition, intepretation, processing, and visualization of 3D data. Except for Chapter 6, the presentation has been confined to image sequences representing information as visible to the human eye. However, the models developed for this purpose hold also for data acquired from other kinds of wave fields than the electromagnetic spectrum of light. It is the purpose of this chapter to show the applicability of the methods presented so far to an entirely different modality: acoustical waves in the sound and ultrasound range. Since the general principles of imaging and rendering hold for both electromagnetic and acoustic waves, this chapter covers topics presented earlier in Chapters 1, 2, 3, and 5 for images taken with optical sensors. However, differences in sensor technology, wave length, reflection properties and wave propagation speed between optics and acoustics require a separate presentation of acoustical sensors and corresponding processing methods.

The chapter starts with a review of methods for 3D acoustical imaging and the respective transducer technologies. This section corresponds to Chapter 1 on optical 3D Sensors. Interestingly, the design methods for electro-acoustic transducers rely not only on physical models for mechanical and electro-magnetical oscillations, but also on volume models for the transducer geometry similar to those presented in Chapter 6.

The second section describes methods for the rendering of 3D acoustic scenes. This process is also called aurealization to emphasize the close methodical connections to visualization. Especially some methods for sound propagation in enclosures are directly derived from their visual counterparts in computer graphics.

The final two sections on fusion of multisensor data and object localization are more than just an acoustical complement of the corresponding section on object tracking in image sequences in Chapter 2. Instead, they cover object localization by processing information from different modalities, namely audio and video. There are two sides of this coin: the aspect of data fusion and the aspect of localization. Each of these is treated in a separate section. The section on multisensor data fusion uses a statistical model (the Kalman filter) already introduced in Chapter 3 and developes a decentralized form suitable for combined audio and video processing. The section on object localization using audio and video signals shows how to obtain suitable object position estimates from each one of these modalities.

7.1 3D ACOUSTICAL IMAGING

R. LERCH, M. KALTENBACHER, H. LANDES

Acoustical imaging is utilized in various fields of technical applications. The most important fields are medical imaging for diagnostic purposes, non-destructive evaluation of material defects, acoustic microscopy for examinations of thin materials as well as the object recognition within automation, manufacturing and process control. The great advantage in comparison to the competitive technical methods, such as optical and x-ray imaging, is that acoustical imaging allows the imaging of non-opaque media without using ionised radiation.

In this article, we first review the basic principles of acoustical imaging. In the following sections, we report the design of acoustical imaging systems. This design has to be mostly concentrated on developing appropriate and optimized ultrasound transducers since the transducer used as transmitter as well as sensor is the most important part of the imaging system. This key role element finally determines the resulting imaging modalities and features as well as the quality of the image. Therefore, we utilize appropriate computer modeling systems for an efficient computer aided design

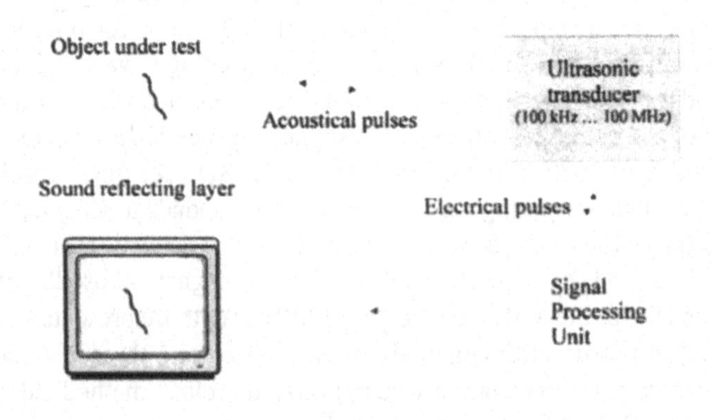

Figure 7.1. Principle of acoustical imaging.

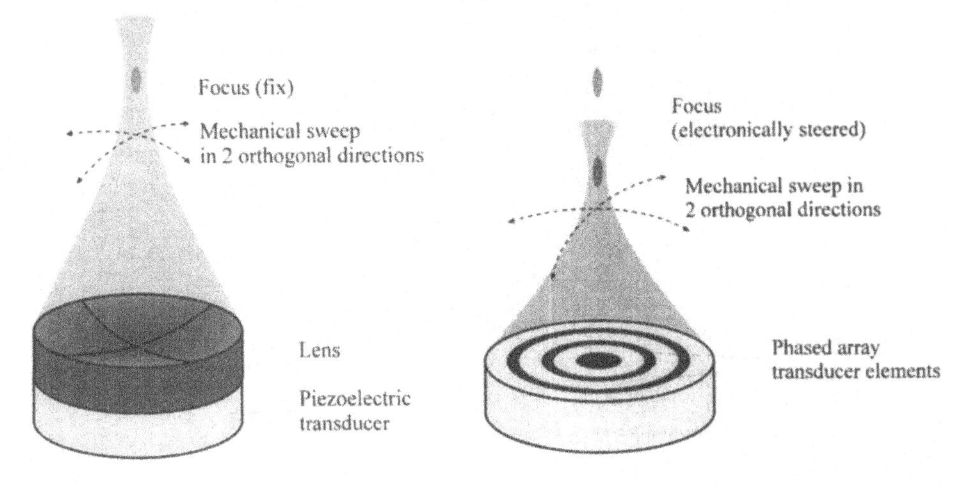

Figure 7.2. Fix focus transducer.

Figure 7.3. Piezoelectric ring antenna.

of these piezoelectric and electrostatic ultrasound transducers. This computer modeling is based on physical models of the transducers and their environment described by the governing partial differential equations. Furthermore, the modeling asks for appropriate geometrical models which are represented by finite element and boundary element meshes. All these models including their generation and use are described in this section.

The basic principle of acoustic imaging is shown in Figure 7.1. The heart and most sophisticated part of the acoustical imaging chain is the ultrasound transducer, which is either based on piezoelectric or electrostatic transducing mechanisms. The operated frequency range of theses transducers strongly depends on the area of application. For ultrasound in air the maximum operating frequency is limited to about 1 MHz due to the high absorption of ultrasound in air. The medical imaging of human tissue as well as the non-destructive evaluation of solid workpieces is mostly performed in the frequency range 1 MHz to 10 MHz. For acoustic microscopy frequencies above 50 MHz are used.

The purpose of the ultrasound transducer is to produce short acoustic pulses which are emitted to the object under test (see Figure 7.1). From reflecting layers (borders of different acoustic impedances within the object), these pulses are reflected towards the transducer which then converts it back to an analog electrical pulse. In operation the ultrasound beam produced by the transducer has to be mechanically or electronically swept over the area or through the volume of interest. Therein, each reflexion is imaged on a TV screen by a grey (or colour) code, which is proportional to the amplitude of the local reflexion. In the following, the different transducer imaging principles used for 3D acoustical imaging are described.

In Figure 7.2, the most simple imaging principle is shown. Here, a piezoelectric disc transducer with a lens attached on its top is used to produce a focussed ultrasound beam. Because the radius of curvature of the lens is fix and given by the manufacturing process, the focus position of the acoustic beam is also fix. In consequence the

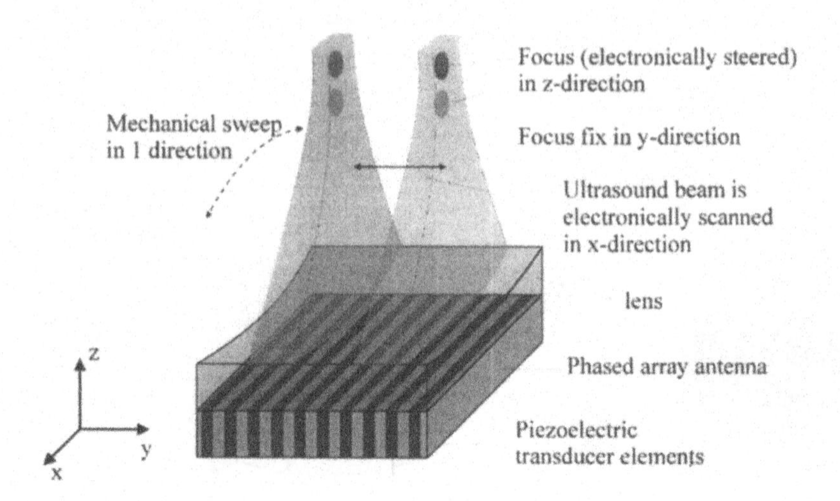

Figure 7.4. Most popular imaging modality within medical imaging by utilizing a 1D phased array antenna.

resolution of the image is low since the focussing has to be weak. Otherwise, the depth of the imaged field would suffer. To come to a 3D image, the region (volume) of interest is scanned by a mechanical sweep in 2 orthogonal local directions (see Figure 7.2). The correlation between the actual position of the transducer (ultrasonic beam) and the position within the image (actual scan line) is performed by position sensors.

The above imaging modality can be improved by substituting the fix focus transducer by a piezoelectric ring antenna (see Figure 7.3). Here, the various ring elements can be separately fed by electric pulses. Especially a pulse can be applied to the ring electrodes in a manner that it is time delayed with a variable time delay for each transducer ring. Especially when the outer rings are served earlier than the inner rings, a focus within the ultrasonic beam will result thereof. By modifying the time delay between the different rings the focus can be (electronically) moved in axial direction. By such an electronic lens, a focussing along the actual image line can be performed. This so-called dynamic focussing (in contrast to the static focussing shown in Figure 7.2) allows much better resolution and sharper images. As a drawback we again have to perform a mechanical sweep of the transducer in two local orthogonal directions to scan the volume of interest and, therewith, come to a 3D image.

In Figure 7.4, the most popular imaging modality of medical imaging for diagnostic purposes is demonstrated. In practice the mechanical sweep is often performed manually by the physician. Nevertheless, the real 3D imaging systems ask for a mechanical sweep in the direction orthogonal to the electronic scan and an appropriate position detecting system for the correlation of local beam and image position. The heart of such an imaging system is a phased array antenna consisting of several hundred separately steered piezoelectric transducer elements (piezoelectric bars). From these, a sub-group of 16, 32, 64 or even 128 elements are fed by an electrical pulse which is time delayed in such a manner that a lens function appears in respect to the

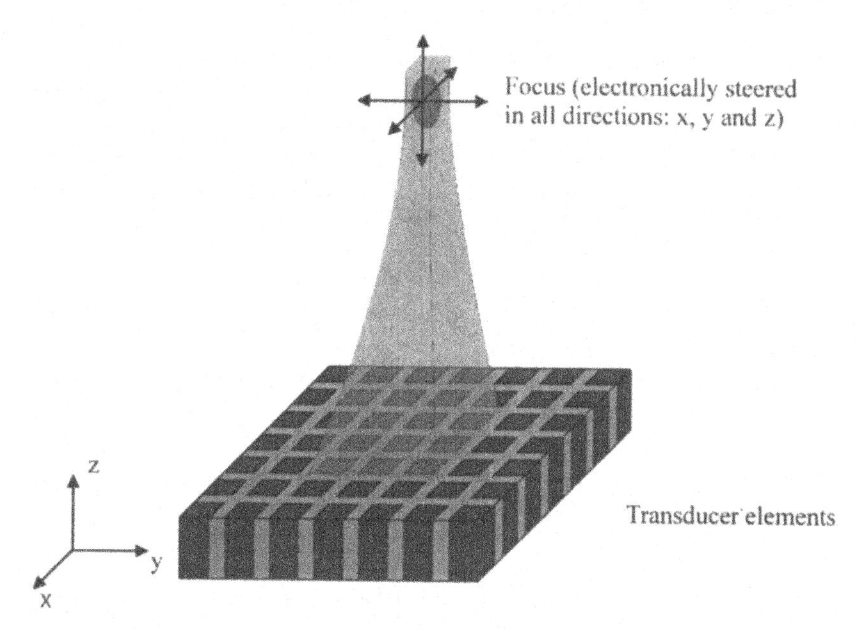

Figure 7.5. 2D phased array antenna.

y-direction. By modifying the delay times for the different transducer elements the focus can be electronically steered in z-direction. For the x-direction we have a fix focus, the position of which is determined by the radius of curvature of the mechanical lens (see Figure 7.4). Furthermore, appropriate electric circuitry allows the stepping of the ultrasonic beam along the y-direction. Therewith, the image scan in y-direction is performed.

The major problem of all above imaging systems is that they need a mechanical sweep of the transducers in one or two directions. Therewith, the imaging is slow, sometimes too slow for omitting movement artefacts within the image. In medical imaging this movement is inherently given by the respiration of imaged patients. This mechanical sweep can only be circumvented by utilizing a 2D phased array antenna, which must be separately steered in each transducer position (see Figure 7.5). Here, an arbitrary positioned sub-group of transducer elements is taken to set up an ultrasonic transducer with electronic lens. The focus position or beam direction of this transducer can be arbitrarily steered to any desired position within the region (volume) of interest. One of the basic problems of 3D acoustical imaging is that it asks for highly sophisticated 2D transducer arrays with several hundred transducer elements in each direction having a total number of transducer elements of 10,000 and more. Such a high number of elements, however, often lead to tremendous problems within transducer design and fabrication on the basis of technologies available today. Therefore, the more rudimentary imaging and transducer principles shown in Figures 7.2–7.4 can be a good alternative to nowadays available 2D phased arrays which exhibit only a small number of elements. In the following sections we describe the design and optimisation process of the above imaging transducers.

7.1.1 Electro-Acoustic Transducers Principles

Piezoelectric Transducers. The development of piezoelectric finite elements is based on three equations, namely the material equations, Newton's law and the potential equation. The material equations, which are the basis of linear piezoelectricity, take into account the piezoelectric effect both in the description of the mechanical stresses as well as the dielectric displacement. According to [471], these equations may be written as

$$\boldsymbol{T} = \mathbf{c}^E \boldsymbol{S} - \mathbf{e}^t \boldsymbol{E}, \tag{7.1}$$

$$\boldsymbol{D} = \mathbf{e}\boldsymbol{S} + \varepsilon^S \boldsymbol{E}. \tag{7.2}$$

Herein, $\boldsymbol{T}$ denotes the mechanical stress, $\mathbf{c}^E$ the mechanical material tensor at constant electric field $\boldsymbol{E}$, $\mathbf{e}$ and $\mathbf{e}^t$ the piezoelectric coupling tensor and its transpose, ε^S the dielectric tensor at constant strain $\boldsymbol{S}$ and $\boldsymbol{D}$ the electrical displacement.

This material constitutive equation has to be combined with the differential equations governing the mechanical and electric behavior. These equations can be written in divergence form as

$$\nabla \cdot \boldsymbol{T} = \rho \frac{\partial^2 \boldsymbol{u}}{\partial t^2}, \tag{7.3}$$

$$\nabla \cdot \varepsilon \nabla \Phi = q, \tag{7.4}$$

respectively. Herein, ρ denotes the density of the material, $\boldsymbol{u}$ the mechanical displacement, and q represents the free electric volume charge.

In many cases the presence of a surrounding fluid strongly influences the behavior of a piezoelectric transducer and, therefore, must also be considered in the numerical simulation. Due to continuity considerations, the normal velocities of the fluid and the solid must coincide at the fluid-structure interface. Using piezoelectric finite elements for the transducer and acoustic finite and infinite elements in the fluid region, the following system of equation results, which fully describes the behavior of a piezoelectric transducer in an unbounded fluid region [429, 430]

$$\begin{pmatrix} \mathbf{M}_{uu} & 0 & 0 \\ 0 & 0 & 0 \\ 0 & 0 & \mathbf{M}_{\psi\psi} \end{pmatrix} \begin{pmatrix} \ddot{U} \\ \ddot{\Phi} \\ \ddot{\Psi} \end{pmatrix} + \begin{pmatrix} \mathbf{C}_{uu} & 0 & \mathbf{C}_{u\psi} \\ 0 & 0 & 0 \\ \mathbf{C}^t_{u\psi} & 0 & \mathbf{C}_I \end{pmatrix} \begin{pmatrix} \dot{U} \\ \dot{\Phi} \\ \dot{\Psi} \end{pmatrix} + \begin{pmatrix} \mathbf{K}_{uu} & \mathbf{K}_{u\phi} & 0 \\ \mathbf{K}^t_{u\phi} & \mathbf{K}_{\phi\phi} & 0 \\ 0 & 0 & \mathbf{K}_{\psi\psi} + \mathbf{K}_I \end{pmatrix} \begin{pmatrix} U \\ \Phi \\ \Psi \end{pmatrix} = \begin{pmatrix} F \\ Q \\ 0 \end{pmatrix}. \tag{7.5}$$

In Equation 7.5 $\mathbf{K}_{uu}$, $\mathbf{C}_{uu}$ and $\mathbf{M}_{uu}$ denote the mechanical stiffness, damping and mass matrix, respectively, and $\mathbf{K}_{\phi\phi}$ and $\mathbf{K}_{u\phi}$ the dielectric stiffness- and the piezoelectric coupling matrix. Furthermore, $\mathbf{K}_{\psi\psi}$, $\mathbf{K}_I$, $\mathbf{C}_I$, and $\mathbf{M}_{\psi\psi}$ denote the stiffness, damping, and mass matrices corresponding to finite and infinite fluid regions, whereas $\mathbf{C}_{u\psi}$ stands for the fluid-solid coupling matrix. $\{F\}$ and $\{Q\}$ denote the external mechanical forces and electric charges, $\{u\}$ the nodal vector of displacement, $\{\Phi\}$ the nodal vector of scalar electric potential, and $\{\Psi\}$ the nodal vector of acoustic velocity potentials.

An alternative to the approach described above, a combined finite-element-boundary-element method has also been developed [431], in which the acoustic infinite elements are replaced by acoustic boundary elements. The application of the combined FEM/BEM modeling scheme to the simulation of piezoelectric ring antennas is described below.

Electrostatic-Mechanical Transducers. In the case of an electrostatic-mechanical transducer the coupling between the electric and the mechanical field is caused by the electrostatic force between the electrodes. This force is calculated based on the electrostatic force tensor $\mathbf{T}_E$, where $\boldsymbol{E} = (E_x, E_y, E_z)$ denotes the electric field

$$\mathbf{T}_E = \begin{bmatrix} \varepsilon E_x^2 - \frac{1}{2}\varepsilon|E|^2 & \varepsilon E_x E_y & \varepsilon E_x E_z \\ \varepsilon E_y E_x & \varepsilon E_y^2 - \frac{1}{2}\varepsilon|E|^2 & \varepsilon E_y E_z \\ \varepsilon E_z E_x & \varepsilon E_z E_y & \varepsilon E_z^2 - \frac{1}{2}\varepsilon|E|^2 \end{bmatrix}. \tag{7.6}$$

The electrostatic force $\boldsymbol{F}_E$ is given by

$$\boldsymbol{F}_E = \iint_A \mathbf{T}_E \boldsymbol{n} \, dS, \tag{7.7}$$

where $\boldsymbol{n}$ is the normal vector on the surface A. The electrostatic force leads to a deformation of the electrodes, which is described by Equation 7.3 and, therefore, introduces a geometric nonlinearity in Equation 7.4.

Splitting the coupled system into mechanical-acoustic and electric part and applying the FE-method analogous to [371] we obtain the following two matrix equations

$$\begin{aligned} \begin{pmatrix} \mathbf{M}_{uu} & 0 \\ 0 & \mathbf{M}_{\psi\psi} \end{pmatrix} \begin{pmatrix} \ddot{u} \\ \ddot{\Psi} \end{pmatrix} + \begin{pmatrix} \mathbf{C}_{uu} & \mathbf{C}_{u\psi} \\ \mathbf{C}_{u\psi}^t & \mathbf{C}_I \end{pmatrix} \begin{pmatrix} \dot{u} \\ \dot{\Psi} \end{pmatrix} & \\ + \begin{pmatrix} \mathbf{K}_{uu} & 0 \\ 0 & \mathbf{K}_\psi + \mathbf{K}_I \end{pmatrix} \begin{pmatrix} u \\ \Psi \end{pmatrix} &= \begin{pmatrix} F_u(\phi) \\ 0 \end{pmatrix} \qquad (7.8) \\ \mathbf{K}_\phi(u)\{\Phi\} &= \{F_\phi(u)\}. \qquad (7.9) \end{aligned}$$

In Equations 7.8 and 7.9, $\{F_u(\phi)\}$ denotes the nodal vector of external mechanical and electrostatic forces, and $\{F_\phi(u)\}$ the electric source vector, which now both depend on the current state of the mechanical displacements and the electric potentials, respectively.

Moving mesh techniques have been successfully applied for the solution of the pure finite element approach described above. Therewith, large mesh deformations in the FE-grid can be avoided. As an alternative a coupled FEM-BEM simulation scheme to coupled electrostatic-mechanical transducers has also been developed [372].

7.1.2 Imaging Transducers

1D and 2D Piezoelectric Ultrasound Phased Array Antennas. The typical setup of a 1D piezoelectric ultrasound phased array antenna, as used in medical diagnostics, is shown in Figure 7.6. Whereas the core of each array element consists of a piezoelectric

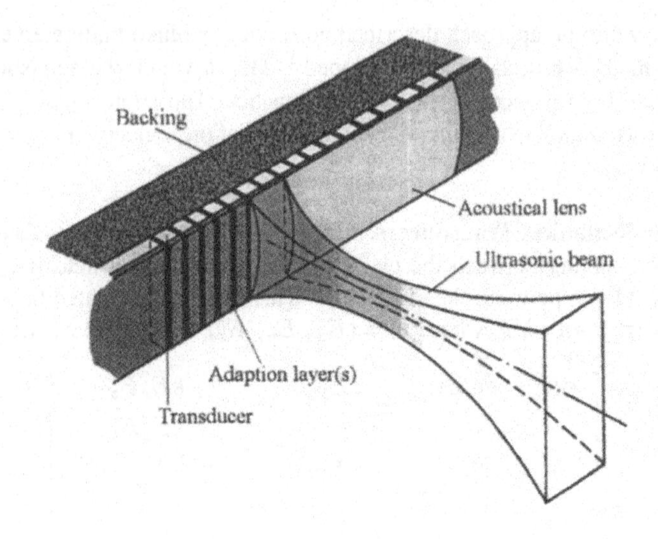

Figure 7.6. Principle setup of a 1D piezoelectric phased array antenna.

transducer, the backing is required for the energy trapping which is neccessary for the generation of short pulses required for a high spatial resolution. The adaption layers perform impedance matching between the high acoustic impedance of the transducer material and the acoustic lense, which in turn is used for mechanical focussing in the third dimension. The numerical simulation of these ultrasound array antennas requires the full description of the interaction of mechanics, electric field and, acoustic field. As shown above, this problem can be efficiently treated by using dedicated finite elements. A detail of a corresponding finite element grid of a phased array antenna is shown in Figure 7.7. For a 1D antenna, a 2D modelling is fully sufficient, since the extension of the model in the third direction is by a factor of 20 and more higher than in the other directions.

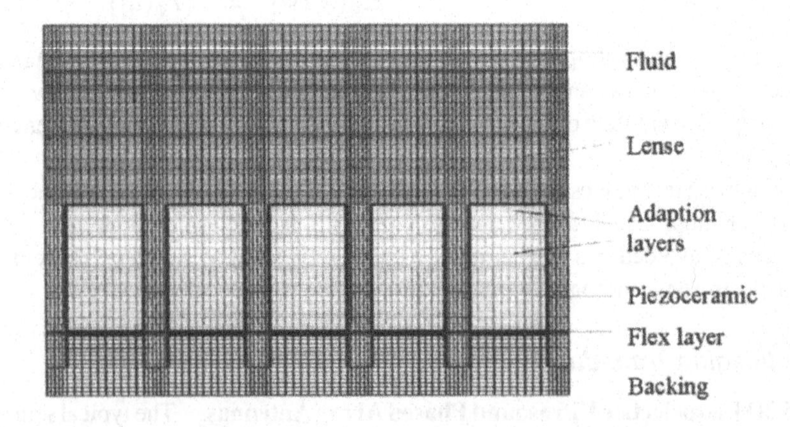

Figure 7.7. Detail of a 2D finite element model of 1D ultrasound phased array antenna.

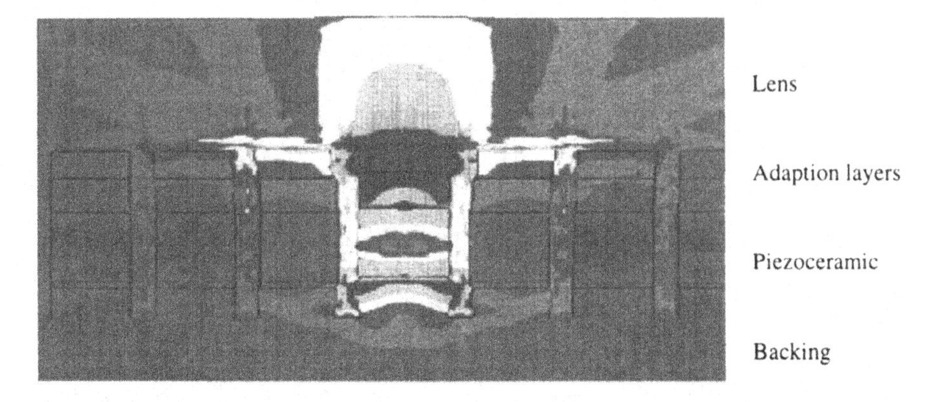

Figure 7.8. Isoplots of the mechanical displacements inside an ultrasound array antenna.

With the aid of these simulations, detailed studies of the interior physical behavior of such antennas can be performed and new insight into the underlying physics may be achieved. Considering, for example, the internal mechanical deformations (see Figure 7.8), which can not be measured, new insight regarding electrical and mechanical crosstalk between neighbored transducer elements can be obtained. Therewith, optimization of the crosstalk can be performed, resulting in an improved farfield or directivity pattern of the antenna. The most important operation mode of these antennas is the pulse-echo mode, in which the antenna acts first as a transmitter and then as a receiver. A straight forward application of the finite element method to this type of problem is not feasible, since the distance between antenna and reflector typcially

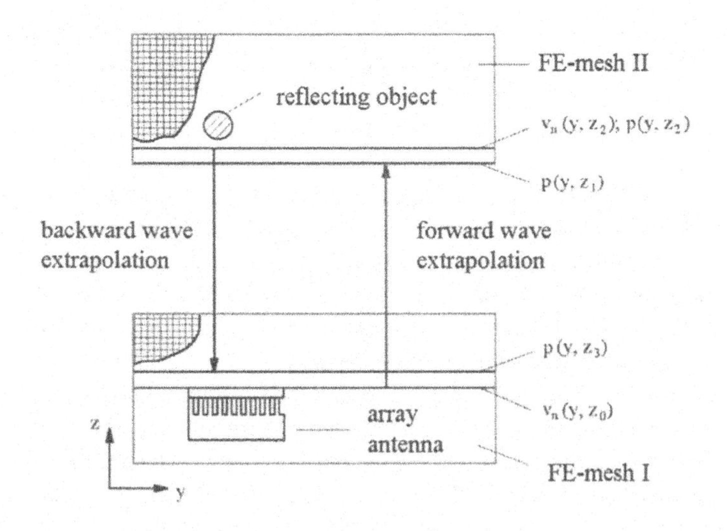

Figure 7.9. Principle of pulse-echo simulation.

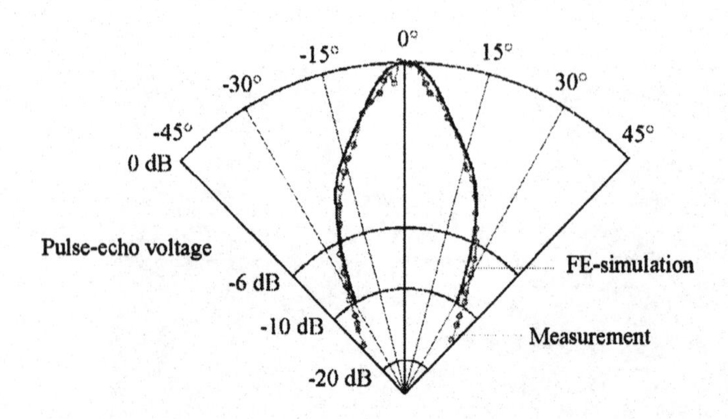

Figure 7.10. Directivity pattern of an ultrasound array antenna (single element excited, plane steel reflector).

is 100 wavelengths and more. Therewith, even for a 2D problem, tremendous grids would result. Furthermore, due to the large bandwith of the considered signals, a FEM/BEM approach would suffer from a very high number of frequencies (spectral lines) which must be considered. Consequently, the simulation of this most important operation mode requires the use of another hybrid method, which is based on the use of finite elements and soundfield calculations by means of the Huygens integral representation [432]. The approach is based on a sequence of calculations, which are shown in Figure 7.9. In a first simulation, the transmit mode is calculated by means of a finite element method. Based on the calculated data on the interface between lense and fluid, a forward wave extrapolation is performed into the region near the scatterer. Now, in a next step, the interaction of the acoustic wave with the scatterer or reflector is simulated using finite elements. Taking the pressure and velocity data of the reflected wave in a plane perpendicular to the wave propagation direction, a backward wave interpolation is performed using Huygens integral respresentation once more. Finally, the simulation of the receive mode concludes the full simulation of the pulse-echo mode.

In Figure 7.10, calculated and measured directivity pattern of a phased array antenna are compared in case of a plane steel reflector and a single excited element. Another essential criterion for the quality of the resulting pictures, besides the directivity pattern, consists of the pulse-echo signal and the spectrum thereof. To achieve a high spatial resolution, a short impulse response of the antenna is required, i.e. the pulse-echo signal due to a Dirac-like exciation should show a high amplitude and a short decay time. In Figure 7.11 calculated and measured pulse-echo signals and spectra for an already optimized 5 MHz antenna are compared.

In recent times, concepts for 2D ultrasound arraysultrasound array, in which the geometric focussing of the lense is replaced by an addititional subdicing of the array in the length direction, have been developed. A finite element model of such a twodimensional array is shown in Figure 7.12. Whereas the simulation of 1D and 1.5D ultrasound arrays could be performed by a two-dimensional finite element simulation, for 2D arrays a full three-dimensional simulation is necessary. In a first simulation

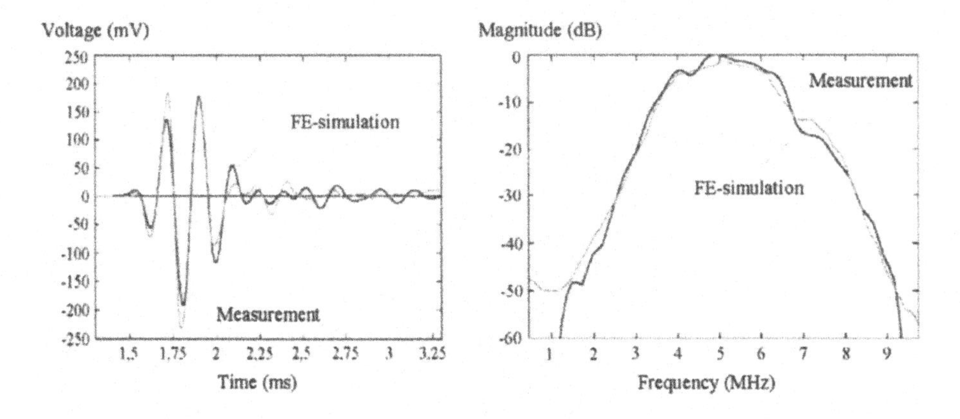

Figure 7.11. Comparison of measured and calculated pulse-echo signal of an ultrasound array antenna.

of such a 2D antenna, a detail of the antenna was modeled and the center element was excited by a short electric pulse. The resulting deformations of the antenna are shown in Figure 7.13. As can already be seen from this simple snapshot, questions of crosstalk, both mechanical and electric, as well as sawcut filling will play an even more prominent role than for the standard antenna configurations.

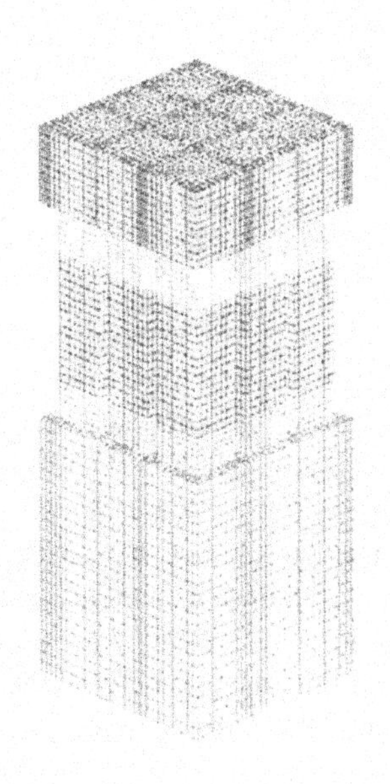

Figure 7.12. Finite element model of 2D piezoelectric ultrasound phased array antenna.

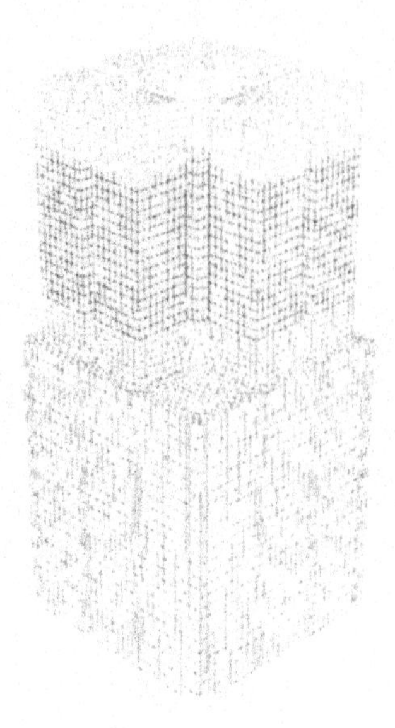

Figure 7.13. Deformed shape of 2D ultrasound array (2nd matching layer and one pillar removed).

Ring Antenna. The simulation of wave radiation in large or unbounded media requires the description of an unbounded medium by means of a discretization process. Using an approach based on finite elements requires the use of infinite elements to account for the unbounded region. However, the standard forms of these elements have to be applied in the far-field of the radiating structure in order to minimize boundary reflections and, therewith, the occurence of standing waves. This problem becomes very significant in the case of piezoelectric annular array antennas, which may lead to very large near-fields.

As a consequence of these considerations, a combination of acoustic finite and infinite elements can hardly be used in the simulation of a piezoelectric annular array antenna (ring antenna). Instead, an approach based on finite elements and acoustic boundary elements has been utilized, the principle of which is shown in Figure 7.14. Here, the array antenna, which consisted of transducer elements, matching layer and backing, as well as a small amount of the surrounding fluid medium is modeled by piezoelectric and acoustic finite elements, respectively. The unbounded acoustic medium, however, is realized by means of acoustic boundary elements, which are located on the boundary of the finite element region. With this approach, only a very small surrounding region of the array antenna must be modeled by means of finite elements.

Since the boundary elements have to be applied in the frequency domain, however, in the case of a non-harmonic excitation a Fourier transform of the excitation must now be performed and the calculation is then performed at a limited number of equally

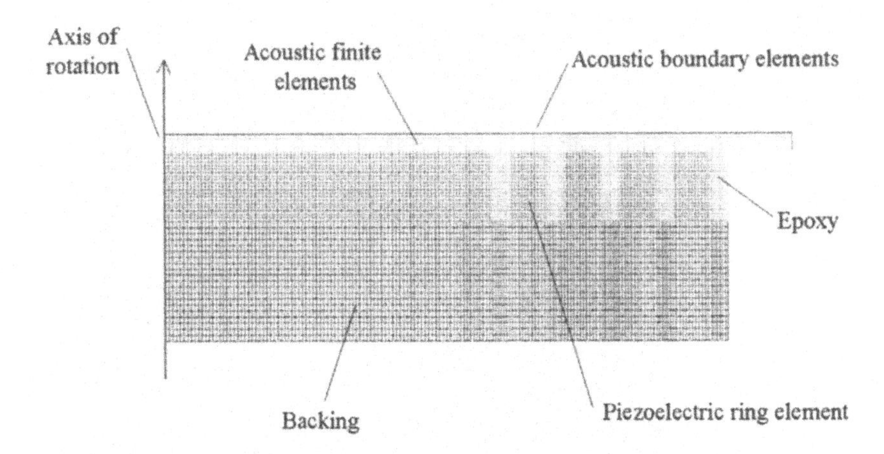

Figure 7.14. FEM/BEM model of piezoelectric annular array antenna.

spaced, discrete frequencies. Finally, an inverse Fourier transform is used to combine the results of the harmonic calculations to get the final results of the transient behavior. A snapshot of the mechanical deformation of the annular array, in case that a single transducer element was excited, is shown in Figure 7.15, whereas the resulting sound field, due to the small band excitation, is shown in Figure 7.16. The calculation of the sound field was performed using the acoustic boundary elements, and, since several postprocessing steps for each field point were involved, considerable calculation time was required.

2D Electrostatic Phased Array Antenna. Since capacitive ultrasound transducers can be fabricated by adding a few technological steps to a standard CMOS process [196] industry's interest in these transducers is rapidly growing. Due to small size and the possibility of integrating signal processing electronics on chip, these transducers are an attractive alternative to standard ultrasound transducers.

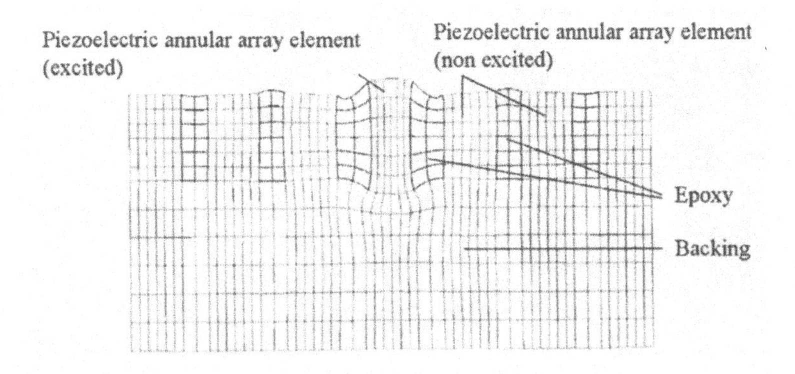

Figure 7.15. Deformation of piezoelectric annular array anrtenna.

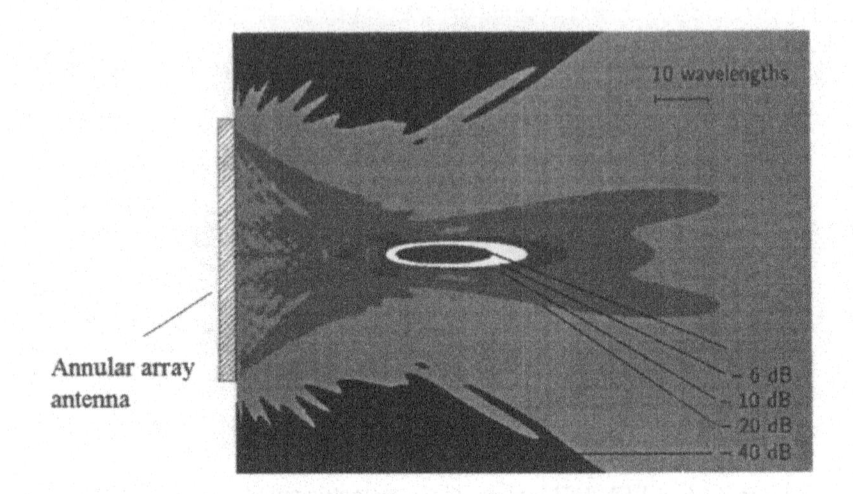

Figure 7.16. Sound field of piezoelectric annular array antenna.

As an application, an array consisting of 19 capacitive transducer cells, as shown in Figure 7.17, was considered [196]. As shown in Figure 7.18, the finite element model used consisted of a quarter of one array. The membranes had a thickness of 1 μm and the gap between the electrodes was 500 nm. A DC voltage of 10 V was applied to the electrodes and a single period of a sine burst with frequency 5 MHz and amplitude

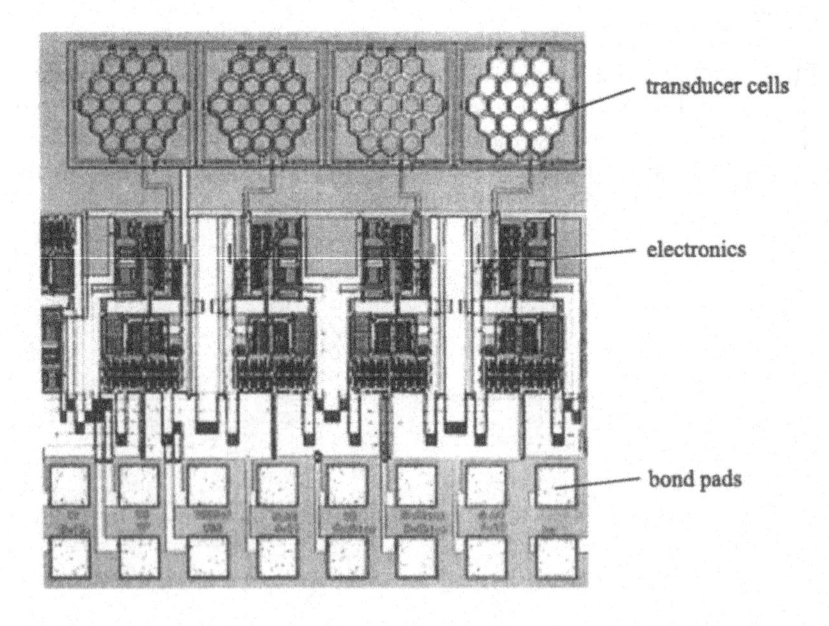

Figure 7.17. Top view of a CMOS chip with 4 arrays, each containing 19 capacitive transducers.

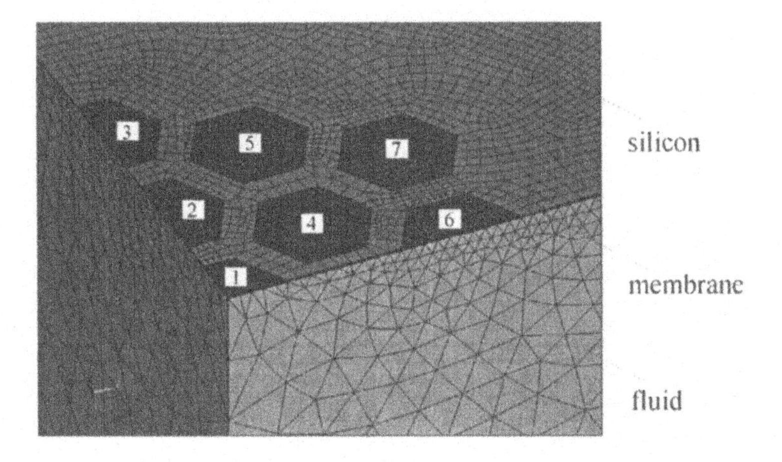

Figure 7.18. Detail of the finite element model; membranes are marked by 1–7.

10 V was used for excitation. During experiments a very long ring down time of the membranes had to be noticed. Furthermore, due to the fluid-solid coupling, a strong crosstalk between the individual membranes was detected. Therefore, investigations focused on these topics have been performed.

In order to decrease the ring down time of the membranes, the transducers were operated in a controlling loop. Therefore, the controller was also included in our finite element model. Due to the quadratic dependency of the electrostatic force on the

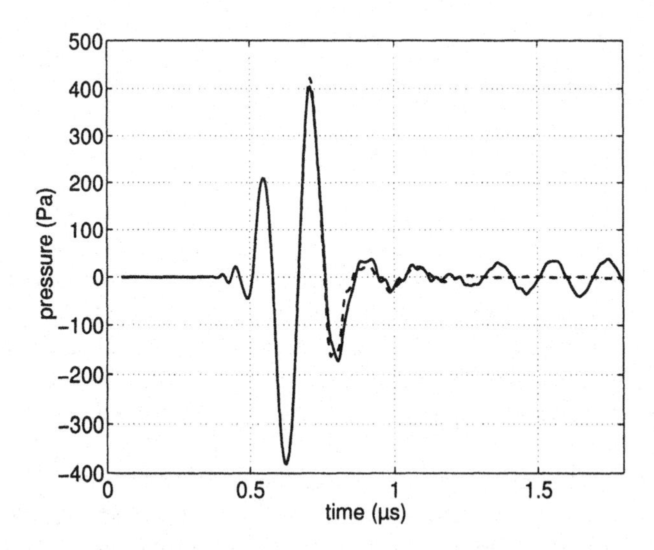

Figure 7.19. Pressure signal of uncontrolled (solid line) and controlled (dashed line) array, when all membranes are excited.

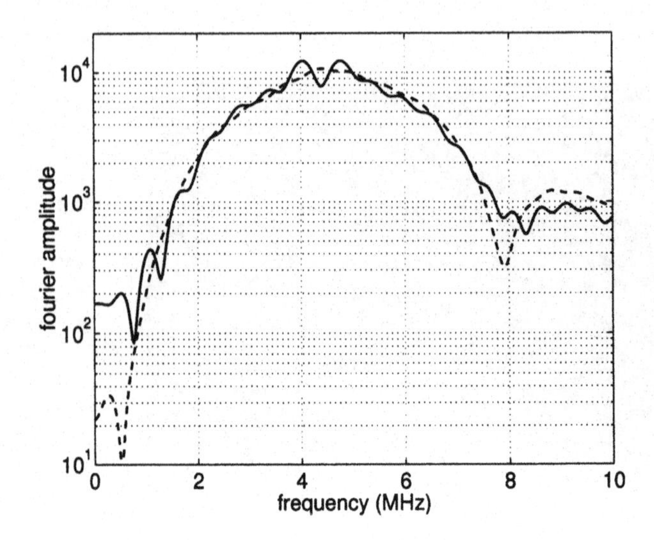

Figure 7.20. Frequency spectrum of uncontrolled (solid line) and controlled (dashed line) array, when all membranes are excited.

deflection of the membranes, a nonlinear controller has been designed. The change of the capacitance of each transducer is computed from the mechanical displacements and used as the input of the controller. The controller algorithm then calculates the voltage for each transducer, which is a direct input value for the electric source vector.

As a consequence of the controller, the ring down time of the membranes and the acoustic crosstalk is strongly reduced. Therewith, the secondary signal in the acoustic pressure, as observed for the uncontrolled case, is no longer present for the controlled membrane array. This is shown in Figure 7.19 for the case, that all membranes are driven in parallel. As a consequence, a smoothing effect due to the controller operation is also observed in the frequency spectrum (cf. Figure 7.20).

7.1.3 Conclusion

In this section we have reported the basic principles of acoustical imaging and corresponding systems which perform this imaging. The electroacoustic transducer which is operated as sensor and transmitter of ultrasound as well is often the weakest part of the imaging chain and, therefore, plays the key role within these mechatronic systems. The transducer features and quality mainly determine the overall performance of the resulting image. To come to an optimized transducer design, we have developed a computer aided design environment which utilized for the efficient development of such imaging transducers.

This computer aided design is mainly based on models: physical models described by partial differential equations, geometrical models determined by finite element meshes as well as appropriate visualization models for the representation of the results of the design process. With this modeling support we achieved an optimization of ultrasound transducers and the overall image performance as well.

7.2 AUREALIZATION

R. RABENSTEIN, L. TRAUTMANN

Visualization denotes the creation of a visual impression of a 3D scene on a graphical output device. In a similar way, aurealization denotes the creation of an audible impression of a sound field by electro-acoustic means. The ultimate goal is to give the listener or even a group of listeners the impression of being immersed in an artificially created sound field. This is why aurealization techniques are also frequently refered to as immersive audio.

To create a true acoustic rendering of a 3D scene requires a three-step procedure:

- (Re-)Production of the acoustic signals emitted by the sound sources in the scene.

  The faithful reproduction from high-fidelity recordings is state of the art. However, this requires such recordings to be available and is thus restricted to real-world scenes, e.g. musical performances.

  On the other hand, acoustical rendering of virtual scenes cannot rely on recordings of real world signals. This situation calls for sound source models from which previously unheard sound signals can be produced. The focus in this contribution will be on sound source models rather than reproduction of recorded signals.

- Rendering of the sound propagation process from the source positions to the listener.

  Once the acoustical source signals are created, the propagation of the sound waves from the source to the listener determines the spatial hearing impression. This is not a major problem under free field conditions. However, most 3D scenes of interest are enclosed by walls and consist of absorbing and reflecting objects. Their influence on the resulting sound field is correctly described by the acoustical wave equation. This partial differential equation is the exact physical model, but its numerical solution is computationally demanding. To avoid this expense, a number of approximate models are in practical use. In one way or the other, they model acoustic wave propagation in analogy to light rays. Thus methods from computer graphics become applicable for the simulation of room acoustics.

- Creation of suitable binaural signals for human sound perception.

  Even the correct reproduction of the sound field at the position of the listener is not enough for the illusion of being immersed in a sound field. The last step is to derive suitable left and right ear signals which enable the listener to actually perceive a sound source at its virtual position. The psychoacoustics of human localization of sound sources in space are not as rigorously understood as the purely physical propagation phenomena. The classical theory of spatial hearing is based on certain cues like interaural time differences and interaural intensity differences. More advanced concepts involve spectral cues in terms of head related transfer functions. The application of psycho-acoustic theory to acoustic scene aurealization largely depends on the sound transducer configuration, e.g. earphones, a pair of loudspeakers or surround sound systems.

The following subsections present an overview of methods and techniques for scene aurealization according to the three-step procedure outlined above.

7.2.1 Sound Source Models

Three methods for sound synthesis will be presented in ascending order of modelling complexity. The discussion focusses on models for musical sounds. The first method, *wavetable synthesis*, is based on samples of recorded sounds with little consideration of their physical nature. The *Spectral synthesis* creates sounds from models of their time-frequency behaviour. The most advanced method, *physical modelling*, is based on models of the physical properties of the vibrating structure which produces the sound.

Wavetable Synthesis. The most widespread method for sound generation in digital musical instruments today is wavetable synthesis, also simply called sampling. Here, the term wavetable synthesis will be used, while sampling strictly denotes time discretization of continuous signals in the sense of signal theory.

Wavetable synthesis does not require a parameterized sound source model. It consists of a database of digitized musical events (the wavetable) and a set of playback tools. The musical events are typically single notes recorded at various instruments and at various frequencies. The recording must be long enough to capture the attack of the tone as well as a portion of the sustain. Capturing the attack is necessary to reproduce the typical sound of an instrument. Recording a sufficiently long sustain period avoids a strict periodicity during playback.

The playback tools consist of various techniques for sound variation during reproduction. The most important components of this toolset are pitch variation, looping, enveloping, and filtering. They are discussed here only briefly. See [367, Chapter 8] and [576] for a more detailed treatment.

Pitch Variation. Recording notes at all possible frequencies for all instruments of interest would require excessive memory. To avoid this situation only a subset of the frequency range is recorded. Missing keys are reconstructed from the closest recorded frequency by pitch variation during playback. Pitch shifting is accomplished by sample rate conversion techniques. Pitch variation is only possible within the range of a few semitones without noticeable alteration of the sound characteristics.

Looping. Memory limitations do not permit a full length recording of each note, which can be many seconds long. As mentioned above, only a certain period is recorded, long enough to capture the richness of the sound. This period is extended by recursive read out during playback to produce the required duration of the tone. Care has to be taken to avoid discontinuities at the loop boundaries.

Enveloping. Tones produced by looping consist of the initial attack and a possibly extended sustain period of constant amplitude. The typical attack-decay-sustain-release envelope of an instrument is destroyed by looping. It can be reconstructed by application of a time varying gain function.

Filtering. While enveloping changes the time pattern of a note, also its spectral content can be modified by filtering. Usually recursive digital filters of low order (up to four) with adjustable coefficients are used. This allows not only more sound variety than present in the originally recorded wavetables but also time-varying effects which are not possible with acoustic instruments.

Despite these playback tools for sound alteration (and others not mentioned here), the sound variation of wavetable synthesis is limited by the recorded material. However, with the availability of cheap memory, wavetable synthesis has become popular for two reasons: Low computing requirements and ease of operation. More advanced synthesis techniques need more processing power and require more skill of the performing musician to fully exploit their advantages.

Spectral Synthesis. While wavetable synthesis is based on sampled waveforms in the time domain, spectral synthesis produces sounds from frequency domain models. There is a variety of methods based on a common generic signal representation: the superposition of basis functions $\psi(t)$ with time-varying amplitudes $F_l(t)$

$$f(t) = \sum_l F_l(t)\psi_l(t) \,. \tag{7.10}$$

Only a short description of the main approaches is given here, based on [367, Chapter 9], [576], and [252]. Practical implementations often consist of combinations of these methods.

Additive Synthesis. In additive synthesis, Equation 7.10 describes the superposition of sinusoids

$$f(t) = \sum_l F_l(t)\sin(\theta_l(t)) + n(t). \tag{7.11}$$

Sometimes a noise source $n(t)$ is added to account for the stochastic character which is not modelled well by sinusoids. In the simplest case, each frequency component $\theta_l(t)$ is given by a constant frequency and phase term $\theta_l(t) = \omega_l t + \phi_l$. In practical synthesis, the time signals in Equation 7.11 are represented by samples and the synthesized sound is processed in subsequent frames. The time variation of the amplitude and the frequency of the sinusoids are considered by changing the values of F_l, ω_l, and possibly ϕ_l from frame to frame. If the spectral lines ω_l in each frame are chosen to be multiples of a fundamental frequency ω_0, then also Discrete Fourier Transform methods or filterbanks can be used. On the other hand, $\theta_l(t)$ may be represented by a cubic polynomial in t, where the coefficients are updated from frame to frame.

Subtractive Synthesis. Additive synthesis builds up signals by adding basis functions. Subtractive synthesis, on the other hand, shapes signals by taking away frequency components from a spectrally rich excitation signal. This is achieved by exciting time-varying filters with noise. This approach is closely related to filtering in wavetable synthesis. However, in subtractive synthesis, the filter input is a synthetic signal rather than a wavetable. Since harmonic tones cannot be well approximated by filtered noise, subtractive synthesis is mostly used in conjunction with other synthesis methods.

Granular Synthesis. In granular synthesis the basis functions $\psi_l(t)$ in Equation 7.10 are chosen to be concentrated in time and frequency. These basis functions are called atoms or grains here. Building sounds from such grains is called granular synthesis. Sound grains can be obtained by various means: from windowed sine segments, from wavetables, from Gabor expansions, or with wavelet techniques.

Frequency Modulation. Making the phase term in the sine function time dependent leads to the frequency modulation (FM) method. In its simplest form, the time function $f(t)$ is given by

$$f(t) = F(t)\sin(\omega_0 t + \phi(t)). \tag{7.12}$$

The implementation consists of at least two coupled oscillators. In Equation 7.12 the carrier $\sin(\omega_0 t)$ is modulated by the time-dependent modulator $\phi(t)$ such that the frequency becomes time-dependent with $\omega(t) = \omega_0 + (\partial/\partial t)\phi(t)$. The inherent nonlinearity produces an artificial sound frequently used in synthesizers and in sound cards for personal computers. However it fails to reproduce natural instruments.

Physical Modelling. Wavetable synthesis and spectral synthesis are based on waveform descriptions in the time and frequency domain. A family of methods called physical modelling goes one step further by modelling directly the vibrating structure from which sound waves emanate. Invoking the laws of acoustics and elasticity theory results in the physical description of woodwind and string instruments by partial differential equations. Most methods are based on the wave equation which describes wave propagation in solids and in air [778].

Finite Difference Methods. The most direct approach is the discretization of the wave equation by finite difference approximations of the partial derivatives with respect to time and space. However, a faithful reproduction of the harmonic spectrum of an instrument requires small step sizes in time and space. The resulting numerical expense is considerably. The application of this aproach to piano strings has for example been shown by [112]. A physical motivation of the space discretization is given by the mass-spring-models described in [549].

Modal Synthesis. Vibrating structures can also be described in terms of their characteristic frequencies or modes and the associated decay rates. This approach allows the formulation of couplings between different substructures. Except for simple cases, the determination of the eigenmodes can only be conducted by experiments [549].

Digital Waveguides. A well known theoretical approach to the solution of the wave equation in one spatial dimension is the d'Alembert solution. It separates the wave propagation process into a pair of waves travelling into opposite directions without dispersion or losses. This separation is the basis of the digital waveguides described in [655], [367, Chapter 10], and [576, Chapter 7]. The digital model consists of a bidirectional delay line with coupling coefficents between the taps approximating losses and dispersion. The digital waveguide method has been refined by proper adjustment of the delay lines using fractional delay filters [721]. Applications to string instruments are found in [720] and to woodwind instruments in [602].

Couplings between sections with different wave impedances are modelled by scattering junctions. They approximate the partial reflection at discontinuities. Waveguide methods have also been extended to two and three spatial dimensions, however with considerable increase in computational demand.

Physical modeling by digital waveguides is incorporated into various commercial musical instruments using appropriate models for excitation (e.g. plucked, struck, and bowed strings) and boundary conditions. Furthermore, it provides a sound basis for the creation of artificial instruments like bowed flutes.

Transfer Function Models. This relatively new approach starts directly at the partial differential equation (PDE) describing the continuous vibrations in a musical instrument. It transforms the PDE with suitable functional transformations into a multidimensional (MD) transfer function model (TFM). For the time variable the Laplace transformation is used. The spatial transformation depends on the PDE and its boundary conditions. This leads to a generalized Sturm-Liouville type problem whose solutions are the eigenfunctions $K(x, \beta_\mu)$ and eigenvalues β_μ. They are used in the spatial transformation as transformation kernel and spatial frequency variable [704].

The physical effects modelled by the PDE like longitudinal and transversal oscillations, loss and dispersion are treated with this method in an exact fashion. Moreover, the TFM explicitely takes initial and boundary conditions, as well as linear and nonlinear excitation functions into account. The discretization of this continuous model for computer implementation based on analog-to-discrete transformations preserves not only the inherent stability, but also the natural frequencies of the oscillating body.

All parameters of this method are strictly based on physical parameters (e.g. the length or the Young's modulus of a string) and the output signal is calculated analytically from these parameters.

7.2.2 Sound Propagation Methods

Sound propagation methods can be divided into geometrical and numerical methods. Geometrical methods usually rely on the plane wave assumption and model sound propagation and reflection with methods from geometrical optics. Numerical models attempt to solve the partial differential equation for 3D wave propagation numerically.

Geometrical Methods. Geometrical methods model sound propagation by sound rays which travel similarly to light rays in optics. Diffraction effects are neglected by this assumption. With this simplification, sound ray propagation can be treated with the same principle methods which are used with great success in visual rendering. However, acoustical rendering requires some additional consideration, which is not necessary in visual rendering. The most important difference is the finite propagation speed of sound waves compared to the almost instantaneous spread of light. Furthermore, damping of light intensity is negligible for small distances, whereas sound waves may undergo severe attenuation. The adaption of visual rendering schemes for acoustics has to take these effects into account.

Image Source Methods. The simplest way of geometric modelling is the method of image or mirror sources. It is a classical method in physics for the treatment of boundary value problems. An early implementation as a computer program for acoustic simulation has been described e.g. by [10].

To describe acoustic ray propagation, the position of the sound source and the receiver have to be fixed. The propagation is modelled as the superposition of direct, single and multiple reflected rays from the source to the receiver. Rather than constructing all the different propagation paths within a room, the reflections are modelled as direct paths from additional imaginary sources outside the room. They are positioned in such a way, that their associated direct paths coincide with the reflections inside the room. These positions can be obtained by symmetry considerations.

For a simple concrete model, we can imagine a room with ideally reflecting walls. The response to an impulse at the source is then simply the sum of delayed impulses, where the delay time is given by the total length of the reflection path divided by the speed of sound.

For acoustic modelling, we have to consider the mirror sources as emitters of spherical pressure waves. The sound pressure at the receiver caused by the source with number i at distance d_i emitting a pulse $\delta(t)$ is then given by

$$p_i(t) = \frac{1}{4\pi d_i} \delta \left(t - \frac{d_i}{c} \right). \tag{7.13}$$

When all walls are ideal reflectors then the total impulse response in terms of sound pressure is given by

$$h(t) = \sum_i p_i(t), \tag{7.14}$$

where the sum is taken over all surrounding image sources within a certain region (ideally infinitely many). Angle and frequency independent wall absorption can be modelled by inclusion of appropriate absorption coefficients multiplied with the source pressure.

With a time dependence of the absorption coefficients the model can be extended to a frequency dependent reflection in which those absorption impulse responses have to be convolved with the source pressure. An angle dependence can also be included by different absorption impulse responses for different angles of incidence.

The image source model is conceptually simple, but it requires a high number of image sources in order to model the complete impulse response of a room. If we want to compute the impulse response of a room with volume V up to a length of t_0 seconds, we have to extend the room in either directions until the image sources fill a sphere of radius ct_0. The number of image sources is roughly given by the relation of the volume of the sphere to the volume of the room

$$N = \frac{\frac{4}{3}\pi (ct_0)^3}{V}. \tag{7.15}$$

For $t_0 = 1$s and $V = 100\text{m}^3$, a total of 1.7 million image sources results. This number can be handled for frequency independent reflections, but the consideration of frequency dependent reflections by multiple convolutions becomes impractical. The numerical expense increases also when multiple sources have to be considered. Additionally, any change of position of the receiver or one of the sources requires a recalculation of the impulse response.

Furthermore, obtaining the locations of the image sources is a tedious procedure if the building geometry is more complicated than a rectangular box. This procedure is described in [78] for polyhedra with arbitrary shape. The locations of possible image sources are computed by a tree-structured algorithm and tested for validity, proximity and visibility. Only valid sources which are visible and sufficiently close to the listener contribute to the impulse response. However, the computing time required to carry out this procedure is a severe limiting factor.

On the other hand, the image source method is also capable of modelling the interference between reflected sounds, if the model is formulated in terms of pressure rather than intensity [150].

Ray and Beam Tracing Methods. The ray tracing method is also based on geometrical acoustics. Its basic concept is similar to the image source method, since it also considers the possible reflection paths by which the sound travels to the receiver.

The source model emits a number of sound rays in all directions. A certain amount of energy is assigned to each ray. The propagation path of each ray is traversed and the energy losses due to absorption in the air and at each reflection are recorded. All rays that hit the receiver within a certain amount of time contribute to the impulse response.

This procedure is very similar to the well known application of ray tracing in visual rendering, except for the different physical effects, which account for the energy losses along the path. The advantage of the ray tracing method over the image source method lies in the treatment of arbitrary complex 3D shapes without requiring excessive computing power. A recent implementation is described in [235].

Radiosity Methods. The image source method and the ray tracing method depend on the position of the receiver. Whenever the receiver changes its position, the impulse response has to be computed anew. The radiosity method is a view independent algorithm. It requires knowledge about the enclosure geometry and about the sound sources. From this input, an approximation of the sound field can be constructed.

The radiosity method has been used for a long time in thermodynamics to calculate the radiation heat transfer between surfaces. Later, it became one of the working horses of computer graphics for visual rendering. Its adaption to auditory rendering is described in [642].

When applied to light waves, the radiosity method describes an equilibrium for the energy distribution within an enclosure that is constructed from a finite set of surfaces (patches). This finite set of patches covers the surface of the enclosure completely, so that all energy flows are considered. Any openings are also described as a partial surface with appropriate properties. The rate B_ℓ at which energy leaves the surface ℓ is derived from the conservation of energy. Since the energy exchange between the surfaces is assumed to happen instantaneously, B_ℓ is always given as the sum of the energy E_ℓ, which is generated by the surface ℓ itself and the reflection of the irradiations from the other surfaces:

$$B_\ell = E_\ell + \rho_\ell \sum_i B_i F_{\ell i}. \tag{7.16}$$

B_i is the rate at which energy leaves the other surfaces, $F_{\ell i}$ is the fraction of B_i which arrives at surface ℓ and ρ_ℓ is the reflection coefficient at surface ℓ. $F_{\ell i}$ depends on the size and the relative orientation of the surfaces ℓ and i. It is also called *form factor*. Their calculation is discussed thoroughly in standard texts on heat radiation and computer graphics.

The two basic differences between radiation transfer and sound propagation are

- damping of the sound intensity by the air,

- finite propagation speed of sound.

Since we are not dealing with real or virtual point sources as in the image source or ray tracing method, the geometrical attenuation with increasing distance is included in the form factors $F_{\ell i}$. The damping considered here is only due to conversion of kinetic to thermal energy by friction and molecular effects in the air given by the air attenuation factor $\Phi_{\ell i}$. Thus $\Phi_{\ell i}$ describes the fraction of the rate of energy $B_i F_{\ell i}$ which arrives at surface ℓ in the presence of transmission losses. Damping of the sound intensity by the air can be modelled by $\Phi_{\ell i}$, which acts on B_i in the same way as the form factor $F_{\ell i}$ does.

A more fundamental change to the equilibrium Equation 7.16 is necessary to take the finite propagation speed into account. Equation 7.16 shows no time dependence, since it is understood that any changes in the system lead instantaneously to a new equilibrium. No dynamical or memory effects have to be considered with respect to the speed of light.

This is no longer true for sound propagation. The energy transport from distant surfaces is not only subject to transmission losses but also to time delay. Sound energy that leaves a surface does not arrive in the same time instant at other surfaces. To account for the travel time of the sound waves, the conservation of energy has to be formulated with time dependent quantities. The time delays are expressed by the distance between the surfaces and the speed of sound. This leads to the following time dependent formulation of the energy balance, which includes propagation losses and delay

$$B_\ell(t) = E_\ell(t) + \rho_\ell \sum_i B_i(t - \frac{d_{\ell i}}{c}) F_{\ell i} \Phi_{\ell i}. \tag{7.17}$$

The energy exchange rates $B_\ell(t)$ and the emission rates $E_\ell(t)$ are now functions of time.

Numerical Methods. The methods for the physical modelling of vibrating structures from Section 7.2.1 can also be extended to the numerical simulation of room acoustics. Two methods will be discussed here: A three-dimensional extension of the digital waveguides and a method based on the wave digital principle. Since an in-depth presentation of the simulation algorithms is rather involved, only a brief description of the general principles is given.

Waveguide Methods. Waveguide methods from physical modelling of musical instruments are also applied to the numerical solution of the 3D wave equation. Detailed accounts are given by [599], [600], and [349].

The basic idea is to discretize the three-dimensional (3D) space by a 3D grid and to consider each node as a lossless junction. The physical nature of wave propagation is taken into account by requiring for each node that

- the sum of all inputs to the junction equals the sum of its outputs,
- all impedances at the junction are equal.

These requirments formaly agree with the basic rules of electrical network theory. Deriving a discrete-time relation for the sound pressures at the grid nodes from these requirements results in

$$p[\boldsymbol{n}, k] = \frac{1}{3} \sum_{\boldsymbol{m}} p[\boldsymbol{m}, k-1] - p[\boldsymbol{n}, k-2]. \tag{7.18}$$

The sum over $\boldsymbol{m}$ denotes summation over all spatial neighbors of the pressure $p[\boldsymbol{n}, k]$ at the node with 3D spatial coordinates $\boldsymbol{n}$ and time index k. This difference equation results also from a finite-difference discretization of the Helmholtz equation [599], with time discretization performed by a forward Euler-step. Variations of this scheme are possible by different spatial sampling grids (e.g. triangular, hexagonal).

A problem associated with this approach is the direction dependent dispersion. [600] show how this problem can be alleviated by interpolation methods.

State Space Methods. A direct method to computational acoustics has been presented by [563]. It leads from the partial differential equations for the sound pressure and the particle velocity directly to a state space description of the sound field simulation algorithm. The method is closely related to the wave digital principle for numerical integration of partial differential equations with its favourable numerical properties as introduced by [222].

The starting point are the the equation of motion and the equation of continuity for the acoustic pressure $p(\boldsymbol{x}, t)$ and the acoustic fluid velocity vector $\boldsymbol{v}(\boldsymbol{x}, t)$ (see [607] and [778]). Under reasonable simplifications for sound propagation in air, they are given by

$$\rho_0 \frac{\partial}{\partial t} \boldsymbol{v}(\boldsymbol{x}, t) + \mathrm{grad} p(\boldsymbol{x}, t) = 0, \tag{7.19}$$

$$\frac{1}{\rho_0 c^2} \frac{\partial}{\partial t} p(\boldsymbol{x}, t) + \mathrm{div} \boldsymbol{v}(\boldsymbol{x}, t) = 0. \tag{7.20}$$

where t denotes time and $\mathbf{x}$ the vector of space coordinates x, y, z. ρ_0 is the static density of the air and c is the speed of sound.

After a series of intermediate steps, the state space representation of a discrete-time and discrete-space algorithm is obtained. These steps are:

- Separation into spatial components

  The PDE description in Equations 7.19 and 7.20 is transformed into a symmetric form and then separated into three different spatial components. Each component contains derivatives with respect to time and only one of the spatial directions x, y, or z.

- Numerical solution for each component

  The numerical integration for the spatial components is performed by the trapezoidal rule in two dimensions (one time and one space dimension).

- Combination of the spatial components

  The discrete-time, discrete-space approximations for each of the three spatial components are combined into a full four-dimensional representation (one time and three space dimensions) of the PDE description in Equations 7.19 and 7.20.

- State space formulation

  A suitable choice of internal states allows to formulate the discrete numerical sound propagation model in the state space context. The resulting discrete-time discrete-space state equation represents directly the structure of the numerical routines of the simulation algorithm.

As an example for the performance of this method, consider the propagation of a sound wave in a building with three interconnected rooms (compare [599]). The resulting pressure field for a fixed time in a plane parallel to the floor is displayed in Figure 7.21. The outer walls (not shown) and the inner walls are highly reflecting, whereas floor and ceiling (not shown) are absorbing. A sound pulse emanates from the corner in front and propagates through all three rooms. Note that the numerical simulation correctly reproduces the diffraction effects at the openings, which could not be obtained by geometric models.

7.2.3 Models for Human Sound Perception

The psychoacoustics of human localization of sounds sources in space are not as rigorously understood as the purely physical propagation phenomena. The classical theory of spatial hearing is based on certain cues like interaural time diffferences, interaural intensity differences, and head movement. More advanced concepts describe the influence of spectral cues in terms of head related transfer functions. The foundations of spatial hearing are given in [71]. The application to virtual reality systems is described in [49] and [243].

We will quote only some of the common concepts to describe what is called spatial hearing. This term can be roughly defined by some questions, which naturally arise, when we try to evaluate the human sound perception capabilities:

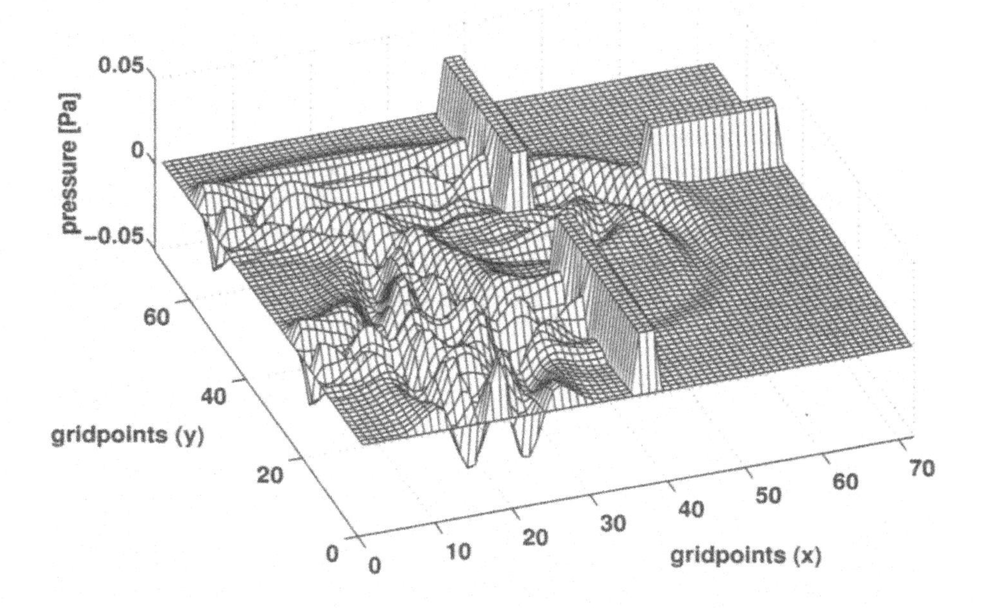

Figure 7.21. Sound propagation in a three-room building.

- How can we tell the direction of a sound source?
- How can we tell the distance of a sound source?
- How can we separate different sound sources, even in the presence of strong background noise?

Interaural Time and Intensity Differences. The ability to tell the direction of a sound source can be traced to two basic differences in the sound perception with both ears:

- The interaural time difference (ITD) is the difference in the arrival time of a wave front between the two ears.
- The interaural intensity difference (IID) is the difference in perceived intensity between the two ears.

Both kinds of differences act together to provide spatial information about the direction of arrival of a wave front. The human hearing system can resolve time differences between both ears for signals from 20 Hz up to frequencies of 1 kHz. The relation between the relative time difference of two signals and the direction of arrival is based on a simple geometrical argument. In this sense, the ears act as a two-microphone-array. For low frequencies, there is no noticeable intensity difference between both ears, due to diffraction at the head. For frequencies with a wavelength shorter than the diameter of the head (typically 4 kHz and higher) diffraction becomes negligible and an intensity difference between both ears is perceived.

There is evidence, that the human hearing system combines ITD and IID for the location of sound sources. ITD dominates for low frequencies, while IID works for high frequencies. In the frequency range between approximately 1 and 4 kHz, time and intensity differences are perceived, which results in several conflicting cues that tend to cause localization errors in virtual systems.

Application of these concepts to acoustic rendering calls for post processing of the receiver signal. Once the response to a sound source is determined by the methods of the previous section, the receiver signal has to be split into two signals for the left and for the right ear respectively. These signals are delayed and filtered in order to produce the interaural time and intensity differences which correspond to the direction of the source signal.

Head-Related Transfer Functions. The ITD and IID just considered are only the starting point for an interpretation of the human hearing system. They allow to discriminate the source position in terms of 'left' and 'right', but they do not explain why we can also locate sound sources above and below or in front or in the back of our head.

These effects can only be explained when the shape of the pinnae, and the influence of head and torso are taken into account. They act as a filter between the sound field outside the ears and the eardrum. Technically, their effect on spatial hearing can be described by a transfer function, the so called *head-related transfer function (HRTF)*. Of course there are infinitely many HRTFs for different source directions and two HRTFs for each sound source, one for each ear. They are defined as the frequency dependend relation of the sound pressure at the eardrum to the sound pressure at the receiver location when the listener is absent. Much effort has been devoted to measurements of the HRTFs and corresponding data sets for typical HRTFs are available.

The drawback of using HRTFs is that they vary significantly from person to person. This fact requires precise measurements of each listener's individual HRTFs which is impractical for widespread use. Data bases with different pinnae shapes are available to minimize the dicrepancy to the individual HRTFs.

The application of HRTFs to acoustical rendering is straightforward. After the receiver signal has been split into a left and right ear signal and both signals have been treated with the appropriate ITD and IID information, the final step is the filtering with the corresponding HRTF for each ear. When these signals are delivered by ear phones directly to the listener's eardrum, they carry the necessary spatial cues to enable spatial hearing of virtual sound sources.

7.2.4 Aurealization Systems

Systems combining sound sources, sound propagation, and models for human sound perception to achieve a spatial hearing effect are called *aurealization systems*. An overview over the implementation of aurealization systems is given in [236] and [411]. The available systems can be roughly categorized in head-related systems, desktop applications, and surround sound systems.

Head-Related Systems. Head-related systems deliver a two-channel signal to one listener by earphones. This has the advantage that there is no cross-talk from one earphone to the contralateral ear, i.e. sending the left and the right channel separately to the listerners ears.

In a first step recordings of real sound fields for head-related systems can easily be done by a dummy head microphone system that is based on average human characteristics or, corresponding to the individualized HRTFs, by probe microphones that are inserted in the listener's ear canal.

To place a monoaural *virtual* source in a 3D sound field, it must first be convolved with the room impulse response with no listener in the room which is then convolved with the corresponding HRTFs for the given directions. The results can be startingly realistic.

However, if the listener turns his head, the HRTFs must be updated because they are given by the relative position of the head to the sound source. This procedure is called head tracking. This procedure is only possible with virtual sound sources and not with recordings of real sound fields. The head can be tracked either by a motion and rotation sensor in the headphone or by vision based methods.

Although there are several advantages in using earphones to obtain a correct lateral sound field impression this method is not widely used. Drawbacks of this method are large errors in sound position perception associated with headphones. These are front-back reversals, the difficulty to externalize the 'inside-the-head' sensation by using earphones and the uncomfortability of headphones for long times [411].

Desktop Applications. Desktop applications attempt to produce a spatial impression for a listener sitting in front of a pair of loudspeakers. This is for example useful in teleconferencing systems or other computer applications where one person sits in front of two loudspeakers placed at both sides of the monitor. In this arrangement some of the drawbacks mentioned at the head-related systems are avoided but here the cross-talk from one loudspeaker to the oppsite-site (contralateral) ear has to be taken into account. Additionally inverse filtering of the room acoustics is required [236].

Several schemes have been studied for cross-talk cancellation. They are all based on preconditioning the signal in each loudspeaker such that the output sound generates the desired binaural sound pressure at each ear. This preconditioning is done by implementing the inverse of a 2×2 system transfer matrix containing the transfer functions from each loudspeaker to each ear. The filter coefficients for the cross-talk canceller are obtained by measurements with a dummy head microphone or with a calibration procedure done by the listener. This method gives good spatial hearing results without a totally immobilized head. Only for rotations of more that ± 10 degrees or for lateral movings of more than 10 cm head tracking as described above is necessary.

In reverberant rooms the ability to distinguish between different elevations and between front and back located sound sources is degraded for this method. This fact requires the implementation of inverse filtering of the room impulse response. To avoid stability problems only the minimum phase portion of the room impulse response is equalized. Combined systems with cross-talk cancellation and inverse filtering of room impulse responses are realized.

Front-back reversals, which are also existent in this system, are reduced by using four instead of only two loudspeakers. They are arranged by two loudspeakers with cross-talk cancellation in front and by two loudspeakers with independent cross-talk cancellation in the back of the listener. This system leads to the sourround sound systems described in the next section.

Surround Sound Systems. Surround sound systems attempt to reproduce the true 3D wave field within an enclosure using a large number of loudspeakers. Although the most expensive solution, it works for more than one listener and it does not require head tracking [328, 411].

Surround sound systems were initially developed in the early 1950's for cinemas. These systems included additionally to the three speakers in front (left, center and right speaker) a monophonic channel that was reproduced over two loudspeakers behind the audience. This so called effect channel increased the spatial hearing but it had the following disadvantages. On the one hand, when sitting on-center the listerner perceived an 'inside-the-head' localization. On the other hand the listeners sitting off-center in the back of the cinema heard the effect speaker next to them as the dictating sound radiator. These difficulties have been overcome by using loudspeaker arrays for the effect channel along the sides of the cinema.

Early surround sound systems for homes used four speakers in a square arrangement with the difficulty in reproducing directions between the speakers. The latest surround sound systems for home theaters are all based on the 3/2 format, with three speakers in front and two speakers in the rear of the listener. This arrangement gives good localization results since the ability of resolving different sound source angles is in front of the listener much better than in the rear. The well known coding schemes for this arrangement reach from the stereo compatible Dolby Surround format with a monophonic and bandlimited rear channel up to five independent full-band channels coded by Dolby AC-3 or by Digital Theater System (DTS) [457].

More advanced but also more expensive methods are based on loudspeaker arrays. They are called wave field synthesis methods. They can simulate the wave field in a whole room with advanced digital signal processing methods controlling the distribution of the sound on different loudspeakers. The method is based on Green's theorem that states that the sound field in an enclosure is totally given by the sound pressure on the surrounding surface. This leads to a loudspeaker array in one plane to realize the soundfield in this plane, or to loudspeaker arrays on all walls of a room to realize the exact soundfield in the whole room. Sound sources can easily be replaced with this method. Furthermore the virtual room ambience the audience is placed in can be changed [727]. In contrast to the other methods the sound field synthesis is a real multi-listener immersive audio method since all listeners feel equally immersed in a virtual constructed room.

7.2.5 Conclusions

In this section we gave a brief overview of different aurealization schemes. We started at the sound source and showed how different sound synthesis methods model the sounds of real or virtual structures. A subdivision was made into classical methods modelling the sound like wavetable and spectral synthesis, and into physical modelling methods modelling the sound production mechanism like finite difference, digital waveguide or transfer function methods. Although physical modelling methods are more flexible and can be used more intuitively they need much more computational power than the classical methods.

The propagation of the sound produced by sound source methods was subdivided into geometrical and numerical methods. The geometrical methods using modified methods of geometrical optics are more useful for real-time implementations because of the lower computational power, whereas the numerical methods solving the 3D wave equation numerically are often preferred when a more precise result is demanded.

Since in aurealization a human listener receives the sound we descibed models for the human sound perception. In the horizontal plane we can distinguish between different sound source angles with interaural time and intensity differences whereas in the median plane the head-related transfer functions filtering the source signal are dominant.

We closed this chapter with some concepts for aurealization systems. Head-related as well as desktop applications are suitable only for one user and are difficult to realize since head-tracking and inverse filtering of the room impulse response must be implemented. If a multi-listener solution is required a first step is the popular 3/2 arrangement with three loudspeakers in front and two loudspeakers in the back of the audience. A more advanced but also the most expensive method where all listerners feel equally immersed in a virtual constructed room is the wave-field synthesis realized with loudspeaker arrays.

7.3 FUSION OF MULTISENSOR DATA

N. STROBEL

This section investigates the problem of combining data acquired from multiple sensor systems. We focus on the process by which data from a multitude of sensors is used to yield an optimal estimate of a specified state vector pertaining to an observed system. This problem is also often referred to as *sensor fusion*.

An emerging application of sensor fusion in scene analysis and interpretation is object tracking using audio and video measurements. In the first case, the object position is estimated from time differences of arrival (TDOAs) of sound waves recorded at a microphone array. Another (complementary) estimate is based on visual object recognition in a camera scene. Our goal is to explain how to optimally combine both estimates.

There are several applications for visual and/or acoustic object localization and tracking. Face localization and tracking, for example, is of particular interest for video communications and coding. Once the face of a speaker has been found in a scene, the associated area can be coded with lower distortion than the remaining background. Recent examples for video coding based on face localization include [223] or [142]. Estimation of the speaker position for video communications using audio information only has been suggested by [719]. The basic idea relies on the estimation of the position of a speaker's mouth based on the received speech signal. Vahedian claims that this approach is much faster and easier to implement than picture-based techniques. So far, however, little attention has been given to multimodal speaker localization using both audio and video information, but the few available publications indicate promising results. A system combining audio and visual cues for locating and tracking of a person in real time was proposed by [546]. There it was shown that combining speech source localization with video-based head tracking provides a more accurate and robust outcome than that obtained using any one of the audio or visual modalities.

To provide a solid basis for the underlying concepts of data fusion, we begin with a brief survey of centralized state estimation techniques. Then by now well known yet still important theoretical results in distributed state estimation are reviewed. In the sequel, we concentrate on recursive state estimation. There the linear (centralized) Kalman filter and the extended Kalman filter are covered. We finally discuss the decentralized Kalman filter and comment on the decentralized extended Kalman filter.

7.3.1 Centralized State Estimation

Let the state of a system of interest be represented by the m-dimensional vector $\mathbf{x}$ and the total set of measured quantities be represented by the n-dimensional vector $\mathbf{y}$. All vectors are assumed to be column vectors. Our problem is to estimate the value of $\mathbf{x}$ from the knowledge of $\mathbf{y}$ according to some specified optimality criterion. To solve this problem, it is necessary (1) to specify a model of the system of interest and the sensor system in the absence of experimental errors, (2) to have some a priori idea about the statistical nature of $\mathbf{x}$ and $\mathbf{y}$, and (3) to know the statistics of associated measurement errors.

Assuming a so-called measurement model, we can write

$$\mathbf{y} = \mathbf{h}(\mathbf{x}) + \mathbf{n}. \tag{7.21}$$

The n-dimensional function $\mathbf{h}(.)$ represents the ideal operation of the sensor system, and $\mathbf{n}$ is a random, n-dimensional vector representing experimental error, model error, background signals, etc. The assumption that the experimental error simplifies to an additive term is clearly an idealization.

The measurement model must include a specification of the a priori statistical properties of $\mathbf{x}$ and $\mathbf{n}$. Otherwise we cannot deduce the a priori statistical properties of $\mathbf{y}$. A common assumption is that the state vector $\mathbf{x}$ and the additive noise $\mathbf{n}$ are statistically independent and that the joint probability density functions (pdfs) of both the state vector, $p_{\mathbf{x}}(\mathbf{x})$, and the process noise vector, $p_{\mathbf{n}}(\mathbf{n})$, are given. It is generally further assumed that $\mathbf{n}$ is a normally distributed random vector with

$$E\{\mathbf{n}\} = \mathbf{0}, \quad \text{and} \quad E\{\mathbf{n}\mathbf{n}^T\} = \mathbf{C}_{\mathbf{nn}}. \tag{7.22}$$

The symbol $E\{.\}$ refers to the expectation operator, the superscript T denotes the transpose, and $\mathbf{C}_{\mathbf{nn}}$ is the covariance matrix of the noise $\mathbf{n}$. The a priori pdf of $\mathbf{x}$, represented by $p_{\mathbf{x}}(\mathbf{x})$, is also assumed to be multivariate Gaussian with

$$E\{\mathbf{x}\} = \boldsymbol{\mu}_{\mathbf{x}}, \quad \text{and} \quad E\{(\mathbf{x} - \boldsymbol{\mu}_{\mathbf{x}})(\mathbf{x} - \boldsymbol{\mu}_{\mathbf{x}})^T\} = \mathbf{C}_{\mathbf{xx}}. \tag{7.23}$$

According to Equation 7.21, the likelihood of $\mathbf{y}$, given $\mathbf{x}$, is the pdf of the random vector $\mathbf{n} = \mathbf{y} - \mathbf{h}(\mathbf{x})$, i.e.,

$$p_{\mathbf{y}|\mathbf{x}}(\mathbf{y}|\mathbf{x}) = p_{\mathbf{n}}(\mathbf{y} - \mathbf{h}(\mathbf{x})). \tag{7.24}$$

The joint pdf $p_{\mathbf{xy}}(\mathbf{x}, \mathbf{y})$ can be written as

$$p_{\mathbf{xy}}(\mathbf{x}, \mathbf{y}) = p_{\mathbf{y}|\mathbf{x}}(\mathbf{y}|\mathbf{x}) p_{\mathbf{x}}(\mathbf{x}), \tag{7.25}$$

and the a posteriori pdf $p_{\mathbf{x}|\mathbf{y}}(\mathbf{x}|\mathbf{y})$ follows from

$$p_{\mathbf{x}|\mathbf{y}}(\mathbf{x}|\mathbf{y}) = \frac{p_{\mathbf{xy}}(\mathbf{x}, \mathbf{y})}{p_{\mathbf{y}}(\mathbf{y})} = \frac{p_{\mathbf{y}|\mathbf{x}}(\mathbf{y}|\mathbf{x}) p_{\mathbf{x}}(\mathbf{x})}{p_{\mathbf{y}}(\mathbf{y})}. \tag{7.26}$$

Often the denominator can be ignored since it does only depend on $\mathbf{y}$, and in the estimation process the random vector $\mathbf{y}$ can normally be regarded as fixed.

Before we can actually compute estimates $\hat{\mathbf{x}}$ of the state vector $\mathbf{x}$ from the observations $\mathbf{y}$, an optimality criterion must be defined. To this end, a cost function $L(\hat{\mathbf{x}}(\mathbf{y}), \mathbf{x})$ is needed. It measures the penalty incurred when we estimate the state vector to be $\hat{\mathbf{x}} = \hat{\mathbf{x}}(\mathbf{y})$ when the actual value of the state is $\mathbf{x}$.

The optimality criterion follows as the average loss over the joint distribution of $\mathbf{x}$ and $\mathbf{y}$. This so-called Bayes risk can be written as

$$R(\hat{\mathbf{x}}(\mathbf{y}), \mathbf{x}) = \iint L(\hat{\mathbf{x}}(\mathbf{y}), \mathbf{x}) p_{\mathbf{xy}}(\mathbf{x}, \mathbf{y}) d\mathbf{x} d\mathbf{y}. \tag{7.27}$$

Using the Bayes rule, we can express the joint pdf $p_{\mathbf{xy}}(\mathbf{x}, \mathbf{y})$ as

$$p_{\mathbf{xy}}(\mathbf{x}, \mathbf{y}) = p_{\mathbf{x}|\mathbf{y}}(\mathbf{x}|\mathbf{y}) p_{\mathbf{y}}(\mathbf{y}). \tag{7.28}$$

Substituting Equation 7.28 into Equation 7.27, we get an expression of the Bayes risk in terms of the conditional pdf $p_{\mathbf{x}|\mathbf{y}}(\mathbf{x}|\mathbf{y})$ and the prior probability of the measurements $p_{\mathbf{y}}(\mathbf{y})$, i.e.,

$$R(\hat{\mathbf{x}}(\mathbf{y}), \mathbf{x}) = \iint L(\hat{\mathbf{x}}(\mathbf{y}), \mathbf{x}) p_{\mathbf{x}|\mathbf{y}}(\mathbf{x}|\mathbf{y}) d\mathbf{x} p_{\mathbf{y}}(\mathbf{y}) d\mathbf{y}. \tag{7.29}$$

Since the marginal density $p_{\mathbf{y}}(\mathbf{y})$ is nonnegative, the Bayes risk may be minimized by minimizing the inner integral for each value of $\mathbf{y}$, i.e.,

$$\hat{\mathbf{x}}(\mathbf{y}) = \arg\min_{\hat{\mathbf{x}}} \left\{ \int L(\hat{\mathbf{x}}(\mathbf{y}), \mathbf{x}) p_{\mathbf{x}|\mathbf{y}}(\mathbf{x}|\mathbf{y}) d\mathbf{x} \right\}. \tag{7.30}$$

To turn this general result into a useful formula, we briefly investigate the quadratic loss function

$$L(\hat{\mathbf{x}}(\mathbf{y}, \mathbf{x})) = (\mathbf{x} - \hat{\mathbf{x}}(\mathbf{y}))^T (\mathbf{x} - \hat{\mathbf{x}}(\mathbf{y})). \tag{7.31}$$

The associated Bayes estimate follows by substituting Equation 7.31 into Equation 7.30 and taking the derivative with respect to $\hat{\mathbf{x}}(\mathbf{y})$. Setting the result equal to zero, we obtain the so-called minimum mean-square error estimate (MMSE)

$$\hat{\mathbf{x}}(\mathbf{y}) = \int \mathbf{x} p_{\mathbf{x}|\mathbf{y}}(\mathbf{x}|\mathbf{y}) d\mathbf{x} = E\{\mathbf{x}|\mathbf{y}\} = \mu_{\mathbf{x}|\mathbf{y}}. \tag{7.32}$$

The symbol $\mu_{\mathbf{x}|\mathbf{y}}$ refers to the (conditional) mean of the a posteriori pdf $p_{\mathbf{x}|\mathbf{y}}(\mathbf{x}|\mathbf{y})$.

7.3.2 Distributed State Estimation

So far we have assumed that a single central processor computes the optimal estimate. In some cases, however, decentralized estimation may be required due to increasing computational loads and reliability concerns. To save in storage and communications, we may choose to compute intermediate estimates of parameters from associated sensor data followed by global fusion of the local estimates. The distributed fusion algorithm attempts to duplicate the results of a centralized approach. That is, the global estimate after fusion is to be identical to the result based on the complete set of observations.

From an abstract point of view, the problem of estimating the state of a system from multiple sets of data produced by separate sensor systems is the same as that involving one set of data. The only difference is that the measurement vector $\mathbf{y}$ is separated into two components $\mathbf{y}_1$ and $\mathbf{y}_2$:

$$\mathbf{y} = \begin{bmatrix} \mathbf{y}_1 \\ \mathbf{y}_2 \end{bmatrix}. \tag{7.33}$$

Nevertheless, significant secondary problems emerge when one investigates the modularization of the estimation process. An example for a configuration with two sensors is shown in Figure 7.22. For simplicity, we will only discuss two-sensor systems in the sequel [569]. They are sufficient to convey the basic ideas, and they appropriately describe the fusion problem for audio and video data.

In our case, optimal distributed state estimation is concerned with computing the joint a posteriori state pdf $p_{\mathbf{x}|\mathbf{y}}(\mathbf{x}|\mathbf{y})$ from two distributed observations $\mathbf{y}_1$ and $\mathbf{y}_2$ as shown in Equation 7.33. Once $p_{\mathbf{x}|\mathbf{y}}(\mathbf{x}|\mathbf{y})$ is known, the optimal global (MMSE) estimate follows from Equation 7.32. Two important questions arise in this context: (1) What requirement is necessary to derive an optimal state estimate based on observations from two sensor systems? (2) What forms of the intermediate estimators $\mathbf{z}_1(\mathbf{y}_1)$ and $\mathbf{z}_2(\mathbf{y}_2)$ are sufficient to compute such an estimate?

Basic Requirement for Optimal Distributed State Estimation. The first requirement for separating the estimation process into two modules can be expressed as

$$p_{\mathbf{y}|\mathbf{x}}(\mathbf{y}_1,\mathbf{y}_2|\mathbf{x}) = p_{\mathbf{y}_1|\mathbf{x}}(\mathbf{y}_1|\mathbf{x})p_{\mathbf{y}_2|\mathbf{x}}(\mathbf{y}_2|\mathbf{x}). \tag{7.34}$$

This means that the observations $\mathbf{y}_1$ and $\mathbf{y}_2$ are statistically independent, given the state vector $\mathbf{x}$.

It does not imply in general that the measurements are independent. That is, Equation 7.34 is not to be confused with

$$p_{\mathbf{y}}(\mathbf{y}_1,\mathbf{y}_2) = p_{\mathbf{y}_1}(\mathbf{y}_1)p_{\mathbf{y}_2}(\mathbf{y}_2). \tag{7.35}$$

In fact, it can be shown that for Equation 7.35 to hold, it is necessary that both

$$p_{\mathbf{y}|\mathbf{x}}(\mathbf{y}_1,\mathbf{y}_2|\mathbf{x}_1,\mathbf{x}_2) = p_{\mathbf{y}_1|\mathbf{x}_1}(\mathbf{y}_1|\mathbf{x}_1)p_{\mathbf{y}_2|\mathbf{x}_2}(\mathbf{y}_2|\mathbf{x}_2), \tag{7.36}$$

and

$$p_{\mathbf{x}}(\mathbf{x}_1,\mathbf{x}_2) = p_{\mathbf{x}_1}(\mathbf{x}_1)p_{\mathbf{x}_2}(\mathbf{x}_2) \tag{7.37}$$

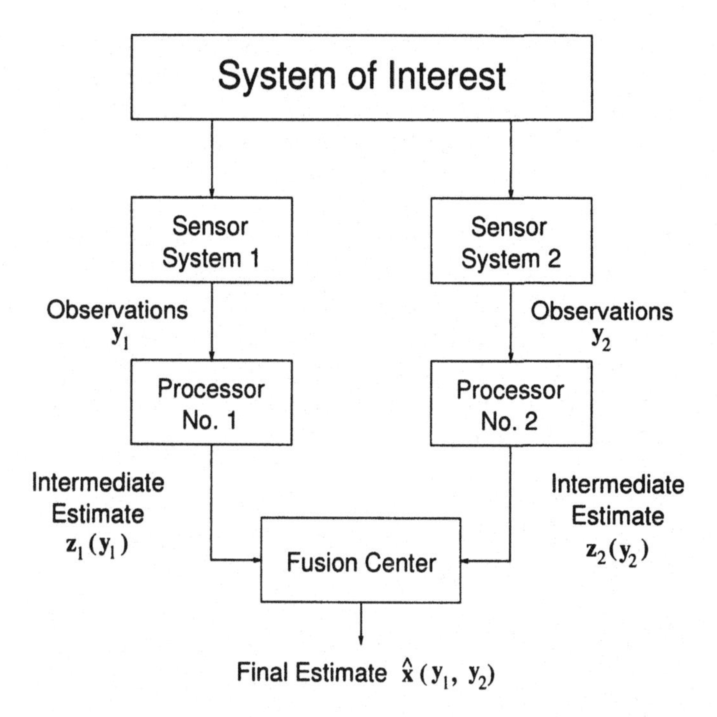

Figure 7.22. System for the fusion of two intermediate estimates $\mathbf{z}_1(\mathbf{y}_1)$ and $\mathbf{z}_2(\mathbf{y}_2)$ based on the associated sensor measurements $\mathbf{y}_1$ and $\mathbf{y}_2$. The output of the fusion center is the final estimate $\hat{\mathbf{x}}(\mathbf{y}_1, \mathbf{y}_2)$.

are satisfied. For more details see [569]. Equation 7.36 states that the two sensor systems observe independent parts of the system of interest. In other words, $\mathbf{y}_1$ only depends on $\mathbf{x}_1$, while $\mathbf{y}_2$ only depends on $\mathbf{x}_2$. Equation 7.37 implies that the components $\mathbf{x}_1$ and $\mathbf{x}_2$ of the underlying state vector

$$\mathbf{x} = \begin{bmatrix} \mathbf{x}_1 \\ \mathbf{x}_2 \end{bmatrix} \tag{7.38}$$

are statistically independent.

Once the general solution for the distributed state estimation is derived below, it will become evident that Equation 7.35 greatly simplifies the fusion problem by separating it into two independent (traditional) estimation problems for $\mathbf{x}_1$ and $\mathbf{x}_2$.

Another instructive interpretation of Equation 7.34 is possible. To this end, let us divide the left hand side (lhs) of Equation 7.34 by $p_{\mathbf{y}_2|\mathbf{x}}(\mathbf{y}_2|\mathbf{x})$. This yields

$$\frac{p_{\mathbf{y}|\mathbf{x}}(\mathbf{y}_1, \mathbf{y}_2|\mathbf{x})}{p_{\mathbf{y}_2|\mathbf{x}}(\mathbf{y}_2|\mathbf{x})} = \frac{p_{\mathbf{y}_1|\mathbf{x},\mathbf{y}_2}(\mathbf{y}_1|\mathbf{x}, \mathbf{y}_2)p_{\mathbf{y}_2|\mathbf{x}}(\mathbf{y}_2|\mathbf{x})}{p_{\mathbf{y}_2|\mathbf{x}}(\mathbf{y}_2|\mathbf{x})} = p_{\mathbf{y}_1|\mathbf{x},\mathbf{y}_2}(\mathbf{y}_1|\mathbf{x}, \mathbf{y}_2). \tag{7.39}$$

Since Equation 7.39 is identical to the remainder of the right hand side (rhs) of Equation 7.34 after dividing it by $p_{\mathbf{y}_2|\mathbf{x}}(\mathbf{y}_2|\mathbf{x})$, we can write

$$p_{\mathbf{y}_1|\mathbf{x},\mathbf{y}_2}(\mathbf{y}_1|\mathbf{x}, \mathbf{y}_2) = p_{\mathbf{y}_1|\mathbf{x}}(\mathbf{y}_1|\mathbf{x}). \tag{7.40}$$

If, for example, $\mathbf{y}_1$ refers to audio time delay measurements used to estimate the speaker position $\mathbf{x}$, and $\mathbf{y}_2$ specifies video measurements, then Equation 7.40 means that the audio measurements only depend on the speaker position $\mathbf{x}$ and not on the video measurements $\mathbf{y}_2$. Of course, the same relation with $\mathbf{y}_1$ and $\mathbf{y}_2$ interchanged also holds.

Sufficient Distributed State Estimates. The second question about sufficient intermediate estimates is studied below for two extreme case. Initially we discuss the general case. There we only assume that the observations $\mathbf{y}_1$ and $\mathbf{y}_2$ are conditionally independent, given the state $\mathbf{x}$. Then we assume that $p_{\mathbf{x}|\mathbf{y}_1}(\mathbf{x}|\mathbf{y}_1)$ and $p_{\mathbf{x}|\mathbf{y}_2}(\mathbf{x}|\mathbf{y}_2)$ are independent and Gaussian. In each case, we determine what kinds of intermediate processors outputs $\mathbf{z}_1(\mathbf{y}_1)$ and $\mathbf{z}_2(\mathbf{y}_2)$ are required such that the fusion module can reproduce an optimal state estimate.

General Case. Here we consider the general situation in which the observations $\mathbf{y}_1$ and $\mathbf{y}_2$ are only assumed to be conditionally independent, given the state $\mathbf{x}$.

For Bayesian estimation under quadratic loss, we recall that the optimal estimate is

$$\hat{\mathbf{x}} = E\{\mathbf{x}|\mathbf{y}_1, \mathbf{y}_2\} = \int \mathbf{x} p_{\mathbf{x}|\mathbf{y}}(\mathbf{x}|\mathbf{y}_1, \mathbf{y}_2) dx = \mu_{\mathbf{x}|\mathbf{y}}. \tag{7.41}$$

The conditional pdf $p(\mathbf{x}|\mathbf{y}_1, \mathbf{y}_2)$ can be computed from

$$p_{\mathbf{x}|\mathbf{y}}(\mathbf{x}|\mathbf{y}_1, \mathbf{y}_2) = \frac{p_{\mathbf{y}|\mathbf{x}}(\mathbf{y}_1, \mathbf{y}_2|\mathbf{x}) p_{\mathbf{x}}(\mathbf{x})}{p_{\mathbf{y}}(\mathbf{y}_1, \mathbf{y}_2)}. \tag{7.42}$$

For conditionally independent observations, we can use Equation 7.34 to rewrite Equation 7.42. The result is

$$p_{\mathbf{x}|\mathbf{y}}(\mathbf{x}|\mathbf{y}_1, \mathbf{y}_2) = \frac{p_{\mathbf{y}_1|\mathbf{x}}(\mathbf{y}_1|\mathbf{x}) p_{\mathbf{y}_2|\mathbf{x}}(\mathbf{y}_2|\mathbf{x}) p_{\mathbf{x}}(\mathbf{x})}{p_{\mathbf{y}}(\mathbf{y}_1, \mathbf{y}_2)}. \tag{7.43}$$

Substituting

$$p_{\mathbf{y}_1|\mathbf{x}}(\mathbf{y}_1|\mathbf{x}) = \frac{p_{\mathbf{x}|\mathbf{y}_1}(\mathbf{x}|\mathbf{y}_1) p_{\mathbf{y}_1}(\mathbf{y}_1)}{p_{\mathbf{x}}(\mathbf{x})} \tag{7.44}$$

and

$$p_{\mathbf{y}_2|\mathbf{x}}(\mathbf{y}_2|\mathbf{x}) = \frac{p_{\mathbf{x}|\mathbf{y}_2}(\mathbf{x}|\mathbf{y}_2) p_{\mathbf{y}_2}(\mathbf{y}_2)}{p_{\mathbf{x}}(\mathbf{x})} \tag{7.45}$$

into Equation 7.43, we finally get

$$p_{\mathbf{x}|\mathbf{y}}(\mathbf{x}|\mathbf{y}_1, \mathbf{y}_2) = \frac{p_{\mathbf{x}|\mathbf{y}_1}(\mathbf{x}|\mathbf{y}_1) p_{\mathbf{x}|\mathbf{y}_2}(\mathbf{x}|\mathbf{y}_2)}{p_{\mathbf{x}}(\mathbf{x})} \frac{p_{\mathbf{y}_1}(\mathbf{y}_1) p_{\mathbf{y}_2}(\mathbf{y}_2)}{p_{\mathbf{y}}(\mathbf{y}_1, \mathbf{y}_2)}. \tag{7.46}$$

To compute the global state estimate by means of Equation 7.41, we need the global a posteriori pdf $p_{\mathbf{x}|\mathbf{y}}(\mathbf{x}|\mathbf{y}_1, \mathbf{y}_2)$. Equation 7.46 now shows that it directly involves the two local pdfs $p_{\mathbf{x}|\mathbf{y}_1}(\mathbf{x}|\mathbf{y}_1)$ and $p_{\mathbf{x}|\mathbf{y}_2}(\mathbf{x}|\mathbf{y}_2)$. As a result, we can take the pdfs $p_{\mathbf{x}|\mathbf{y}_1}(\mathbf{x}|\mathbf{y}_1)$ and $p_{\mathbf{x}|\mathbf{y}_2}(\mathbf{x}|\mathbf{y}_2)$ as intermediate estimates $\mathbf{z}_1(\mathbf{y}_1)$ and $\mathbf{z}_2(\mathbf{y}_2)$ computed at the local processors. The pdf $p_{\mathbf{x}}(\mathbf{x})$ is known a priori, and the second term on the rhs of Equation 7.46 is just a normalization factor.

Above we promised to revisit Equation 7.35 once the general result for distributed state estimation had been established. To show that Equation 7.35 greatly simplifies the fusion problem, we first cancel the last term on the rhs of Equation 7.46 by substituting Equation 7.35. This leaves us with a product in the numerator composed of $p_{\mathbf{x}|\mathbf{y}_1}(\mathbf{x}|\mathbf{y}_1)$ and $p_{\mathbf{x}|\mathbf{y}_2}(\mathbf{x}|\mathbf{y}_2)$. The first term can be rewritten according to

$$\begin{aligned} p_{\mathbf{x}|\mathbf{y}_1}(\mathbf{x}_1,\mathbf{x}_2|\mathbf{y}_1) &= p_{\mathbf{x}_1|\mathbf{x}_2,\mathbf{y}_1}(\mathbf{x}_1|\mathbf{x}_2,\mathbf{y}_1)p_{\mathbf{x}_2}(\mathbf{x}_2) \\ &= p_{\mathbf{x}_1|\mathbf{y}_1}(\mathbf{x}_1|\mathbf{y}_1)p_{\mathbf{x}_2}(\mathbf{x}_2). \end{aligned} \tag{7.47}$$

To derive the last term in Equation 7.47, we used the fact that for Equation 7.35 to hold it is necessary that the two state components $\mathbf{x}_1$ and $\mathbf{x}_2$ are statistically independent. Applying the same logic to $p_{\mathbf{x}|\mathbf{y}_2}(\mathbf{x}|\mathbf{y}_2)$, we obtain

$$p_{\mathbf{x}|\mathbf{y}_2}(\mathbf{x}_1,\mathbf{x}_2|\mathbf{y}_2) = p_{\mathbf{x}_2|\mathbf{y}_2}(\mathbf{x}_2|\mathbf{y}_2)p_{\mathbf{x}_1}(\mathbf{x}_1). \tag{7.48}$$

Next, Equations 7.47 and 7.48 are substituted into Equation 7.46, and Equation 7.37 is used to simplify the denominator. This yields

$$p_{\mathbf{x}|\mathbf{y}}(\mathbf{x}_1,\mathbf{x}_2|\mathbf{y}_1,\mathbf{y}_2) = p_{\mathbf{x}_1|\mathbf{y}_1}(\mathbf{x}_1|\mathbf{y}_1)p_{\mathbf{x}_2|\mathbf{y}_2}(\mathbf{x}_2|\mathbf{y}_2), \tag{7.49}$$

which is equivalent to Equation 7.46 if Equation 7.35 is true. Equation 7.49 implies that the optimal overall estimate corresponds to the direct combination of the results of two separate estimation problems.

Returning to Equation 7.46, one might ask the question: "If the values of $\mathbf{x}$ that maximize $p_{\mathbf{x}|\mathbf{y}_1}(\mathbf{x}|\mathbf{y}_1)$ and $p_{\mathbf{x}|\mathbf{y}_2}(\mathbf{x}|\mathbf{y}_2)$, respectively, are sufficiently close together, can we average them to get a good approximation to the value of $\mathbf{x}$ that maximizes $p_{\mathbf{x}|\mathbf{y}}(\mathbf{x}|\mathbf{y}_1,\mathbf{y}_2)$?" The answer is no, although this question has a certain relevance to the Gaussian case discussed next.

Gaussian Case. For the more restricted Gaussian case in which the independent pdfs $p_{\mathbf{x}|\mathbf{y}_1}(\mathbf{x}|\mathbf{y}_1)$ and $p_{\mathbf{x}|\mathbf{y}_2}(\mathbf{x}|\mathbf{y}_2)$ are both Gaussian in $\mathbf{x}$, knowledge of the a posteriori means and covariance matrices of $\mathbf{x}$ with $\mathbf{y}_1$ and $\mathbf{y}_2$ separately suffices to determine $p_{\mathbf{x}|\mathbf{y}_1}(\mathbf{x}|\mathbf{y}_1)$ and $p_{\mathbf{x}|\mathbf{y}_2}(\mathbf{x}|\mathbf{y}_2)$. The optimal estimate $\hat{\mathbf{x}}(\mathbf{y}_1,\mathbf{y}_2)$ is the mean of the a posteriori pdf $p_{\mathbf{x}|\mathbf{y}}(\mathbf{x}|\mathbf{y}_1,\mathbf{y}_2)$ as shown in Equation 7.46.

Let us simplify matters by assuming linear measurement models involving a normally distributed state vector $\mathbf{x}$. Given constant output matrices $\mathbf{H}_1$ and $\mathbf{H}_2$, we can write

$$\mathbf{y}_1 = \mathbf{H}_1\mathbf{x} + \mathbf{n}_1 \quad \text{and} \quad \mathbf{y}_2 = \mathbf{H}_2\mathbf{x} + \mathbf{n}_2. \tag{7.50}$$

The measurement noise components $\mathbf{n}_1$ and $\mathbf{n}_2$ are assumed to be independent zero-mean Gaussian random vectors with covariance matrix $\mathbf{C}_{\mathbf{n}_1\mathbf{n}_1}$ and $\mathbf{C}_{\mathbf{n}_2\mathbf{n}_2}$, respectively.

Under these conditions, the joint pdfs $p_{\mathbf{x},\mathbf{y}_i}(\mathbf{x},\mathbf{y}_i)$, $i = 1, 2$, are Gaussian. The mean and covariances of the individual (Gaussian) posteriori probability density functions

$p_{\mathbf{x}|\mathbf{y}_i}(\mathbf{x}|\mathbf{y}_i), i = 1, 2$, can be computed using the Gauss-Markov theorem following, e.g., [606]. The conditional means are

$$\boldsymbol{\mu}_{\mathbf{x}|\mathbf{y}_i} = \boldsymbol{\mu}_{\mathbf{x}} + \mathbf{C}_{\mathbf{x}|\mathbf{y}_i}\mathbf{H}_i^T\mathbf{C}_{\mathbf{n}_i\mathbf{n}_i}^{-1}(\mathbf{y}_i - \mathbf{H}_i\boldsymbol{\mu}_{\mathbf{x}}), \quad i = 1, 2, \tag{7.51}$$

and the associated covariances follow from

$$\mathbf{C}_{\mathbf{x}|\mathbf{y}_i} = \left[\mathbf{C}_{\mathbf{xx}}^{-1} + \mathbf{H}_i^T\mathbf{C}_{\mathbf{n}_i\mathbf{n}_i}^{-1}\mathbf{H}_i\right]^{-1}, \quad i = 1, 2. \tag{7.52}$$

According to Equation 7.32, the conditional means $\boldsymbol{\mu}_{\mathbf{x}|\mathbf{y}_i}$ are the optimal MMSE estimates $\hat{\mathbf{x}}(\mathbf{y}_i)$ based on the individual observations $\mathbf{y}_i$, i.e.,

$$\hat{\mathbf{x}}(\mathbf{y}_i) = \boldsymbol{\mu}_{\mathbf{x}|\mathbf{y}_i}, \quad i = 1, 2. \tag{7.53}$$

Similarly, the optimal global estimate $\hat{\mathbf{x}}(\mathbf{y}_1, \mathbf{y}_2)$ is the conditional mean of the normal distribution $p_{\mathbf{x}|\mathbf{y}}(\mathbf{x}|\mathbf{y}_1, \mathbf{y}_2)$. It can be written as

$$\hat{\mathbf{x}}(\mathbf{y}_1, \mathbf{y}_2) = \boldsymbol{\mu}_{\mathbf{x}|\mathbf{y}} = \mathbf{C}_{\mathbf{x}|\mathbf{y}}\left[\mathbf{C}_{\mathbf{x}|\mathbf{y}_1}^{-1}\boldsymbol{\mu}_{\mathbf{x}|\mathbf{y}_1} + \mathbf{C}_{\mathbf{x}|\mathbf{y}_2}^{-1}\boldsymbol{\mu}_{\mathbf{x}|\mathbf{y}_2} - \mathbf{C}_{\mathbf{xx}}^{-1}\boldsymbol{\mu}_{\mathbf{x}}\right]. \tag{7.54}$$

The covariance matrix of the normal distribution $p_{\mathbf{x}|\mathbf{y}}(\mathbf{x}|\mathbf{y}_1, \mathbf{y}_2)$ is called $\mathbf{C}_{\mathbf{x}|\mathbf{y}}$, and it follows from

$$\mathbf{C}_{\mathbf{x}|\mathbf{y}} = \left[\mathbf{C}_{\mathbf{x}|\mathbf{y}_1}^{-1} + \mathbf{C}_{\mathbf{x}|\mathbf{y}_2}^{-1} - \mathbf{C}_{\mathbf{xx}}^{-1}\right]^{-1}. \tag{7.55}$$

Similar results have also been reported by [569], and by [440].

Since the mean and the covariance matrix of the state $\mathbf{x}$ are assumed to be known a priori, the quantities $\mathbf{z}_1(\mathbf{y}_1) = \boldsymbol{\mu}_{\mathbf{x}|\mathbf{y}_1}$ and $\mathbf{z}_2(\mathbf{y}_2) = \boldsymbol{\mu}_{\mathbf{x}|\mathbf{y}_2}$ together with their associated covariance matrices provide sufficient information for the fusion module in Figure 7.22 to compute the optimal estimate.

If the second set of measurements cannot be received ($\mathbf{H}_2 = \mathbf{0}$) or if it does not contain any meaningful information due to noise ($\mathbf{C}_{\mathbf{n}_2\mathbf{n}_2}^{-1} = \mathbf{0}$), then Equations 7.51 and 7.52 collapse to

$$\mathbf{C}_{\mathbf{x}|\mathbf{y}_2} = \mathbf{C}_{\mathbf{xx}} \quad \text{and} \quad \boldsymbol{\mu}_{\mathbf{x}|\mathbf{y}_2} = \boldsymbol{\mu}_{\mathbf{x}}. \tag{7.56}$$

Substituting these results into Equation 7.54, we get

$$\boldsymbol{\mu}_{\mathbf{x}|\mathbf{y}} = \boldsymbol{\mu}_{\mathbf{x}|\mathbf{y}_1}. \tag{7.57}$$

This result is a meaningful special case of Equation 7.54.

So far we only considered the stationary case. There we are concerned with estimating a constant (yet unknown) state vector. For the problem of acoustic object localization from audio and video data, the techniques described in the previous sections provide an optimal position estimate only if the speaker does not move. In general, however, the system state, i.e., the speaker position, changes with time. As a consequence, we consider the problem of recursively estimating a time-varying state vector from associated measurement data to better adapt to changing system conditions next.

7.3.3 Recursive State Estimation

A recursive linear estimator computes a new estimate on the basis of a linear combination of a previous estimate and a new observation. A very popular example of a recursive linear estimator is the Kalman filter discussed in the sequel.

Linear Kalman Filter. The Kalman filter is based on recursive evaluation, an internal model of the system dynamics, and a comparison of incoming measurements with ongoing estimates. It produces estimates of the state of the system observed. The basic (centralized) Kalman filter loop is shown in Figure 7.23.

The input to the Kalman filter is a n-dimensional measurement vector recorded at time instant k, $\mathbf{y}[k]$, expressed as

$$\mathbf{y}[k] = \mathbf{H}\mathbf{x}[k] + \mathbf{n}[k]. \tag{7.58}$$

The state-space model for the system dynamics is formulated as

$$\mathbf{x}[k+1] = \mathbf{A}\mathbf{x}[k] + \mathbf{B}\mathbf{u}[k] + \mathbf{v}[k], \tag{7.59}$$

where $\mathbf{x}[k]$ is the $m \times 1$ state vector at time instant (or step) k, $\mathbf{A}$ is the $m \times m$ nonsingular state matrix, and $\mathbf{B}$ denotes the $m \times l$ input matrix. The initial mean and covariance of the state are assumed to be known. The vector $\mathbf{u}[k]$ denotes a $l \times 1$ deterministic control. Both measurement noise $\mathbf{n}[k]$ and process noise $\mathbf{v}[k]$ are assumed to be normal, zero-mean, and white. They are further supposed to be independent of each other and also independent of the initial state $\mathbf{x}[0]$. The associated measurement noise and process noise covariance matrices are called $\mathbf{C_{nn}}$ and $\mathbf{C_{vv}}$, respectively.

The useful outputs of the Kalman filter are the difference between the predicted and observed measurements, usually referred to as innovation $\boldsymbol{\nu}[k]$, and the *a posteriori* state estimate at step k, $\hat{\mathbf{x}}[k|k]$, given the vector of measurements, $\mathbf{y}[k]$, at the same time instant.

We further define $\hat{\mathbf{x}}[k|k-1]$ to be the *a priori state estimate* at step k, based on the knowledge of the process up to step $k-1$.

Given the a priori state estimate, $\hat{\mathbf{x}}[k|k-1]$, and the a posteriori state estimate, $\hat{\mathbf{x}}[k|k]$, we can define a priori and a posteriori estimate errors as

$$\mathbf{e}[k|k-1] = \mathbf{x}[k] - \hat{\mathbf{x}}[k|k-1] \tag{7.60}$$

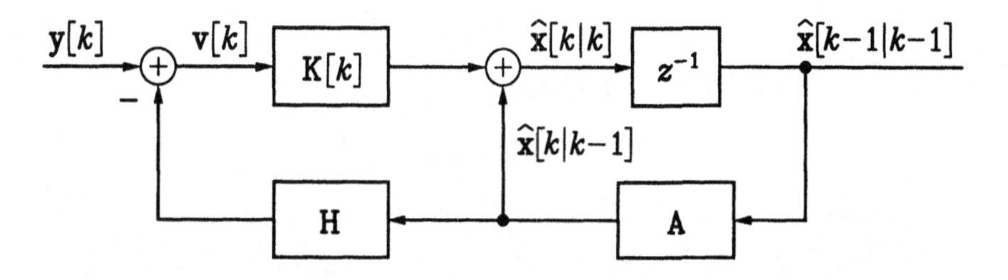

Figure 7.23. The basic Kalman filter loop.

and

$$\mathbf{e}[k|k] = \mathbf{x}[k] - \hat{\mathbf{x}}[k|k]. \tag{7.61}$$

The associated a priori estimate error covariance is

$$\mathbf{P}[k|k-1] = E\{\mathbf{e}[k|k-1]\mathbf{e}[k|k-1]^T\}, \tag{7.62}$$

and the a posteriori estimate error covariance follows from

$$\mathbf{P}[k|k] = E\{\mathbf{e}[k|k]\mathbf{e}[k|k]^T\}. \tag{7.63}$$

The conventional Kalman filter equations for state prediction $\mathbf{x}[k|k-1]$, covariance prediction $\mathbf{P}[k|k-1]$, state update $\mathbf{x}[k|k]$, and variance update $\mathbf{P}[k|k]$ may be found in [40], or [606]. They have the following form:

1. **Prediction-Update Equations**:

   The a priori state at the time of the next measurement may be predicted as

   $$\hat{\mathbf{x}}[k|k-1] = \mathbf{A}\hat{\mathbf{x}}[k-1|k-1] + \mathbf{B}\mathbf{u}[k-1]. \tag{7.64}$$

   The associated a priori estimate error covariance $\mathbf{P}[k|k-1]$ can be computed from

   $$\mathbf{P}[k|k-1] = \mathbf{A}\mathbf{P}[k-1|k-1]\mathbf{A}^T + \mathbf{C_{vv}}, \tag{7.65}$$

   where $\mathbf{C_{vv}}$ is the a posteriori estimate error covariance at step $k-1$.

   The next measurement is predicted using the characteristics of the measurement process and the predicted (or a priori) state

   $$\hat{\mathbf{y}}[k|k-1] = \mathbf{H}\hat{\mathbf{x}}[k|k-1]. \tag{7.66}$$

2. **Measurement-Update Equations**:

   The innovation is defined as

   $$\boldsymbol{\nu}[k] = \mathbf{y}[k] - \hat{\mathbf{y}}[k|k-1]. \tag{7.67}$$

   It reflects the discrepancy between the actual measurement and the predicted measurement. The associated innovation covariance $\mathbf{S}[k] = E\{\boldsymbol{\nu}[k]\boldsymbol{\nu}[k]^T\}$ is

   $$\mathbf{S}[k] = \mathbf{H}\mathbf{P}[k|k-1]\mathbf{H}^T + \mathbf{C_{nn}}. \tag{7.68}$$

   It can be used to compute the Kalman gain

   $$\mathbf{K}[k] = \mathbf{P}[k|k-1]\mathbf{H}^T\mathbf{S}^{-1}[k]. \tag{7.69}$$

   The innovation weighted by the filter gain and the a priori state estimate yield the a posteriori state estimate

   $$\hat{\mathbf{x}}[k|k] = \hat{\mathbf{x}}[k|k-1] + \mathbf{K}[k]\boldsymbol{\nu}[k]. \tag{7.70}$$

The associated a posteriori estimate error covariance is

$$\mathbf{P}[k|k] = \mathbf{P}[k|k-1] - \mathbf{K}[k]\mathbf{S}[k]\mathbf{K}^T[k]. \tag{7.71}$$

The initial conditions are

$$\hat{\mathbf{x}}[0|-1] = E\{\mathbf{x}[0]\} \tag{7.72}$$

and

$$\mathbf{P}[0|-1] = E\{(\mathbf{x}[0] - \hat{\mathbf{x}}[0|-1])(\mathbf{x}[0] - \hat{\mathbf{x}}[0|-1])^T\}. \tag{7.73}$$

The first task during the update is to compute the Kalman gain $\mathbf{K}[k]$. The next step is to actually measure the process to obtain $\mathbf{y}[k]$ and the innovation $\boldsymbol{\nu}[k]$. Then an a posteriori state estimate is computed using Equation 7.70. The final step is to obtain an a posteriori error covariance estimate via Equation 7.71. Both a posteriori estimates are then used to predict a priori estimates $\hat{\mathbf{x}}[k|k-1]$ and $\mathbf{P}[k|k-1]$. These a priori estimates are fed back to compute a posteriori estimates again. This recursive nature is one of the very appealing features of the Kalman filter, since it facilitates practical implementations.

Extended Kalman Filter. The extended Kalman filter (EKF) is a version of the Kalman filter that deals with nonlinear system dynamics or nonlinear measurement equations. The system dynamics are represented by the nonlinear plant equation

$$\mathbf{x}[k+1] = \mathbf{f}(\mathbf{x}[k], \mathbf{u}[k], \mathbf{v}[k]). \tag{7.74}$$

The EKF linearizes the problem by employing the best estimates of the state vector as reference values used at each stage for the linearization. This approximation applied to Equation 7.64 yields

$$\hat{\mathbf{x}}[k|k-1] = \mathbf{f}(\hat{\mathbf{x}}[k-1|k-1], \mathbf{u}[k]). \tag{7.75}$$

The basic control loop of Figure 7.23 still applies, but measurements are predicted using the nonlinear measurement equation

$$\mathbf{y}[k] = \mathbf{h}(\mathbf{x}[k], \mathbf{n}[k]). \tag{7.76}$$

The measurement model $\mathbf{h}(.)$ is linearized about the current predicted state vector $\hat{\mathbf{x}}[k|k-1]$ using the Jacobian, $\mathbf{h}_{\mathbf{x}}$, of the nonlinear measurement function $\mathbf{h}(.)$. The calculations for filter gain, state update, and covariance update are similar to the linear case with the Jacobian $\mathbf{h}_{\mathbf{x}}$ replacing the matrix $\mathbf{H}$. Likewise, state prediction is accomplished by using the nonlinear state equation $\mathbf{f}(.)$, and the plant model is linearized about the current estimated state vector $\hat{\mathbf{x}}[k|k]$. The state prediction covariance, $\mathbf{P}[k|k-1]$, is computed using the plant Jacobian $\mathbf{f}_{\mathbf{x}}$ instead of the matrix $\mathbf{A}$.

The detailed recursion formulas for the EKF together with a tracking example can be found in [85].

Decentralized Kalman Filter. The parallelized or decentralized Kalman filter (DKF) is a multisensor Kalman filter that has been divided up into modules, each one associated with a particular sensor system. The example shown in Figure 7.22 becomes a DKF when the two the local processors and the fusion center become Kalman filters. In a DKF with two sensors, each node computes a local a posteriori estimate, $\hat{\mathbf{x}}_i[k|k], i = 1, 2$, of the environment. These partial estimates are finally assimilated to provide a global a posteriori estimate $\hat{\mathbf{x}}[k|k]$.

The prediction-update equations and measurement-update equations of a DKF with M sensors can, e.g., be found in the paper by [291].

For a two-sensor system, we can partition the observation vector at step k, $\mathbf{y}[k]$, into two components, i.e.,

$$\mathbf{y}[k] = \begin{bmatrix} \mathbf{y}_1[k] \\ \mathbf{y}_2[k] \end{bmatrix}. \tag{7.77}$$

Partitioning the measurement noise component $\mathbf{n}[k]$ conformably, we obtain

$$\mathbf{y}_i[k] = \mathbf{H}_i\mathbf{x}[k] + \mathbf{n}_i[k], \quad i = 1, 2. \tag{7.78}$$

Provided the measurement noise components $\mathbf{n}_1[k]$ and $\mathbf{n}_2[k]$ are independent, the centralized Kalman filter can be parallelized.

If both local sensors observe the same state, the local system dynamics follow as

$$\mathbf{x}[k+1] = \mathbf{A}\mathbf{x}[k] + \mathbf{B}\mathbf{u}[k] + \mathbf{v}_i[k], \quad i = 1, 2, \tag{7.79}$$

and the two local state-space models are identical to the global state-space model formulated in Equation 7.59. Only the additive process noise components differ.

Among others, Hashemipour et al. showed that for this case the global a posteriori state estimate can be expressed as

$$\begin{aligned} \hat{\mathbf{x}}[k|k] &= \mathbf{P}[k|k]\Big(\mathbf{P}^{-1}[k|k-1]\hat{\mathbf{x}}[k|k-1] \\ &\quad + \sum_{i=1}^{2}\{\mathbf{P}_i^{-1}[k|k]\hat{\mathbf{x}}_i[k|k] - \mathbf{P}_i^{-1}[k|k-1]\hat{\mathbf{x}}_i[k|k-1]\}\Big). \end{aligned} \tag{7.80}$$

The matrices $\mathbf{P}[k|k-1]$ and $\mathbf{P}[k|k]$ denote the global a priori and a posteriori error estimate covariances, respectively, while $\mathbf{P}_i[k|k-1]$ and $\mathbf{P}_i[k|k], i = 1, 2$, are their counterparts at the two local processors. The vector $\hat{\mathbf{x}}[k|k-1]$ is the global a priori state estimate, and $\hat{\mathbf{x}}_i[k|k-1]$ together with $\hat{\mathbf{x}}_i[k|k], i = 1, 2$, denote the local a priori and local a posteriori state estimates, respectively. The second term on the right hand side in Equation 7.80 involving the intermediate state estimates can be viewed as a *state error information* vector. Since the local observations $\mathbf{y}_1[k]$ and $\mathbf{y}_2[k]$ are usually different, their associated a priori and a posteriori state estimates differ as well.

The global a posteriori error covariance is given by

$$\mathbf{P}^{-1}[k|k] = \mathbf{P}^{-1}[k|k-1] + \sum_{i=1}^{2}\{\mathbf{P}_i^{-1}[k|k] - \mathbf{P}_i^{-1}[k|k-1]\}. \tag{7.81}$$

The term involving the local covariance matrices represents the *variance error information matrix.*

The prediction-update equation for the a priori global state equals Equation 7.64, and the associated a priori global error covariance is the same as in Equation 7.65. The prediction-update and measurement-update equations for the local processors can be found in the paper by [291]. They are basically identical to the centralized Kalman filter equations as presented in Equation 7.64 through Equation 7.71. The only difference is the presence of different local process noise and measurement noise covariance matrices.

Equations 7.80 and 7.81 summarize the parallel Kalman filter algorithm. In the measurement-update Equation 7.80, the central processor needs the central a priori state estimate, $\hat{\mathbf{x}}[k|k-1]$, the associated global a priori covariance matrix, $\mathbf{P}[k|k-1]$, the a posteriori covariance matrix, $\mathbf{P}[k|k]$, and the state error information vector together with variance error information matrix. There is no need for communications from the local processors to the fusion center during the prediction-update, provided the central coordinator can store the matrices $\mathbf{A}$, $\mathbf{B}$, and $\mathbf{C_{vv}}$.

Note that the estimate generated by the central processor coincides with the global centralized estimate as initially presented in Equation 7.70. As a result, there is no performance loss in the decentralized system. However, the algorithm does assume that the local processors work in sync at the same speed. In general, this may not be the case. A solution to the problem of asynchronous operation can be found in [566].

Decentralized Extended Kalman Filter. In the previous section, we described the decentralized (or parallelized) Kalman filter for linear systems. Similar to the extended centralized Kalman filter, a decentralized extended Kalman can be defined. To this end, we assume identical nonlinear plant equations $\mathbf{f}(.)$, mutually independent process noise components, $\mathbf{v}_i[k]$, and identical control inputs, $\mathbf{u}[k]$ at all local processors. The result is

$$\mathbf{x}[k+1] = \mathbf{f}(\mathbf{x}[k], \mathbf{u}[k], \mathbf{v}_i[k]), \quad i = 1, 2. \tag{7.82}$$

Again, we only used two local processors for simplicity. The measurement models of the distributed sensors need not be identical, i.e., different nonlinear measurement equations, $\mathbf{h}_i(.)$, are possible. Then we get

$$\mathbf{y}_i[k] = \mathbf{h}_i(\mathbf{x}[k], \mathbf{n}_i[k]), \quad i = 1, 2. \tag{7.83}$$

In Equation 7.83, the measurement noise components, $\mathbf{n}_i[k]$, are assumed to be Gaussian and mutually independent.

Although we assumed identical nonlinear plant equations, the local a priori estimates, $\hat{\mathbf{x}}_i[k|k-1]$, and a posteriori estimates, $\hat{\mathbf{x}}_i[k|k]$, will not be the same, since the observations, $\mathbf{y}_i[k], i = 1, 2$, usually differ. Ideally, the final state estimate after fusing all individual nonlinear estimates should be identical to the centralized state estimate. Due to the nonlinear equations, a general answer to this problem appears difficult, and, at least to the knowledge of the authors, no solution has been presented so far. Nevertheless, some related work has been presented by [567]. He tackles the problem of crowding estimation and localization of groups of people in underground stations.

There, an extended Kalman filter is used to estimate multiple states each representing the number of people within a region of interest. The different state estimates depend on each other, and each state follows its own associated plant equation. However, no attempt is made to centrally combine the individual state estimates into a single overall composite description of the environment.

7.3.4 *Conclusions*

We have analyzed the problem of how to decompose a centralized state estimation system into meaningful stand-alone modules each providing separate estimates of the overall (global) state. We showed that this problem can only be solved if the local observations are statistically independent given the overall state vector. Further, we discussed the question of how much processing can be accomplished by the separate sensor systems before the final fusion takes place. In the general case no meaningful local estimates can be computed at the local processors, and they have to transmit their entire a posteriori conditional pdfs $p_{\mathbf{x}|\mathbf{y}_i}(\mathbf{x}|\mathbf{y}_i)$ to the fusion center. At the other extreme, i.e., in the case of linear Gaussian measurement models (which imply that the a posteriori conditional pdfs $p_{\mathbf{x}|\mathbf{y}_i}(\mathbf{x}|\mathbf{y}_i)$ are Gaussian), it is only necessary to use the conditional means and the associated conditional covariance matrices as inputs to the fusion module. After discussing the stationary case, we entered the domain of time-sequential estimation problems. There we first discussed the centralized linear Kalman filter. Then we briefly summarized the extended Kalman filter. Finally, we highlighted the decentralized linear Kalman filter and shortly mentioned the concept of a decentralized extended Kalman filter.

The target tracking methods described in the section on recursive state estimation allow noisy audio and video measurements to be recursively combined into more reliable position estimates. To this end, a model of the system dynamics and a common coordinate system is required. A system design could, for example, start with tracking the positional estimates in the operation region of an acoustic microphone array as presented in [676]. An interesting system for detecting and tracking speakers using a video camera is described in [205]. Their face detection algorithm is hierarchically organized and incorporates coarse scanning, fine scanning, and ellipse fitting. Although some more work is required to arrive at a joint audio-video position estimator, the ideas presented in this section outline the theoretical framework for such a system.

7.4 OBJECT LOCALIZATION USING AUDIO AND VIDEO SIGNALS

N. STROBEL, R. RABENSTEIN

This section presents object localization using audio and video signals. In some cases the knowledge of the position of an acoustic source can greatly simplify scene analysis and understanding. In video conferencing applications, for example, we commonly find a stationary background, and the only unknown objects are speakers who are positioned at random locations to be estimated. The estimated speaker positions may be used to automatically focus the camera and the speech acquisition system to obtain a clearer sound and a better field of view. This, in turn, could simplify speech or face recognition.

Audio and video observations are independent of each other, given the source position. This is to say that the audio measurements only depend on the speaker position and not on the video measurements and vice versa. As a result, we can decompose the joint estimation problem for the source position into two separate object localization problems. For acoustic object localization, we need a wave propagation model. Otherwise no progress can be made. Video-based source localization, on the other hand, requires a visual model of the object whose position is to be estimated. Once acoustic and visual position estimates are available, they can be combined using the *sensor fusion* techniques presented in Section 7.3.

To this end, we proceed as follows. First, a maximum likelihood (ML) estimator yielding a position estimate based on acoustic measurements is derived. We show that it can be configured as a steered filter-and-sum beamformer. In the next step, we suggest an algorithm for face localization in video sequences. It is based on (1) background image subtraction, (2) color filtering, and (3) a statistical method for estimating the face position and size. Assuming a stationary speaker, we finally show how to fuse the audio and video position estimates into a more robust overall result.

7.4.1 Acoustic Object Localization

For acoustic object localization, we use a number of spatially distributed microphone sensors to capture incoming sound waves emitted at a distant point source. If the amplitude gradient across the microphone array is negligible, essentially all of the geometric information is encoded in the time differences of arrival (TDOAs) of the wavefront at the microphone signals. In this case, the time delays between different microphone signals can be expressed in terms of the unknown source location parameters.

There are *indirect* and *direct* acoustic source localization methods. Indirect techniques detect a set of time differences of arrival of the wavefront between the different microphone sensors first. Then geometrical properties are used to infer the source position. Periodicity in the signal, room reverberations, and noise, however, often result in peak-picking errors yielding wrong TDOA estimates. Since they are subsequently used to compute the source location, unreliable position estimates are not uncommon. The direct approach, on the other hand, is based on summing the systematically delayed microphone signals and observing the power of the overall output signal. This strategy is usually implemented using a steered filter-and-sum beamformer. Our experimental results indicated that the beamformer method is preferable for the localization of acoustic sources producing non-stationary signals such as speech, since it's position estimates were less sensitive to the kind of acoustic signal received.

Among others, [35], [274], or [108] provide many insights into the sonar target detection problem which we applied to speaker localization. We also benefited from ideas presented by [523] and [275].

Our goal is to derive the maximum likelihood estimator for the position of an acoustic source. To this end, we introduce a simple acoustic signal model first. Next, the ML estimator is derived. Finally, we show that the ML estimator can be configured as a steered filter-and-sum beamformer provided some conditions are met.

The Receiver Model. For simplicity, only the planar problem will be considered. In particular, it will be assumed that acoustic energy arrives at each microphone through only one propagation path in the same plane with all the receivers and the source. More sophisticated propagation modelling and associated signal processing is considered by [523].

In the planar case, a simple mathematical model for the received signal at the i-th microphone is

$$y_i(t) = s(t - \tau_i) + n_i(t). \tag{7.84}$$

In Equation 7.84, both the signal, $s(t)$, and the additive noise component, $n_i(t)$, are zero-mean, wide sense stationary (WSS), and uncorrelated Gaussian random processes. We further assume that the noise components have identical power spectra. The variable τ_i represents the unknown time delay associated with the signal propagation from the acoustic source to the i-th microphone.

The position, Θ_a, of the acoustic source is defined as

$$\Theta_a = [B_a, R_a]^T, \tag{7.85}$$

where B_a denotes the bearing of the source and R_a refers to the source range with respect to the origin of some suitably chosen coordinate system. The time delay, τ_i, follows as the distance from the source position to the i-th microphone divided by the speed of sound, c_s. The array comprises M microphones.

For further analysis, we express the received waveform in the frequency domain. Taking the short-time Fourier transform of $y_i(t)$, we get

$$Y_i(\omega) = \int_{-T/2}^{T/2} y_i(t) e^{-j\omega t} dt. \tag{7.86}$$

Here the integration time, T, is large compared to the correlation times of the random processes and also to the time needed for the wavefront to traverse the array. Since $y_i(t)$ is assumed to be a WSS Gaussian random process, $Y_i(\omega)$ is a WSS Gaussian random process as well.

For simplicity, we first estimate the location vector, Θ_a, based on evaluating $Y_i(\omega)$ at one frequency, ω_l, only. This result is extended to the whole frequency range later. To this end, we first define a column vector, $\boldsymbol{Y}(\omega_l)$, by stacking up the frequency components of all M microphone signals at frequency ω_l. This yields

$$\boldsymbol{Y}(\omega_l) = [Y_0(\omega_l) \ldots Y_{M-1}(\omega_l)]^T. \tag{7.87}$$

The superscript, T, indicates the transpose of a vector. Substituting the additive noise model defined in Equation 7.84 into Equation 7.87, we obtain

$$\boldsymbol{Y}(\omega_l) = \boldsymbol{V}(\omega_l) S(\omega_l) + \boldsymbol{N}(\omega_l). \tag{7.88}$$

In Equation 7.88, $S(\omega_l)$ denotes the Fourier transform of $s(t)$, $\boldsymbol{V}(\omega_l)$ represents the delay or steering vector

$$\boldsymbol{V}(\omega_l) = \left[e^{-j\omega_l \tau_0} \ldots e^{-j\omega_l \tau_{M-1}}\right]^T, \tag{7.89}$$

and the vector $\boldsymbol{N}(\omega_l)$ comprises the Fourier transforms of the M microphone noise components:

$$\boldsymbol{N}(\omega_l) = [N_0(\omega_l) \dots N_{M-1}(\omega_l)]^T . \tag{7.90}$$

To proceed further, let us assume that an acoustic source is present, i.e., that $\boldsymbol{Y}(\omega_l)$ is available. Then we can subject the vector $\boldsymbol{Y}(\omega_l)$ to maximum likelihood (ML) estimation with respect to the assumed deterministic, but unknown, location vector Θ_a.

The Maximum-Likelihood Estimator. Since we assumed zero-mean Gaussian random processes for both signal and noise, the conditional probability density function (pdf) for complex $\boldsymbol{Y}(\omega_l)$, given Θ_a, can be written as

$$p(\boldsymbol{Y}(\omega_l)|\Theta_a) = \frac{1}{\pi^M \det \boldsymbol{P}(\omega_l)} \exp\{-\boldsymbol{Y}^H(\omega_l)\boldsymbol{P}^{-1}(\omega_l)\boldsymbol{Y}(\omega_l)\}. \tag{7.91}$$

In Equation 7.91, the superscript H is used to denote complex conjugation and transposition, respectively. Note that the pdf for complex random variables differs from its real-valued counterpart as, e.g., explained in [366]. The symbol $\boldsymbol{P}(\omega_l)$ refers to the cross-spectral density matrix evaluated at the frequency ω_l. It is defined as

$$\boldsymbol{P}(\omega_l) = E\{\boldsymbol{Y}(\omega_l)\boldsymbol{Y}^H(\omega_l)\}. \tag{7.92}$$

Substituting Equation 7.88 into Equation 7.92, we obtain

$$\boldsymbol{P}(\omega_l) = R_S(\omega_l)\boldsymbol{V}(\omega_l)\boldsymbol{V}^H(\omega_l) + \boldsymbol{R}_N(\omega_l). \tag{7.93}$$

Here we introduced the spectral density of the source signal, $R_S(\omega_l)$, at the frequency ω_l. It is defined as

$$R_S(\omega_l) = E\{S^2(\omega_l)\}. \tag{7.94}$$

For WSS random processes we have

$$E\{S^2(\omega_l)\} = 2\pi \int r_{ss}(\tau)e^{-j\omega_l\tau}d\tau, \tag{7.95}$$

where $r_{ss}(\tau)$ represents the autocorrelation function of $s(t)$.

We also introduced the noise cross-spectral density matrix, $\boldsymbol{R}_N(\omega_l)$, taken at the frequency ω_l. It is defined as

$$\boldsymbol{R}_N(\omega_l) = E\{\boldsymbol{N}(\omega_l)\boldsymbol{N}^H(\omega_l)\}. \tag{7.96}$$

Since our noise components are uncorrelated, $\boldsymbol{R}_N(\omega_l)$ is a diagonal matrix. As we assumed identical noise power spectra, we can further simplify $\boldsymbol{R}_N(\omega_l)$ to

$$\boldsymbol{R}_N(\omega_l) = R_N(\omega_l)\boldsymbol{I}, \tag{7.97}$$

where

$$R_N(\omega_l) = E\{N_i^2(\omega_l)\} \quad i = 0, \ldots, M-1, \tag{7.98}$$

and $\boldsymbol{I}$ is the identity matrix.

The maximum likelihood estimator selects the parameter Θ_a which maximizes the likelihood function shown in Equation 7.91. However, in the Gaussian case it is generally preferable to work with the log-likelihood function. Then the ML estimate, $\hat{\Theta}_a$, follows from

$$\hat{\Theta}_a = [\hat{B}_a, \hat{R}_a]^T = \arg\max_{\Theta_a}\{\ln\left(p(\boldsymbol{Y}(\omega_l)|\Theta_a)\right)\}, \tag{7.99}$$

while the log-likelihood function evaluated at ω_l can be written as

$$\ln\left(p(\boldsymbol{Y}(\omega_l)|\Theta_a)\right) = -\ln\left(\pi^M \det \boldsymbol{P}(\omega_l)\right) - \boldsymbol{Y}^H(\omega_l)\boldsymbol{P}^{-1}(\omega_l)\boldsymbol{Y}(\omega_l). \tag{7.100}$$

The second term on the rhs of Equation 7.100 depends on Θ_a. This is because Θ_a constrains the time delays, τ_i, of the steering vector, $\boldsymbol{V}(\omega_l)$, which is an important parameter for both $\boldsymbol{Y}(\omega_l)$ and $\boldsymbol{P}(\omega_l)$. It requires a little bit more effort to see that this is really the only term in Equation 7.100 depending on Θ_a. To this end, we use the fact that the determinant of a matrix is equal to the product of eigenvalues. Multiplying the rhs of Equation 7.93 with $\boldsymbol{V}$ from the right, we see that $\boldsymbol{V}$ is a eigenvector. Its eigenvalue is $\lambda_0 = MR_S(\omega_l) + R_N(\omega_l)$. Since the remaining eigenvectors are orthogonal to $\boldsymbol{V}$, all other eigenvalues must be equal to $R_N(\omega_l)$. Thus, in our case the determinant of $\boldsymbol{P}(\omega_l)$ can be expressed as

$$\det \boldsymbol{P}(\omega_l) = (1 + M\frac{R_S(\omega_l)}{R_N(\omega_l)}) \det \boldsymbol{R}_N(\omega_l). \tag{7.101}$$

It does not depend on the steering vector and, therefore, not on Θ_a.

To obtain the ML estimate, $\hat{\Theta}_a$, it is, hence, sufficient to maximize the quadratic form

$$Q(\omega_l) = -\boldsymbol{Y}^H(\omega_l)\boldsymbol{P}^{-1}(\omega_l)\boldsymbol{Y}(\omega_l). \tag{7.102}$$

Equation 7.102 can be simplified by applying the matrix inversion lemma to $\boldsymbol{P}^{-1}(\omega_l)$. The frequency arguments of the functions discussed will from now on be suppressed when necessary to avoid cumbersome notation. Then the inverse of $\boldsymbol{P}$ can be expressed as

$$\boldsymbol{P}^{-1} = \boldsymbol{R}_N^{-1} - \boldsymbol{R}_N^{-1}\boldsymbol{V}\left[R_S^{-1} + \boldsymbol{V}^H\boldsymbol{R}_N^{-1}\boldsymbol{V}\right]^{-1}\boldsymbol{V}^H\boldsymbol{R}_N^{-1}. \tag{7.103}$$

Substituting Equation 7.103 into Equation 7.102, we obtain

$$Q = -\boldsymbol{Y}^H\boldsymbol{R}_N^{-1}\boldsymbol{Y} + \left(\boldsymbol{V}^H\boldsymbol{R}_N^{-1}\boldsymbol{Y}\right)^H\left[R_S^{-1} + \boldsymbol{V}^H\boldsymbol{R}_N^{-1}\boldsymbol{V}\right]^{-1}\boldsymbol{V}^H\boldsymbol{R}_N^{-1}\boldsymbol{Y}. \tag{7.104}$$

The first term on the right-hand side of Equation 7.104 can be dropped, and we obtain an optimal ML estimate for Θ_a by maximizing the spectral density of a postfiltered steered filter-and-sum beamformer. The spectral density can be written as

$$P(\omega_l) = Z^*(\omega_l)|H(\omega_l)|^2 Z(\omega_l) = |H(\omega_l)|^2|Z(\omega_l)|^2. \tag{7.105}$$

In Equation 7.105, the term $Z(\omega_l)$ represents

$$Z(\omega_l) = \boldsymbol{V}^H(\omega_l)\boldsymbol{R}_N^{-1}(\omega_l)\,\boldsymbol{Y}(\omega_l), \tag{7.106}$$

and it can be interpreted as the output of a steered filter-and-sum beamformer. The squared amplitude response of the postfilter at ω_l is

$$|H(\omega_l)|^2 = \left[R_S^{-1}(\omega_l) + \ \boldsymbol{V}^H(\omega_l)\boldsymbol{R}_N^{-1}(\omega_l)\,\boldsymbol{V}(\omega_l)\right]^{-1}. \tag{7.107}$$

In other words, instead of maximizing the log-likelihood function derived in Equation 7.100, or its quadratic part, $Q(\omega_l)$, as shown in Equation 7.102, it is possible to find an ML estimate for Θ_a at some focusing frequency ω_l by maximizing $P(\omega_l)$ provided our assumptions are met.

Equation 7.106 indicates that the ML estimator first pre-whitens the input vector, $\boldsymbol{Y}(\omega_l)$, using the matrix filter $\boldsymbol{R}_N^{-1}(\omega_l)$. In the next stage, the filtered input vector components are properly delayed and summed by evaluating the inner product of $< \boldsymbol{V}, \boldsymbol{R}_N^{-1}\boldsymbol{Y} >$. Finally this sum is filtered with a function $F(\omega)$ and squared. It is required that $|F(\omega)|^2 = |H(\omega)|^2$.

Up to now we only dealt with a (partial) ML estimate for Θ_a based on the focusing frequency, ω_l, only. The overall ML estimate for Θ_a, though, follows by maximizing $P(\omega)$ no longer at one but at all frequencies. Since $P(\omega)$ is nonnegative, this is equivalent to maximizing $\int P(\omega)d\omega$, and thanks to Parseval's theorem, this integral can also be evaluated in the time domain. There we have to maximize the output energy of the steered beamformer, $\int p(t)dt$, where $p(t) = F^{-1}\{P(\omega)\}$.

For uncorrelated noise components only the diagonal elements of $\boldsymbol{R}_N(\omega)$ remain. Since the noise power spectra are all assumed to be equal, these diagonal elements are

$$R_N(\omega) = E\{N_i^2(\omega)\} \quad i = 0, \ldots, M-1, \tag{7.108}$$

where the subscript i is reminiscent of the i-th microphone. Now $|H(\omega)|^2$ simplifies to

$$|H(\omega)|^2 = \frac{R_S(\omega)}{1 + M\frac{R_S(\omega)}{R_N(\omega)}}. \tag{7.109}$$

At this point all components for a physically realizable ML estimator for Θ_a are in place. Such a steered filter-and-sum beamformer is shown in Figure 7.24. A ML estimate for the source position is obtained by forming a 'beam' by the array and 'sweeping' it over the area of interest. The ML estimate for the source position follows for that set of steering delays that yields the maximum output energy.

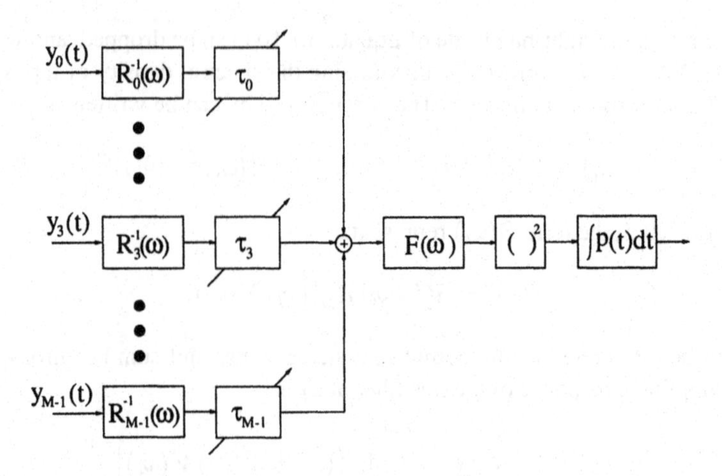

Figure 7.24. Maximum likelihood estimator for the location vector Θ_a. A time-domain implementation of a steered filter-and-sum beamformer is shown. A 'beam' is formed by the array and 'swept' over the area that is of interest to the observer. The ML estimate for the source position is obtained for that set of steering delays that yields the maximum output energy.

A very interesting special case arises when there are only two microphones with identical power spectral densities $R_1(\omega) = R_2(\omega) = R_N(\omega)$. Merging each prefilter R_N^{-1} with the postfilter, $F(\omega)$, we obtain two identical equivalent prefilters with magnitude

$$|G(\omega)| = \frac{|H(\omega)|}{R_N(\omega)} = \sqrt{\frac{R_S(\omega)}{R_N^2(\omega) + 2R_N(\omega)R_S(\omega)}}. \tag{7.110}$$

Both operate directly on the microphone signals $y_0(t)$ and $y_1(t)$. In this case the square of $|G(\omega)|$ is identical to the optimal signal window, W_1, presented by [292].

Whether or not there is only one unique global maximum with respect to Θ_a strictly being a function of the sensor locations in the array. The higher the percentage of sensors which are located at distances on the order of one-half a wavelength at the focusing frequency, i.e., at

$$\frac{\lambda_l}{2} = \frac{\omega_l}{4\pi c_s}, \tag{7.111}$$

the more distinguishable the global maximum will be from the local ambiguous maxima. Conversely, widely spaced sensors may lead to significant estimator ambiguity unless some form of a priori knowledge on the parameters to be estimated can be exploited.

7.4.2 Object Localization in Video Sequences

Acoustic object localization can only work if sound waves can be reliably received at the microphone array. When faced with low signal-to-noise ratios or room reverberations, the steered beamformer may fail to return an accurate source position estimate. Then it is beneficial to use a complementary sensor system to obtain another estimate for the source position. A video camera may be a practical choice for such a complementary sensor system.

To find and identify an object in a video sequence, visual object characteristics must be used. Since the visual appearance of various objects can vary widely, we restrict ourselves to localizing a speaker's face. It serves as a good example to discuss some of the problems encountered when searching for the position of an object in a video sequence. A good presentation of the state-of-the-art in face recognition can be found in [116]. It also covers the face detection problem in video sequences.

We proceed in two steps: First, the face of a speaker is coarsely localized by segmenting the speaker from the background. This important step has already been suggested by [361]. Areas which are found to be different from the background are potential face regions. They are further analyzed using color-based methods to arrive at a more accurate estimate for the face position. To this end, a technique similar to the algorithm proposed by [561] is applied.

Reference Image Subtraction. The basic idea presented relies on computing difference pictures. A difference picture is a matrix with entries representing the value of a difference metric between the current frame and a reference image. It should be computed for every frame within a certain interval of frames. This interval should be small enough so that no important scene changes are missed, yet it should be large enough to ensure a reasonable processing time. In video conferencing applications an image of the background may be taken as a reference. In applications with unknown background we may take the first frame of a video sequence which is then hypothesized to represent a stationary background component. In either case, there are two hypotheses a sample area can satisfy:

1. H_0 : sample area represents background, or

2. H_1 : sample area is taken from foreground.

Once it has been recognized that a subarea of the initial background estimate corresponds to the image of a moving object, the values in this subarea are replaced by estimates of the stationary background at this position.

Our input signal comprises color images, but for simplicity it is desirable to use gray-scale difference images. To this end, a color transform is applied first generating a luminance (or brightness) value, $I[r, c]$, for each pixel with row index, r, and column index c. To compute a difference image, we interprete the hypotheses H_0 and H_1 as two classes whose features are brightness values drawn from a closed neighborhood, $W[r, c]$, with height H and width W centered at the pixel location (r, c). Now, we model both background pixel distribution and foreground pixel distributions as normal distributions, respectively. Then they can be expressed as

$$p(I[r,c]|H_i) = \frac{1}{\sqrt{2\pi}\sigma_i[r,c]} \exp\left(-\frac{1}{2}(\frac{I[r,c]-\mu_i[r,c]}{\sigma_i[r,c]})^2\right), \quad i = 0,1. \quad (7.112)$$

The background pixel distribution within $W[r,c]$ is assumed to have mean $\mu_0[r,c]$ and variance $\sigma_0^2[r,c]$ associated with it, and the ensemble of foreground pixels within $W[r,c]$ is modeled by substituting the mean $\mu_1[r,c]$ and variance $\sigma_1^2[r,c]$ into Equation 7.112.

A useful measure of separability between the two classes H_0 and H_1 associated with $W[r,c]$ is the Mahalanobis distance

$$G[r,c] = \frac{(\mu_0[r,c] - \mu_1[r,c])^2}{\sigma_0^2[r,c] + \sigma_1^2[r,c]}. \tag{7.113}$$

Each Mahalanobis distance, $G[r,c]$, is computed for a neighborhood $W[r,c]$ centered at the pixel location (r,c) in both current and background image. It represents the normalized distance between the background mean, $\mu_0[r,c]$, and the foreground mean, $\mu_1[r,c]$, in standard deviation units. In Equation 7.113, $\sigma_0^2[r,c]$ denotes the variance within the region $W[r,c]$ found in the background image centered around $I[r,c]$, and $\sigma_1^2[r,c]$ is the associated variance in the foreground region.

In what follows, we use a skin color model to map an input color image into a representation, called *skin map*. Finally, we multiply the skin map by the associated values of $G[r,c]$ to highlight (1) areas differing from the background, and (2) showing skin color. These regions are assumed to be faces.

Skin Color Filtering. In the sequel, we derive a skin-color-filter effectively evaluating a statistical distance measure. To this end, we again assume that two hypotheses are possible:

1. H_0 : pixel shows no skin color, or

2. H_1 : pixel represents skin color.

Let $p(\boldsymbol{X}|H_0)$ describe the probability density function (pdf) of pixel $\boldsymbol{X}$ representing any other color but skin color, while $p(\boldsymbol{X}|H_1)$ refers to the pdf of $\boldsymbol{X}$ showing skin color. Then we can define the likelihood ratio

$$\lambda(\boldsymbol{X}) = \frac{p(\boldsymbol{X}|H_1)}{p(\boldsymbol{X}|H_0)}. \tag{7.114}$$

If the measurement vectors $\boldsymbol{X}$ are both characterized by Gaussian densities with means $\mu_i, i = 0,1$, and covariance matrices $C_{ii}, i = 0,1$, then $\lambda(\boldsymbol{X})$ can be expressed as

$$\begin{aligned} \lambda(\boldsymbol{X}) &= \sqrt{\frac{\det(C_{00})}{\det(C_{11})}} \exp\Big(-\frac{1}{2}(\boldsymbol{X} - \mu_1)^T C_{11}^{-1} (\boldsymbol{X} - \mu_1) \\ &\quad + \frac{1}{2}(\boldsymbol{X} - \mu_0)^T C_{00}^{-1} (\boldsymbol{X} - \mu_0)\Big). \end{aligned} \tag{7.115}$$

The subscripts of μ_i and C_{ii} in Equation 7.115 denote the mean and the covariance matrix resulting from the hypothesis carrying the same subscript.

Using the log-likelihood ratio $\Lambda(\boldsymbol{X}) = -2\ln(\lambda(\boldsymbol{X}))$, Equation 7.115 simplifies to

$$\begin{aligned}\Lambda(\boldsymbol{X}) &= (\boldsymbol{X}-\mu_1)^T C_{11}^{-1}(\boldsymbol{X}-\mu_1) - (\boldsymbol{X}-\mu_0)^T C_{00}^{-1}(\boldsymbol{X}-\mu_0) \\ &\quad + \ln\left(\frac{\det(C_{00})}{\det(C_{11})}\right).\end{aligned} \tag{7.116}$$

Equation 7.116 can be interpreted as a quadratic discriminant function. It computes the difference between the Mahalanobis distances of an input pixel to the means of each class.

If we are just interested in a measure describing the distance of an input pixel to the face-color-class, we may take

$$d(\boldsymbol{X}) = (\boldsymbol{X}-\mu_1)^T C_{11}^{-1}(\boldsymbol{X}-\mu_1), \tag{7.117}$$

where we neglected the term $\ln(\det(C_{11}))$, since it is independent of $\boldsymbol{X}$.

To evaluate Equation 7.117, a suitable color space must be chosen. A practical choice is the (r, g) space. It is based on the chromaticity transform with

$$\begin{aligned} r &= \frac{R}{R+G+B}, \\ g &= \frac{G}{R+G+B}. \end{aligned} \tag{7.118}$$

The purpose of employing a transformed color space in contrast to the initial (R, G, B) color components is to reduce the dependency on intensity.

Let $\boldsymbol{X} = [r, g]^T$ denote a skin color pixel. Then the mean of the associated skin color can be expressed as

$$\mu_1 = E\{\boldsymbol{X}\} = [\mu_r, \mu_g]^T, \tag{7.119}$$

while the covariance matrix follows as

$$C_{11} = E\{(\boldsymbol{X}-\mu_1)(\boldsymbol{X}-\mu_1)^T\}. \tag{7.120}$$

Both may be obtained by training based on skin color samples. [561] claim that their skin color model remains valid under a wide range of conditions, once it has been established.

Face Localization. Based on the result from skin color filtering and weighting by the background-foreground separability measure $G[r, c]$, we construct two 1D histograms by projecting the weighted skin map along its x and y direction, respectively. The instantaneous center position and size of a face in the image can be estimated based on statistical measurements derived from the two 1D projection histograms. A simple method estimates the face position using the sample means and standard deviations of the 1D histograms. More specifically, let $h_x(i), i = 0, 1, \ldots,$ and $h_y(i), i = 0, 1, \ldots,$ denote the elements in the projection histograms along the x and y direction, respectively. Then the face center position (x_c, y_c) can be estimated as

$$\hat{x}_c = \frac{\sum_i x_i h_x(i)}{\sum_i h_x(i)}, \qquad \hat{y}_c = \frac{\sum_i y_i h_y(i)}{\sum_i h_y(i)}. \tag{7.121}$$

Face width and height (w_f, h_f) may follow from

$$\hat{w}_f = \alpha\left[\frac{\sum_i (x_i - \hat{x}_c)^2 h_x(i)}{\sum_i h_x(i)}\right]^{0.5}, \qquad \hat{h}_f = \beta\left[\frac{\sum_i (y_i - \hat{y}_c)^2 h_y(i)}{\sum_i h_y(i)}\right]^{0.5}, \tag{7.122}$$

where α and β are constant scaling factors which follow from camera calibration. If more than one speaker and/or skin-color-like objects appear in the foreground, then more robust statistical estimation techniques must be employed as explained in [561]. Alternatively, more intelligent pattern recognition techniques may be employed such as template matching.

Using the position estimate $(\hat{x}_c, \hat{y}_c)$ together with the size of the face and assuming a correctly calibrated camera, it is possible to establish an estimate of the face position in polar coordinates

$$\hat{\Theta}_v = [\hat{B}_v, \hat{R}_v]^T. \tag{7.123}$$

Here $\hat{B}_v$ and $\hat{R}_v$ refer to the estimated azimuth and the estimated range of the face, respectively. The underlying camera coordinate system should be chosen such that it coincides with the coordinate system of the microphone array. Each pixel location in a camera image represents a different azimuth (and elevation) with respect to the camera orientation. This knowledge can be used to compute $\hat{B}_v$. One way to estimate the target distance, $\hat{R}_v$, is by considering the targets size in the image, and incorporating a priori knowledge about the actual size of the object. Since this calculation is highly sensitive to sensor noise and variations in target size, the expected measurement error associated with $\hat{R}_v$ can be expected to be fairly large.

7.4.3 Joint Object Localization

As of now, little attention has been paid to multimodal speaker localization using both audio and video sensors. In related work, [96] suggested a visual tracking technique for steering an acoustic beamformer to achieve robust speech recognition in noisy environments. The opposite approach was taken by [734] who presented a voice activated automatic camera pointing system. A system combining audio and visual cues for locating and tracking of a person in real time was proposed by [546]. There it was shown that combining speech source localization with video-based head tracking provides a more accurate and robust outcome than that obtained using any one of the audio or visual modalities alone.

Common Coordinates. Both the acoustic source position estimate, $\hat{\Theta}_a$, and the visual position estimate, $\hat{\Theta}_v$, involve the angle (bearing) and the range with respect to a common coordinate system. For simplicity it is also assumed that camera and microphone array are adjusted to the height of the speaker. We require that the target's angle of elevation is small enough such that the estimation of the sound azimuth, $\hat{\Theta}_a$, equals the visually measured azimuth in polar coordinates, $\hat{\Theta}_v$.

Sensor Fusion. To better illustrate the underlying concepts, let us focus on a stationary speaker and a simple measurement model for $\hat{\Theta}_a$ and $\hat{\Theta}_v$. They are assumed to be normally distributed, i.e.,

$$\begin{aligned} \hat{\Theta}_a &= \Theta + \boldsymbol{n}_a, \\ \hat{\Theta}_v &= \Theta + \boldsymbol{n}_v. \end{aligned} \tag{7.124}$$

The additive noise components $\boldsymbol{n}_a$ and $\boldsymbol{n}_v$ are zero-mean Gaussian random vectors with independent components. The associated covariance matrices can be expressed as

$$\begin{aligned} C_{n_a n_a} &= \begin{bmatrix} \sigma^2_{B_a} & 0 \\ 0 & \sigma^2_{R_a} \end{bmatrix}, \\ C_{n_v n_v} &= \begin{bmatrix} \sigma^2_{B_v} & 0 \\ 0 & \sigma^2_{R_v} \end{bmatrix}, \end{aligned} \tag{7.125}$$

respectively. The variances associated with bearing and range for each sensor modality express the measurement uncertainties. The vector $\Theta = [B, R]^T$ represents the actual source position to be estimated. Its mean is called μ_Θ, and its covariance matrix is denoted $\boldsymbol{C}_{\Theta\Theta}$. To see what happens if there is no available prior information about Θ, let us assume maximum uncertainty, i.e., $\boldsymbol{C}_{\Theta\Theta}^{-1} = \boldsymbol{0}$.

Similar measurement models have already been used in Section 7.3. There the joint estimation problem for the restricted Gaussian case was discussed. In this case, it was noted that the joint pdfs $p(\hat{\Theta}_a, \Theta)$ and $p(\hat{\Theta}_v, \Theta)$ are Gaussian. The means and covariances of the individual (Gaussian) posteriori probability density functions $p(\Theta|\hat{\Theta}_a)$ and $p(\Theta|\hat{\Theta}_v)$ can be computed using the Gauss-Markov theorem. They are

$$\mu_{\Theta|\hat{\Theta}_a} = \mu_\Theta + (\hat{\Theta}_a - \mu_{\hat{\Theta}_a}) = \hat{\Theta}_a, \tag{7.126}$$

since $\mu_{\hat{\Theta}_a} = E\{\hat{\Theta}_a\} = \mu_\Theta$ due to Equation 7.124. Similarly, we obtain

$$\mu_{\Theta|\hat{\Theta}_v} = \hat{\Theta}_v. \tag{7.127}$$

The associated conditional covariance matrices are

$$C_{\Theta|\hat{\Theta}_a} = C_{n_a n_a}, \tag{7.128}$$
$$C_{\Theta|\hat{\Theta}_v} = C_{n_v n_v}. \tag{7.129}$$

The optimal joint estimate $\hat{\Theta}(\hat{\Theta}_a, \hat{\Theta}_v)$ is the conditional mean of the normal distribution $p(\Theta|\hat{\Theta}_a, \hat{\Theta}_v)$. It can be written as

$$\hat{\Theta}(\hat{\Theta}_a, \hat{\Theta}_v) = \left[C^{-1}_{\Theta|\hat{\Theta}_a} + C^{-1}_{\Theta|\hat{\Theta}_v}\right]^{-1} \left[C^{-1}_{\Theta|\hat{\Theta}_a} \hat{\Theta}_a + C^{-1}_{\Theta|\hat{\Theta}_v} \hat{\Theta}_v\right]. \quad (7.130)$$

Substituting Equations 7.128 and 7.129 into Equation 7.130, we finally obtain

$$\hat{\Theta}(\hat{\Theta}_a, \hat{\Theta}_v) = \begin{bmatrix} \dfrac{\sigma^2_{B_v} \hat{B}_a + \sigma^2_{B_a} \hat{B}_v}{\sigma^2_{B_a} + \sigma^2_{B_v}} \\ \dfrac{\sigma^2_{R_v} \hat{R}_a + \sigma^2_{R_a} \hat{R}_v}{\sigma^2_{R_a} + \sigma^2_{R_v}} \end{bmatrix}. \quad (7.131)$$

This result is intuitively appealing. To see this, assume that the audio sensor fails when the speaker pauses. The resulting measurements are unreliable, i.e., the measurement uncertainty become $\sigma^2_{B_a} \to \infty$ and $\sigma^2_{R_a} \to \infty$. In this case, Equation 7.131 reduces to

$$\hat{\Theta}(\hat{\Theta}_a, \hat{\Theta}_v) = \begin{bmatrix} \hat{B}_v \\ \hat{R}_v \end{bmatrix}. \quad (7.132)$$

That is, in case of uncertain audio measurements, the system defaults to the estimates based on the video sensor.

7.4.4 Conclusions

We have discussed some important issues related to speaker localization based on the fusion of audio and video measurements. Both sensor modalities yield independent position estimates which can be combined using joint estimation techniques.

A steered filter-and-sum beamformer is applied to obtain an estimate of the speaker position based on acoustic measurements. Speakers present in video objects are localized in three steps. First, background image subtraction is carried out. Then facial color characteristics are exploited to further emphasize areas potentially associated with speakers. In the third step, two 1D projections are used to obtain an estimate of the speaker position. Audio and video position estimates are combined assuming a stationary speaker. We presented an example where a simple measurement model is used, and any a-priori statistical knowledge about the joint estimate is neglected. For this case, it was shown that sensor fusion still provides a robust estimate even when one sensor fails. However mathematically convenient, these simplifications need not be appropriate for practical applications. In particular, it does not seem advisable to neglect the covariance matrix associated with the joint position estimate. If there is some knowledge about its statistical behaviour, the overall estimate should improve. Unfortunately, speakers need not remain stationary but may move. In this case, object tracking methods have to be used as, e.g., presented by [566].

Although our discussion implies that decentralized audio and video sensor fusion is feasible in principle, many open issues are yet to be resolved. The most immediate problem is the lack of accurate system models. They are essential to reliably express measurement uncertainties. Just knowing a position estimate obtained either from audio or video measurements is not enough. To successfully fuse these estimates, their associated statistical properties must be known.

8 SELECTED APPLICATIONS

D. PAULUS

INTRODUCTION

In the previous six chapters we have introduced the methodology for model based image interpretation, computer graphics, as well as model based data visualization. In this chapter we apply subsets of the proposed methods to a variety of scenarios.

The expression of human faces are a major source of information in human communication. For image communication as well as for multi-media application it is of high importance to analyze human facial mimics. The appearance of a face is of course greatly influenced by the shape of the bones, such as skull and jaw. In Section 8.5 we describe how facial expressions can be synthesized on a graphical display for a simulated change of the bones. This will be used for surgery planning in medical applications. Due to spasms or paralysis, human facial expression is degraded severely in pathological cases. In Section 8.4 we outline how computer vision techniques can help the physicians to evaluate the degree of facial asymmetry. Model based techniques are used for training and rehabilitation. Section 8.6 describes how facial expressions can be modelled and how the parameters of the model can be used to generate mimics after image and parameter transmission.

One example of a solution for a complex visualization problem is given in Section 8.3 where flow fields are visualized in computational fluid dynamics. From the huge amount of data, that part has to be selected, which is of relevance for the observer.

At the present state of the art, images are presented to the observer on a flat display which provides a two-dimensional view of the data. Images captured by conventional cameras also record information on a two-dimensional grid. Since the real world has a higher dimension, projection methods are used. Models of real-world objects often represent the three-dimensional structure by surface information. In the previous chapters we have learned how three-dimensional information can be recovered even from two-dimensional projected images. In the final two sections of this volume we describe how three-dimensional models are used in automotive design and how three-dimensional surface models can be created from several range images recorded from different directions. We describe reverse engineering and automatic object model generation that can be used in virtual reality environments. We also show how these techniques can be used for manufacturing of real objects in automotive design where esthetic criteria have to fulfilled. As in the application in Section 8.5, physical models which were used formerly are now replaced by computer simulations. By computer simulation, effects of changes in the shape of a car can be inspected which reduces the costs of car design.

8.1 DIGITIZING 3D OBJECTS FOR REVERSE ENGINEERING AND VIRTUAL REALITY

S. KARBACHER, N. SCHÖN, H. SCHÖNFELD, G. HÄUSLER

Optical 3D sensors are used as tools for *reverse engineering* and *virtual reality* to digitize the surface of real 3D objects. Common interactive surface reconstruction is used to convert the sensor point cloud data into a parametric CAD description (e.g. NURBS). We discuss an almost fully automatic method to generate a surface description based on a mesh of curved or flat triangles.

Multiple range images, taken with a calibrated optical 3D sensor from different points of views, are necessary to capture the whole surface of an object and to reduce data loss due to reflexes and shadowing. The raw data are not directly suitable for import into CAD/CAM systems, as these range images consist of millions of single points and each image is given in the sensor's coordinate system. The data usually are distorted, due to the imperfect measuring process, by outliers, noise and aliasing. To generate a valid surface description from this kind of data, several problems need to be solved:

- the *registration* (transformation) of the single range images into one common coordinate system (see Section 4.2),
- the *surface reconstruction* from the point cloud data to regain the object topology (see Section 4.4),
- the *modeling of the surface geometry* to eliminate errors introduced by the measurement process and to reduce the amount of data (see Section 4.3),
- and the *reconstruction of the texture* of the object from additional video images for virtual reality applications.

Commonly used software directly fit *tensor product surfaces* to the point cloud data [186, 180] (see Section 4.6). This approach requires permanent interactive control by the user. For many applications however, e.g. the production of dental prosthesis or

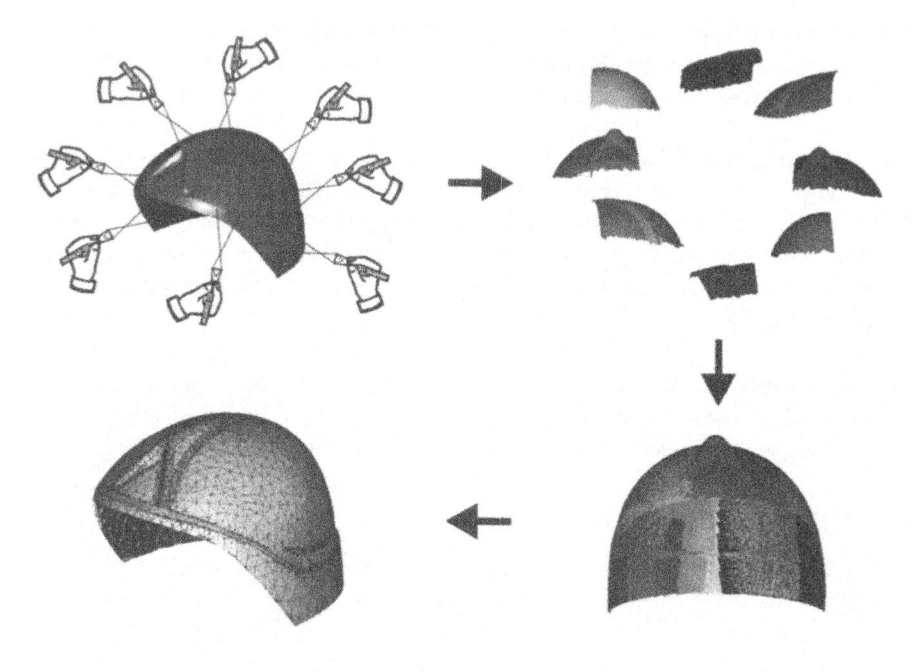

Figure 8.1. Data acquisition (upper left), registration (right) and mesh reconstruction (lower left) for a fire fighter's helmet.

the '3D copier', this kind of technique is not necessary. In this case the generation of meshes of triangles as surface representation is adequate [326, 199, 716, 586, 726, 145].

This section introduces a nearly automatic procedure covering the complex task from gathering data by an optical 3D sensor to generating meshes of triangles [617]. The whole surface of an object can be scanned by registering range images taken from arbitrary positions. The procedure includes the following steps (see Figure 8.1):

Data acquisition: Usually multiple range images of one object are taken to acquire the whole object surface. These images consist of range values arranged in a matrix as on the camera's CCD chip.

Calibration: By measuring a standard with an exactly known shape, a calibration function for transforming the pixel coordinates into metrical coordinates is computed. The coordinate triplets of the calibrated range image have the same structure as the original pixel matrix. This method calibrates each measurement individually. As a result each view has its own coordinate system.

Surface registration: The various views are transformed into a common coordinate system and are adjusted to each other. Since the sensor positions of the different views are not known, the transformation parameters must be determined by an accurate localization method. First the surfaces are coarsely aligned one to another with a feature based method. Then a fine tuning algorithm minimizes the deviations between the surfaces (global optimization).

Mesh reconstruction: The views are merged into a single object model. A surface description is generated using a mesh of curved triangles. The order of the original data is lost, resulting in scattered data.

Data modeling and smoothing: A new modeling method for triangle meshes allows to interpolate curved surfaces only from the vertices and the vertex normals of the mesh (see Section 4.3). Using curvature dependent mesh thinning it provides a compact description of the curved triangle meshes. Measurement errors, such as sensor noise, aliasing, calibration and registration errors, can be eliminated without ruining the object edges.

Texture reconstruction: For each range image a congruent normalized texture map is computed. In order to eliminate artifacts caused by illumination effects, two images of the same object view with illumination from different directions are merged.

8.1.1 Data Acquisition

Depending on the object to be digitized, that is its size and surface characteristics, the appropriate 3D sensor needs to be chosen. For details refer to Chapter 1.

8.1.2 Sensor Calibration

Optical sensors generate distorted coordinates $\boldsymbol{x}' = [x', y', z']^T$ because of perspective, aberrations and other effects. For real world applications a calibration of the sensor that transforms the sensor raw data $\boldsymbol{x}'$ into the metrical Euclidean coordinates $\boldsymbol{x} = [x, y, z]^T$ is necessary.

Our method uses an arbitrary polynomial $\boldsymbol{p}_c$ as a calibration function [304]. The coefficients are determined by a series of measurements of a calibration standard with exactly known geometry. The standard consists of three tilted flat surface patches which intersect virtually at an exactly known position $\boldsymbol{x}_s$. Due to aberrations, the digitized surface patches of the standard are not flat. Therefore *interpolating polynomials* $\boldsymbol{p}_k$, $k \in \{1, 2, 3\}$, (not to be mistaken with the above mentioned calibration polynomial $\boldsymbol{p}_c$)[1] are approximated to find the virtual intersection point $\boldsymbol{x}'_s$ with

$$\boldsymbol{x}_s = \boldsymbol{p}_k(\boldsymbol{x}'_s) \quad \forall\, k. \tag{8.1}$$

In order to fill the whole measuring volume with such calibration points $\boldsymbol{x}_i$, $1 \leq i \leq n$, the standard is moved on a translation stage and measured in n different positions. The intersection points can be calculated with high accuracy, since the information of thousands of data points is averaged by surface approximation. As a result, the accuracy of the calibration method is not limited by the sensor noise. This method is usable for any kind of sensor and, in contrast to other methods, requires no mathematical model of the sensor and no localization of small features, like circles or crosses.

[1] The calibration polynomial $\boldsymbol{p}_c$ transforms arbitrary data points into real world coordinates, the interpolating polynomials $\boldsymbol{p}_k$ only the data points of the k-th calibration plane.

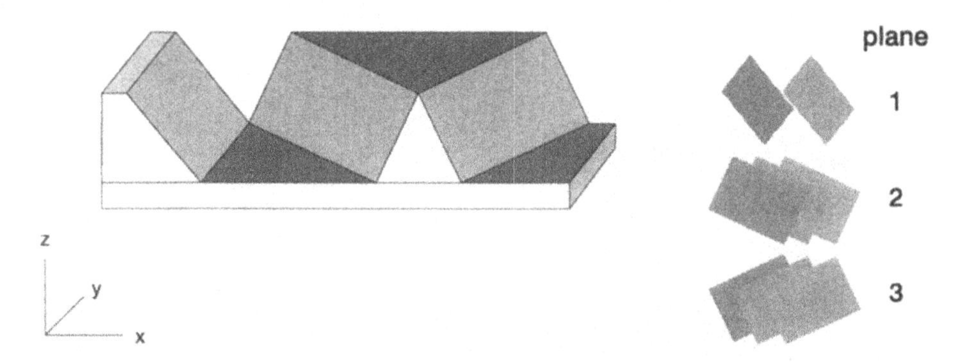

Figure 8.2. Calibration standard with three tilted planes (left) and 8 range images of these planes (right).

Figure 8.2 shows a block of aluminum with three tilted planes, which is used for calibration of our phase measuring (pmt) sensor. It is moved by small steps along y-direction. After each displacement three range images, one for each plane, are taken. Images of the same plane form a class of parallel calibration surfaces. Interpolating polynomials of order 5 are fitted to the deformed plane data. Each polynomial is intersected with polynomials of the other classes (see Figure 8.3). The intersection points are scattered throughout the whole field of view. Their real positions $\boldsymbol{x}_i$ are computed from the geometry of the standard and from its actual translation. The *calibration polynomial* $\boldsymbol{p}_c$ that transforms the measured intersection positions $\boldsymbol{x}_i'$ to the known positions $\boldsymbol{x}_i$ by

$$\boldsymbol{x}_i = \boldsymbol{p}_c(\boldsymbol{x}_i') \tag{8.2}$$

is approximated by polynomial regression.

Calibration of a range image with 512×540 points takes about 3 seconds on an Intel Pentium® 166 CPU. The calibration error is less than 50% of the measurement uncertainty of the sensor. Since coefficients of higher order are usually very close to zero, it is sufficient to use a calibration polynomial of order 4.

8.1.3 Registration

Usually multiple range images from different views are taken. If the 3D sensor is moved mechanically to definite positions, this information can be used to transform the images into a common coordinate system. Some applications require sensors that are placed manually in arbitrary positions, e.g. if large objects like monuments are digitized. Sometimes the object has to be placed arbitrarily, e.g. if the top and the bottom of the same object are to be scanned. In these cases the transformation must be computed solely from the image data.

Coarse Registration. The following procedure for registration of multiple views is based on feature extraction. It is independent of the sensor that is used. Thus it may be applied to small objects like teeth, and to large objects like busts.

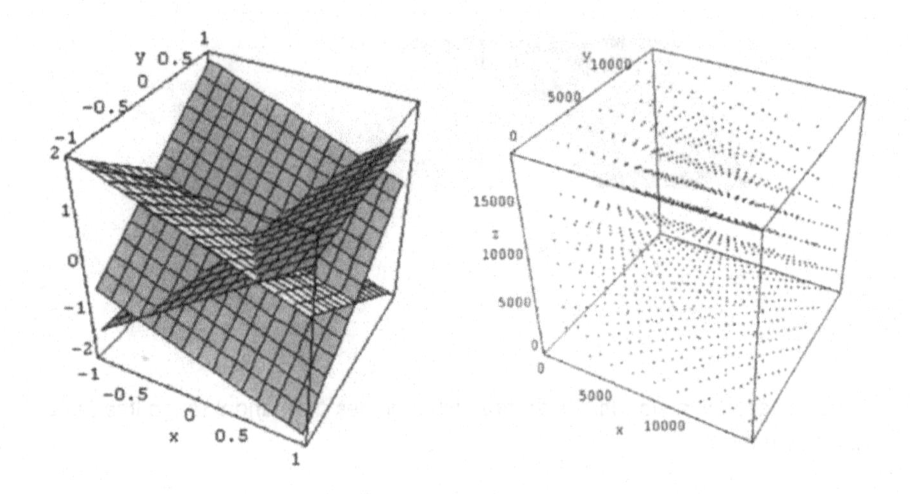

Figure 8.3. Intersection of three polynomial surfaces (one from each class) in arbitrary units (left) and field of view with all measured intersection points in metric coordinates (right).

Zero-dimensional intrinsic features, e.g. corners, are extracted from the range images or congruent gray scale images (see Section 4.2). The detected feature locations are used to calculate the translation and rotation parameters of one view in relation to a master image. Simultaneously, the unknown correlations between the located features in both views are determined by Hough methods. To allow an efficient use of Hough tables, the six-dimensional parameter space is separated into two- and one-dimensional sub-spaces (see [574]).

If intrinsic features are hard or impossible to detect (e.g. on the fireman's helmet in Figures 8.1 and 8.9), artificial markers, which can be detected automatically or manually, are applied to the surface. The views are aligned to each other by pairs. Due to the limited accuracy of feature localization, deviations between the different views still remain.

Fine Registration. A modified Iterated Closest Point (ICP) algorithm is used to minimize the remaining deviations between pairs of views (see Section 4.2.5). Error minimization is done by *simulated annealing*, so that in contrast to classic ICP, local minima of the cost function may be overcome. Since simulated annealing leads to a slow convergence, computation time tends to be rather large. A combination of simulated annealing and the Levenberg-Marquardt algorithm however shows even smaller remaining errors in much shorter time: The registration of one pair of views takes about 15 seconds on an Intel Pentium® 166 CPU. If the data is calibrated correctly, the accuracy (defined by the standard deviation between corresponding data points of both surfaces) is only limited by sensor noise.

Figure 8.4 shows the result of the registration for two views with 0.5% noise. One was rotated by approximately 50°. The final error standard deviation was approximately σ_{noise} (standard deviation of the sensor noise).

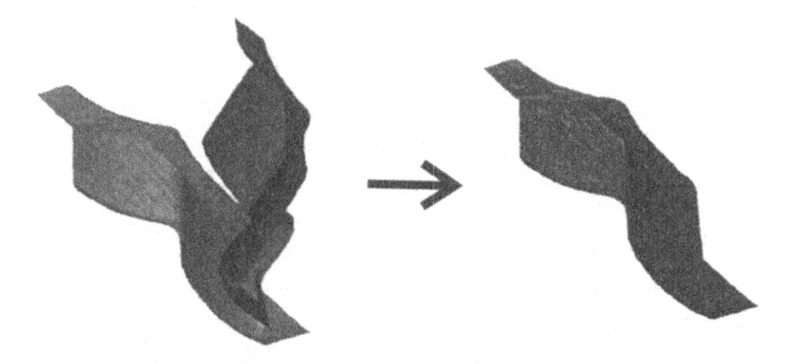

Figure 8.4. Fine registration of two noisy views with very high initial deviation ($\alpha = -30°$, $\beta = 50°$, $\gamma = 40°$, $x = 30$ mm, $y = 60$ mm, $z = -40$ mm).

Global Registration. Solely using registration by pairs, closed surfaces can not be registered satisfactorily. Due to the accumulation of small remaining errors (caused by noise and miscalibration) frequently a chink develops between the surface of the first and last registered view. In such cases the error must be minimized globally over all views. One iteration fixes a single view and minimizes the error of all overlapping views simultaneously. About 5 of these global optimization cycles are necessary to reach the minimum or at least an evenly distributed residual error. Global registration of an object that consists of 10 views takes approximately 30 minutes on an Intel Pentium® II 300 CPU.

8.1.4 Mesh Reconstruction

The object surface is reconstructed from multiple registered range images. These consist of a matrix of coordinate triples $\boldsymbol{x}_{n,m} = [x, y, z]^T_{n,m}$. The object surface may be incompletely sampled and the sampling density may vary, but should be as high as possible. Beyond that, the object may have arbitrary shape, and the field of view may even contain several objects. The following steps are performed to turn this data into a single mesh of curved or flat triangles:

Mesh generation: Because of the matrix-like structure of the range images, it is easy to separately turn each of them into triangle meshes with the data points as vertices. For each vertex the vertex normals are calculated from the normals of the surrounding triangles.

First smoothing: In order to utilize as much of the sampled information as possible, smoothing of measuring errors like noise and aliasing is done before mesh thinning (see Section 4.3.2).

First mesh thinning: Merging dense meshes usually requires too much memory, so that mesh reduction must often be carried out in advance (see Section 4.1.3). The permitted approximation error should be chosen as small as possible, as ideally thinning should be only done at the end of the processing chain.

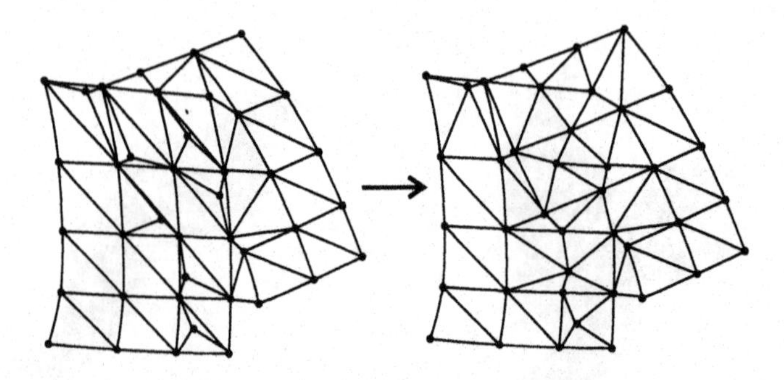

Figure 8.5. Topological optimization of the mesh from Figure 4.26 on page 178 using edge swap operations. Triangles that are as equilateral as possible were aspired.

Merging: The meshes from different views are merged by pairs using local mesh operations like *vertex insertion*, *gap bridging* and *mesh growing* (see Figure 4.26 and Section 4.1.7). Initially a master image is chosen. The other views are merged into it successively. Only those vertices are inserted whose absence would cause an approximation error larger than a given threshold.

Final smoothing: Due to calibration and registration errors, the meshes do not match perfectly. Thus after merging the mesh is usually distorted and has to be smoothed again.

Final mesh thinning: Mesh thinning is continued until the given approximation error is reached. For thinning purposes also a classification of the surfaces according to curvature characteristics is carried out.

Geometrical mesh optimization: Thinning usually causes awkward distributions of the vertices, so that elongated triangles occur. Geometrical mesh optimization moves the vertices along the curved surface in order to produce a better balanced triangulation (see Section 4.3.3).

Topological mesh optimization: At last the surface triangulation is reorganized using *edge swap* operations, in order to optimize certain criteria (see Figure 8.5). The interpolation error is usually minimized.

The result of this process is a mesh of curved triangles. A new modeling method is able to interpolate curved surfaces solely from the vertex coordinates and the assigned normal coordinates (see Section 4.3). This enables a compact description of the mesh, as modern data exchange formats like Wavefront OBJ, Geomview OFF, or VRML support this data structure.

8.1.5 Texture Reconstruction

The previous sections describe methods to reconstruct the shape of three dimensional objects. For visualization purposes it is often desirable to acquire the color texture of the object as well. Color video images of the object usually show artifacts caused by

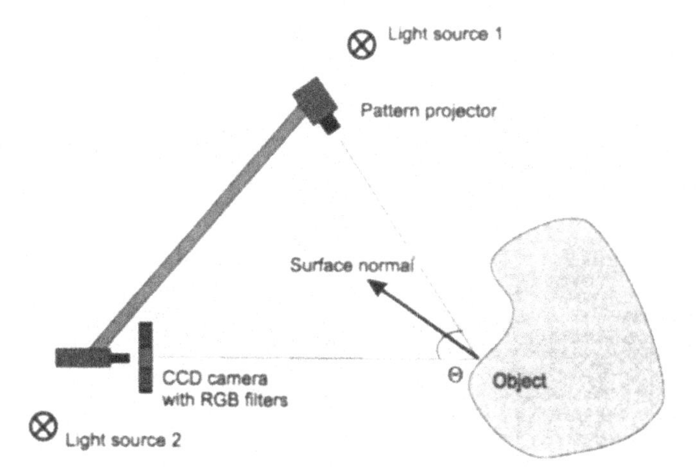

Figure 8.6. A 3D sensor with two additional light sources for texture reconstruction.

illumination effects, which are not part of the surface color itself, for example white spots due to shiny surfaces and brightness variations of diffusely reflected light. These artifacts can be eliminated if two images of the same object view with illumination from different directions are taken [616] (see Figures 8.6 and 8.7). The 3D sensor acquires a range image of the same view, so that for each camera pixel the 3D coordinates are known.

Variations of the orientation of the surface normals change the brightness in diffusely reflecting parts of the image. Assuming Lambertian reflection this effect can be compensated if the normals are known. Compensation factors for each pixel are

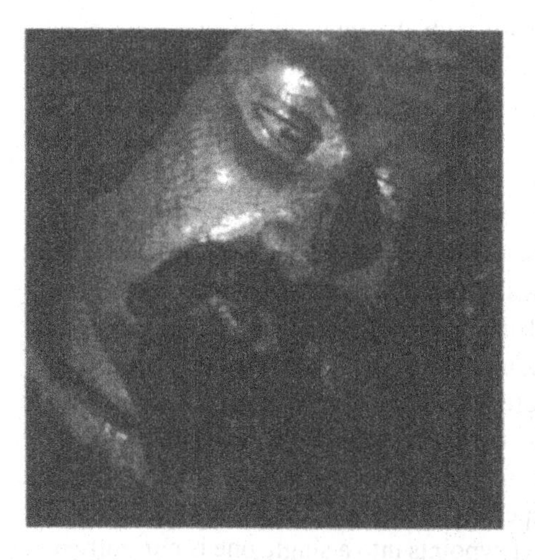
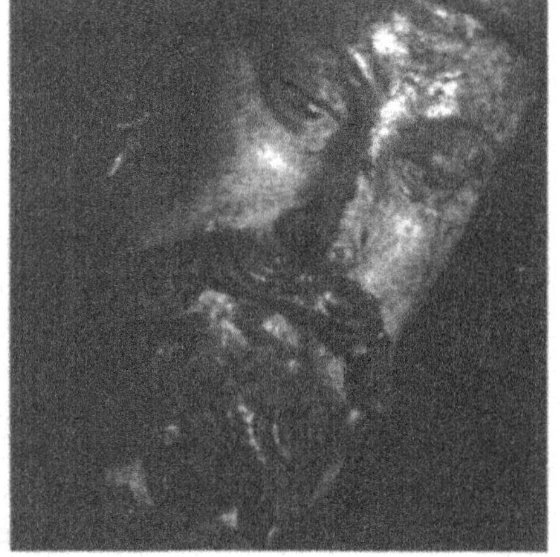

Figure 8.7. Images of St. George ('Germanisches Nationalmuseum') taken from the same view and illuminated from left and right.

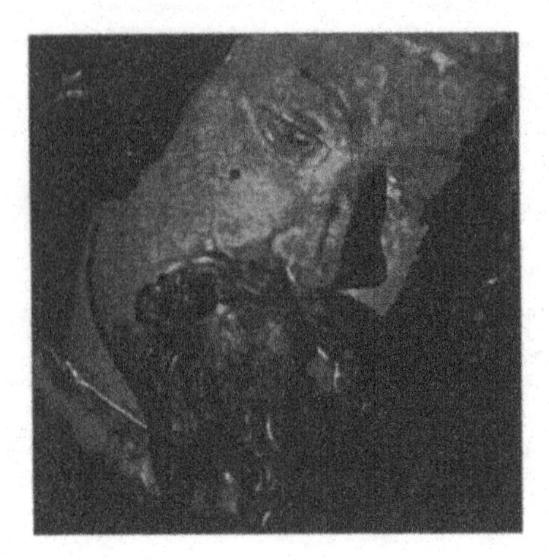

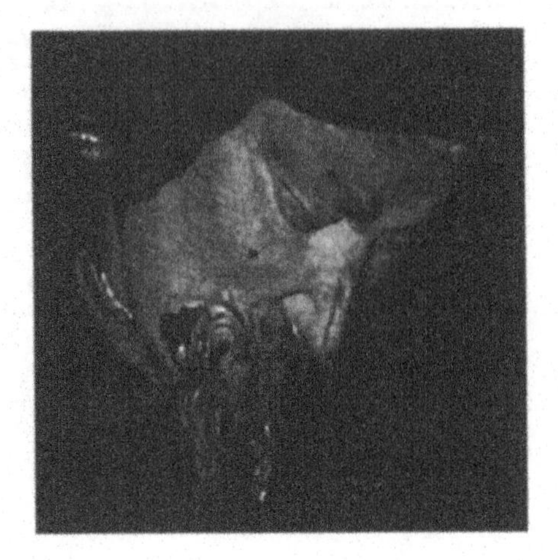

Figure 8.8. A normalized texture image is computed from the two images of Figure 8.7 (left) and mapped onto the triangle mesh of the corresponding range image (right).

calculated from the angles between the normals and direction of illumination. The normals are computed from the range data and therefore are usually distorted by camera noise. Since the influence of this noise depends on the normal orientations, *confidence values* are assigned to each pixel which quantify the accuracy of that pixel. The following observations are considered:

- The lower the intensity the higher is the relative influence of the noise.

- The noise is amplified by the correction factor.

As a result, the confidence values increase with the intensity of the corresponding pixel and decrease with the deviation of the surface normal from the direction of illumination.

In order to remove shiny regions from the texture images, additional confidence values are defined. In contrast to the previous ones, these increase with the deviation of the observation direction from the directions of specular reflection. A mathematical model for specular reflection is used for that purpose. It depends on the spatial angles of the shiny regions, which are automatically computed by masking the highlights.

Finally, the two images with corrected diffuse reflection are merged by weighting the color of each pixel with its confidence value. The result is a normalized texture image without shading and highlights (see Figure 8.8). This texture can be mapped onto the triangle mesh that was generated from the corresponding range image (see Figure 8.8) or may be used for coloring the vertices of the final mesh (see Figure 8.11). The latter method requires meshes with texture dependent densities. Hence, a method to merge normalized texture maps from different viewpoints into a single one is currently being designed. The texture image can be mapped onto a mesh with arbitrary density without loss of details.

8.1.6 Results

The described digitizing method was tested with many different objects, technical and natural as well. The errors reinduced by the modeling process were smaller than the errors in the original data (measuring, calibration and registration errors). The smoothing method is specifically adapted to the requirements of geometric data, as it minimizes curvature variations. Undesirable surface waviness is avoided. Surfaces of high quality for visualization, NC milling and 'real' reverse engineering are reconstructed. The method is well suited for metrology purposes, where high accuracy is desired.

Figure 8.9 shows the results for a design model of a fire fighter's helmet. The reconstructed CAD-data is used to produce the helmet. For that purpose, the triangular mesh was translated into a mesh of Bézier triangles, so that small irregularities on the boundary could be cleaned manually. This work was carried out in collaboration of the Chair for Optics and the Computer Graphics Group of the University of Erlangen-Nürnberg. Eight range images containing 874,180 data points (11.6 MB) were used for surface reconstruction. The standard deviation of the sensor noise is 0.03 mm (10% of the sampling distance), the mean registration error is 0.2 mm. On a machine with an Intel Pentium® II 300 processor, the surface reconstruction took 7 minutes allocating 22 MB of RAM. The resulting surface consists of 33,298 triangles (800 kB) and has a mean deviation of 0.07 mm from the original (unsmoothed) range data.

For a virtual catalogue of teeth for medical applications we measured a series of 26 human tooth models of twice the natural size [417]. Figure 8.10 shows a reconstructed canine tooth and a molar. Ten views (between 15 resp. 13 MB) were processed. Surface reconstruction took 6 resp. 7 minutes.

For the 'Germanisches Nationalmuseum' in Nürnberg we measured parts of a statue of Saint George (15th Century). The results are used to improve the restoration by virtually visualizing each step. The 3D data allow to distinguish between relief and painted structures (texture) of the surface. For the visitors of the museum it is

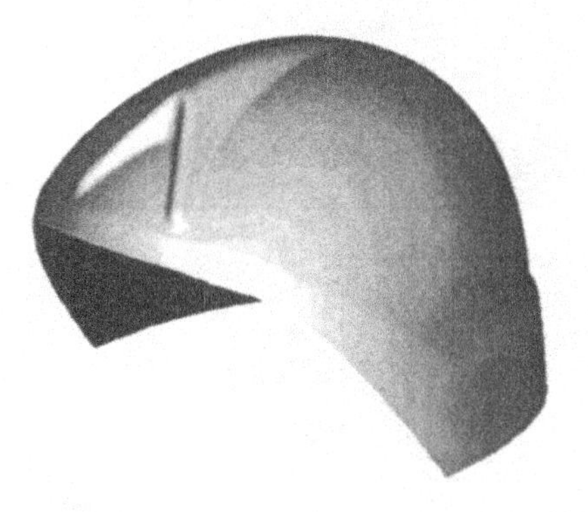

Figure 8.9. Reconstructed mesh of a fire fighter's helmet (left) and the helmet that was produced from the CAD data (right).

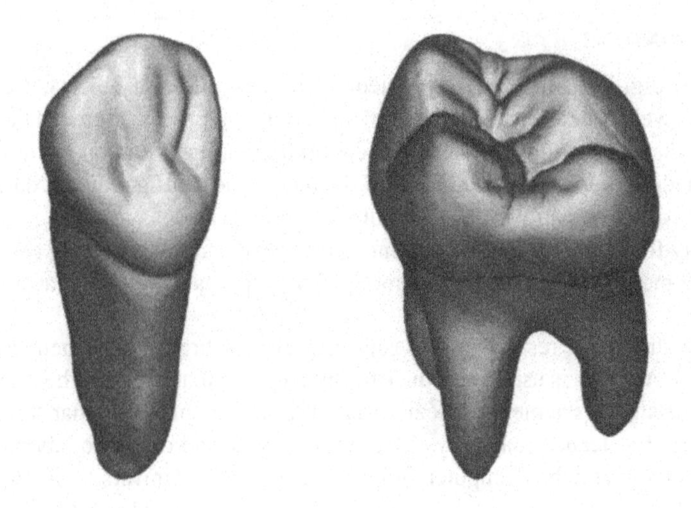

Figure 8.10. Reconstructed surfaces of plaster models of a human canine (left, 35,642 triangles, 870 kB) and a molar (right, 64,131 triangles, 1.5 MB).

interesting to watch a 3D simulation of the restoration process. Moreover, the obtained 3D model enables to view the statue in a virtual (e.g. ancient) environment. The 3D measurements were done using a phase measuring triangulation (pmt) sensor. Since the statue is partially very dark (e.g. the hair), 16 measurements for each view were

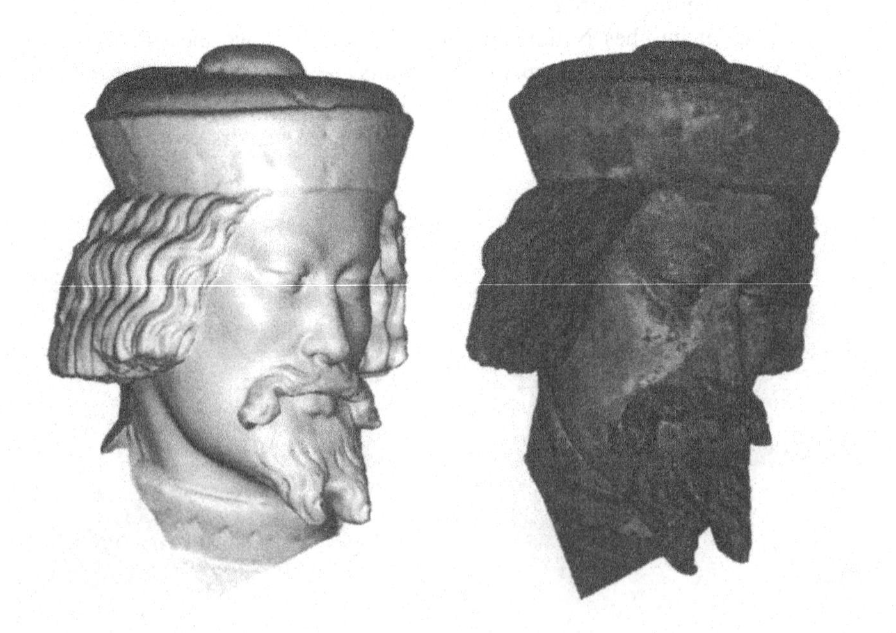

Figure 8.11. Reconstruction of the head of St. George: with plain surface (left) and with textured faces (right).

overlayed. For each of altogether 26 views a normalized texture map was reconstructed. This was used to color the vertices of the final mesh (see Figure 8.11). In order to avoid loss of details, mesh thinning was carried out very carefully. The final mesh consists of 400,000 triangles. Hence, surface reconstruction took 6 hours. The final mesh approximates the raw sensor data with a mean deviation of 140μm.

Acknowledgement

Parts of this work were supported by 'Deutsche Forschungsgemeinschaft (DFG)' #1319/8-2 and by 'Studienstiftung des deutschen Volkes'.

8.2 SURFACE INTERROGATION IN AUTOMOTIVE DESIGN

G. GREINER

The competition on the global market forces car manufacturer to offer a great variety of different models. Moreover, redesign and updates of the models have to be done in shorter periods. The necessary speed-up of the development cycle of a car can be accomplished only by massive computer support. The slogan *digital car* has gained great popularity in the upper management of all car manufacturers. It expresses the desire to use CAx-technologies in order to achieve better quality, in less time and for less costs. In several areas of this complex process geometric modeling and computer graphics play a decisive role. For instance, car body design is based on free form surface modeling. In former time clay models have been used but for quite some time, the *master model*, which serves as reference for all steps in the construction is a digital model. It contains 3D-CAD-data of all components of the vehicle.

A historic fact is that many of the fundamental ideas of free form surface design have been developed in the sixties in the car industry. The theory of Bézier curves and surfaces (see Section 4.5) been established independently by de Casteljau at Citroën and Bézier at Renault. Coons at Ford and Gordon at General Motors used *transfinite interpolation methods* to interpolate a network of curves with fair surfaces. Extending the ideas of de Casteljau and Bézier naturally leads to B-splines and further on to non-uniform rational B-splines (NURBS), which are the de-facto standard of todays CAD systems [77, 215].

Free form surface design plays an important role in several stages of the development of a new car. It is particularly important in the design and construction of the car body. The latter can be subdivided in the following steps (see Figure 8.12):

Design. Stylists design new cars. At this stage none or only little computer support is used. Stylists still use traditional techniques, like hand drawings and modeling with clay.

Construction. Draftsmen use powerful computer hardware and sophisticated software to create a digital model. Reverse engineering techniques are used to build a 3D-CAD model of the stylists' concepts. In addition, fine detail as well as the overall appearance has to be checked for quality.

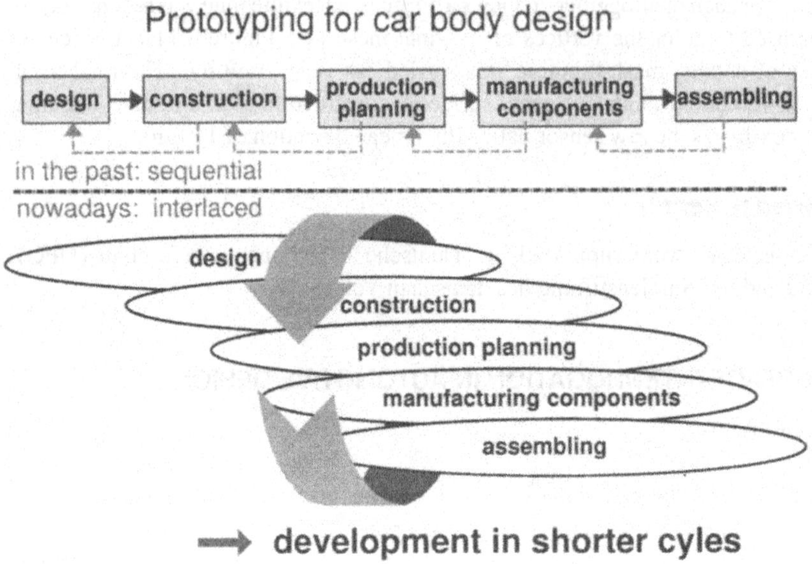

Prototyping for car body design

design → construction → production planning → manufacturing components → assembling

in the past: sequential

nowadays: interlaced

design

construction

production planning

manufacturing components

assembling

→ development in shorter cyles

Figure 8.12. Car body: design and construction of prototypes.

Production planning. Possibility of manufacturing has to be checked, *digital mock-up* will widely replace *physical mock-up*. As a result of this stage, the master model is specified which is a reference for all further steps in the development and also later on in the regular production.

Manufacturing of components. At this stage geometric modeling is used for designing the tools that are necessary for the manufacturing.

Assembling of the components. Computer technology is used for process simulation. Finally prototypes are manufactured.

In order to speed-up the development cycle these steps are no longer executed sequentially but interlaced. This requires a lot of interaction between the different groups. Fast and meaningful visualization of the partial steps is indispensable. Moreover, adequate measures for checking quality and advanced tools for optimizing components are necessary. In the sequel we describe one of these procedures in more detail. In automotive industry there is a long tradition to judge the quality of the car body by inspecting reflection lines on the object. Formerly these lines were generated by placing the real object in front of a set of parallel light lines. Nowadays, reflection lines are simulated on the computer and then will be mapped as texture during display of the 3D-CAD model.

Besides reflection lines (see [388, 380] for more details), there are several other methods for surface interrogation, e.g. displaying color encoded curvatures on the surface [183], focal surfaces [271], isolines of curvatures [514] and isophotes (isolines of brightness) [548].

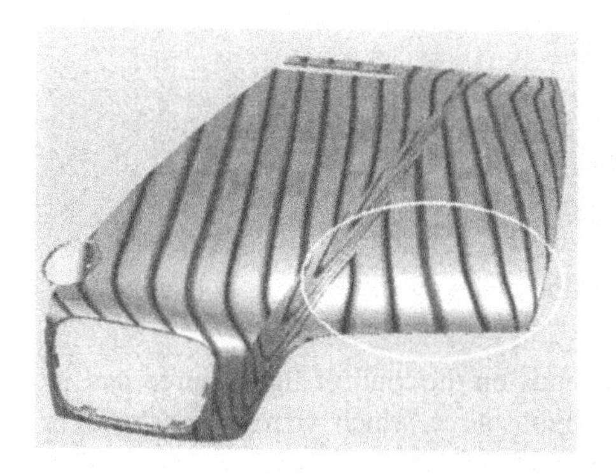
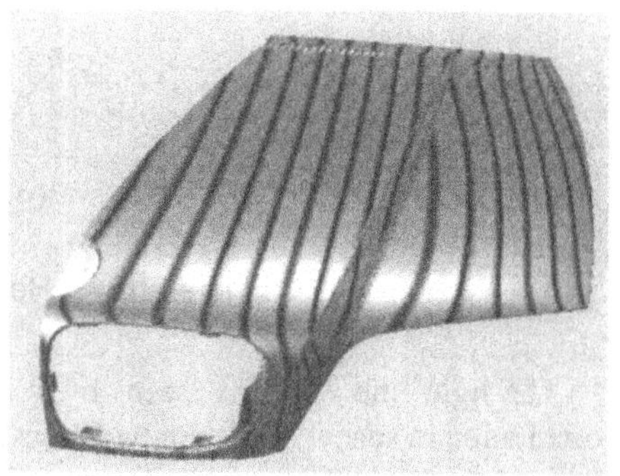

Figure 8.13. Part of a car (front hood): the right version has a more appealing reflection lines pattern.

Reflection lines are very popular in automotive industry probably because they are familiar to the designer from 'real' world experiments. Moreover, they are very sensitive to small shape deviations and, last but not least, modern graphics hardware allows rendering of reflection lines at interactive rates. In Figure 8.13 two version of a front hood are shown. The reflection lines are pre-calculated and then mapped by environment texture (see Section 5.3) on the Gouraud-shaded geometric model. Reflection lines are created by simulating the reflection of a family of parallel light lines in 3D-space at the surface. They are viewer dependent and assume a specular reflecting surface.

Due to the popularity of reflection lines, there is a need for a surface design tool that directly operates on these features, allowing the user to specify the desired lines and automatically generate the corresponding surface shape. In this section we outline a method which has been presented in [451] and constructs a surface geometry for a given reflection line pattern. Exact matching to the user specified data is in most cases inherently impossible (see [15, 16]). Therefore one tries to find an approximation which comes as close as possible to the desired surface shape.

8.2.1 Determination of Reflection Lines

For simplicity we assume that the surface F under consideration (describing a component of the car) is represented as a graph of a function, i.e. $F(u, v) = (u, v, f(u, v))$. Reflection lines are viewer dependent. Therefore we have to fix a viewer position. For simplicity, we assume that viewer and light source are located at infinity. Let $\boldsymbol{s}$ be the vector pointing towards the viewer.

An eye ray emerging from the viewer is reflected at the surface into a direction $\boldsymbol{r}$. Since the viewer position lies in infinity, $\boldsymbol{s}$ is constant. Thus $\boldsymbol{r}$ depends only on the surface normal $\boldsymbol{n}$ at the point of reflection (see Figure 8.14). A simple calculation shows

$$\boldsymbol{r} = 2\langle \boldsymbol{s} | \boldsymbol{n} \rangle \boldsymbol{n} - \boldsymbol{s} \tag{8.3}$$

where we assume $||\boldsymbol{n}|| = ||\boldsymbol{s}|| = 1$ hence $||\boldsymbol{r}|| = 1$ as well.

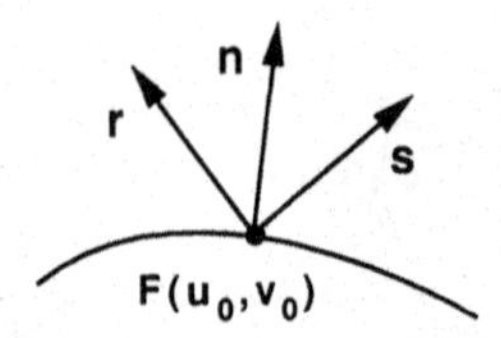

Figure 8.14. Reflection of the eye ray.

The light line which is 'seen' by $\boldsymbol{r}$ depends on the kind of light source and its orientation in space. There are two types of light source, which we refer to as *axial* or *radial* light lines. The axial light source is a family of straight (light)-lines, all parallel to a vector $\boldsymbol{z} \in \mathbb{R}^3$, which determines the orientation. Moving these lines to infinity (perpendicular to $\boldsymbol{z}$), the projection onto the unit sphere assigns a longitude to each light line (see Figure 8.15). By completing $\boldsymbol{z}$ to an orthonormal coordinate system $\{\boldsymbol{x}, \boldsymbol{y}, \boldsymbol{z}\}$ we can parameterize the longitudes from 0 to 2π as shown in the left image of Figure 8.15. Thus, every light line is determined by a unique angle $\phi \in [0, 2\pi[$. The radial light source is composed of circles in infinity (however, we still call them light 'lines'), all circles being perpendicular to $\boldsymbol{z}$. The projection onto the sphere maps them onto latitudes, which we number from 0 to π (right image in Figure 8.15). In this case each light line is determined uniquely by an angle $\theta \in]0, \pi[$. A different interpretation of this situation is the following statement: Light lines are isolines of the scalar function $\phi = \phi(u, v)$ and $\theta = \theta(u, v)$ respectively. These functions are obtained by inserting Equation 8.3 into the following identities

$$\tan\phi = \frac{\langle \boldsymbol{r}|\boldsymbol{y}\rangle}{\langle \boldsymbol{r}|\boldsymbol{x}\rangle} \quad \text{or} \quad \phi = \arctan\frac{\langle \boldsymbol{r}|\boldsymbol{y}\rangle}{\langle \boldsymbol{r}|\boldsymbol{x}\rangle}$$

and

$$\cos\theta = \langle \boldsymbol{r}|\boldsymbol{z}\rangle \quad \text{or} \quad \theta = \arccos\langle \boldsymbol{r}|\boldsymbol{z}\rangle \ .$$

These functions are the *height field of reflection lines* subsequently abbreviated by HFR. For a given surface F and specified viewing direction s, the HFR is considered as

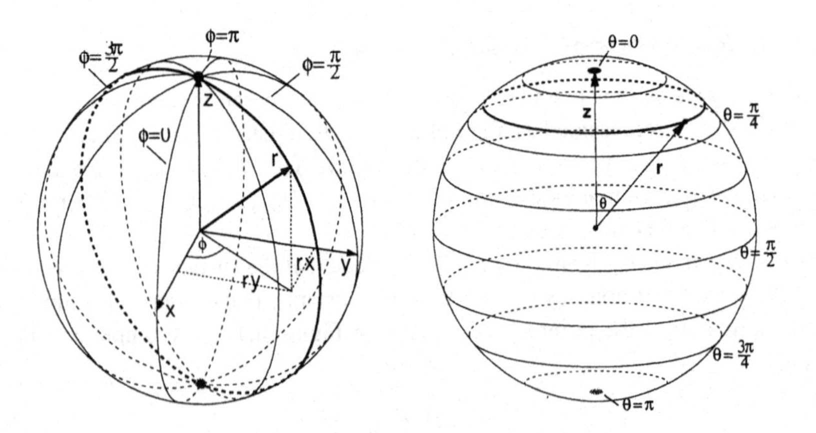

Figure 8.15. Model for axial (left) and radial (right) sources.

function defined on the parametric domain of F. To express specifically the dependence on F we will subsequently mark the HFR with a subscript $_F$, i.e. $I_F = I_F(u, v)$.

In order to manipulate reflection lines or to fit a geometry to a prescribed pattern of reflection lines we only consider the HFR. We outline the procedure by considering two examples: Fairing of reflection lines and surface reconstruction based on reflection lines.

8.2.2 *Fairing of Reflection Lines.*

If reflection lines are used as surface interrogation tool, it is more natural to perform the fairing procedure on the reflection lines or the corresponding HFR and to adapt the surface to the resulting HFR. The user smoothes the HFR first, obtaining better reflection lines while retaining the global look. Then the system computes the appropriate surface modification automatically. Figure 8.16 illustrates the fairing strategy. First the user specifies a region where the reflection lines and the surface should be faired. The outside region remains fixed.

For a given geometry F and a specified viewing direction s we proceed as follows:

- Compute the HFR of the surface F, thus obtaining a scalar function I_F. In order to obtain a B-spline representation of I_F, we evaluate I_F on a regular grid and interpolate the sampled data (see Section 4.6).

- Smooth the HFR I_F by a standard procedure, e.g. by minimizing an energy functional $J_{HFR}(I)$.

$$J_{HFR}(I) = \alpha J_{fair}(I) + (1 - \alpha) J_{fit}(I) \ ,$$

where J_{fair} is a simple quadratic fairing functional, e.g.

$$J_{fair}(I) = \int I_{uu}^2 + 2I_{uv}^2 + I_{vv}^2 \ .$$

The functional

$$J_{fit}(I) = \int (I - I_F)^2$$

measures the deviation of the new HFR obtained by minimizing J_{HFR} to the existing one. The weight $\alpha \in [0, 1]$ allows to control to which extent the lines should be smoothed. The result will be a scalar function I^*.

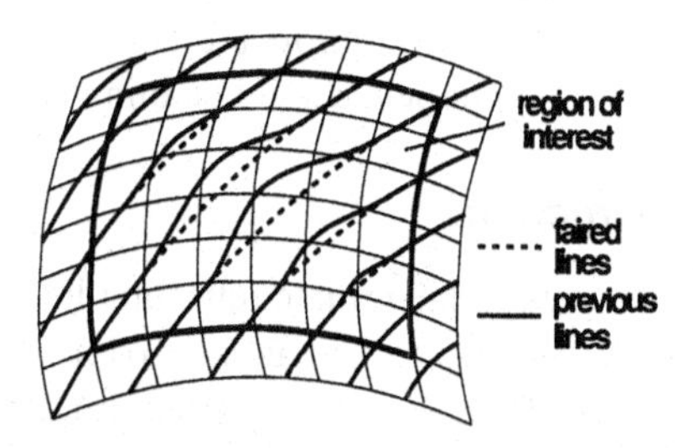

Figure 8.16. Local fairing of reflection lines for surface.

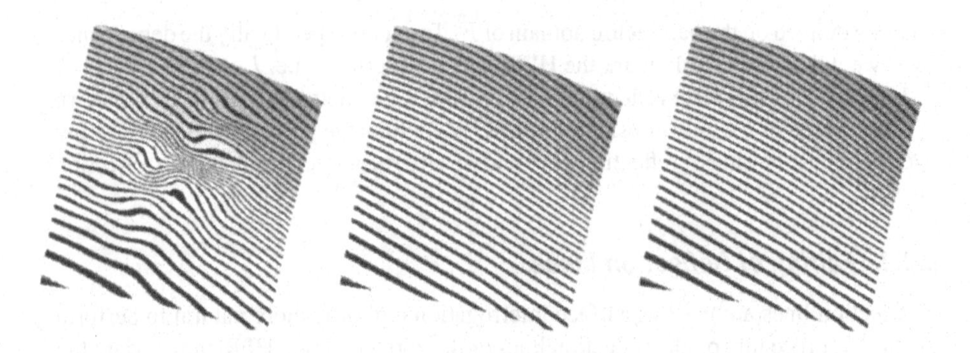

Figure 8.17. Initial surface (left), faired lines (middle), faired surface (right).

- Determine a new surface F^* such that the HFR I_{F^*} satisfies $I_{F^*} \approx I^*$. This will also be achieved by minimizing an appropriate error functional. Good results will be obtained using the following functional

$$\begin{aligned} J(F) &= \int (\text{grad} I_F - \text{grad} I^*)^2 \\ &= \int \left((I_F)_u - (I^*)_u\right)^2 + \left((I_F)_v - (I^*)_v\right)^2. \end{aligned} \tag{8.4}$$

  We additionally impose F^* to interpolate at the boundary the derivatives up to order two of the initial surface F. This ensures that the reflection lines of F^* connect tangent continuous to the lines in the outside region.

An example for the described fairing method is given in Figure 8.17. The left picture shows the initial surface with obvious curvature perturbations. The picture in the middle shows the smoothed lines, being texture mapped onto the initial surface. The right picture shows the faired surface and its reflection line pattern, which has been computed by approximating the line pattern of the middle picture.

The reason for incorporating gradient deviation in the error functional in Equation 8.4 is the following. Isolines of a scalar function (which represent the reflection lines in our situation) are closely related to the gradient of the function: isolines and gradient are orthogonal at every point. Thus in order to make sure that the reflection lines of the resulting surface F^* closely follow the faired isolines of F the gradient deviation of the corresponding height field should be minimal.

8.2.3 Surface Reconstruction

In the previous example we tried to interpolate reflection lines which may not correspond to a surface geometry. We had to find an approximation. If however, a set of reflection lines, which is created from an existing surface is used as input, the algorithm is capable to reconstruct the original surface. Since mechanical measurements often do not provide satisfying accuracy, it seems reasonable to use optical measuring based

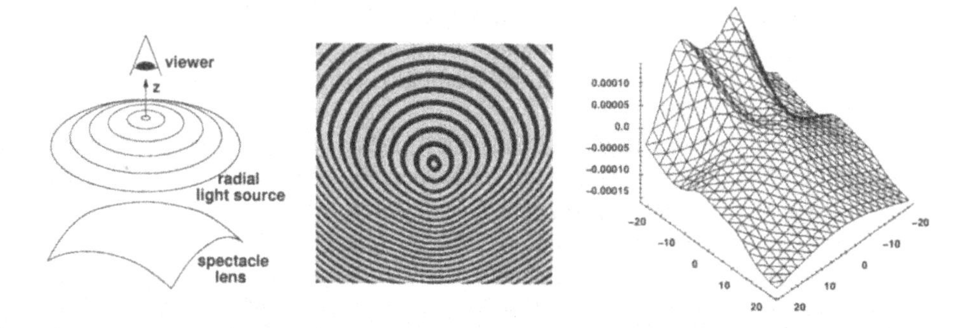

Figure 8.18. Configuration of radial light lines and viewing position (left), reflection line pattern of a progressive lens (middle), error plot (right).

on specular reflection. As an example (see Figure 8.18) the geometry of a progressive addition lens is reconstructed. We used this type of lens, because it has an increasing mean curvature along its vertical axis, and therefore exhibits a more interesting reflection pattern than uni-focal lenses. As input we computed the HFR from CAD data of such a lens (see [465, 450]).

The arrangement of camera, light lines and lens is illustrated in the left image of Figure 8.18. The orientation of the radial light lines is given by the direction $z = (0, 0, 1)$, which also points to viewer position. The arrangement is adopted from Halstead et.al. [278], where it was used to reconstruct the topography of a human cornea. The reconstruction algorithm used there is based on fitting a set of normal vectors which are obtained by backward ray tracing. We use the following strategy to reconstruct the geometry of the lens.

- For the known HFR I we determine a surface F such that the error measured by the functional

$$J(F) = \int (\mathrm{grad} I_F - \mathrm{grad} I^*)^2$$

is minimal.

The error plot in the right image of Figure 8.18 shows the difference between the original lens geometry and the surface which was reconstructed by adaption to the reflection line pattern shown in the middle image.

8.3 FLUID DYNAMICS

F. SCHÄFER, M. BREUER, C. TEITZEL

In Sections 6.4 and 6.5 of this book, methods for the visualization of three-dimensional scalar data (volume data) and vector fields have been described. An important field of application of these methods is the visualization of fluid dynamical data, which can be obtained by computer simulations of flows or detailed flow measurements.

Fluid dynamical data can contain three-dimensional, time-dependent scalar fields as well as vector fields and second-order tensor fields (e.g. pressure, velocity and stress, respectively). Hence, a variety of different methods is needed for the visualization of these different types of flow data.

In the following we want to indicate exemplarily how data visualization can be used in the field of fluid dynamics. We use the term 'data visualization' to distinguish the subject of this section from 'experimental visualization', where, for example, smoke is photographed while being advected in a flow. In Section 8.3.1 the purpose of data visualization in fluid dynamics is discussed, followed by a brief summary of data visualization techniques suitable for flow field visualization (see Section 8.3.2). The application of selected techniques to two different flow problems of actual research will be discussed in Sections 8.3.3 and 8.3.4. Here the main focus will be on computational fluid dynamics, since the spatial and temporal resolution of the flow data is usually higher for simulations than for measurements. But all of the visualization techniques are, in principle, applicable to experimental data as well, provided that the data resolution in space and time is sufficiently high.

To reveal the time-dependent behavior of fluid flows, animations are especially important. For the stirred vessel example discussed in Section 8.3.4 video sequences are available in the world wide web. Details about the online resources can be found in Section 8.3.5.

8.3.1 Purpose of Data Visualization in Fluid Dynamics

The progress in computational fluid dynamics (CFD) and experimental measuring techniques has led to increasingly detailed information about the investigated flow fields such as velocity, pressure, temperature and more. Experimental methods such as laser-Doppler anemometry [193], hot-wire anemometry [95] and particle-image velocimetry [3] have been developed allowing detailed measurements of the velocity field in a flow. Moreover, by solving the governing equations of fluid dynamics numerically, CFD can provide values of the flow field variables at a high resolution in space and time [221, 321].

In particular for CFD this has been possible due to increasing computer power and the development of efficient numerical algorithms. The most general description of a fluid flow, on which computer simulations can be based, is obtained from the full system of Navier-Stokes equations expressing the conservation of mass, momentum and energy. These equations may be written as

$$\begin{aligned}
\frac{\partial \rho}{\partial t} + \operatorname{div}(\rho \boldsymbol{v}) &= 0 \\
\frac{\partial (\rho \boldsymbol{v})}{\partial t} + \operatorname{div}(\rho \boldsymbol{v}\boldsymbol{v} + \tau) &= -\operatorname{grad} p + \boldsymbol{F} \\
c_p \left(\frac{\partial (\rho T)}{\partial t} + \operatorname{div}(\rho \boldsymbol{v} T) \right) &= \operatorname{div}(k \operatorname{grad} T) + \frac{\partial p}{\partial t} + \boldsymbol{v} \operatorname{grad} p + S,
\end{aligned}$$

where ρ denotes the density, $\boldsymbol{v}$ the velocity, p the pressure, T the temperature, τ the stress tensor, c_p the specific heat at constant pressure and k the thermal conductivity of the fluid. $\boldsymbol{F}$ is a volume specific force acting on the fluid (for instance gravity), and S is a volume specific source term for heat energy. Additional equations may be formulated for further scalar quantities such as the concentration of chemical species.

In addition to the Navier-Stokes equations, also lower approximation levels are in use such as Euler equations or potential flow models. Regardless of the underlying set of equations, numerical flow simulations are generally based on a spatial and temporal discretization of the computational domain, using different approaches such as finite elements, finite volumes or finite differences methods. In combination with different grid arrangements such as structured, block-structured and unstructered grids, this leads to a variety of different CFD methods. They all have in common that an approximate solution of the governing equations is determined. A special problem arises in the case of turbulent flows, since the resolution of the smallest vortices in the flow requires such a fine discretization in space and time that the direct numerical simulation (DNS) of the governing equations leads to prohibitive long computation times. To cope with this problem, often the Reynolds-averaged Navier-Stokes equations are used for the simulation of turbulent flows together with statistical turbulence models.

Depending on the flow problem and the computational method applied, a wide range of spatial and temporal discretizations can be used in CFD applications. Nowadays, three-dimensional grids containing 10^4 to more than 10^7 grid points are quite common, and the resulting output data sets contain just as many velocity values, pressure values, etc. The information output is even larger in the most general case of a time-dependent simulation, where for every time step such a voluminous data set is produced.

Having this in mind, the question arises how to get useful information from the columns of numbers produced by CFD applications and high resolution measurements. Of course, the numerical values of the calculated or measured flow field variables are important for comparison with theoretical data, as far as they are available, and for comparison between simulations and measurements. Furthermore, integral quantities such as drag or lift coefficients can be derived from the flow field data, leaving details of the flow information unconsidered. But the trend is to analyse flow field data in more detail, and if one wants to *understand* the physical behavior of the flow, pure columns of numbers are not very useful. Here visualization offers the possibility for a detailed exploration of flow field data. By using the visual pattern recognition facility of the human brain, visualization can provide a deeper insight into the dynamical behavior of fluid flows. Therefore, visualization is an important research tool for improving our physical understanding of fluid flows and for communicating these physical insights within the scientific community [162].

8.3.2 Data Visualization Methods

The field variables that are of interest in fluid dynamics, such as pressure, temperature, velocity and stress, are quite different with respect to their information content, since they include scalar, vector and second-order tensor quantities. Going from scalar to vector and from vector to tensor fields, there is an increase of information contained in the data, and its visualization becomes more and more sophisticated. Accordingly, there is a need for different visualization techniques depending on the kind and amount of information that has to be visualized.

The following list is a compilation of methods suitable for the visualization of scalar, vector and second-order tensor fields. It is certainly not complete, but it is thought to contain the most important techniques used for flow field visualization. For details about the methods we refer to Sections 6.4 and 6.5 in this book and to the cited references. A survey of visualization methods frequently used in fluid dynamics is also given in [555].

- Visualization of scalar fields
  - Contour plots [589, 99]
  - Isosurfaces [452]
  - Volume rendering [433]
  - Color coding (see Section 6.5.3)
- Visualization of vector fields
  - Arrow plots (see Section 6.5.1)
  - Glyphs [426]
  - Line integral convolution (LIC) [103]
  - Particle tracing
    - Basic trace types: stream lines, streak lines, path lines, time lines [23, 418]
    - Variants: lines, bands [51], tubes [177]; time surfaces [649]; stream surfaces [347]; flow volumes [475]
  - Vector field topology [316]
- Visualization of second-order tensor fields
  - Hyperstreamlines [165]

It is worth noticing that some methods for vector field visualization such as glyphs, stream bands, stream surfaces and flow volumes are also capable of displaying information about tensor quantities (for instance shear). Depending on the application this may be sufficient, but the more specialized methods for tensor field visualization are more powerful.

Whatever method is used for visualization, special attention has to be paid to the accurate and careful interpretation of the visualization results. One always has to bear in mind what kind of information is displayed by a specific visualization method, otherwise the obtained images can be misleading. For example, stream lines are well suited for depicting the fluid's path of motion in a steady flow, but applied to an unsteady flow only the local flow directions are displayed. The same is true for other methods generating a representation of the instantaneous velocity field (e.g. LIC, arrow plots). Other visualization methods have their specific limitations as well. Nevertheless, they can be useful even in cases where their limitations take effect, as long as the restrictions are taken into account during interpretation.

8.3.3 Simulated Flow in a Melting Crucible

The first application example we want to discuss is the flow in a melting crucible filled with pure silicon. The crucible is part of an arrangement used for the synthesis of monocrystalline silicon according to the Czochralski process, which is the most popular method in the semiconductor industry for producing large crystals for electronic and photonic devices. In this process a rotating seed crystal is placed on top of the melt's surface so that an almost cylindrical single crystal starts growing and can be dragged from the melt. The quality of this crystal depends on many parameters, in particular on those determining the silicon flow in the crucible. In order to investigate the influence of different parameters on the flow, three-dimensional time-dependent simulations of the turbulent flow and heat transfer in the crucible have been performed within the scope of a research project [45, 209]. The calculation considered here is a direct numerical simulation using a very fine discretization in space and time. Thus, we can expect that very small flow structures are resolved by the simulation.

For visualization we have selected a data set representing the flow field after a simulated time of two minutes, with zero velocity as initial condition. Besides Coriolis and centrifugal forces, the flow in the crucible is driven by buoyancy forces due to the temperature gradients in the heated melt. From the contour plot of the temperature field in Figure 8.19, it can be seen that there is an increase in temperature from the inner top of the crucible to the outer bottom. The reason for this temperature profile is twofold. First, the crucible is heated radiatively from the sides and the vertical heating elements extend below the crucible. This causes that the highest temperature is reached at the outer bottom. On the other hand, the cylindrical crystal is placed at the inner top of the crucible so that the fluid is cooled down there to allow crystallization of the silicon. The diameter of the single crystal is almost 60 percent of the crucible diameter, and its 'fingerprint' can be seen clearly in the contour plot.

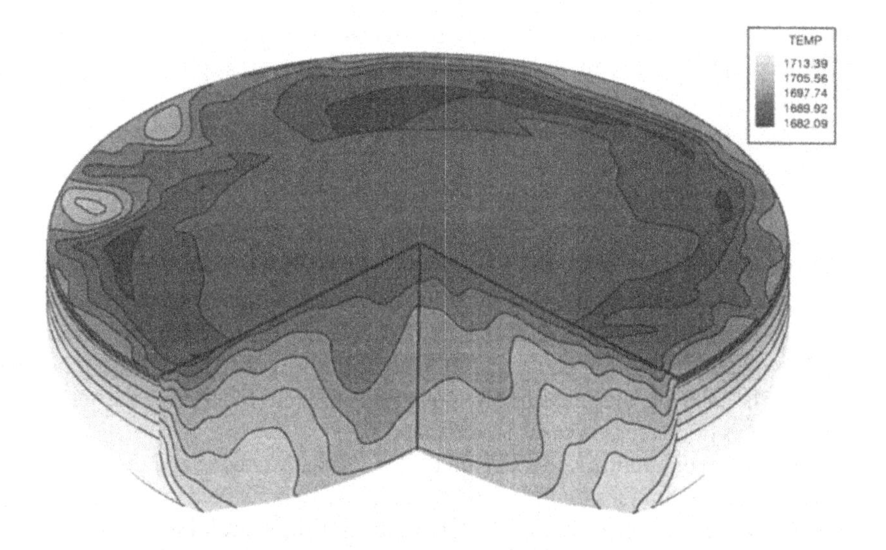

Figure 8.19. Contour plot of the temperature field in the crucible.

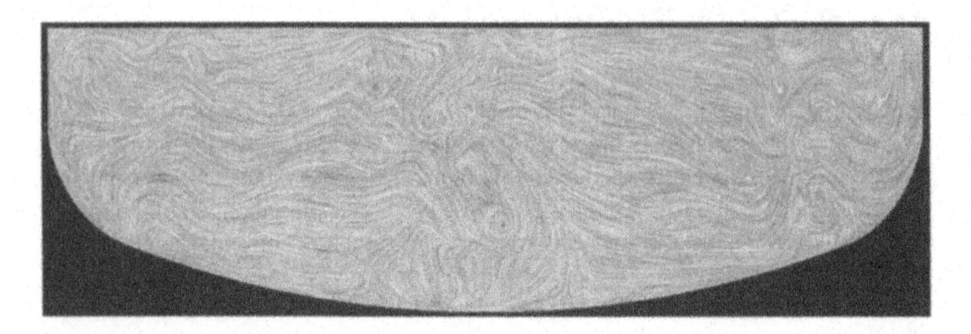

Figure 8.20. Line integral convolution applied to the turbulent flow in the crucible: LIC of the velocity field in a vertical cutting plane.

We can get an idea of the fluid flow from the LIC image in Figure 8.20, which is a visualization of the velocity field in a vertical cutting plane (compare Section 6.5.1 for a description of the method). It gives a very detailed impression of the instantaneous velocity field parallel to the plane without regarding the velocity component perpendicular to the plane or the evolution of the velocity field in time. However, it can be observed that there are very small vortical structures in the flow, as we had to expect for the direct numerical simulation. To take a closer look at the flow field, a magnified clipping of the cutting plane is depicted in Figure 8.21. Three different visualization techniques have been applied to the velocity field: LIC, arrow plot, and stream lines restricted to the cutting plane. Again, the LIC image shows very clearly the structures in the flow, whereas the magnitude and direction of the velocity cannot be determined. On the other hand, the arrow plot is particular good at depicting the magnitude and direction of the velocity components parallel to the plane, whereas the flow structures cannot be recognized as good as in the LIC image. Hence, combining LIC and arrow plot can give a good idea of the instantaneous velocity field in the cutting plane.

The stream line plot in Figure 8.21 is also capable of displaying the main flow structures in the plane, but it is much less detailed than the LIC image. This is because only a few stream lines can be used in order to avoid cluttering of the plot. Since the starting points of the stream lines have to be selected manually, small flow structures are only found by chance and can be missed easily.

Taking a closer look at the arrow plot of the velocity and the contour plot of the temperature in Figure 8.21, the interaction between both fields can be understood. Most clearly the upwards moving stream along the right wall of the crucible is caused by buoyancy due to the high temperature in this region compared to the surrounding, confirming that the flow is partly driven by temperature gradients. On the other hand, the flow is carrying thermal energy along and is likely to deform the temperature contour lines in this region to an S-like shape.

Concluding Figure 8.21, this example shows the advantages and disadvantages of three different techniques for the visualization of the velocity field in a cutting plane. In order to extract more information from the data, it is often useful to combine different visualization techniques revealing different information about the field quantities and their interaction. Furthermore, it is found that the two-dimensional cutting plane

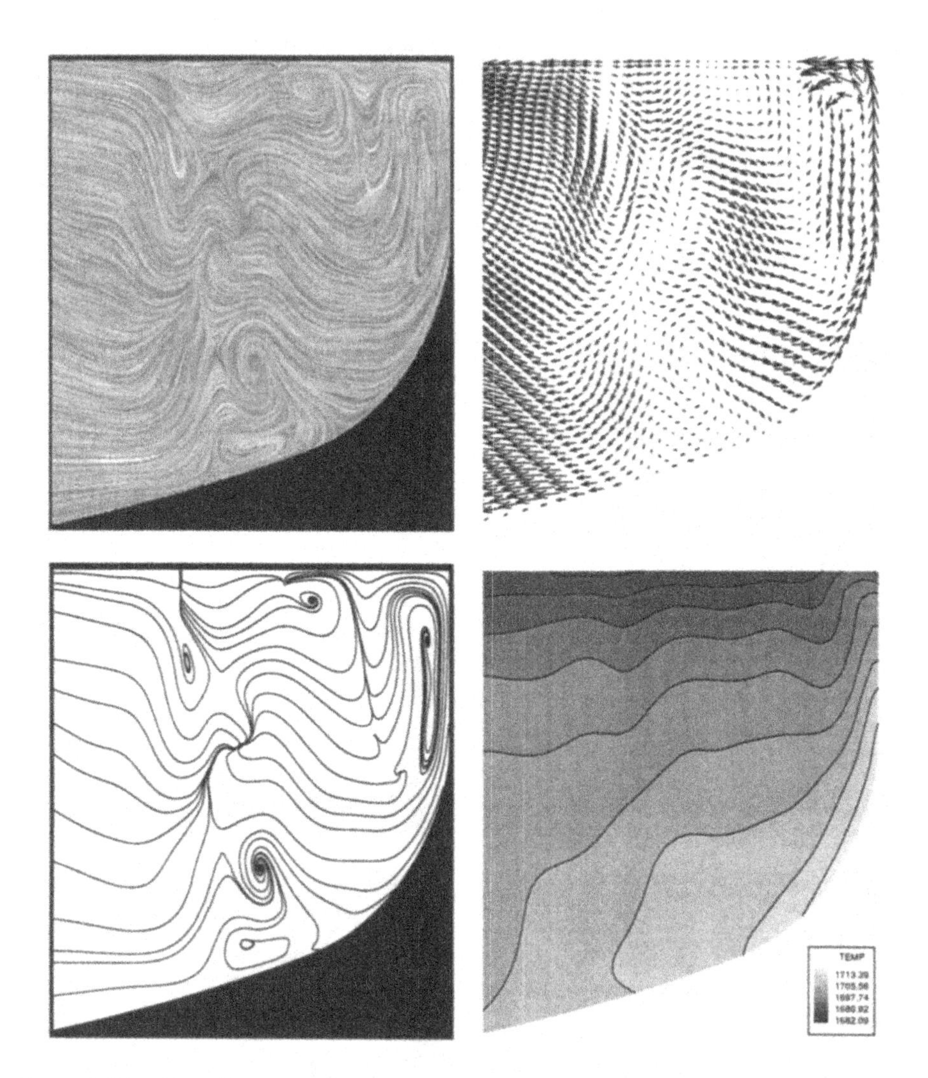

Figure 8.21. Visualization of the velocity and the temperature field in the crucible. The region shown is a magnified clipping of the cutting plane in Figure 8.20. Top left: LIC of the velocity field. Top right: velocity field depicted as an arrow plot. The arrow length is proportional to the magnitude of the velocity component parallel to the plane. Bottom left: 2D stream lines restricted to the cutting plane. Bottom right: contour plot of the temperature.

visualization is able to resolve very small vortical structures in the three-dimensional flow. Similar investigations can be made to examine the influence of the velocity component perpendicular to the cutting plane and the evolution of the vortical structures in time. This can be done by using several cutting planes with data sets representing different time steps. However, there are more specialized methods for the visualization of time-dependent flows in three space dimensions, as will be shown in the next example.

8.3.4 Simulated and Measured Flow in a Stirred Vessel

Impeller stirred reactors play an important role for production processes in the chemical and process engineering industry. In order to permit reliable process optimization, a lot of experimental and numerical research studies were carried out to increase the understanding of the mixing process in stirred vessels. Owing to several problems encountered in experimental investigations, CFD has gained increasing interest during the past few years.

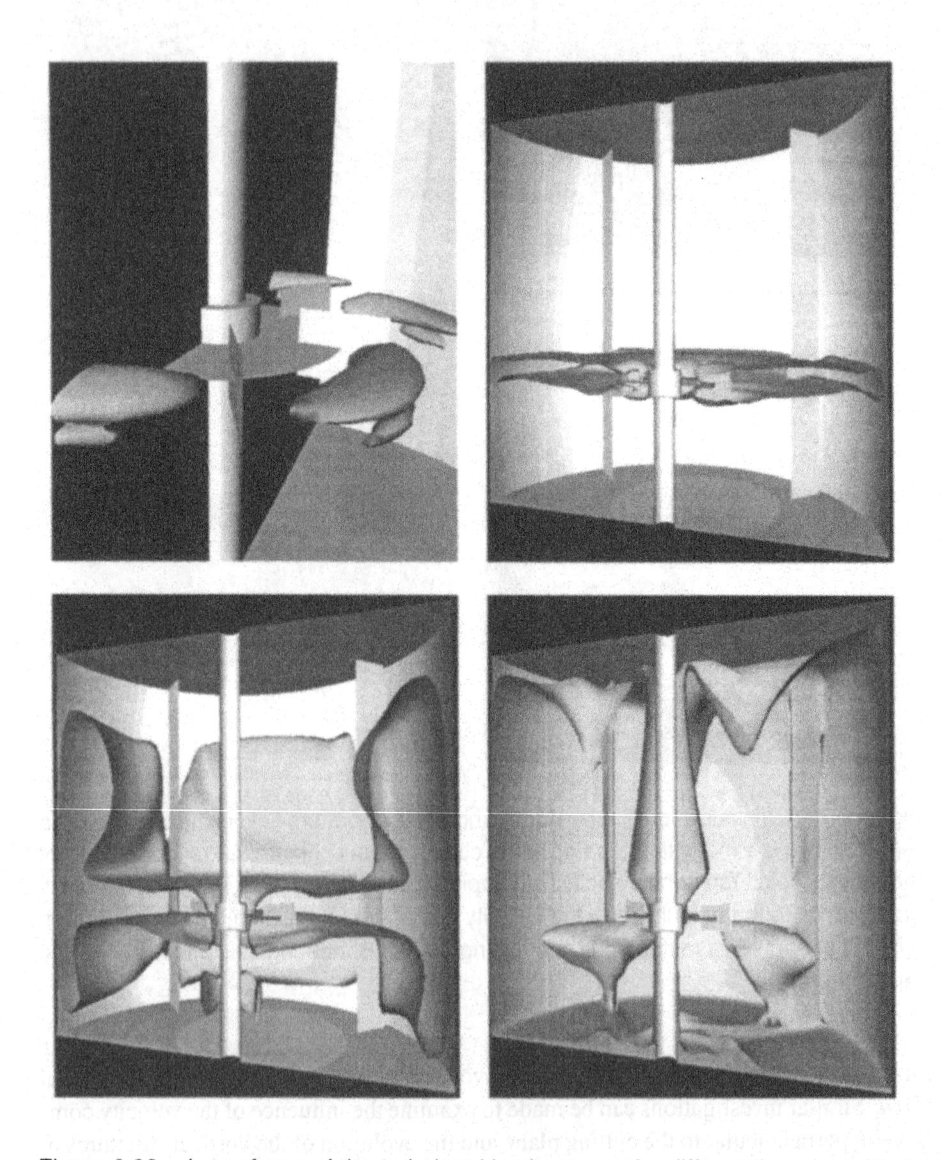

Figure 8.22. Isosurfaces of the turbulent kinetic energy for different isovalues (4, 1, 0.25, 0.1 J/kg, from top left to bottom right, respectively). Flow in a vessel stirred by a Rushton turbine (only half of the vessel is depicted).

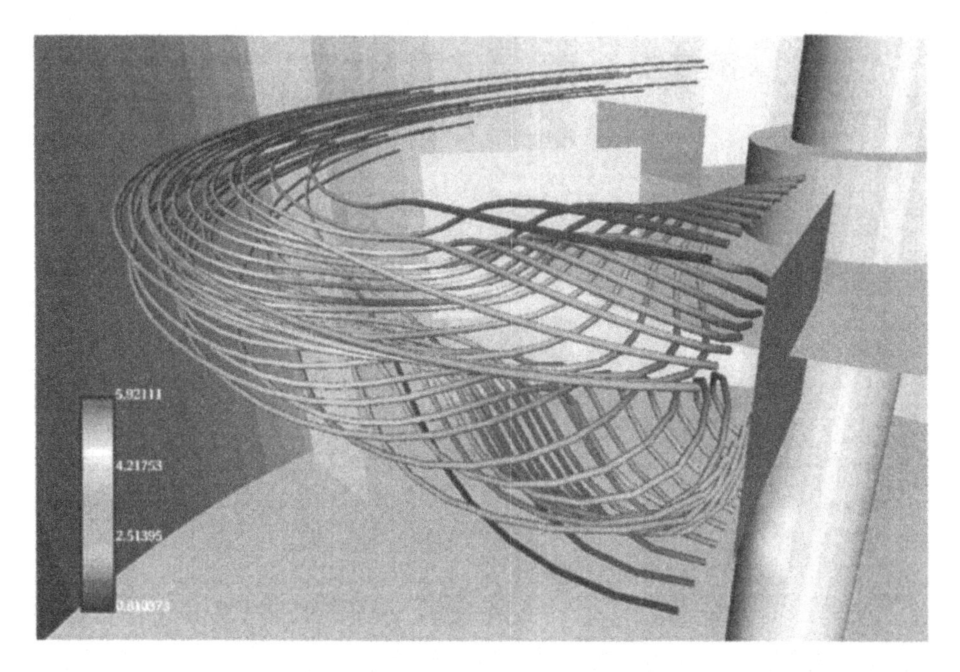

Figure 8.23. Streak lines starting behind a blade of the Rushton turbine (the impeller is rotating counterclockwise). Color coding has been used to colorize the lines according to the turbulent kinetic energy at the particle positions. The lines are represented by pipes in order to obtain a three-dimensional depth cue.

The flow field in a stirred vessel is complex and three-dimensional with vortical structures and high turbulence levels in the vicinity of the impeller requiring large computational efforts. However, based on efficient numerical algorithms and fast supercomputers, it is nowadays possible to predict the flow in the vessel accurately. Within a joint European research programme detailed numerical and experimental investigations have been carried out for different impeller types [603, 604, 739]. The vessel considered here for visualization is stirred by a Rushton turbine, i.e. the impeller consists of a disc carrying six vertical blades.

The underlying numerical simulations of the vessel flow are three-dimensional and time-dependent. They are based on the Reynolds-averaged Navier-Stokes equations combined with a statistical two-equation turbulence model, which does not resolve the small scale vortices but introduces the so-called turbulent kinetic energy and the dissipation to represent their kinetic energy content and their dissipation rate, respectively. Since turbulence comes along with a very efficient mixing of the fluid, the turbulent kinetic energy is of great interest for the investigation of the flow in the stirred vessel. Isovalue surfaces of constant turbulent kinetic energy are a very powerful tool for exploring the turbulence distribution in the vessel. Figure 8.22 shows a sequence of isosurfaces for different isovalues of the turbulent kinetic energy. As it can be seen from the first image, the highest values occur at a short distance behind the blades of the Rushton turbine. Furthermore, the distribution of the turbulent kinetic

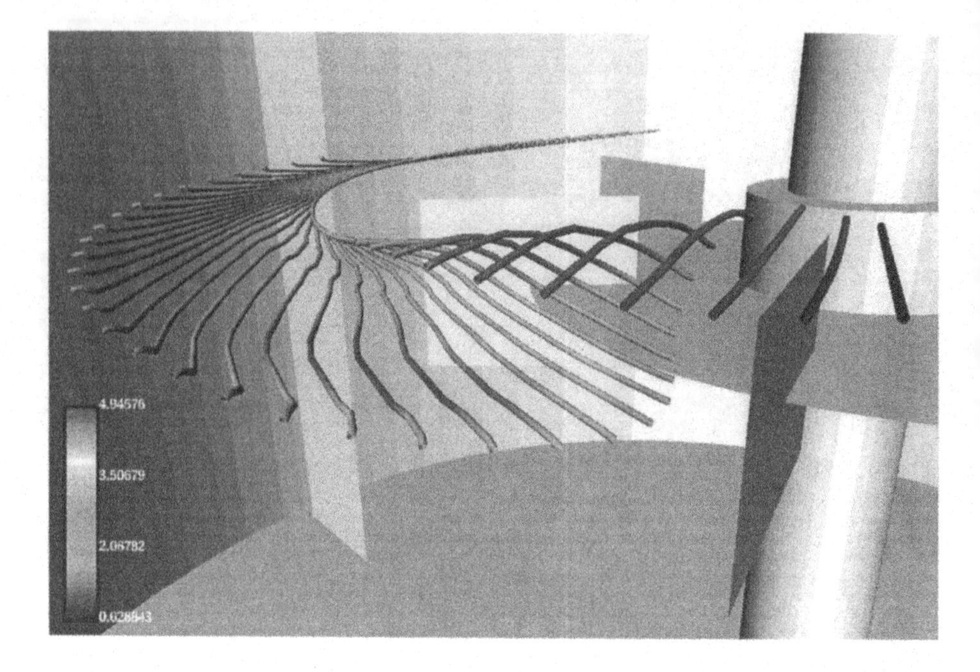

Figure 8.24. Time lines starting in front of a blade of the Rushton turbine. Color coding according to the turbulent kinetic energy.

energy is clearly asymmetric with respect to the horizontal impeller plane. Below the plane there is only a small volume enclosed by the isosurface, indicating that the turbulence level is lower than above the plane. Going to a four times lower value of the turbulent kinetic energy, the isosurface encloses almost the complete vessel cross-section in the height of the impeller. Just the innermost region between the blades and the rotating axle is spared out. Decreasing the isovalue by a factor of four again, the isosurface extends upwards and downwards along the vessel walls. Only for even smaller isovalues the isosurface reaches out into the outer regions of the vessel volume. As a result, the isosurface sequence shows that the turbulent fluid motion is concentrated more or less in the horizontal plane around the Rushton turbine.

An investigation of the three-dimensional time-dependent velocity field can be performed by using a particle tracing technique such as streak lines or time lines. Streak lines are generated by starting particles at each time step at some fixed seed positions and connecting all particles coming from the same seed position. Since this visualization method is time-dependent, animations are most powerful. However, for animations we must refer to our web-pages (see Section 8.3.5), since here we can only show single snapshots depicting the streak lines at one instant in time. This is the case in Figure 8.23, with the seed positions located behind an impeller blade. In order to investigate the vortices being dragged along with the blade, the seed positions are fixed with respect to the blade, i.e. they are rotating with the same angular velocity. The streak lines reveal two vortices behind the blade, one above and one below the impeller plane. This vortical region is restricted to a relatively small distance behind

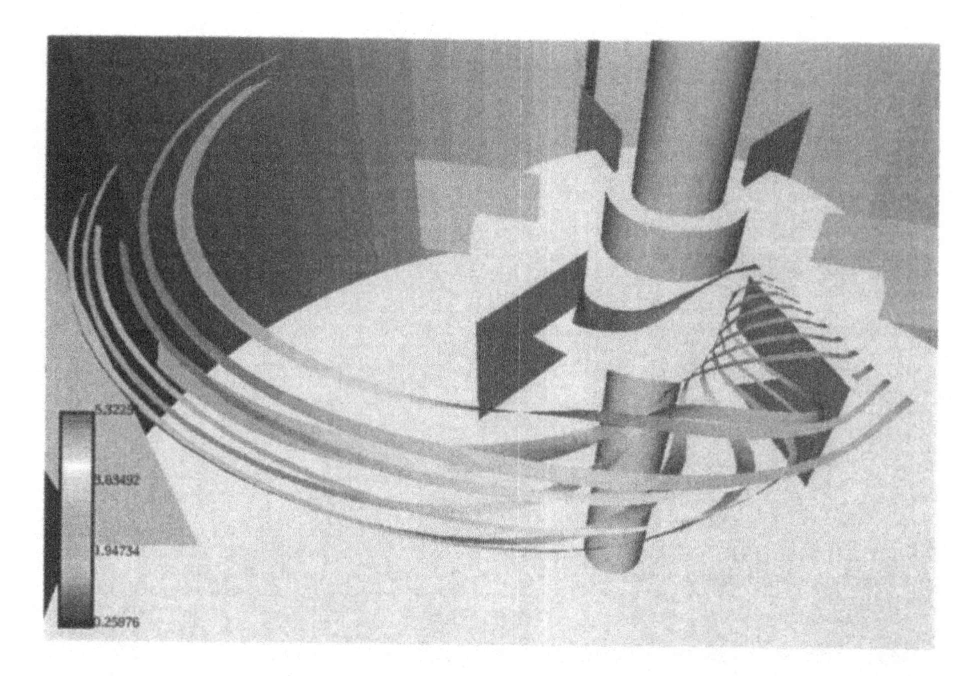

Figure 8.25. Streak bands starting in front of a blade of the Rushton turbine. Color coding according to the turbulent kinetic energy.

the blade. In addition, it can be seen that the streak lines are converging towards the impeller plane when passing through the vortices.

To take a more detailed look at the vortex region trailing the blade, a time line visualization is shown in Figure 8.24. Time lines are constructed by connecting the particles released at the same time step. To avoid cluttering of the image we restrict ourselves to the upper vortex by starting the particles only above the impeller plane. In the vortex region the time lines are stretched and rotated while their front end is turned to the back and the back end to the front. Furthermore, the time lines are moving vertically towards the impeller plane. We will come back to this vertical movement later, since it is part of a large scale vortex in the vessel.

Another possibility to reveal rotation in the flow is shown in Figure 8.25. The streak bands in this image are contructed by connecting two neighboring streak lines. When passing through the vortex behind the impeller blade, the bands are twisted and stretched, in accordance with the time line visualization in Figure 8.24. Moreover, from this perspective it can be seen very clearly that the particles are moving in a spiral outwards to the vessel wall after leaving the vortex region.

Besides the small vortices trailing the blades also a large scale vortical motion exists in the vessel. As has been shown in Figures 8.23 and 8.24, the fluid is moving vertically towards the impeller plane when passing through the vortices behind the blades. In addition, Figure 8.25 has demonstrated that there is a trailing spiral behind the vortex region. Both movements are part of a large scale flow, which can be seen in the LIC image in Figure 8.26. This is a velocity field visualization in a vertical cutting

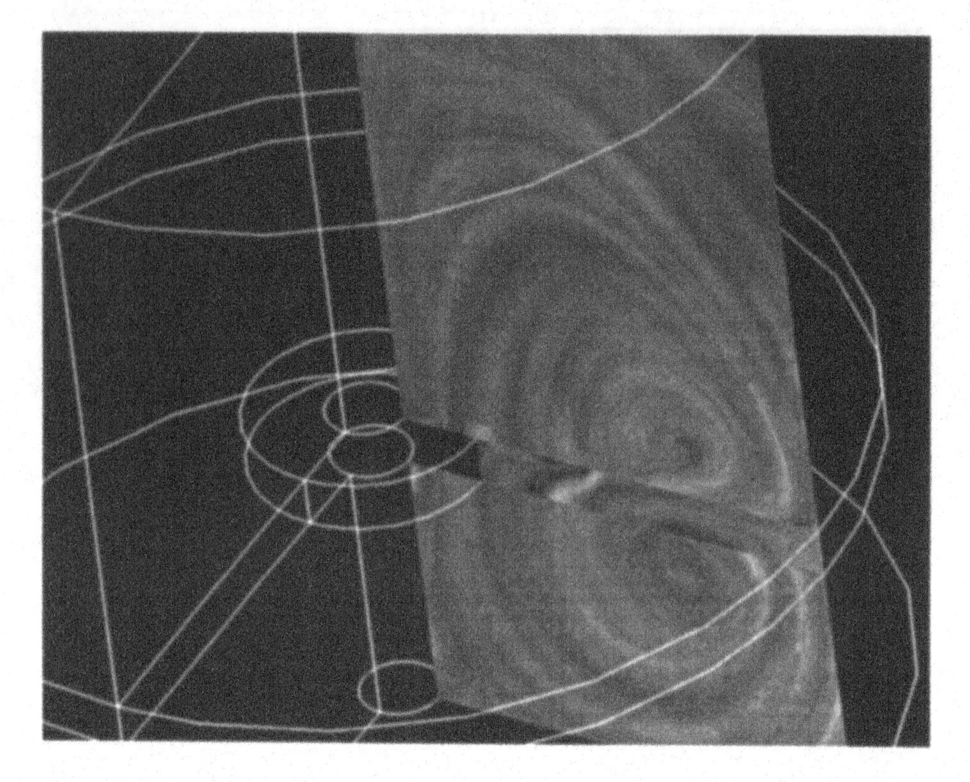

Figure 8.26. Line integral convolution applied to experimental velocity data obtained for the stirred vessel [689]. The white lines are representing the vessel geometry. The impeller is indicated by the white circles next to the gap in the LIC image. There are discontinuities between the middle and the adjacent top and bottom parts of the LIC image, because the LIC algorithm has been applied to the three parts independently.

plane of the vessel based on experimental data. The velocity has been measured using laser-Doppler anemometry. There are two large vortices filling the vessel volume, one above the impeller plane and one below. These vortices can be found in CFD data as well.

To conclude this example, the isosurface visualization has proved to be powerful for the investigation of the spatial distribution of some scalar field in the vessel. Some main flow structures have been revealed by using particle tracing methods in combination with line integral convolution applied to CFD data and experimental data, respectively. However, we want to emphasize that time-dependent visualization methods such as streak lines, streak bands and time lines are most powerful when they are animated.

8.3.5 Online Resources

For the stirred vessel example discussed in Section 8.3.4, video sequences are available in the world wide web (http://www.lstm.uni-erlangen.de/SFB603_C3/applications/). All images presented in this section can be found there as well, a colored version is provided where available. Moreover, links have been added to lead you to further material created by our and other research groups.

Acknowledgements

Without data there is no data visualization. We would like to thank the following people for making the data sets available to us which have been visualized in this section. The crucible simulation has been performed by Sven Enger at the Institute of Fluid Mechanics (LSTM), University of Erlangen-Nürnberg. Klaus Wechsler at LSTM provided the stirred vessel simulation data, which have been generated within the scope of the BRITE EURAM project "Chemical Reactor Modeling for Fast Exothermic and Mixing Sensitive Reactions". The experimental stirred vessel data have been provided by Marcus Schäfer at LSTM.

8.4 DIAGNOSIS SUPPORT OF PATIENTS WITH FACIAL PARESIS

A. GEBHARD, D. PAULUS

Computers are used more and more in medical applications. Well-known examples are the analysis of radiographs or MR images [536, 445]. A special field is the analysis of human faces and facial features. Already realized applications exist e.g. for face recognition [89, 90] or face analysis [594, 317].

With the presented system we consider the problem of diagnosis support of patients with facial paresis. Approximately 350 patients per year are registered in the Department of Otorhino-Larygology of the University of Erlangen with new occurrences of this type of paresis (cf. [763] as an overview). The current way to diagnose the paresis is a subjective judgement of the functionality of the face muscles by a physician. The patient performs mimic exercises such as closing the eyes or showing the teeth, while he is observed by the physician. The subjective observations of the physician are then graded by means of two medical indexing systems [343, 671].

One part of the rehabilitation of the patients is to perform specific mimic exercises with the face's musculature. The success of the therapy is also observed by a physician or educated clinic personal. For every observation the patient has to travel to the clinics. Furthermore, educated personal is needed for the diagnosis and rehabilitation observation of the paresis.

The applications of our system are on the one hand diagnosis support in the clinics. We want to improve the subjective judgments of a physician by objective measurements and numerical features from the face. On the other hand the supervision of the rehabilitation process of the patient will be enhanced by placing the system to the patient's home. The patient can use the system in a convenient environment and does not have to travel to the clinics. The patient does *not* have to pay for the more convenient observation environment by wearing any artificial markers inside the face and he is allowed to move in front of the system's camera.

We present two different approaches for the analysis of facial features which will measure the face during the performance of mimic exercises. The result is a value for every picture with the level of asymmetry of the eye and mouth region which can be used to classify the facial paresis. The asymmetry of the face can be used as we handle just patients with paresis in one half of the face. This class of patients is the most frequent class. Patients with double sided pareses are approximately 1% of the face paresis patients.

This contribution is organized as follows: In Section 8.4.1 we compare the use of different data sources (2D vs. 3D images). A survey about related work on the field of facial image processing is given in Section 8.4.2. The localization of facial features will be described in detail in Section 8.4.3. It is followed by the description of the analysis approaches (see Section 8.4.4). In Section 8.4.5 we show how the analysis results can be used to classify the facial paresis inside the observed regions. Results are shown in Section 8.4.6. Finally, Section 8.4.7 will recapitulate the contents of this paper with the main results in a short way.

8.4.1 Sensor Selection

The analysis of facial images can be performed using either the frontal view in a 2D projection, or taking into account the 3D structure of the head. The first approach can be motivated by human perception. For human interaction and human interpretation of mimics, 2D information seems to be sufficient. We can clearly tell whether a person smiles, frowns, or whether his eyes fixate ours, even when we close one eye. Paralyzed parts in the face are less perceived when they are in the lateral part of the cheeks. In the following we concentrate on 2D images and use a model of the projected face.

Three-dimensional information on the face is available when stereo information is taken into account. However, it is disparity is difficult to estimate for facial images, since these images exhibit neither significant texture—except for areas with facial hair—for area based matching, nor many lines or prominent points. Traditional stereo algorithms on the face thus result in sparse or coarse depth data.

8.4.2 Related Work

In this section we give a survey of related areas of the field of face image processing.

Face Localization and Tracking. One of the basic components of many systems which process face images is a face localization module. Different approaches can be found in the literature. In the following we present exemplary two different approaches.

One approach was introduced by De Silva et al. [651]. A person is expected to sit in front of a homogenous background. A gray-level image of the head and the shoulders is captured. The edge strength representation of the image (cf. Section 3.1) is used to find an elliptic region containing a high amount of edge strength which is supposed to be the face.

Another approach was introduced by Oliver et al. [520] and bases on the segmentation of *blobs*. A blob is a "compact set of pixels that share a visual property that is not shared by the surrounding pixels" (see [520, p. 123]), and is thus a special kind of region segmentation (see Section 3.1.2). Oliver et al. use the normalized color as the visual property of a pixel. Every image coordinate is combined with color and brightness information by generating a vector $(i, j, \frac{r}{r+g+b}, \frac{g}{r+g+b})^T$. A model for skin color was trained by thousands of samples that is valid for a broad spectrum of users. With an adaptive strategy, face color regions are grown on the complemented image data. The classification accuracy is close to 100%.

Face Recognition. There are two major classes of face recognition operations. On the one hand there are methods using geometrical features or template-matching and on the other hand there is the processing on grey-level information.

The techniques basing on features or templates where analyzed in detail by Brunelli and Poggio [91]. They show the extraction of features or the template-matching with the goal of recognition. The results achieved by template-matching were better.

A prominent technique which bases on the processing of grey-level information is the *Eigenface* approach (e.g. [717]). The basic idea of the Eigenface approach is to encode a face image, and compare one face encoding with a database of models encoded similar. The encoding is done in the following way: A face image of size $N \times N$ is interpreted as a point in an N^2-dimensional space. A set of training images of faces with similar overall configuration (e.g. face as the dominant image region) can be described by a relatively low dimensional subspace of the huge image space. Principal Component Analysis used with different images of one person gives vectors that best account for the distribution of face images within the entire image space. Those vectors are called *eigenfaces* (comp. the *eigenspaces* approach in Section 3.4.3).

For recognition, a new face image is transferred into its eigenface components and assigned to the eigenface with minimal distance.

Face Coding. The principal component representation of face images can also be used to encode faces, transmit the encoded information via a relative slow line, and decode and view then at the destination. Moghaddam and Pentland [485] propose this proceeding for the video telephony task.

Another model-based approach for this task was presented by Tao and Huang [684]. A basic articulation model can be influenced by articulation parameters such as facial action coding system (FACS) parameters of MPEG-4 facial animation parameters (FAPs).

Medical Applications. In [318] a system for the automatic diagnosis of craniofacial dysmorphic signs is presented. Different multi-layer perceptrons were trained with a training set of 31 images using back-propagation as learning algorithm. The classification of a face whether craniofacial dysmorphic signs were present or not was done with a correct classification rate of 95%.

8.4.3 Localization and Tracking of Faces and Face Features

In the application presented in this contribution facial asymmetries are judged by means of asymmetries in the eyes, nose and mouth regions. The approaches we chose for the analysis of the facial paresis base on the localization of the mentioned facial features and of their surrounding (in case of the eyes the surrounding covers eyebrows and zygomatic). In our system the localization is done with a parametric face model, which is fitted to the image data by an energy-minimization process.

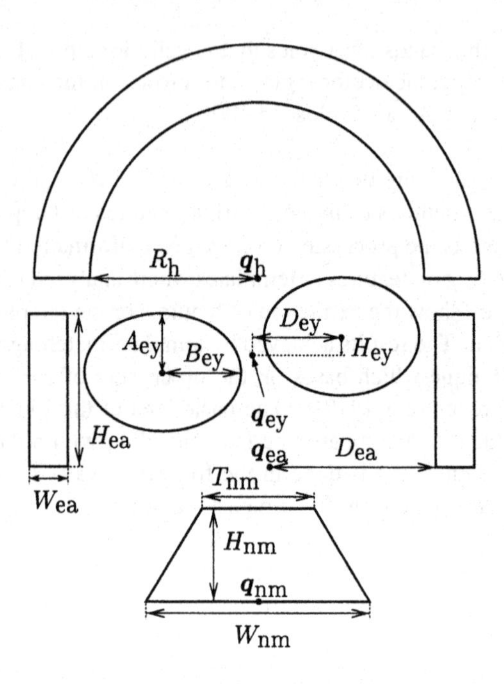

Figure 8.27. Parametric Face Model. All shown parameters are optimized during face localization.

Face Model. We assume that the patient's face is the dominant region inside the image. The background is expected to be either homogeneous or a background image is captured prior to the analysis which is used to part foreground from background. The localization is executed by calculating parameters of the face model shown in Figure 8.27. All calculations are performed on the edge strength part e_s of a Sobel filtered image $\boldsymbol{f}$ (see Section 3.1).

Localization. The localization is performed as a four-step process. The *first* step is to localize the upper arc of the head. A ring segment with fixed width C_R is found in the image which is expected to be the top of the patient's head. There are three parameters which have to be estimated: the x-coordinate x_h and the y-coordinate y_h of the origin q_h and the radius R_h of the arc. We expect much edge energy between patient and background. That lets us find the parameters of the circle and of the following part of the model in the edge strength representation of the original image. Equation 8.5 gives the edge-energy inside the head arc in the image:

$$E_h = \int_{0}^{\pi} \int_{R_h}^{R_h+C_R} e_s(x_h + r\cos\phi, y_h + r\sin\phi)\,dr\,d\phi. \tag{8.5}$$

To get the optimal model parameters the ratio of edge energy inside the head arc and the area of the head arc must be optimized. This optimization is done by the following maximization process:

$$(x_{\mathrm{h}}^*, y_{\mathrm{h}}^*, R_{\mathrm{h}}^*) = \underset{(x_{\mathrm{h}}, y_{\mathrm{h}}, R_{\mathrm{h}})}{\operatorname{argmax}} \frac{E_{\mathrm{h}}}{\left(\frac{(R_{\mathrm{h}}+C_{\mathrm{R}})^2 - R_{\mathrm{h}}^2}{2}\right)^{C_{\mathrm{h}}}}. \tag{8.6}$$

The constant C_{h} influences the energy to area ratio. The values of the constants are $C_{\mathrm{R}} = 20$ and $C_{\mathrm{h}} = 1.4$.

The optimization is performed by an adaptive random search with a subsequent local simplex method [210]. To speed up the optimization there are restrictions for plausible parameters: The parameter x_{h} can vary form $N_{\mathrm{y}}/4$ to $3N_{\mathrm{y}}/4$, y_{h} form $M/4$ to $M/2$, and R_{h} from $N_{\mathrm{y}}/10$ to $N_{\mathrm{y}}/3$.

The *next* step is to localize the ears which are modeled as rectangles positioned below the arc of the head. The parameters which have to be determined are the position of the origin q_{ea}, the width W_{ea} and height H_{ea} of the ears, and the distance $2D_{\mathrm{ea}}$ between the two ears. The equation to calculate the edge-energy inside the ears' region is

$$\begin{aligned} E_{\mathrm{ea}} &= \int_0^{W_{\mathrm{ea}}} \int_0^{H_{\mathrm{ea}}} \{e_{\mathrm{s}}(x_{\mathrm{ea}} + w + D_{\mathrm{ea}}, y_{\mathrm{ea}} + h) \\ &\quad + e_{\mathrm{s}}(x_{\mathrm{ea}} + w - D_{\mathrm{ea}}, y_{\mathrm{ea}} + h)\} dh dw \end{aligned} \tag{8.7}$$

and similar to Equation 8.6 the calculation of the optimal ear parameters bases on the ratio of edge energy inside the ears regions and the area of this regions. They can be determined as

$$(x_{\mathrm{ea}}^*, y_{\mathrm{ea}}^*, D_{\mathrm{ea}}^*, W_{\mathrm{ea}}^*, H_{\mathrm{ea}}^*) = \underset{(x_{\mathrm{ea}}, y_{\mathrm{ea}}, D_{\mathrm{ea}}, W_{\mathrm{ea}}, H_{\mathrm{ea}})}{\operatorname{argmax}} \frac{(E_{\mathrm{ea}} D_{\mathrm{ea}})^{C_{\mathrm{ea}}}}{(W_{\mathrm{ea}} H_{\mathrm{ea}})} \tag{8.8}$$

with the ratio influencing constant $C_{\mathrm{ea}} = 1.4$. Additionally the distance between the left and right ear region is involved in the optimization as they are used as the horizontal boundaries of the face and therefore the distance D_{ea} should be as big as posible.

There are anatomic restrictions for the ear parameters. The origin of the ears q_{ea} must have a lower horizontal distance than $R_{\mathrm{h}}/3$, the vertical position of q_{ea} must be below the origin of the head arc q_{h}, but with a lower distance than $R_{\mathrm{h}}/2$. D_{ea} must be between $0.9R_{\mathrm{h}}$ and $N_{\mathrm{y}}/3$.

The eyes, which are found in the *third* step, are modeled as ellipses. The parameter $q_{\mathrm{ey}} = (x_{\mathrm{ey}}, y_{\mathrm{ey}})^T$ is the position of the origin of the eyes, A_{ey} and B_{ey} the length of the ellipses axis, $2H_{\mathrm{ey}}$ the vertical and $2D_{\mathrm{ey}}$ the horizontal distance of the eye centers to each other. We use the following equation to calculate the edge energy inside the eyes' regions:

$$\begin{aligned} E_{\mathrm{ey}} &= \int_0^{2\pi} \int_0^{1} e_{\mathrm{s}}(x_{\mathrm{ey}} + A_{\mathrm{ey}} r \cos\phi + D_{\mathrm{ey}}, y_{\mathrm{ey}} + B_{\mathrm{ey}} r \sin\phi + H_{\mathrm{ey}}) \\ &\quad + e_{\mathrm{s}}(x_{\mathrm{ey}} + A_{\mathrm{ey}} r \cos\phi - D_{\mathrm{ey}}, y_{\mathrm{ey}} + B_{\mathrm{ey}} r \sin\phi - H_{\mathrm{ey}}) dr d\phi. \end{aligned} \tag{8.9}$$

With the following optimization we get the eye parameters:

$$(x_{\rm ey}^*, y_{\rm ey}^*, D_{\rm ey}^*, A_{\rm ey}{}^*, B_{\rm ey}{}^*, H_{\rm ey}^*) = \operatorname*{argmax}_{(x_{\rm ey}, y_{\rm ey}, D_{\rm ey}, A_{\rm ey}, B_{\rm ey})} \frac{E_{\rm ey}}{\left(A_{\rm ey}{}^2 + B_{\rm ey}{}^2\right)^{C_{\rm ey}}} \tag{8.10}$$

with a constant $C_{\rm ey} = 1.4$ influencing the ratio of energy to area of the eyes' regions.

The anatomic restrictions here are: The horizontal distance of the origin of the eyes $q_{\rm ey}$ must be less than $R_{\rm h}/3$, the vertical position must be greater than $y_{\rm h}$ but lower than $y_{\rm h} + R_{\rm h}$. $A_{\rm ey}$ and $B_{\rm ey}$ must be lower than $0.4R_{\rm h}$ and $D_{\rm ey}$ lower than $D_{\rm ea}$. The two eye regions must not overlap and the eye regions must not overlap the ears' regions.

Finally the nose/mouth region is to be found. It is modeled as a triangle stump with parameters: origin $q_{\rm nm} = (x_{\rm nm}, y_{\rm nm})^T$, height $H_{\rm nm}$, length of base line $W_{\rm nm}$ and the length of top line $T_{\rm nm}$.

The amount of edge strength in the nose mouth region can be determined by the following equation:

$$E_{\rm nm} = \int_0^{H_{\rm nm}} \int_{-1}^{1} e_{\rm s}(x_{\rm nm} + w(W_{\rm nm} - \frac{h}{H_{\rm nm}}(W_{\rm nm} - T_{\rm nm})), y_{\rm nm} + H_{\rm nm}) dw dh. \tag{8.11}$$

The optimal parameters are found as

$$(x_{\rm nm}^*, y_{\rm nm}^*, W_{\rm nm}{}^*, T_{\rm nm}{}^*, H_{\rm nm}{}^*) = \operatorname*{argmax}_{(x_{\rm nm}, y_{\rm nm}, W_{\rm nm}, T_{\rm nm}, H_{\rm nm})} \frac{(E_{\rm nm})^{C_{\rm nm}} H_{\rm nm}}{W_{\rm nm} + T_{\rm nm}}. \tag{8.12}$$

Here the constant to influence the ratio of edge energy to area is $C_{\rm nm} = 1.2$. The height of the nose/mouth region appears in the numerator of the ratio in Equation 8.12 to avoid that the optimization result is a region that covers just the nostrils or the mouth and not both of them.

The horizontal distance of the origin $q_{\rm nm}$ must be lower than $D_{\rm ey}/3$ from $q_{\rm ey}$. The vertical position must be between $y_{\rm ey} + 2D_{\rm ey}$ and $y_{\rm ey} + 3D_{\rm ey}$. $W_{\rm nm}$ must be between $2D_{\rm ey}$ and $3D_{\rm ey}$, $T_{\rm nm}$ between $D_{\rm ey}$ and $2D_{\rm ey}$, and $H_{\rm nm}$ between $D_{\rm ey}$ and $3D_{\rm ey}$.

All the restrictions to the optimized parameters were imposed because of observations of anatomic facts. The constants $C_{\rm R}$, $C_{\rm h}$, $C_{\rm ea}$, $C_{\rm ey}$, and $C_{\rm nm}$ which were used for the calculations were determined experimentally by localization and tracking of patient faces and facial features in 1000 images.

Tracking. When the face and the facial features are localized in one image (i.e. the parameters of the face model are calculated) the face and features can be found in the following image in approximately the same position as we postprocess a video stream with 25 frames per second and a relatively slow moving patient.

We initialize the simplex optimization with the parameters from image i and a set of parameters which are normally distributed with mean old parameter and variance depending on the expected parameter variation.

That reduces the search area very much and we initialize the simplex optimization with less parameter sets (instead of 5000 we use 500) to get the optimal parameter in a reliable way. The reduction of the initialization set also results in a noticeable decrease of calculation time (instead of 40 sec we need 8 sec per image).

8.4.4 Analysis of Facial Paresis

As written in the introduction we operate with patients with single sided facial pareses. In this case asymmetries can occur in the eye and mouth region when specific mimic exercises are performed. This asymmetries are considered to be symptoms for the present paresis. We will give two different data-driven analysis methods for the analysis of the eye region and mouth region of a human face.

The first mimic exercise is to lift up the eyebrows such that wrinkles will appear on the forehead. In the following we will call this exercise 'frowning'. Depending on the grade of paresis, some patients are not able to lift the eyebrow. This will result in certain asymmetries in the eye/eyebrow-region. Also the second exercise, the closing of the eyes, can generate asymmetries in this face area, as some patient are not able or have severe problems to close their eyes.

The other two exercises tell us something about the patient's mouth region. Asymmetries can arise when patients try to point their mouth or when they show the teeth. We record images of the patient while he is performing the mentioned exercises. To grade the asymmetries we need an additional view, the relaxed face.

Comparison of Intensity Values and Edge Strength. The first attempt is the direct comparison of gray-level values of mouth and eyes. To get a feature of the eye region, we take the gray-values of the left eye, mirror the single lines and match the gray-levels with the right eye by varying the x- and y-coordinate to find the minimal absolute sum D_1^* of the pixel differences.

$$D_1^* = \min_{x_t, y_t} \int_{right\ Eye} |(f(x,y) - f(2(B_{\text{ey}} - D_{ey}) - x + x_t, y - 2H_{\text{ey}} + y_t)| dx dy. \quad (8.13)$$

The absolute sum divided by the area of the right eye (in our face model both eyes have the same size) is taken as feature c_1.

$$c_1 = \frac{D_1^*}{(A_{\text{ey}} B_{\text{ey}})}. \quad (8.14)$$

To analyze the mouth region we find the row index r_{min} inside the mouth region that will give the minimal sum of absolute differences when matching the pixels on the left of row r_{min} with those on the right side. Row r_{min} represents the vertical symmetric axis of the mouth which is the elongation of the nasal labial fold

$$D_2^* = \min_r \int_{Mouth\ left}^{r} \int_0^{H_{\text{nm}}} |(f(x,y) - f(2r - x, y)| dx dy. \quad (8.15)$$

D_2^* is devided by the area of the analyzed region to get a feature for the asymmetry of the mouth region.

$$c_2 = \frac{D_2^*}{r_{min} H_{\text{nm}}}. \quad (8.16)$$

To keep the following more predictable: Odd-indexed features c_{2i+1} result from comparisons of the eye regions, even-indexed features c_{2i} result from the mouth region. As mentioned before interesting areas inside the face are often regions where changes of the gray-values occur. Such regions appear emphasized in the edge-strength representation of the image. For that reason we additionally applied the methods not only on the gray-levels f but also on the edge-strength part e_s of the Sobel-filtered image. That gives us the next two features c_3 for the eye region and c_4 for the mouth region.

Using Averaging Filter Responses for Face Analysis. The second class of approaches for the analysis of facial asymmetries arises from the theory of steerable filters [770]. We use the response of averaging rotated wedge filters to characterize the direction information in the environment of certain key points (see Figure 8.28). The key points here are the corners of the eyes and the mouth and the extracted information contains the opening angle of those facial features (see Figure 8.29). The disadvantage of this approach is that the positions of the angles of the eyes and the mouth have to be determined as exact as possible. This additional localization is of course another source for errors. The localization of these features can be very hard even for a exercised person and it is often the case that just an unprecise estimation can be performed.

To get an estimate for the position of the angles of eyes and mouth we use the columns of the surrounding boxes of eyes and the mouth. We noticed that the angles often appear noticeable darker than the surrounding intensity values. The search for the angle positions is started at the columns of the outer borders of the localized facial feature regions (cf. Section 8.4.3). We calculate the average intensity value of one

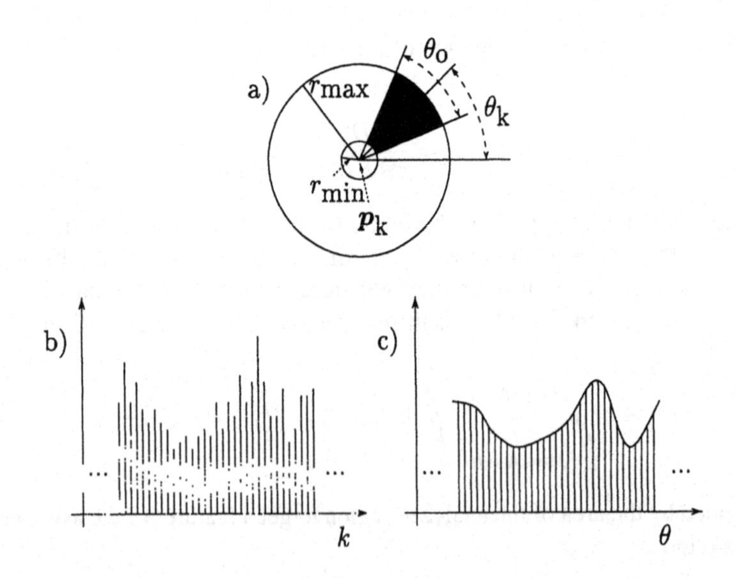

Figure 8.28. a) an averaging mask centered at key point p_k, b) the response of the individual wedge filters, c) the reconstructed (Gaussian smoothed) filter response.

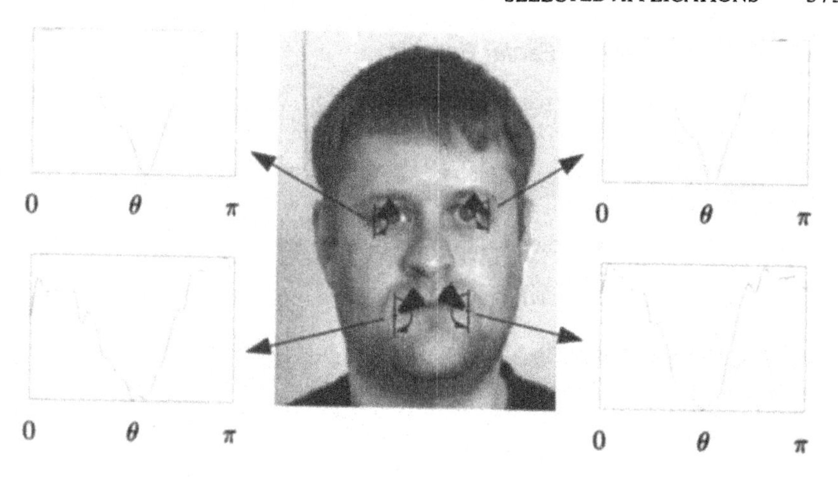

Figure 8.29. Smoothed responses of the wedge filters applied to the angles of the eyes and the mouth.

column, and compare the minimum of the column to it. If the quotient is below a threshold θ_e of θ_m, we stop the process and consider the angle as found. This simple method gives useful results which are used in the analysis process.

We take the localized eye angles and mouth angles as the key points of the wedge filters. The filters are rotated not the whole 2π but only π over the face feature in 2 degree steps which gives 91 averaged gray-values for every localized four facial feature angle. The opening angle of the wedge is 6 degree. The results are the four vectors $\boldsymbol{w}_i, i = 1, \ldots, 4$ of filter responses with 91 entries each $\boldsymbol{w}_i = [w_{i,1}, \ldots, w_{i,91}]^T$. Feature $\boldsymbol{w}_1$ results from the filtering of the corner of the right eye, $\boldsymbol{w}_1$ from the left eye, $\boldsymbol{w}_3$ from the right corner of the mouth, and $\boldsymbol{w}_4$ from the left corner. The filters are also applied to the edge-strength part e_s of a Sobel-filtered image. That gives another 4 vectors $\boldsymbol{w}_5$ to $\boldsymbol{w}_8$, resulting from the respective facial feature corners. With these eight vectors we can calculate another eight features values to analyze the facial asymmetry. c_5 is the absolute component difference sum of $\boldsymbol{w}_1$ and $\boldsymbol{w}_2$

$$c_5 = \sum_{i=1}^{91} |w_{1,i} - w_{2,i}|. \tag{8.17}$$

The same is done with $\boldsymbol{w}_3$ and $\boldsymbol{w}_4$ to calculate the feature c_6, c_7 with $\boldsymbol{w}_5$ and $\boldsymbol{w}_6$, and c_8 with $\boldsymbol{w}_7$ and $\boldsymbol{w}_8$. The feature c_9 is generated after matching $\boldsymbol{w}_1$ to $\boldsymbol{w}_2$. This is done by translations (t_1) and scalings (s_1) of the vector components of $\boldsymbol{w}_2$

$$c_9 = \min_{s_1, t_1} \sum_{i=1}^{91} |w_{1,i} - s_1 w_{2,i+t_1}|. \tag{8.18}$$

In analogy to the feature c_9 the features c_{10}, c_{11} and c_{12} can be calculated.

8.4.5 *Classification of Facial Paresis*

To classify a face whether a paresis is present or not, we proceed as follows. First we calculate the features c_1 to c_{12} while the person's face is in the following states:

1. relaxed face: all face muscles in a relaxed state,
2. lifting up the eyebrows: the result of this motion are wrinkles on the forehead,
3. closing the eyes: patients with paresis in the eye region have problems to do this,
4. pointing of the mouth,
5. showing the teeth: the last two exercises give information about a potential paresis in the mouth region.

That gives five sets of the twelve parameters each which are used for classification:

1. $c_{1,\text{norm}}, \ldots, c_{12,\text{norm}}$: extracted parameters while the face is in a relaxed state,
2. $c_{1,\text{frown}}, \ldots, c_{12,\text{frown}}$: extracted parameters while frowning,
3. $c_{1,\text{close}}, \ldots, c_{12,\text{close}}$: extracted parameters while closing the eyes,
4. $c_{1,\text{point}}, \ldots, c_{12,\text{point}}$: extracted parameters while pointing the mouth,
5. $c_{1,\text{teeth}}, \ldots, c_{12,\text{teeth}}$: extracted parameters while showing the teeth.

To analyze the facial paresis we are interested in asymmetries which are caused by the performation of the mimic exercises. To subtract the asymmetry information from the extracted features which is caused not by the paresis but other factors like the illumination or anatomic reasons, all the parameters $c_{1,\text{frown}}, \ldots, c_{12,\text{frown}}$, $c_{1,\text{close}}, \ldots, c_{12,\text{close}}$, $c_{1,\text{point}}, \ldots, c_{12,\text{point}}$, and $c_{1,\text{teeth}}, \ldots, c_{12,\text{teeth}}$ are normalized by the parameters $c_{1,\text{norm}}, \ldots, c_{12,\text{norm}}$, which mainly contain asymmetry information not generated by the mimic exercises.

This normalization is done by calculating the ratios

1. $\dfrac{c_{1,\text{frown}}}{c_{1,\text{norm}}}, \ldots, \dfrac{c_{12,\text{frown}}}{c_{12,\text{norm}}}$.
2. $\dfrac{c_{1,\text{close}}}{c_{1,\text{norm}}}, \ldots, \dfrac{c_{12,\text{close}}}{c_{12,\text{norm}}}$.
3. $\dfrac{c_{1,\text{point}}}{c_{1,\text{norm}}}, \ldots, \dfrac{c_{12,\text{point}}}{c_{12,\text{norm}}}$.
4. $\dfrac{c_{1,\text{teeth}}}{c_{1,\text{norm}}}, \ldots, \dfrac{c_{12,\text{teeth}}}{c_{12,\text{norm}}}$.

The four sets of normalized parameters with 12 values each are finally used to detect a facial paresis. Again, the odd-indexed normalized parameters contain information about the eye region, even-indexed normalized parameters about the mouth region. In the present state of the system the normalized features are thresholded to get the information whether facial paresis exists or not. For every set of features we calculate if facial paresis is present. E.g. if $\frac{c_{1,\text{frown}}}{c_{1,\text{norm}}}$, $\frac{c_{1,\text{close}}}{c_{1,\text{norm}}}$, $\frac{c_{1,\text{point}}}{c_{1,\text{norm}}}$, or $\frac{c_{1,\text{teeth}}}{c_{1,\text{norm}}}$ are greater than the thresholds $\theta_{1,\text{frown}}$, $\theta_{1,\text{close}}$, $\theta_{1,\text{frown}}$, or $\theta_{1,\text{close}}$ we consider a paresis in the eye region. If $\frac{c_{2,\text{frown}}}{c_{2,\text{norm}}}$, $\frac{c_{2,\text{close}}}{c_{2,\text{norm}}}$, $\frac{c_{2,\text{point}}}{c_{2,\text{norm}}}$, or $\frac{c_{2,\text{teeth}}}{c_{2,\text{norm}}}$ are over the thresholds $\theta_{2,\text{frown}}$, $\theta_{2,\text{close}}$ $\theta_{2,\text{point}}$, or $\theta_{2,\text{teeth}}$ facial paresis in the mouth region is supposed.

This gives six ways to detect facial paresis in the eyes region and another six ways to diagnose the mouth region. The classification results of all twelve feature sets are presented in Section 8.4.6.

8.4.6 Experimental Results

The evaluation of the whole diagnosis support system was performed with 16 patients with different grades of paralysis and 4 healthy persons. In this section we present the results of all our experiments.

We started with the generation of image sequences of the 20 persons. The persons performed mimic exercises in front of a homogeneous background. and the image series including 20 images were taken when the extremal positions of the exercises described in the last section were reached. That gives five image sequences of length 20 of every person.

In every first image of the series the face was located by means of the parametric face model and tracked in the remaining 19 images. In Figure 8.30 the localization results are shown graphically. In that example the input image was of size 384×288. The calculated parameters are shown in Table 8.1.

Totally the localization was successful in 82% of the eyes and in 73% of the mouth and whenever the facial features where localized correctly the tracking was done error-free.

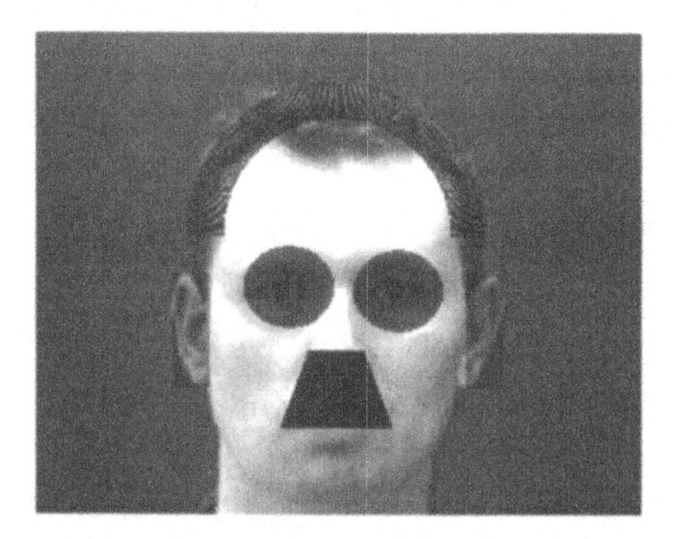

Figure 8.30. Localization example of face and facial features.

Head: $(x_{\mathrm{h}}^*, y_{\mathrm{h}}^*, R_{\mathrm{h}}^*)$	(191,123,65)
Ears: $(x_{\mathrm{ea}}^*, y_{\mathrm{ea}}^*, D_{\mathrm{ea}}^*, W_{\mathrm{ea}}^*, H_{\mathrm{ea}}^*)$	(192,144,20,62,93)
Eyes: $(x_{\mathrm{ey}}^*, y_{\mathrm{ey}}^*, D_{\mathrm{ey}}^*, A_{\mathrm{ey}}{}^*, B_{\mathrm{ey}}{}^*, H_{\mathrm{ey}}^*)$	$(194, 153, 31, 26, 22, 1)$
Nose/Mouth $(x_{\mathrm{nm}}^*, y_{\mathrm{nm}}^*, W_{\mathrm{nm}}{}^*, T_{\mathrm{nm}}{}^*, H_{\mathrm{nm}}{}^*)$	$(190{,}229, 63, 31, 43)$

Table 8.1. Computed parameters (see Figure 8.27) for real face.

The localization of the eye and mouth angles (cf. Figure 8.31) was performed correctly with a rate of 50% for the eye angles and 45% for the mouth angles. In the case of an error the correct position was hand-segmented.

With the localized facial features the classification of the facial paresis was performed.

In Figure 8.32 characteristical evaluations of the features c_1 (eye region) and c_2 (mouth region) are shown. The features were calculated while the observed person performed the five mimic expressions described in the last section. The numbers at the x-axis show the exercise which was performed to calculate the feature. A patterned entry shows a change of asymmetry that might be originated by a facial paresis.

Then we generate the ratio of the parameters belonging to the relaxed face with the corresponding other ones. That produces for every of the twelve features c_1 to c_{12} four ratios. The ratios are compared with a threshold and if a certain number of ratios are greater than the threshold the face is classified to contain paresis or not.

The extracted features were used to classify face images into healthy persons and patients with facialis paresis. The selected threshold for all ratios was $\theta = 0.7$. Table 8.2 shows the classification results. The correctness of a classification was detected by comparation with the grading of a medical specialist.

8.4.7 Conclusion

We presented the basics of a system for diagnosis support and rehabilitation supervision of patients with facial paresis. The advantages of this system are an objective method to diagnose facial paresis. This is performed by means of measurements and numerical features from the face.

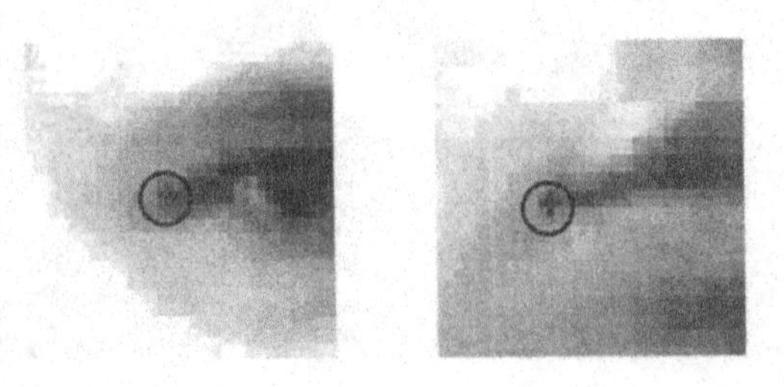

Figure 8.31. Localization example of the eye angle and mouth angle.

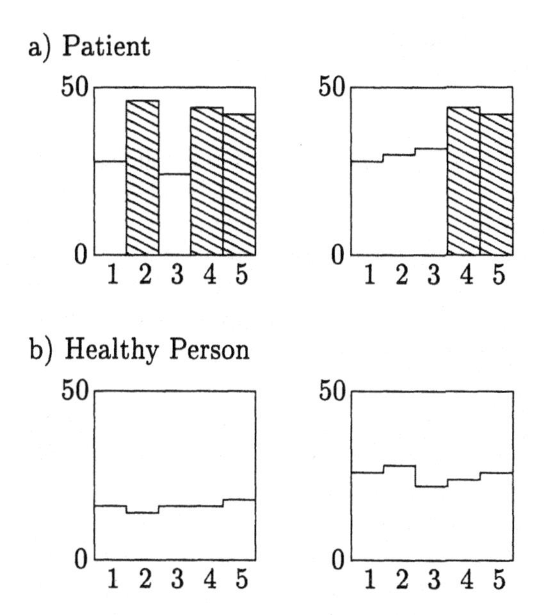

Figure 8.32. Results of face analysis: a) Patient, left: feature c_1 (eye), right: feature c_2 (mouth); b) Healthy Person, left: feature c_1 (eye), right: feature c_2 (mouth).

In a first step we localized the face and facial features using a dynamic face model and a energy maximization Simplex approach. The segmented model parameters where used to analyze the eye and mouth regions of the face towards asymmetries as asymmetries are symptoms for facial paresis. The extracted numerical features made us grade the facial paresis in mouth and eye regions. We evaluated the system by 15 patients and 5 healthy persons.

Eye						
	c_1	c_3	c_5	c_7	c_9	c_{11}
healthy	0.75	0.75	0.75	0.50	0.75	0.25
paresis	0.6	0.6	0.67	0.67	0.73	0.73
Mouth						
	c_2	c_4	c_6	c_8	c_{10}	c_{12}
healthy	1.0	1.0	0.25	0.25	0.50	0.50
paresis	0.87	0.87	0.73	0.73	1.0	0.93

Table 8.2. Classification results.

8.5 SURGICAL PLANNING

M. TESCHNER, S. GIROD, B. GIROD

Although 'virtual aircrafts' are commonly used as a training method for pilots, there exists no 'virtual surgery room' to educate physicians. When physicians finish their theoretical studies they gain surgical experiences on real patients. This fact is a strong motivation to develop computer-based simulation methods for surgical procedures. These methods can be used in medical education, they can improve diagnosis, surgical planning, and surgical treatment by providing the ability to perform 'virtual surgeries'.

The development of simulation methods for surgical procedures is a very complex task and only very specific problems have been solved in recent years. In order to describe a 'virtual patient' huge data sets consisting of different sensory modalities are required and interdisciplinary methods have to be applied to the data sets in order to build a surgical simulation system.

The basis for developing simulation methods for surgical procedures has been given in 1973, when Hounsfield invented computed tomography (CT) [342]. This technology enabled the generation of three-dimensional models that might be used in surgical simulation. There are two principal methods of simulating surgeries based on models:

One method is to build physical models. The idea of manufacturing lifelike models has been introduced in 1980 [9]. These models are generated from CT data using stereolithography and milling. In order to segment the bone structure a semiautomatic contour detection is applied to each CT slice. Further processing of contour data is done by a milling machine or by a stereolitography machine. While the milling machine builds the model from a block of polyurethane, the stereolitography machine builds the model slice by slice in a basin filled with liquid, photosensitive resin. Realistic, lifelike models improve osteotomy planning and allow accurate manufacturing of transplants. Moreover, these models can be used for educational purposes and demonstrations. However, lifelike models are expensive and their production can take several days. Furthermore, stereolitography machines have problems to reconstruct small bone structures that are not connected to the main part of the model. Stereolitographic models shrink up to six months after their production. Milling machines cannot reproduce hollows, undercuts and small holes. While physical models provide information on the bone structure, they do not contain knowledge of other structures.

The second method of simulating surgical procedures is to utilize medical imaging for generating computer models. The generation of computer models takes less time and is cheaper compared to stereolitographic and milled models. Additional inaccuracies caused by the manufacturing process of physical models are avoided. Computer models are more flexible than physical models and are able to provide more information by integrating several sensory modalities. A multimodal computer model of the bone structure and soft-tissue can be used for simulating osteotomies as well as for assessing the resulting soft-tissue changes. Various simulations can be performed with less additional effort compared to physical models. This is especially helpful in cases where various surgical options are possible.

The improvement of three-dimensional visualization techniques of tomographic data sets in the late 80's promoted the development of surgical procedures based on computer models. The *Marching Cubes algorithm* by Lorensen, 1987 [452] provided the ability to generate iso-surface triangle meshes from volume data sets. Based on this algorithm research focussed on simulating the manipulation of bone structure. In 1984 first results on modeling of deformable objects were published by Barr [43], in 1986 modeling of skin deformation using finite element models was proposed by Larrabee [420]. Deng (1988) [166] attempted to simulate wound closing using a three-layer soft-tissue model. Yasuda (1990) [768] and Pieper (1991) [545] introduced first craniofacial surgery systems. These systems were able to perform simple facial surgical simulations and to estimate the corresponding soft-tissue changes roughly. The idea of estimating soft-tissue changes due to bone realignment was formulated by Vannier in 1983 [722]. In 1992 further approaches to craniofacial surgery simulation were introduced by Kikinis [387], followed by Delingette 1994 [164], Bohner 1996 [75], Koch 1996 [398], and Bro-Nielsen 1998 [82].

At the Telecommunications Laboratory, University of Erlangen-Nürnberg, methods for craniofacial surgery simulation based on 3D computer models are investigated since 1993 [245, 383, 382]. In this section, an overview of components of the current surgical planning system is given [693].

The system uses an optimization approach for fast soft-tissue simulation. The section is organized as follows. In the next subsection the generation of the 3D computer models of the bone structure and the face surface is described. In Section 8.5.2 mesh simplification is described. In Section 8.5.3 the simulation of bone realignment is explained. In Section 8.5.4 the structure and parametrization of the soft-tissue model is described. In Section 8.5.5 optimization methods are compared that are used to estimate the soft-tissue deformation due to bone realignment. Simulation results are presented in Section 8.5.6.

8.5.1 Data Acquisition

In order to plan surgeries various sensory modalities are commonly used, e.g. radiographs, CT, MRI. Figure 8.33 illustrates the process of acquisition and model generation in case of craniofacial surgery simulation. Triangle meshes that describe the surface of the face and the bone structure of the head are the basic elements of the craniofacial simulation process. These meshes are built using two different sensory modalities.

A CT scan provides the anatomically correct representation of the bone structure and a laser scanner records a photorealistic, 3D model of the patient's face. The triangle mesh that represents the surface of the bone structure is generated by segmenting bone from the CT scan and applying the *Marching Cubes algorithm* [452] to the result. Although the *Marching Cubes algorithm* extracts isosurfaces from volume data sets it is useful to segment the bone structure in advance to reduce artifacts of the CT scan. The triangle mesh that represents the face surface is computed from the depth and color map of the laser scan.

Both modalities are registered by exploiting corresponding cephalometric landmarks of the laser scan and the skin surface taken from the CT scan [382].

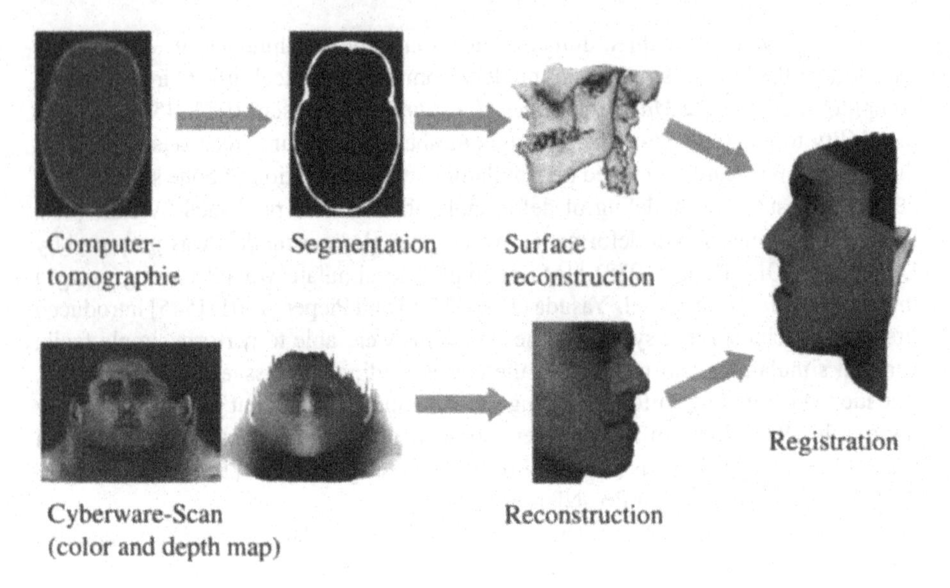

Figure 8.33. Data acquisition for craniofacial surgery simulation.

8.5.2 Data Reduction

Models represented as triangular meshes consist of a large number of triangles. Extraction of isosurfaces from volume data sets (MRI, CT) or reconstruction of surfaces from range scanner data sets (Cyberware) easily generates hundreds of thousands of triangles. In order to enable interactive visualization and handling of these triangle meshes they must be decimated. Several triangle mesh decimation algorithms have been developed in recent years. A classification of these algorithms is given in [130]. Basic operations of simplification algorithms are:

- *Vertex removal*: Vertices are removed and the resulting polygon is retriangulated.
- *Vertex clustering*: Vertices are grouped in clusters. These clusters are replaced by new representative vertices.
- *Edge / face collapsing*: Successive collapse of edges into vertices or triangles into edges or vertices.
- *Coplanar triangle merging*: Coplanar or nearly coplanar triangles are merged and the resulting polygon is retriangulated into fewer triangles.

In most cases mesh simplification is an iterative process that generates a sequence of meshes $M_0 \rightarrow M_1 \rightarrow \cdots \rightarrow M_{i-1} \rightarrow M_i$. A criterion for choosing an item for the reduction step $M_{i-1} \rightarrow M_i$ can consider the current mesh M_{i-1} (*local criterion*) or the initial mesh M_0 (*global criterion*) to compute a certain error measure. These error measures are defined as distances, angles or as a combination of various measures described by an energy function.

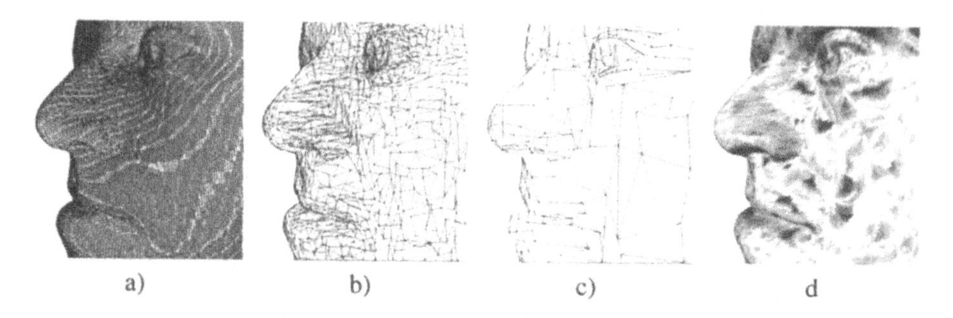

Figure 8.34. Mesh simplification. a) Original mesh, 100.0%. b) Simplified mesh, Hausdorff distance 0.1mm, 7.8%. c) Simplified mesh, Hausdorff distance 0.5mm, 1.8%. d) Hausdorff distances between a) and c). White: 0mm deviation. Black: 0.5mm deviation.

Due to the fact that local error criteria consider the current, partially simplified mesh, they do not describe the difference of the initial and the simplified mesh. However, in case of medical applications a desired accuracy of the decimated mesh compared to the initial mesh should be guaranteed by the algorithm.

In [105] a simplification method is proposed that incorporates the one-sided Hausdorff distance as global error criterion. The one-sided Hausdorff distance is defined by

$$d_h(X, Y) = \max_{\boldsymbol{x} \in X}(\min_{\boldsymbol{y} \in Y}(\|\boldsymbol{x} - \boldsymbol{y}\|)), \tag{8.19}$$

with X denoting the set of vertices of the simplified mesh and Y denoting the surface of the original mesh. The decimation algorithm guarantees that the global deviation of the original and the decimated mesh is not larger than the given Hausdorff distance. Figure 8.34 illustrates an original mesh generated from a laser range scan, two reduced triangle meshes and the Hausdorff distance between the original and a reduced mesh.

8.5.3 Simulation of Bone Movement

The process of surgery planning requires a realistic simulation of the transformation of certain bone structures, e.g. craniofacial surgery, knee surgery, hip surgery. While graphical environments provide the functionality to transform bone models they do not check for penetration of these models. This problem is addressed by collision detection algorithms. Basically, these algorithms work on surface models and check for interferences between triangles of different geometric objects. To speed up the collision test some approaches are restricted to a class of triangle meshes, e.g. convex meshes. However, these algorithms are not suitable for medical applications. Another approach to speed up the collision test and to avoid considering all triangles in the interference check is to employ hierarchical object representations.

In [255] a collision detection algorithm is presented that can be applied to objects represented as unstructured triangle meshes. The objects do not have to be convex and they are not considered to have any special properties. As an initialization step the

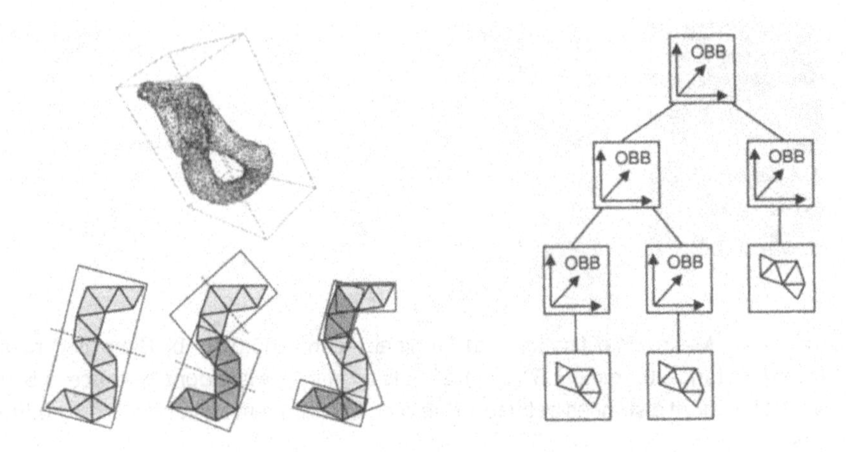

Figure 8.35. OBB-tree representation.

algorithm computes a hierarchical tree-structure of oriented bounding boxes (OBB). In contrast to axis-aligned bounding boxes (AABBs), OBBs are aligned with the principal axes of an object and grant a better approximation of the object. Although the computation of OBBs is more time-consuming than the computation of AABBs OBBs do not have to be recomputed if the object is transformed. In case of transformation OBBs only have to be transformed with the object. AABBs would have to be recomputed in case of rotation. Following certain rules a generated OBB is separated into two areas with quite the same number of triangles. For each of these areas a new OBB is generated and so on. This leads to a hierarchical OBB tree-representation (see Figure 8.35).

One important reason for the speed of the interference check of this algorithm is that only the OBBs are transformed instead of the whole object. Only if two OBBs are not separated, the next layer of OBBs is transformed and tested. If leaves are reached in both structures then all triangles that are represented by these leaves are transformed and tested for interference. This method optimizes the amount of transformation as well as the amount of interference tests (see Figure 8.36).

Collision detection based on this algorithm can be performed in real-time and enables interactive physiological bone movement and bone realignment.

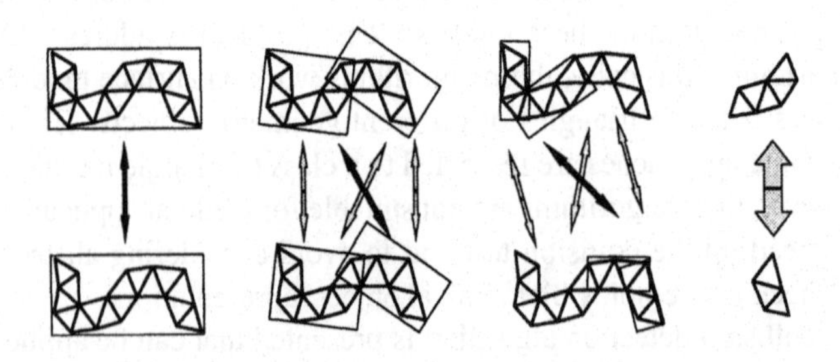

Figure 8.36. Collision detection based on an OBB-tree.

8.5.4 Soft-Tissue Model

Given the triangle meshes that represent the skull and the face, the soft-tissue model is generated. In recent years, several soft-tissue models based on springs or finite elements have been developed [164, 82, 382, 474, 163, 398]. As computational costs for finite-element methods are high and these methods seem to be less suitable for interactive applications, in our work, a mesh of springs is utilized. The springs are categorized according to their location and function (see Figure 8.37):

- *Layer springs* represent soft-tissue layers. In order to model differentiated elasto-mechanical properties of soft-tissue layers, each layer is represented by a particular class of springs. The number and the thickness of soft-tissue layers are variable (see Figure 8.38). Simulations have been performed with one, three, and five layers.

- *Bone springs* represent connections between bone and soft-tissue. Only some regions of the soft-tissue are connected to the underlying bone structure. To mimic sliding contact, these connections are modeled using springs.

- *Boundary springs* prevent the soft-tissue model from undergoing global transformation. Due to the fact that the face model does not include the complete head surface but only the facial region, these springs anchor the face and the underlying soft-tissue in space.

A spring is characterized by a spring constant k, which describes its stiffness and by a length l (see Figure 8.38). Every spring class is parametrized with a particular spring constant to model the elasto-mechanical properties of the corresponding soft-tissue layer. Bone springs and boundary springs are parametrized with a comparatively large spring constant. The length of bone springs and boundary springs is zero. The springs that represent the skin surface are given a certain strain. This strain corresponds to the skin turgor. Setting the natural length of all surface springs to $c_{\text{turgor}} \cdot l$, with $0 < c_{\text{turgor}} < 1$, introduces a certain strain. The difference of l and $c_{\text{turgor}} \cdot l$ corresponds to the desired skin turgor.

A *soft-tissue position* is characterized by a location $P \in \mathbb{R}^3$ and a mass m in order to enable simulation of gravity (see Figure 8.38). Every soft-tissue layer is parametrized by an overall mass, which is distributed according to the topology of the representing soft-tissue positions.

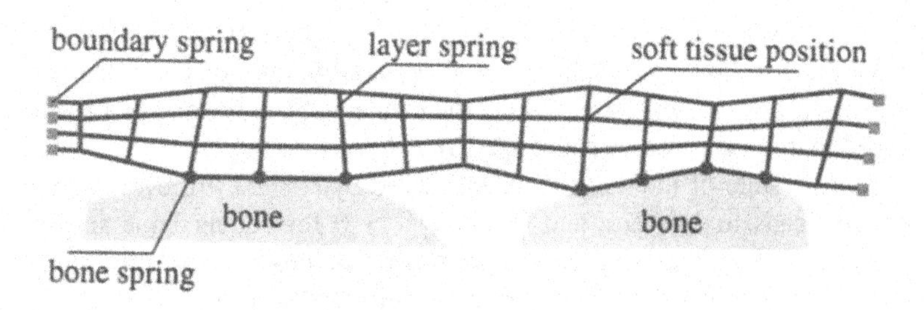

Figure 8.37. Soft-tissue model.

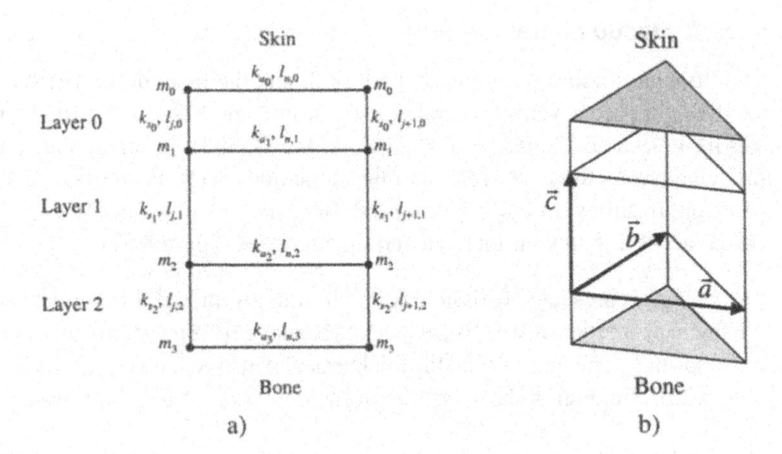

Figure 8.38. Soft-tissue representation. a) Parametrization. b) Basic element.

Due to the masses and the strain of the surface there are forces at each soft-tissue position. In order to prevent the model from changing without performing any bone realignment and obtaining a stable equilibrium of the mesh, the sum of these forces has to be zero. This is achieved by determining appropriate strains for all springs, given the strain of the skin surface.

8.5.5 Soft-Tissue Deformation

The described soft-tissue model is used to estimate soft-tissue deformation due to simulated bone realignment. Basically, the soft-tissue deformation is computed by applying an optimization method that minimizes the energy of the spring mesh. In the initial state of the simulation process the energy is zero. The energy is increased by performing bone transformation. An optimization process deforms the spring mesh in order to minimize the energy. The energy function

$$g(P_0, P_1, \ldots, P_{N-1}) = \lambda \sum_i k_i (l_{0i} - l_i)^2 + (1 - \lambda) \sum_j (v_{0j} - v_j)^2 \tag{8.20}$$

depends on $3 \cdot N$ independent variables determining N soft-tissue positions $P_i \in \mathbb{R}^3$. It mainly captures differences between initial spring lengths l_{0i} and current spring lengths l_i and differences between initial volumes v_{0i} and current volumes v_i. The values k_i are spring constants, and λ $(0 < \lambda < 1)$ weights the influence of both terms of the function. The values v_i and v_{0i} are volumes of basic elements of the soft-tissue. Figure 8.38 illustrates the basic elements. The volume of a basic element is approximated by a cross product $0.5 \cdot (\boldsymbol{a} \times \boldsymbol{b}) \cdot \boldsymbol{c}$.

The initial state of the mesh is characterized by $l_i = l_{0i}$ for every spring. Bone movement leads to $l_i \neq l_{0i}$ and $(l_{0i} - l_i)^2 > 0$ for certain bone springs and to $g(P_0, P_1, \ldots, P_{N-1}) > 0$. Now, new soft-tissue positions P^* are computed by minimizing g. These values P^* describe the deformed soft-tissue:

$$P_0^*, P_1^*, \ldots, P_{N-1}^* = \operatorname{argmin} g(P_0, P_1, \ldots, P_{N-1}). \tag{8.21}$$

Optimization method	Order of additional memory	Requires partial deriva-tives	Time $[s]$	min g
Conjugate gradient, parabolic interpolation	N	yes	0.60	4.33
Conjugate gradient, derivative based	N	yes	2.21	4.33
Direction set (Powell)	N^2	no	81.43	4.33
Variable metric (quasi-Newton)	N^2	yes	2.91	4.30

Table 8.3. Comparison of optimization methods using a synthetic data set, the energy function in Equation 8.20, and performing an exemplary bone movement. 138 soft-tissue positions P_i, 986 springs, 144 volumes (SGI O2, R10000, 175 *MHz*). N is the number of parameters of the energy function.

During the minimization process there are no additional restrictions applied to the soft-tissue positions P apart from the energy function. All soft-tissue positions are considered in the minimization process, regardless of the simulated bone realignment.

Four optimization methods have been compared with regard to computational costs and robustness of the result (see Table 8.3). All optimization methods are iterative processes. They terminate if the difference of two P^* or the difference of two evaluations of g in successive steps is tolerably small. This tolerance can be chosen. On one hand, it influences the accuracy of the minimum, on the other hand it has an effect on the computation time. The slightly different minima found by the optimization algorithms (see Table 8.3) are due to this tolerance. Some methods require the calculation of partial derivatives. The methods differ in the amount of allocated memory. The order

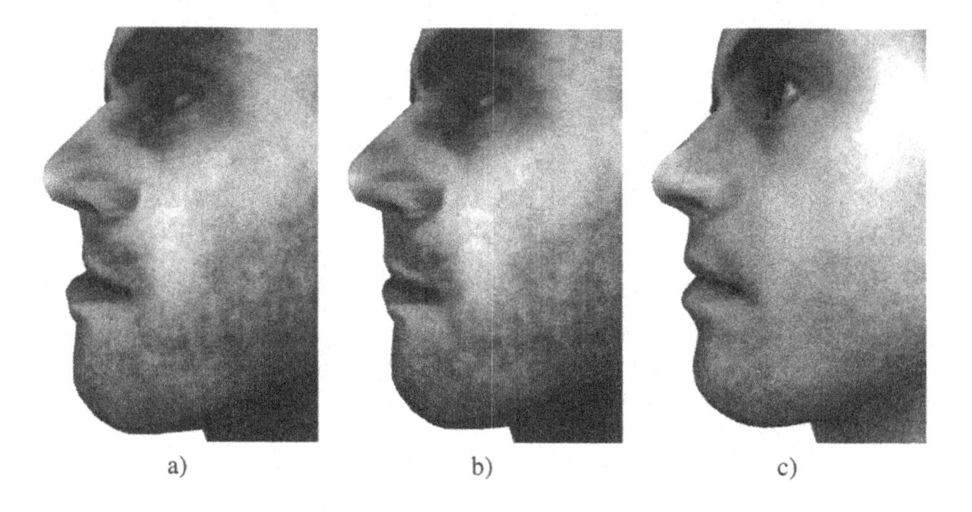

Figure 8.39. Craniofacial surgery simulation for a patient. a) Preoperative appearance. b) Simulated postoperative appearance. c) Postoperative appearance. Figure 8.40 shows the corresponding bone realignment.

Figure	Number of soft-tissue positions	Number of springs	Number of volumes	Max. simulation time [s]
8.39, 8.40	954	6103	874	1.1
8.40, 8.41	1838	11763	1696	4.5

Table 8.4. Model parameters and simulation speed.

of additional memory that is needed by an optimization method is important due to the fact that its amount is dependent on the number of soft-tissue positions. If the model consists of 3000 positions, then the energy function in Equation 8.20 has 9000 parameters and an optimization method that requires memory in order of N^2 would need a multiple of 81 *MByte* memory instead of a multiple of 9 *kByte* for an algorithm with order of N. All optimization methods are described in [560].

Tests have shown that the conjugate gradient method provides reliable results and is very efficient with regard to memory and computational complexity. Parabolic interpolation is used for 1D sub-minimization due to the quadratic form of g. Although partial derivatives of the energy function are calculated by this optimization method, its computational expense is comparatively low because of the similarity of the energy function and its partial derivatives. The partial derivatives are responsible for fast convergence of the optimization process and fast convergence reduces the number of function evaluations.

In addition to computational costs another important criterion of an minimization algorithm is the quality of the minimum found. It cannot be guaranteed that the

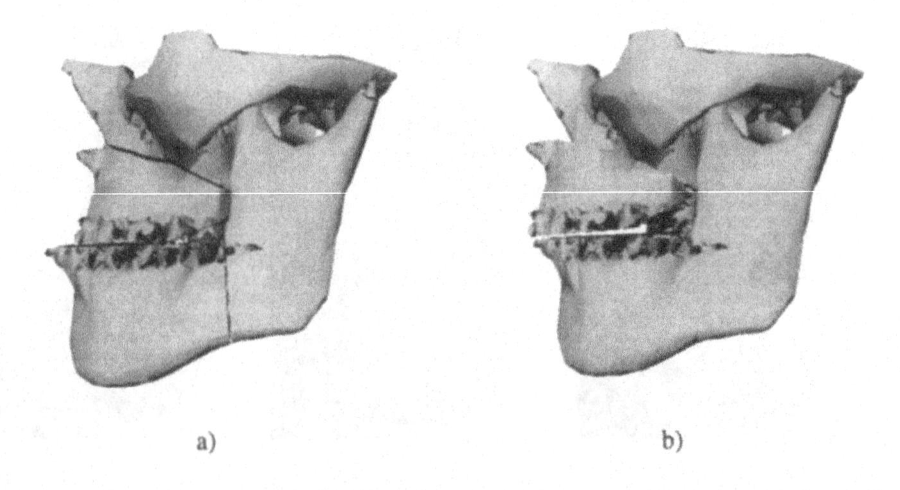

Figure 8.40. Craniofacial surgery simulation for a patient. a) Preoperative bone structure. b) Simulated postoperative bone structure. The upper jaw is repositioned 4mm forward and 2mm upward anteriorly and 4mm posteriorly. The lower jaw is moved backwards 5mm. The corresponding soft-tissue changes are illustrated in Figure 8.39.

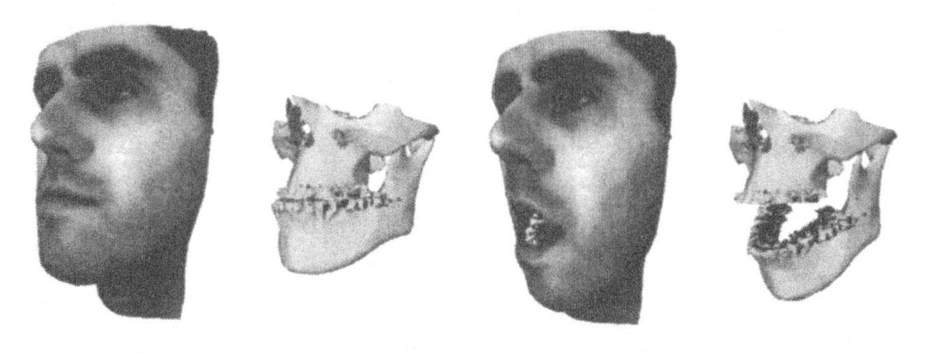

Figure 8.41. Simulated physiological lower jaw movement for a patient.

minimum P^* is the global minimum, and it is difficult to prove that fact in a space with $> 10^3$ dimensions. A method to check the robustness of the minimum P^* is to perform a certain bone movement in different ways and to compare the results. If only a small movement is performed, the distance of the initial soft-tissue positions P and P^* is small, g is comparatively small, and the global minimum is likely to be found. For example, translating a bone by 0.1mm ten times or translating a bone by 1mm once should lead to the same P^*. Several tests using the multidimensional conjugate gradient method have been performed and all minima have been reliable.

8.5.6 Results

Table 8.4 and Figures 8.39–8.44 show examples of simulations performed. The soft-tissue prediction is tested with three individual patient data sets. Several simulations of bone movement have been applied to each model. The last column of Table 8.4 shows the maximum time needed by the optimization process. All tests have been performed on a standard workstation SGI O2, R10000, 175MHz. Figure 8.45 shows parts of the simulated planning process in case of a craniosynostosis. In this case, only cutting of the bone structure and its realignment has been simulated.

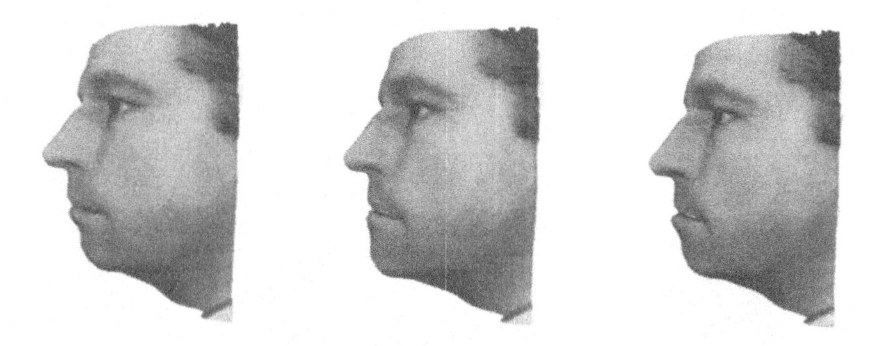

Figure 8.42. Simulated jaw movement for a patient. Figure 8.43 shows the corresponding bone realignment.

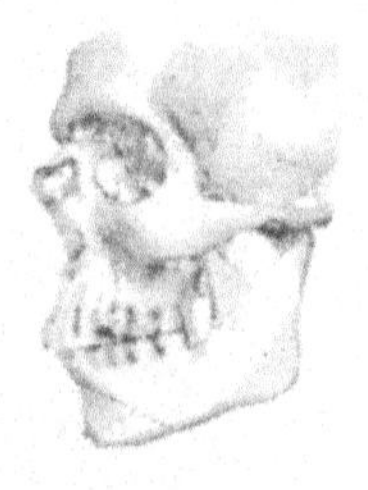
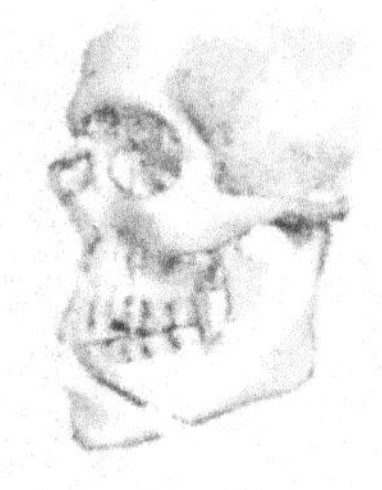
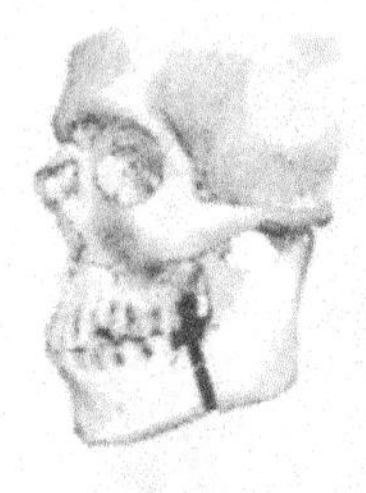

Figure 8.43. Simulated bone realignment for a patient. The corresponding soft-tissue changes are illustrated in Figure 8.42.

8.5.7 Ongoing Work

In this section, a craniofacial surgery simulation system has been presented. It simulates bimaxillary osteotomies, physiological jaw movement and predicts the soft-tissue changes caused by bone realignment. The soft-tissue prediction is based on an optimization method and has been tested with several individual patient data sets. Interactive collision detection and collision response has been integrated into the system to enable a realistic simulation of bone movement. Ongoing work focuses on realistic simulation transplants. As well as estimating the patient's static postoperative appearance and simulating physiological bone movement, the visualization of the patient's post-operative facial expressions is very useful. Therefore, it is planned to add muscles to the existing soft-tissue model. Furthermore, it is intended to register very accurate measurements of the jaws with the CT scan, in order to consider the occlusion of the jaws in the planning process.

Acknowledgement

This work is supported by the Deutsche Forschungsgemeinschaft DFG (SFB 603, Project C4). Patient data have been provided by the Maxillofacial Surgery Department of the University of Cologne, by the Maxillofacial Surgery Department of the University of Erlangen-Nürnberg, and by the Childrens' Hospital of the University of Erlangen-Nürnberg.

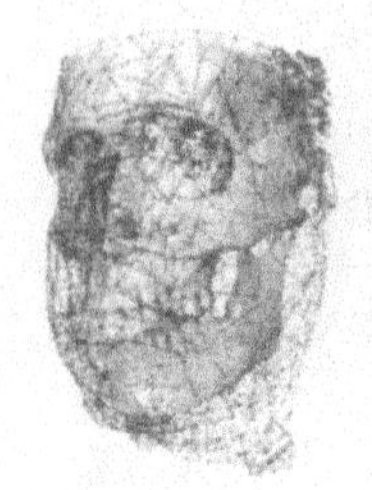
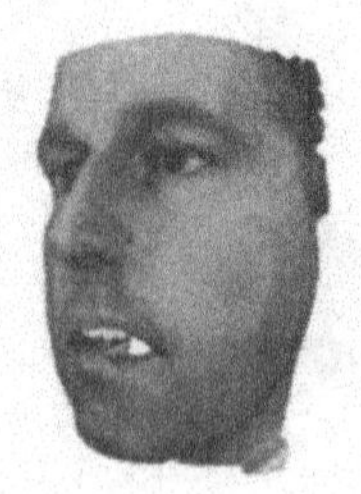
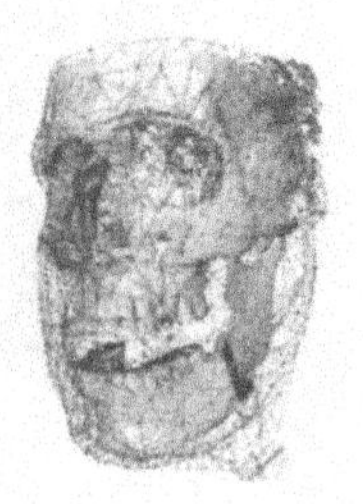

Figure 8.44. Simulated jaw movement for a patient.

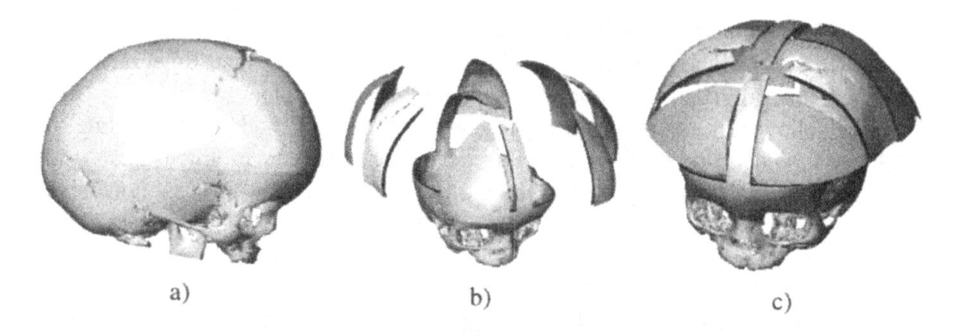

Figure 8.45. Planning process in case of a craniosynostosis. a) Reconstructed pre-operative skull. b) Segmentation of the Cranial Vault. c) Reconstruction of the Cranial Vault.

8.6 IMAGE COMMUNICATION

P. EISERT, B. GIROD

Source models play an important role in image coding. Knowledge that is available a priori and that can be represented appropriately need not be transmitted. Rate distortion theory allows us to calculate a lower bound for the average bitrate of any coder, if a maximum permissible average distortion may not be exceeded. Many of today's sophisticated coding schemes probably operate very close to their rate distortion theoretical bound. It may not be concluded, however, that this fundamentally prevents us from inventing even more efficient coding schemes. Rate distortion theoretical bounds are valid only for a given source model. Another source model might result in a lower rate at a given distortion. Better source models are the key to more efficient image compression schemes.

The majority of images are the result of a camera pointing to a three-dimensional scene. The scene consists mostly of surfaces reflecting the illumination towards the camera according to well understood physical laws. *Three-dimensional models* throughout this chapter are models capturing the three-dimensional spatial structure of a scene in front of the camera.

The attempt to explicitly recover 3D structure for a still image and use this information for coding is not very promising. The projection of the 3D scene on the image plane is certainly an enormous data reduction, and a 3D reconstruction has to overcome many ambiguities. How, e.g., to encode a flat photograph in a 3D scene? We can nevertheless hope for enormous data reduction using 3D models for image sequences. Even sequences with a lot of change from one frame to the next can often be described by only a few parameters in the domain of 3D objects and 3D motion. Using such *model-based coding* techniques, extremely low bitrates can be achieved for particular video sequences. For example, for an image sequence resulting from a camera moving through a static room, we would ideally transmit a texture-mapped 3D model of the room once, and then only update the 3D motion parameters of the camera. The decoder includes a 3D computer graphics renderer in this scenario. The problem of rendering photorealistic images from 3D models is basically solved today. The difficult part which is addressed in this section is the automatic estimation of the object's deformation and 3D motion from the original images.

In the following sections, we first present the concept of model-based coding of head-and-shoulder video sequences. This technique uses a three-dimensional head model to describe the appearance of a talking person sitting in front of a camera. The head model together with the modeling of facial expressions are discussed in the next section. We then describe how the 3D facial motion and deformation can be estimated from 2D images. Finally, experimental results are shown that demonstrate the efficiency of model-based coding techniques.

8.6.1 Model-Based Video Coding of Head-and-Shoulder Scenes

In model-based coding of image sequences [8, 437, 533, 744], 3D models are needed for all objects in the scene describing their shape and texture. Therefore, we have to put some restrictions on the scene content. In this context, we focus on head-and-shoulder scenes, which are typical for video telephone and video conferencing applications. Encoding of other classes of sequences makes basically no difference provided that an explicit 3D model is available at the codec.

For videotelephony, we want to transmit the head-and-shoulder view of a talking person recorded with a single camera. Model-based coding is used as illustrated in Figure 8.46. The encoder analyzes the incoming frames and estimates the 3D motion and the facial expressions of the person. A set of facial expression parameters is obtained that describes (together with the 3D model) the current appearance of the person. Only a few parameters have to be encoded and transmitted, resulting in very low bitrates, typically less than 1 kbit/s [201]. At the decoder, the parameters are decoded and then used to deform the head model according to the person's facial expressions. The original video frame is finally approximated by simply rendering the 3D model at the new position.

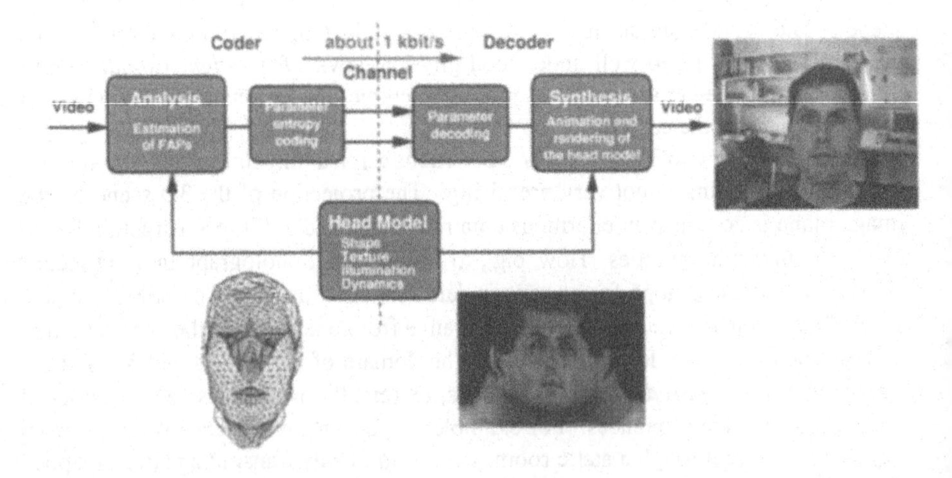

Figure 8.46. Basic structure of a model-based video codec.

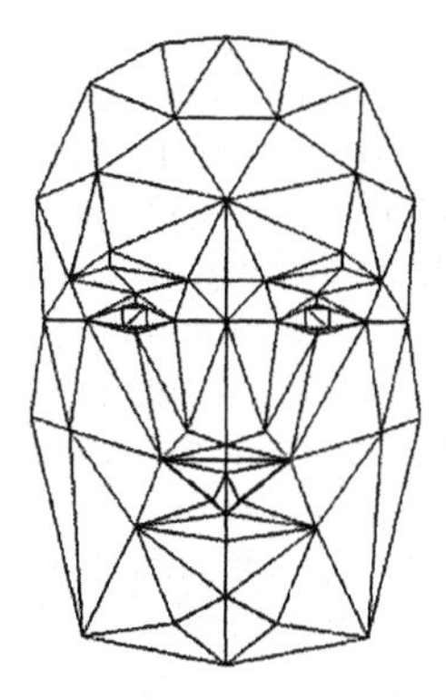

Figure 8.47. Wireframe of the CANDIDE face model.

8.6.2 Head-and-Shoulder Models

The modeling of the objects in the scene is an important part of model-based coding, since both encoder and decoder strongly depend on the shape, color and motion information from the 3D model. Many researchers have made progress modeling head and shoulders of a talking person in the past years [526]. One of the first parameterized head model was the CANDIDE model [587] that is shown in Figure 8.47. Similar to the face model developed by Parke [525] it consists of a three-dimensional triangular mesh that can be deformed to show facial expressions. Geometric details like eyes and the mouth are visible but due to the use of shaded single color triangles they typically lack reality.

A large step towards photorealistic modeling of human heads is the use of texture mapping [109, 73] and today almost all models have textured surfaces [8, 122, 201]. Figure 8.48 shows a head model in wireframe representation and the corresponding textured version.

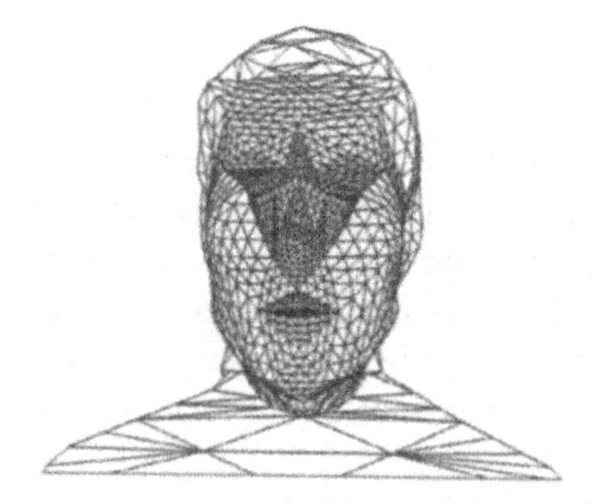

Figure 8.48. left: wireframe representation of a head model, right: corresponding textured version.

Eyes, teeth and the interior of the mouth, including the tongue, can be modeled similarly with textured polygonal meshes but a realistic representation of hair is still not available. A lot of work has been done in this field to model the fuzzy shape and reflection properties of the hair. For example, single hair strands are modeled with polygonal meshes [737] and the hair dynamics are incorporated to model moving hair [17]. However, these algorithms are computational intensive and are not practical for real-time applications in the next future. Image-based rendering techniques [435] might provide new chances for realistic hair modeling.

Modeling a human head with polygonal meshes results in a representation with a large number of triangles and vertices, which have to be moved and deformed to show facial expressions. To reduce the large number of degrees of freedom in the shape, splines [227] can be used to describe the surface, exploiting the smoothness of the facial tissue. Hoch et al. [322] have used B-splines with about 200 control points for this purpose. To allow a local refinement for the spline topology in areas that are more curved, triangular B-splines [201] or hierarchical splines [229] are used.

Rather than defining the shape of a human head by the position of the vertices or control points of a polygonal mesh or spline surface, a parameterized view-based modeling technique can be used [728]. A large database with the shape and texture of different people is established that spans a high-dimensional space of faces. Using the knowledge about the point correspondences between all 3D models, new shape and texture can be created by linearly interpolating or morphing between the models in the database. Even if not all individual characteristics of a person can be reconstructed exactly, the advantage of this approach is the small number of parameters (coordinates in the face space) that are sufficient to describe the shape and texture of a person.

8.6.3 Facial Expression Modeling

Once a 3D head model is available, new views can be generated by rotating and translating the 3D object. However, for the synthesis of facial expressions the model can no longer be static. In general, two different classes of facial expression modeling can be distinguished: the clip-and-paste method and algorithms based on the deformation of the 3D model.

Clip-and-Paste Method. For the clip-and-paste method [7, 744], templates of facial features like eyes and the mouth are extracted from previous frames and mapped onto the static 3D shape model. The model is not deformed according to the facial expression but remains rigid and is only used to compensate the global motion given by head rotation and translation. All local changes in the face must therefore be compensated by changing the texture of the model. During the video sequence, a codebook containing templates for the different facial expressions is built. A new expression can then be synthesized by combining several feature templates that are specified by their position on the model and their template index from the codebook.

Deformation Method. With the clip-and-paste method, a discrete set of facial expression can be synthesized. However, the transmission of the template codebook to the decoder consumes a large number of bits which makes the scheme unsuitable for coding [744]. Beyond that, the localization of the facial features in the frames is a hard task. The pasting of templates extracted at inaccurate positions leads to unpleasening jitter in the resulting synthetic sequence.

The deformation method avoids these problems by using the same model for all facial expressions. The texture remains basically constant and the facial expressions are generated by deforming the 3D surface. In order to avoid the transmission of the changed position of all vertices of the triangular mesh, the facial expressions are parameterized using high-level expression parameters. Deformation rules associated to the 3D head model describe how certain areas in the face are deformed if a parameter is changed. The superposition of many of these local deformations is then expected to lead to the desired facial expression.

One system for facial expression parameterization that is widely used today [7, 122, 322, 438] is the facial action coding system (FACS) developed by the psychologists Ekman and Friesen [204]. Any facial expression results from the combined action of the 268 muscles in the face. Ekman and Friesen discovered that there are only 46 possible basic actions performable on the human face. Each of these basic actions, which they call *action units*, consists of a set of muscles that cannot be controlled independently. To obtain the deformation of the facial skin that is caused by a change of an action unit, the motion of the muscles and their influence on the facial tissue can be simulated using soft-tissue models. Head models which exploit the properties of the tissue and the muscles are called muscle-based models [692]. Due to the high computational complexity of the simulation, in many applications the surface deformation is modeled directly [7, 322] using heuristic transforms between action units and surface motion.

Very similar to the FACS is the parameterization in the recently determined *synthetic and natural hybrid coding* (SNHC) part of the MPEG-4 video coding standard [493]. SNHC allows the transmission of a 3D face model that can be animated to generate different facial expressions. Rather than specifying groups of muscles that can be controlled independently and that sometimes leads to deformations in larger areas in the face, in this system the single parameters directly correspond to locally limited deformations of the facial surface. There are 65 different facial animation parameters (FAPs) that control both global and local motion. Examples of different facial expression synthesized using this scheme are depicted in Figure 8.49.

 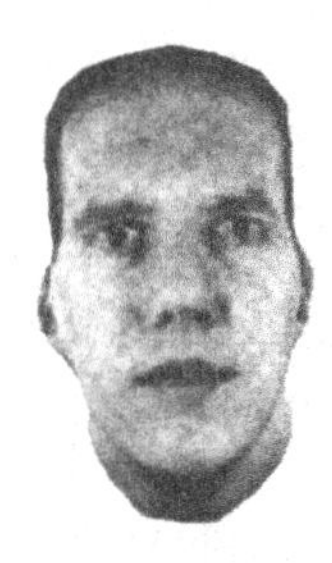  

Figure 8.49. Images synthesized using the MPEG-4 facial expression modeling scheme.

8.6.4 Analysis

The analysis of facial expressions from the image data itself is a hard task. Three-dimensional motion and deformation in the face have to be estimated. Due to the loss of one dimension in the camera projection, this ill-posed problem can only be solved by using additional assumptions and restrictions on the motion characteristics. In the case of head-and-shoulder scenes, a parameterized 3D head model provides us with information about shape, color and motion constraints. The motion constraints are defined by the particular facial expression modeling. If, for example, the FACS system is used, the complete motion in the face is described by the superposition of 46 well defined 3D displacement fields each controlled by a single parameter. Instead of estimating the motion vector of every surface point on the face, only 46 parameters corresponding to the action units have to be determined.

The motion constraints are assigned to the objects using 3D and 2D motion models as described in Section 2.2. Global head motion, characterized by head translation and rotation, is often estimated prior to the local motion [397, 438]. For the global motion, the rigid body model in Equation 2.13 can be used that leads to a motion constraint in the 2D image plane as shown in Equation 2.34. Equivalently, the local motion in the face caused by facial expressions is modeled using a flexible body motion model as in Equation 2.17 with the corresponding 2D representation given by Equation 2.38. Assuming that 2D displacements $(x_2 - x_1,\ y_2 - y_1)$ of object points between two frames recorded at the time instants t_1 and t_2 are given, we can solve for the unknown motion and deformation parameters. Since the underlying motion model is valid for the whole object and each displacement provides two additional equations, the number of point correspondences must be at least half the number of parameters to be estimated.

However, the displacements in the 2D images have to be determined first. Based on the way this is accomplished, the algorithms for estimating motion and deformation parameters from 2D images are typically classified into two groups. First, methods that are based on the tracking of discrete feature points and, second, algorithms that use the complete image incorporating gradient-based or optical-flow-based techniques.

Facial Parameter Estimation using Feature Points. One common way for determining the motion and deformation in the face between two frames of a video sequence is the use of feature points [7, 4, 346, 373]. Highly discriminant areas with large spatial variations, e.g. like areas containing the eyes, nostrils, or mouth corners are searched and tracked over the frames. If the corresponding features are found in two frames the change in position can be taken as the displacement.

How the features are searched depends on their properties like color, size, or shape. For facial features, extensive research has been performed especially in the area of face recognition [116]. Templates [91], often used for finding facial features, are small reference images of typical features. They are compared at all positions in the frame to find a good match between the template and the current image content. The best match is then said to be the desired feature. Problems with templates arise from the wide variability of captured images due to illumination changes or different viewing position. To compensate these effects, Eigen-features [486], which span a space of possible feature variations, or deformable templates [771], that reduce the features to their contours, can be utilized.

Rather than estimating single feature points, also the whole contour of features can be tracked [346, 533] using snakes. Snakes [378] are parameterized active contour models that are composed of several energy terms. Internal energy terms account for the shape of the feature and smoothness of the contour while the external energy attracts the snake towards feature contours in the image.

Optical-Flow Based Estimation. When using feature based algorithms, single features like the eyes can be found quite robustly. Dependent on the image content, however, only a small number of feature correspondences is typically determined. As a result, the estimation of the 3D motion and deformation parameters from the displacements is completely wrong, if only one feature is associated to a different feature in the second frame.

In contrast, optical flow based techniques [160, 201, 397, 438] utilize the complete image information leading to a large number of point correspondences. The single correspondences are not as reliable as the ones for feature-based methods, but due to the large number of equations, their influence is not critical. Additionally, possible outliers can generously be removed without obtaining an under-determined system of equations.

Gradient-based algorithms utilize the optical flow constraint [333]

$$\frac{\partial I(x,y)}{\partial x}(x_2 - x_1) + \frac{\partial I(x,y)}{\partial y}(y_2 - y_1) = I_1(x,y) - I_2(x,y), \tag{8.22}$$

where $\frac{\partial I}{\partial x}$ and $\frac{\partial I}{\partial y}$ are the spatial derivatives of the image intensity at image position (x, y). This equation, obtained by Taylor series expansion up to first order of the image intensity, can be set up anywhere in the image. It relates the unknown 2D motion displacement $(x_2 - x_1, y_2 - y_1)$ with the content of the images.

The solution of this problem is under-determined since each equation has two new unknowns for the displacement coordinates. For the determination of the optical flow or motion field, additional constraints are required [333]. Exploiting knowledge about the shape and motion characteristics of the object, any 2D motion model of Section 2.2.3 can be used as an additional motion constraint. Inserting the 2D motion model into Equation 8.22 reduces the number of unknowns to the number of motion parameters of the corresponding model. In that case, it is assumed that the motion model is valid for the complete object. An over-determined system of equations is obtained that can be solved robustly for the unknown motion and deformation parameters in a least-squares sense. The computational complexity is moderate, since the 2D motion models are linear in the parameters.

In the case of facial expression analysis, the motion models are taken from the motion characteristics of the head model description according to Equation 2.15. Global motion can be estimated first, followed by local motion refinement [122, 397] or both global and local motion are determined in a uniform framework [160, 201].

Since the optical flow constraint Equation 8.22 is derived assuming the image intensity to be linear, it is only valid for small motion displacements between two

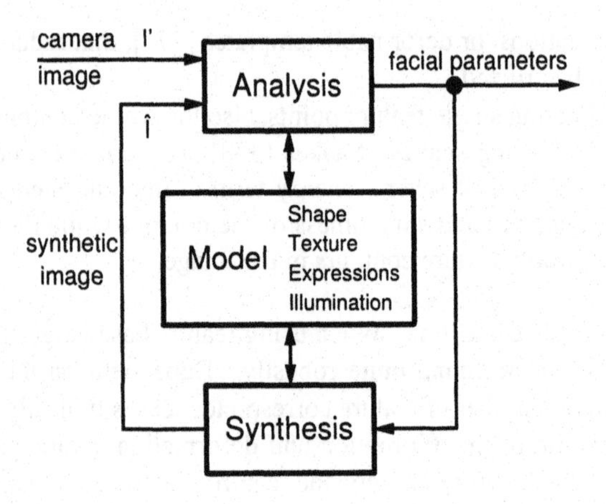

Figure 8.50. Analysis-synthesis loop of the model-based coder.

successive frames. To overcome this limitation, a hierarchical framework can be used [201]. First, a rough estimate of the facial motion and deformation parameters is determined from sub-sampled and low-pass filtered images, where the linear intensity assumption is valid over a wider range. The 3D model is motion compensated and the remaining motion parameter errors are reduced on frames having higher resolutions.

The hierarchical estimation can be embedded into an analysis-synthesis loop [438] as shown in Figure 8.50. In the analysis part, the algorithm estimates the parameter changes between the previous synthetic frame $\hat{I}_1$ and the current frame I' from the video sequence. The synthetic frame $\hat{I}_1$ is obtained by rendering the 3D model (synthesis part) with the previously determined parameters. This approximative solution is used to compensate the differences between the two frames by rendering the deformed 3D model at the new position. The synthetic frame now approximates the camera frame much better. The remaining linearization errors are reduced by iterating through different levels of resolution. By estimating the parameter changes with a synthetic frame that corresponds to the 3D model, an error accumulation over time is avoided.

8.6.5 Experimental Results

As an example for facial expression analysis we present some results from the algorithm proposed in [201] that uses the model-based concept for encoding of head-and-shoulder video sequences.

Video sequences are recorded of a talking person at a frame rate of 25 Hz and CIF resolution (352 x 288 pixels). The camera is calibrated using Tsai's camera calibration technique [709]. The obtained internal camera parameters like aspect ratio of the image and viewing angle form the parameters of the perspective projection model that describes the relation between 3D and 2D coordinates.

A triangular B-spline based 3D head model as depicted in Figure 8.48 provides us with information about shape and color of the person. To individualize the generic model, a 3D laser scan of the person is utilized. The control points of the splines, defining the shape of the surface, are optimized to adapt the model to the measured

Figure 8.51. Original (left) and corresponding synthetic frame (right) of the sequence 'Peter'.

data of the scan. The texture for the 3D model is extracted from the first video frames of the corresponding sequence.

For the estimation of the facial expression parameters that represent the 3D motion and deformation of the 3D head model, a hierarchical gradient-based algorithm is used as described above. In our experiments, 23 parameters are estimated. These parameters include global head rotation and translation (6 parameters), shoulder motion (4 parameters), movement of the eyebrows (4 parameters), two parameters for eye blinking, and 7 parameters for the motion of the mouth and the lips. These parameters are transmitted to the decoder, where they determine the position and deformations of the 3D head model. The original images are reconstructed by simply rendering the model. Figures 8.51 and 8.52 show examples of the resulting synthetic frame for two video sequences.

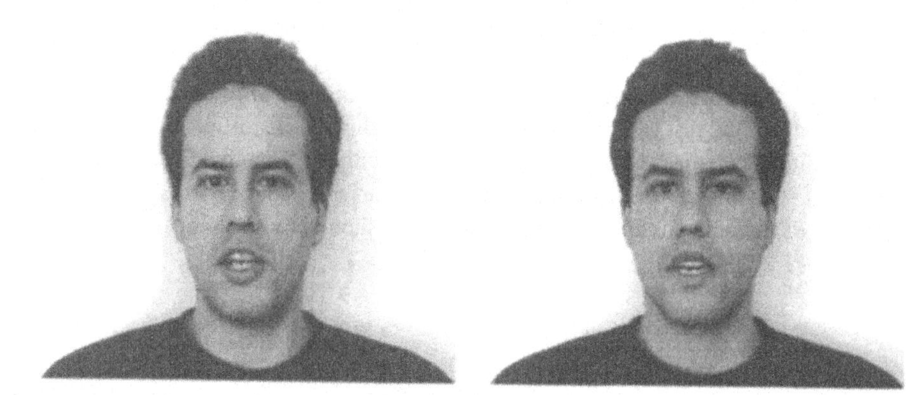

Figure 8.52. Original (left) and corresponding synthetic frame (right) of the sequence 'Eckehard'.

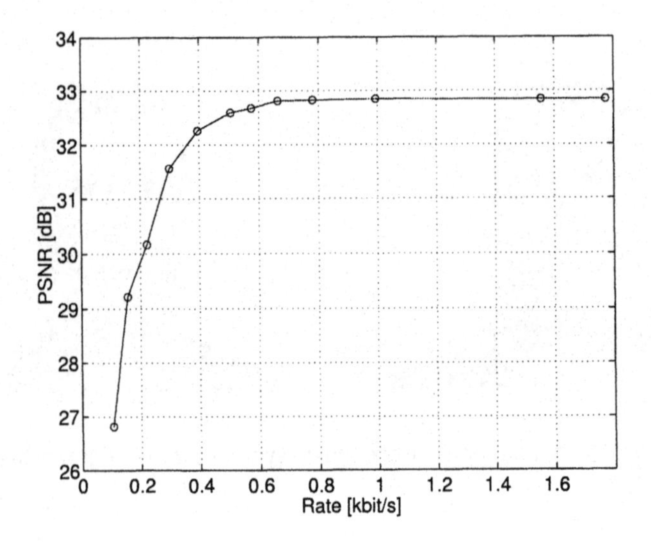

Figure 8.53. Rate-distortion plot for the model-based coder.

To measure the quality of the resulting synthetic sequences we use the PSNR (peak signal to noise ratio) as an error measure which is defined as

$$PSNR = 10 \log \left(\frac{255^2}{\frac{1}{N} \sum_{i=0}^{N-1} (I_{orig,i} - I_{synth,i})^2} \right). \tag{8.23}$$

In this equation, $I_{orig,i}$ is the luminance component of pixel i of the original camera image with values from 0 to 255, and $I_{synth,i}$ is the corresponding value in the synthetic image. N denotes the number of pixels in the image. Evaluating this error measure in the facial area of the two sequences, having a length of 230 and 220 frames, leads to an average value of 32.8 dB and 32.6 dB, respectively.

To determine the bitrate necessary to transmit the 23 parameters over a channel, we predict the current values from the previous frame, scale and quantize them. In this experiment, only every third frame is encoded leading to a frame rate of 8.33 Hz. An arithmetic coder that is initialized with experimentally determined probabilities is then used to encode the quantized values. The training set for the arithmetic coder is separate from the test set. Figure 8.53 illustrates the rate-distortion curve, where the average image quality is plotted over the average bitrate. Note that the model failures of the 3D head model lead to a saturation of the curve at about 32.8 dB. However, the plot shows that it is possible to transmit head-and-shoulder video sequences at bitrates of below 1 kbit/s.

References

[1] S. Abramowski and H. Müller. *Geometrisches Modellieren.* BI, Mannheim, Germany, 1991.

[2] E. H. Adelson and J. R. Bergen. The plenoptic function and the elements of early vision. In M. Landy and J. A. Movshon, editors, *Computational Models of Visual Processing.* MIT Press, Cambridge, MA, 1991.

[3] R. Adrian. Particle-imaging-techniques for experimental fluid mechanics. *Ann. Rev. Fluid Mechanics*, 23:261–304, 1991.

[4] H. Agawa, G. Xu, Y. Nagashima, and F. Kishino. Image analysis for face modeling and facial image reconstruction. *Proceedings of the SPIE*, 1360:1184–1197, 1990.

[5] J. K. Aggarwal and N. Nadhakumar. On the computation of motion from sequences of images—a review. *Proceedings of the IEEE*, 76(8):917–935, 1988.

[6] U. Ahlrichs, B. Heigl, D. Paulus, and H. Niemann. Wissensbasierte aktive Bildanalyse. In *Von der Informatik zu Computational Science und Computational Engineering. Abschlußkolloquium des Sonderforschungsbereichs 182 Multiprozessor- und Netzwerkkonfigurationen*, pages 97–113, University of Erlangen-Nürnberg, 1998.

[7] K. Aizawa, H. Harashima, and T. Saito. Model-based analysis synthesis image coding (MBASIC) system for a person's face. *Signal Processing: Image Communication*, 1(2):139–152, October 1989.

[8] K. Aizawa and T. S. Huang. Model-based image coding: Advanced video coding techniques for very low bit-rate applications. *Proceedings of the IEEE*, 83(2):259–271, February 1995.

[9] C. Alberti. Three-dimensional CT and structure models. *British Journal of Radiology*, 53:261–262, 1980.

[10] J. B. Allen and D. A. Berkley. Image method for efficiently simulating small-room acoustics. *Journal of the Acoustical Society of America*, 65(4):943–950, 1979.

[11] Y. Aloimonos. Purposive and qualitative active vision. In *Proceedings of DARPA Image Understanding Workshop*, pages 816–828, 1990.

[12] Y. Aloimonos. *Active Perception*. Lawrence Erlbaum, Hillsdale, New Jersey, London, 1993.

[13] Y. Aloimonos. What I have learned. *Computer Vision, Graphics, and Image Processing*, 60(1):74–85, 1994.

[14] Y. Aloimonos, I. Weiss, and A. Bandyopadhyay. Active vision. *International Journal of Computer Vision*, 2(3):333–356, 1988.

[15] E. Andersson, R. Andersson, M. Boman, B. Dahlberg, T. Elmroth, and B. Johansson. Automatic construction of surfaces with prescribed shape. *Computer Aided Design*, 20(6):317–324, 1988.

[16] R. Andersson. Surface design based on brightness intensity or isophotes-theory and practice. In J. Hoschek and P. Kaklis, editors, *Advanced Course on FAIRSHAPE*, pages 131–143, Stuttgart, 1996. B. G. Teubner.

[17] K. Anjyo, Y. Usami, and T. Kurihara. A simple method for extracting the natural beauty of hair. *Computer Graphics (SIGGRAPH)*, 26(2):111–120, July 1992.

[18] E. Arge, M. Dæhlen, and A. Tveito. Approximation of scattered data using smooth grid functions. Technical report, SINTEF-SI, Oslo, 1994.

[19] F. Arman and J. K. Aggarwal. Model-based object recognition in dense-range images—a review. *ACM Computing Surveys*, 25(1):5–43, March 1993.

[20] V. I. Arnold. *Ordinary Differential Equations*. MIT University Press, Cambridge, MA, 1985.

[21] K. S. Arun, T. S. Huang, and S. D. Blostein. Least-squares fitting of two 3D point sets. *IEEE Transactions on Pattern Analysis and Machine Intelligence*, 9(5):698–700, September 1987.

[22] J. Arvo. Transfer equations in global illumination. In *SIGGRAPH '93 Course Notes*. ACM, August 1993.

[23] T. Asanuma and Y. Tanida. Fluid dynamics. In W.-J. Yang, editor, *Handbook of Flow Visualization*, pages 7–28. Hemisphere Publishing Corporation, New York, 1989.

[24] I. Ashdown. Near-Field Photometry: A New Approach. *Journal of the Illuminating Engineering Society*, 22(1):163–180, Winter 1993.

[25] I. Ashdown. Near-Field Photometry: Measuring and Modeling Complex 3D Light Sources. In *ACM SIGGRAPH '95 Course Notes—Realistic Input for Realistic Images*, pages 1–15, 1995.

[26] L. Aupperle and P. Hanrahan. A hierarchical illumination algorithm for surfaces with glossy reflection. In *Computer Graphics (SIGGRAPH '93 Proceedings)*, pages 155–162, August 1993.

[27] L. Aupperle and P. Hanrahan. Importance and discrete three point transport. In *Fourth Eurographics Workshop on Rendering*, pages 85–94, Paris, June 1993.

[28] S. Avidan and A. Shashua. Novel view synthesis by cascading trilinear tensors. *IEEE Transactions on Visualization and Computer Graphics*, 4(4), October 1998.

[29] N. Ayache and C. Hansen. Rectification of images for binocular and trinocular stereovision. *International Conference on Pattern Recognition*, pages 11–16, 1988.

[30] C. L. Bajaj, V. Pascucci, and D. R. Schikore. Fast isocontouring for improved interactivity. In *ACM Symposium on Volume Visualization*, pages 39–46, 1996.

[31] R. Bajcsy. Active perception. *Proceedings of the IEEE*, 76(8):996–1005, 1988.

[32] R. Bajcsy and M. Campos. Active and exploratory perception. *Computer Vision, Graphics, and Image Processing*, 56(1):31–40, 1992.

[33] H. Baker. Surface representation through sequential tracking. *Proceedings 22nd Asilomar Conference on Signals, Systems and Computer*, ANOV 1988.

[34] D. H. Ballard and C. M. Brown. Principles of animate vision. *Computer Vision, Graphics, and Image Processing*, 56(1):3–32, 1992.

[35] W. J. Bangs and P. M. Schultheiss. Space-time processing for optimal parameter estimation. In J. W. R. Griffiths, P. L. Stocklin, and C. Van Schooneveld, editors, *Signal Processing*, pages 577–591. New York, Academic Press, 1973.

[36] R. E. Bank. Hierarchical bases and the finite element method. *Acta Numerica*, pages 1–43, 1996.

[37] R. E. Bank, A. H. Sherman, and A. Weiser. Refinement algorithms and data structures for regular local mesh refinement. In R. Stepleman, editor, *Scientific Computing*, pages 3–17, Amsterdam, 1983. IMACS/North Holland.

[38] D. C. Banks. Illumination in diverse codimensions. In *Computer Graphics (Proceedings of SIGGRAPH '94)*, pages 327–334, July 1994.

[39] D. C. Banks and B. A. Singer. Vortex tubes in turbulent flows: Identification, representation, reconstruction. In R. D. Bergeron and A. Kaufman, editors, *Visualization '94*, pages 132–139, Washington, D.C., October 1994. IEEE Computer Society Press.

[40] Y. Bar-Shalom and T. E. Fortmann. *Tracking and Data Association*. Academic Press, Boston, San Diego, New York, 1988.

[41] R. Baribeau, M. Rioux, and G. Godin. Color reflectance modeling using a polychromatic laser sensor. *IEEE Transactions on Pattern Analysis and Machine Intelligence*, 14(2):263–269, 1992.

[42] R. E. Barnhill. Representation and approximation of surfaces. In J. D. Rice, editor, *Mathematical Software III*, pages 69–120. Academic Press, 1977.

[43] A. H. Barr. Global and local deformations of solid primitives. *Computer Graphics*, 18(3):21–30, 1984.

[44] H. Bartels, C. Beatty, and A. Barsky. *Splines for Use in Computer Graphics & Geometric Modeling*. Morgan Kaufmann Publishers Inc., Los Altos, 1987.

[45] B. Basu, S. H. Enger, and M. Breuer. Three-dimensional simulation of the flow field in the czochralski melt using a block-structured finite volume method. Journal of Crystal Growth, submitted, 2000.

[46] B. G. Baumgart. Winged edge polyhedron representation. Technical Report STAN-CS-320, Computer Science Department, Stanford University, Palo Alto, California, 1972.

[47] G. A. Baxes. *Digital Image Processing*. John Wiley & Sons, Inc., New York, 1994.

[48] P. Beckmann and A. Spizzichino. *The Scattering of Electromagnetic Waves from Rough Surfaces*. McMillan, 1963.

[49] D. R. Begault. *3D Sound for Virtual Reality and Multimedia*. Academic Press, Cambridge, USA, 1994.

[50] P. N. Belhumeur, J. P. Hespanha, and D. J. Kriegman. Eigenfaces vs. Fisherfaces: Recognition using class specific linear projection. *IEEE Transactions on Pattern Analysis and Machine Intelligence*, 19(7):711–720, July 1997.

[51] R. G. Belie. Some advances in digital flow visualization. In *AIAA Aerospace Sciences Conference*, Reno, Nevada, January 1987.

[52] C. Bennis, J.-M. Vézien, and G. Iglésias. Piecewise surface flattening for non-distorted texture mapping. In *ACM Computer Graphics (SIGGRAPH '92 Proceedings)*, pages 237–246, 1991.

[53] J. R. Bergen, P. Anandan, K. J. Hanna, and R. Hingorani. Hierarchical model-based motion estimation. *Proceedings European Conference on Computer Vision*, pages 237–252, 1992.

[54] M. Berger. Tracking rigid and non polyhedral objects in an image sequence. In *Scandinavian Conference on Image Analysis*, pages 945–952, Tromsø, Norway, 1993.

[55] R. Bergevin, M. Soucy, H. Gagnon, and D. Laurendeau. Towards a general multi-view registration technique. *IEEE Transactions on Pattern Analysis and Machine Intelligence*, 18(5):540–547, May 1996.

[56] A. Berler and S. E. Shimony. Bayes networks for sonar sensor fusion. In *Proceedings Thirteenth Conference on Uncertainty in Artificial Intelligence*. Morgan Kaufmann, 1997.

[57] S. Bernstein. Démonstration du théorème de Weierstrass, fondée sur le calcul des probabilités. *Communication Societé Mathématique Kharkov*, 13:1–2, 1912.

[58] P. J. Besl. Active, optical range imaging sensors. *Machine Vision and Application*, 1:127–152, 1988.

[59] P. J. Besl and R. C. Jain. Invariant surface characteristics for 3D object recognition in range images. *Computer Vision, Graphics, and Image Processing*, 33:33–80, 1986.

[60] P. J. Besl and N. D. McKay. A method for registration of 3D shapes. *IEEE Transactions on Pattern Analysis and Machine Intelligence*, 14(2):239–256, February 1992.

[61] J. R. Beveridge. *Local Search Algorithms for Geometric Object Recognition: Finding the Optimal Correspondence and Pose*. PhD thesis, University of Massachusetts, Boston, 1993.

[62] J. Bey. Tetrahedral grid refinement. *Computing*, 55(4):355–378, 1995.

[63] P. Bézier. *The Mathematical Basis of the UNISURF CAD System*. Butterworth, London, 1986.

[64] G. J. Bierman. *Factorization Methods for Discrete Sequential Estimation*. Academic Press, New York, 1977.

[65] C. M. Bishop. *Neural Networks for Pattern Recognition*. Oxford University Press, 1995.

[66] M. J. Black and P. Anandan. The robust estimation of multiple motions: Parametric and piecewise-smooth flow fields. *Computer Vision and Image Understanding*, 63(1):75–104, AJAN 1996.

[67] M. J. Black and Y. Yacoob. Tracking and recognizing rigid and non-rigid facial motions using local parametric models of image motion. *International Conference on Computer Vision*, pages 374–381, June 1995.

[68] A. Blake, R. Curwen, and A. Zisserman. A framework for spatiotemporal control in the tracking of visual contours. *International Journal of Computer Vision*, 11(2):127–145, 1993.

[69] A. Blake, M. Isard, and D. Reynard. Learning to track the visual motion of contours. *Artificial Intelligence*, 78(1–2):179–212, 1995.

[70] A. Blake and A. Yuille. *Active Vision*. MIT Press, Cambridge, Massachusetts, London, England, 1992.

[71] J. Blauert. *Spatial Hearing*. MIT Press, Cambridge, USA, 1983.

[72] J. F. Blinn. Simulation of wrinkled surfaces. In *Computer Graphics (SIGGRAPH '78 Proceedings)*, pages 286–292, August 1978.

[73] J. F. Blinn and M. E. Newell. Texture and reflection in computer generated images. *Communications of the ACM*, 19:542–546, 1976.

[74] M. I. Bloor, M. J. Wilson, and H. Hagen. The smoothing properties of variational schemes for surface design. *Computer Aided Geometric Design*, 12:381–394, 1995.

[75] P. Bohner, P. Pokrandt, and S. Haßfeld. Simultaneous planning and execution in cranio-maxillo-facial surgery. In *Medicine Meets Virtual Reality 4 (MMVR4)*, 1996.

[76] J. D. Boissonnat. Representing 2D and 3D shapes with the Delaunay triangulation. In *ICPR '84*, Seventh International Conference on Pattern Recognition, pages 745–748, July 1984.

[77] C. de Boor. *A Practical Guide to Splines*. Springer, New York, 1987.

[78] J. Borish. Extension of the image model to arbitrary polyhedra. *Journal of the Acoustical Society of America*, 75(6):1827–1836, 1984.

[79] M. Born and E. Wolf. *Principles of Optics*. Pergamon Press, Oxford, 6th edition, 1993.

[80] F. Bornemann, B. Erdmann, and R. Kornhuber. Adaptive multilevel methods in three space dimensions. *International Journal of Numererical Methods in Engineering*, 36:3187–3203, 1993.

[81] M. Brill, H. Hagen, H.-C. Rodrian, W. Djatschin, and S. V. Klimentko. Streamball techniques for flow visualization. In R. D. Bergeron and A. Kaufman, editors, *Visualization '94*, pages 225–231, Washington, D.C., October 1994. IEEE Computer Society Press.

[82] M. Bro-Nielsen. Finite element modeling in surgery simulation. *Proceedings of the IEEE: Special Issue on Virtual & Augmented Reality in Medicine*, 86(3):524–530, March 1998.

[83] K. Brodlie and P. Mashwama. Controlled interpolation for scientific visualization. In G. M. Nielson, H. Hagen, and H. Müller, editors, *Scientific Visualization: Overviews, Methodologies, and Techniques*, pages 253–276. IEEE Computer Society Press, Los Alamitos, California, 1997.

[84] T. Broida and R. Chellappa. Performance bounds for estimating 3D motion parameters from a sequence of noisy images. *Journal of the Optical Society of America*, 6:879–889, AJUN 1986.

[85] C. Brown, H. Durrant-Whyte, J. Leonard, et al. Distributed data fusion using Kalman filtering. In M. A. Abidi and R. C. Gonzales, editors, *Data Fusion in Robotics and Machine Intelligence*, pages 267–309. Academic Press, Boston, 1992.

[86] C. M. Brown. Issues in selective perception. In *Proceedings of International Conference on Pattern Recognition*, pages 21–30, 1992.

[87] C. M. Brown. Toward general vision. *Computer Vision, Graphics, and Image Processing*, 60(1):89–91, 1994.

[88] L. G. Brown. A survey of image registration techniques. *ACM Computing Surveys*, 24(4):325–376, 1992.

[89] V. Bruce, P. Hancock, and M. Burton. Comparison between human and computer recognition of faces. In *ICAFGR*, pages 408–413, Nara, Japan, 1998.

[90] R. Brunelli and T. Poggio. Face recognition through geometrical features. In G. Sandini, editor, *Computer Vision—ECCV '92*, Lecture Notes in Computer Science, pages 792–800, Santa Margherita Ligure, 1992. Springer.

[91] R. Brunelli and T. Poggio. Face recognition: Features versus templates. *IEEE Transactions on Pattern Analysis and Machine Intelligence*, 15(10):1042–1052, 1993.

[92] G. Brunnett, H. Hagen, and P. Santarelli. Variational design of curves and surfaces. *Surveys on Mathematics for Industry*, 3:1–27, 1993.

[93] K. Brunnström, T. Lindeberg, and J. Eklundh. Active detection and classification of junctions by foveation with a head-eye system guided by the scale-space primal sketch. In G. Sandini, editor, *Computer Vision—ECCV '92*, Lecture Notes in Computer Science, pages 701–709, Berlin, Heidelberg, New York, London, 1992.

[94] K. Brunnström and A. J. Stoddart. Genetic algorithms for free-form surface matching. In *13th Int. Conference on Pattern Recognition*, pages D673–689, Vienna, Austria, 1996.

[95] H. H. Bruun. *Hot-Wire Anemometry*. Oxford University Press, Oxford, 1995.

[96] U. Bub, M. Hunke, and A. Waibel. Knowing who to listen to in speech recognition: Visually guided beamforming. In *Proceedings of the 1995 IEEE International Conference on Acoustics, Speech, and Signal Processing*, volume 1, pages 848–851, 1995.

[97] B. Buchanan and E. Shortliffe. *Rule-Based Expert Systems*. Addison-Wesley, Reading, Massachusetts, 1984.

[98] P. Bui-Tuong. Illumination for computer generated pictures. *Communications of the ACM*, 18(6):311–317, June 1975.

[99] P. Buning. Numerical algorithms in CFD post-processing. In *Computer Graphics and Flow Visualization in Computational Fluid Dynamics*, number 1989-07 in Lecture Series, Brüssel, Belgium, 1989. Von Karman Institute for Fluid Dynamics.

[100] P. Burt. Smart sensing with a pyramid vision machine. *Proceedings of the IEEE*, 76(8):1006–1015, 1988.

[101] H. Buxton and S. Gong. Visual surveillance in a dynamic and uncertain world. *Artificial Intelligence*, 78:431–459, 1995.

[102] B. Cabral, N. Cam, and J. Foran. Accelerated volume rendering and tomographic reconstruction using texture mapping hardware. In A. Kaufman and W. Krüger, editors, *1994 Symposium on Volume Visualization*, pages 91–98. ACM SIGGRAPH, 1994.

[103] B. Cabral and L. Leedom. Imaging vector fields using line integral convolution. In *Computer Graphics Proceedings (SIGGRAPH '93 Proceedings)*, Annual Conference Series, pages 263–270, Los Angeles, California, July 1993. ACM SIGGRAPH, Addison-Wesley Publishing Company, Inc.

[104] S. Campagna. *Polygonreduktion zur effizienten Speicherung, Übertragung und Darstellung komplexer polygonaler Modelle*. PhD thesis, University of Erlangen-Nürnberg, 1999.

[105] S. Campagna, L. Kobbelt, and H.-P. Seidel. Directed edges—a scalable representation for triangle meshes. Technical report, IMMD IX, University of Erlangen-Nürnberg, 1998.

[106] J. Canny. A computational approach to edge detection. *IEEE Transactions on Pattern Analysis and Machine Intelligence*, 8(3):679–698, 1986.

[107] M. P. do Carmo. *Differential Geometry of Curves and Surfaces*. Prentice-Hall Inc., 1976.

[108] G. C. Carter. Time delay estimation for passive sonar signal processing. *IEEE Transactions on Acoustics, Speech, and Signal Processing*, 29(3):463–470, 1981.

[109] E. E. Catmull. *A Subdivision Algorithm for Computer Display of Curved Surfaces*. PhD thesis, Department of CS, University of Utah, December 1974.

[110] G. Celniker and D. Gossard. Deformable curve and surface finite-element for free-form shape design. In *ACM Computer Graphics (SIGGRAPH '91 Proceedings)*, pages 257–266, 1991.

[111] A. Certain, J. Popović, T. DeRose, T. Duchamp, D. Salesin, and W. Stuetzle. Interactive multiresolution surface viewing. In H. Rushmeier, editor, *SIGGRAPH*

'96 Conference Proceedings, Annual Conference Series, pages 91–98. ACM SIGGRAPH, Addison Wesley, August 1996.

[112] A. Chaigne and A. Askenfelt. Numerical simulations of piano strings. I: A physical model for a struck string using finite difference methods. *Journal of the Acoustical Society of America*, 95(2):1112–1118, 1994.

[113] C.-F. Chang, G. Bishop, and A. Lastra. LDI tree: A hierarchical representation for image-based rendering. Technical Report TR98-030, Department of Computer Science, University of North Carolina—Chapel Hill, October 1998.

[114] S. Chaudhuri and S. Chatterjee. Robust estimation of 3D motion parameters in presence of correspondence mismatches. *Proceedings International Symposium Intelligent Robotics, Bangalore, India*, 1991.

[115] S. Chaudhuri, S. Sharma, and S. Chatterjee. Recursive estimation of motion parameters. *Computer Vision and Image Understanding*, 1996.

[116] R. Chellappa, C. W. Wilson, and A. Sirohey. Human and machine recognition of faces: A survey. *IEEE Proceedings*, 83(5):705–740, 1995.

[117] S. E. Chen. Quicktime VR—an image-based approach to virtual environment navigation. In R. Cook, editor, *SIGGRAPH '95 Conference Proceedings*, Annual Conference Series, pages 29–38. ACM SIGGRAPH, Addison Wesley, August 1995.

[118] S. E. Chen, H. E. Rushmeier, G. Miller, and D. Turner. A progressive multi-pass method for global illumination. *Computer Graphics (SIGGRAPH '91 Proceedings)*, 25(4):165–174, July 1991.

[119] J. Cheng and T. Huang. Image registration by matching relational structures. *Pattern Recognition*, 17(1):149–159, 1984.

[120] Y. T. Chien and K. S. Fu. Selection and ordering of feature observations in a pattern recognition system. *Information And Control*, 12:395–414, 1968.

[121] Z. H. Cho, J. P. Jones, and M. Singh. *Foundations in Medical Imaging*. J. Wiley & Sons, Inc., 1993.

[122] C. Choi, K. Aizawa, H. Harashima, and T. Takebe. Analysis and synthesis of facial image sequences in model-based image coding. *IEEE Transactions on Circuits and Systems for Video Technology*, 4(3):257–275, June 1994.

[123] H. J. Christensen and C. B. Madsen. Purposive reconstruction. *Computer Vision, Graphics, and Image Processing*, 60(1):103–108, 1994.

[124] P. H. Christensen, D. Lischinski, E. Stollnitz, and D. H. Salesin. Clustering for glossy global illumination. *ACM Transactions on Graphics*, 16(1):3–33, January 1997.

[125] P. H. Christensen, D. H. Salesin, and T. DeRose. A continuous adjoint formulation for radiance transport. In *Fourth Eurographics Workshop on Rendering*, pages 95–104, Paris, June 1993. Eurographics.

[126] C. S. Chua and R. Jarvis. Point signatures: a new representation for 3D object recognition. *International Journal of Computer Vision*, 25(1):63–85, 1997.

[127] P. Cignoni, L. de Floriani, C. Montani, E. Puppo, and R. Scopigno. Multiresolution modeling and visualization of volume data based on simplicial complexes. In *1994 Symposium on Volume Visualization*, pages 19–26. ACM SIGGRAPH, 1994.

[128] P. Cignoni, P. Marino, C. Montani, E. Puppo, and R. Scopigno. Speeding up isosurface extraction using interval trees. *IEEE Transactions on Visualization and Computer Graphics*, 3(2):158–170, 1997.

[129] P. Cignoni, C. Montani, E. Puppo, and R. Scopigno. Optimal isosurface extraction from irregular volume data. In *1996 Symposium on Volume Visualization*, pages 31–39, 1996.

[130] P. Cignoni, C. Montani, and R. Scopigno. A comparison of mesh simplification algorithms. *Computers & Graphics*, 22(1):37–54, 1998.

[131] R. Clouard, A. Elmoataz, C. Porquet, and M. Revenu. Borg: A knowledge-based system for automatic generation of image processing programs. *IEEE Transactions on Pattern Analysis and Machine Intelligence*, 21:128–144, 1999.

[132] L. Cloutot, X. Laboureux, and G. Häusler. Some medical applications of high speed PMT. *Chair for Optics, Annual report*, 1998.

[133] J. Cohen, M. Olano, and D. Manocha. Appearance-preserving simplification. In M. Cohen, editor, *SIGGRAPH '98 Conference Proceedings*, Annual Conference Series, pages 115–122. ACM SIGGRAPH, Addison Wesley, July 1998.

[134] J. Cohen, A. Varshney, D. Manocha, G. Turk, H. Weber, P. Agarwal, F. P. Brooks, Jr., and W. Wright. Simplification envelopes. In H. Rushmeier, editor, *SIGGRAPH '96 Conference Proceedings*, Annual Conference Series, pages 119–128. ACM SIGGRAPH, Addison Wesley, August 1996.

[135] M. Cohen, S. E. Chen, J. R. Wallace, and D. P. Greenberg. A progressive refinement approach to fast radiosity image generation. *Computer Graphics (SIGGRAPH '88 Proceedings)*, 22(4):75–84, August 1988.

[136] M. F. Cohen and J. R. Wallace. *Radiosity and Realistic Image Synthesis*. Academic Press, 1993.

[137] R. L. Cook. Shade trees. In H. Christiansen, editor, *Computer Graphics (SIGGRAPH '84 Proceedings)*, volume 18, pages 223–231, July 1984.

[138] R. L. Cook, T. Porter, and L. Carpenter. Distributed ray tracing. *Computer Graphics (SIGGRAPH '84 Proceedings)*, 18(3):137–145, July 1984.

[139] R. L. Cook and K. E. Torrance. A reflectance model for computer graphics. In *Computer Graphics (SIGGRAPH '81 Proceedings)*, pages 307–316, August 1981.

[140] R. Courant and D. Hilbert. *Methods of Mathematical Physics*, volume 1. Wiley, New York, 1953.

[141] F. C. Crow. Summed-area tables for texture mapping. In *Computer Graphics (SIGGRAPH '84 Proceedings)*, pages 207–212, July 1984.

[142] J. L. Crowley and P. Berard. Multi-modal tracking of faces for video communications. In *Proceedings of the 1997 IEEE Computer Society Conference on Computer Vision and Pattern Recognition*, pages 640–645, 1997.

[143] J. L. Crowley and H. I. Christensen. *Vision as Process*. Springer Erlbaum, Berlin, 1995.

[144] S. Culhane and J. Tsotsos. An attentional prototype for early vision. In G. Sandini, editor, *Computer Vision—ECCV '92*, Lecture Notes in Computer Science, pages 551–560, Berlin, Heidelberg, New York, London, 1992. Springer.

[145] B. Curless and M. Levoy. A volumetric method for building complex models from range images. In H. Rushmeier, editor, *SIGGRAPH '96 Conference Proceedings*, Annual Conference Series, pages 303–312. Addison Wesley, August 1996.

[146] H. B. Curry and I. J. Schoenberg. On Pólya frequency functions. IV: The fundamental spline functions and their limits. *Journal d'Analyse Mathématique*, 17:71–107, 1966.

[147] Cyberware laboratory incorporated. Cyberware model 3030RGB digitizer manual, 1995.

[148] W. Dahmen. Subdivision algorithms converge quadratically. *Journal of Computational and Applied Mathematics*, 16:145–158, 1986.

[149] W. J. Dally, L. McMillan, G. Bishop, and H. Fuchs. The delta tree: An object-centered approach to image-based rendering. Technical Memo AIM-1604, MIT, May 1996.

[150] S. M. Dance, J. P. Roberts, and B. M. Shield. Computer prediction of sound distribution in enclosed spaces using an interference pressure model. *Applied Acoustics*, 44:53–65, 1995.

[151] P.-E. Danielsson and O. Seger. Generalized and separable Sobel operators. In *Machine Vision for Three-Dimensional Scenes*, pages 347–380. Academic Press, San Diego, 1990.

[152] K. Daniilidis, M. Hansen, C. Krauss, and G. Sommer. Auf dem Weg zum künstlichen aktiven Sehen: Modellfreie Bewegungsverfolgung durch Kameranachführung. In *DAGM '95, Bielefeld*, pages 277–284, 1995.

[153] J. Danskin and P. Hanrahan. Fast algorithms for volume ray tracing. In *1992 Symposium on Volume Visualization*, pages 91–98. ACM SIGGRAPH, 1992.

[154] I. Daubechies. *Ten Lectures on Wavelets.* Society for Industrial and Applied Mathematics, 1992.

[155] R. Davis. Meta-rules: Reasoning about control. *Artificial Intelligence*, 15:241–254, 1980.

[156] P. E. Debevec and J. Malik. Rendering synthetic objects into real scenes: Bridging traditional and image-based graphics with global illumination and high dynamic range photography. In *Computer Graphics (SIGGRAPH '98 Proceedings)*, pages 189–198, July 1998.

[157] P. E. Debevec, C. J. Taylor, and J. Malik. Modeling and rendering architecture from photographs: A hybrid geometry- and image-based approach. In H. Rushmeier, editor, *SIGGRAPH '96 Conference Proceedings*, Annual Conference Series, pages 11–20. ACM SIGGRAPH, Addison Wesley, August 1996.

[158] P. E. Debevec, Y. Yu, and G. Boshokov. Efficient view-dependent image-based rendering with projective texture-mapping. Technical Report CSD-98-1003, University of California, Berkeley, May 1998.

[159] D. DeCarlo and D. Metaxas. The integration of optical flow and deformable models with applications to human face shape and motion estimation. *Computer Vision and Pattern Recognition*, pages 231–238, 1996.

[160] D. DeCarlo and D. Metaxas. Deformable model-based shape and motion analysis from images using motion residual error. *International Conference on Computer Vision*, pages 113–119, 1998.

[161] M. Deering. Geometry compression. In *Computer Graphics (SIGGRAPH '95 Proceedings)*, pages 13–20, 1995.

[162] T. A. DeFanti, M. D. Brown, and B. H. McCormick. Visualization: Expanding scientific and engineering research opportunities. In G. M. Nielson and B. Shrivers, editors, *Visualization in Scientific Computing*, pages 32–47. IEEE Computer Society Press, Los Alamitos, California, 1990.

[163] H. Delingette. Toward realistic soft-tissue modeling in medical simulation. *Proceedings of the IEEE: Special Issue on Virtual & Augmented Reality in Medicine*, 86(3):524–530, March 1998.

[164] H. Delingette, G. Subsol, S. Cotin, and J. Pignon. A craniofacial surgery testbed. *Technical Report 2119, Institut National de Recherche en Informatique et Automatique, (France)*, 1994.

[165] T. Delmarcelle and L. Hesselink. Visualizing second-order-tensor fields with hyperstreamlines. *IEEE Computer Graphics and Applications*, 13(7):25–33, July 1993.

[166] X. Q. Deng. *A Finite Element Analysis of Surgery of the Human Facial Tissue*. PhD thesis, Columbia University, New York, 1988.

[167] J. Denzler. *Aktives Sehen zur Echtzeitobjektverfolgung*, volume 163 of *Dissertationen zur künstlichen Intelligenz*. infix, St. Augustin, 1997.

[168] J. Denzler, B. Heigl, and H. Niemann. An efficient combination of 2D and 3D shape description for contour based tracking of moving objects. In H. Burkhardt and B. Neumann, editors, *Computer Vision—ECCV '98*, Lecture Notes in Computer Science, pages 843–857, Berlin, Heidelberg, New York, London, 1998. Springer.

[169] J. Denzler, B. Heigl, and D. Paulus. Farbsegmentierung für aktives Sehen. In V. Rehrmann, editor, *Erster Workshop Farbbildverarbeitung*, volume 15 of *Fachberichte Informatik*, pages 9–12, Universität Koblenz-Landau, 1995.

[170] J. Denzler and H. Niemann. Combination of simple vision modules for robust real-time motion tracking. *European Transactions on Telecommunications*, 5(3):275–286, 1995.

[171] J. Denzler and H. Niemann. Active rays: A new approach to contour tracking. *International Journal of Computing and Information Technology*, 4(1):9–16, 1996.

[172] J. Denzler and H. Niemann. Real-time pedestrian tracking in natural scenes. In G. Sommer, K. Daniilidis, and J. Pauli, editors, *Computer Analysis of Images and Patterns, (CAIP '97, Kiel)*, Lecture Notes in Computer Science, pages 42–49, Berlin, Heidelberg, New York, London, 1997. Springer.

[173] J. Denzler and H. Niemann. Active rays: Polar-transformed active contours for real-time contour tracking. *Journal on Real-Time Imaging*, 1999. to appear.

[174] R. Deriche, V. Gouet, and P. Montesinos. Differential invariants for color images. In *ICPR '98*, page CV21, 1998.

[175] P. Deuflhard and F. Bornemann. *Numerische Mathematik II: Integration gewöhnlicher Differentialgleichungen*. Walter de Gruyter, Berlin, New York, 1994.

[176] DGZfP. *Handbuch OF1: Verfahren für die optische Formerfassung*. Deutsche Gesellschaft für zerstörungsfreie Materialprüfung e. V., 1995.

[177] R. R. Dickinson. A unified approach to the design of visualization software for the analysis of field problems. In *Three-Dimensional Visualization and Display Technologies*, volume 1083, pages 173–180, Washington, January 1989. SPIE.

[178] E. D. Dickmanns and B. Mysliwetz. Recursive 3D road and relative Ego-state recognition. *IEEE Transactions on Pattern Analysis and Machine Intelligence*, 14:199–214, 1992.

[179] N. Diehl. Object-oriented motion estimation and segmentation in image sequences. *Signal Processing: Image Communication*, pages 23–56, 1991.

[180] P. Dierckx. *Curve and Surface Fitting with Splines*. Oxford Clarendon Press, 1995.

[181] U. Dietz. B-spline aproximation with energy constraints. In J. Hoschek and P. Kaklis, editors, *Advanced Course on FAIRSHAPE*, pages 229–240. B. G. Teubner, Stuttgart, 1996.

[182] U. Dietz. *Geometrierekonstruktion aus Meßpunktwolken mit glatten B-Spline-Flächen*. PhD thesis, Technische Universität Darmstadt, 1998.

[183] J. C. Dill. An application of color graphics to the display of surface curvature. In *SIGGRAPH '81 Conference Proceedings*, volume 15, pages 153–161. ACM SIGGRAPH, 1981.

[184] P. Dombrowski and C. F. Gauß. *150 Years after Gauss' "Disquisitiones Generales Circa Superficies Curvas"*. Paris Societé Mathématique de France, 1979.

[185] R. Dorsch, J. M. Herrmann, and G. Häusler. Laser triangulation: Fundamental uncertainty of distance measurement. *Applied Optics*, 33(7):1306–1314, 1994.

[186] E. Dotzauer. *Mathematische Modellierung von 3D-Freiformobjekten*. Hanser, München, Wien, 1992.

[187] D. Dovey. Vector plots for irregular grids. In G. M. Nielson and D. Silver, editors, *Visualization '95*, pages 248–253, Atlanta, Georgia, 1995. IEEE Computer Society Press.

[188] B. Drebin, L. Carpenter, and P. Hanrahan. Volume rendering. *Computer Graphics*, 22(4):65–74, August 1988.

[189] T. Dresel, G. Häusler, and H. Venzke. 3D sensing of rough surfaces by coherence radar. *Applied Optics*, 33:919–925, 1992.

[190] R. C. Dubes and A. K. Jain. Random field models in image analysis. In *Statistics and Images*, volume 1 of *Advances in Applied Statistics*, pages 121–154. Carfax Publishing Company, Abingdon, 1993.

[191] M. Dubuisson and A. K. Jain. Object contour extracting using color and motion. In *Proceedings of IEEE Conference on Computer Vision and Pattern Recognition*, pages 471–476. IEEE Society, Ney York City, 1993.

[192] M. Dubuisson and A. K. Jain. Contour extraction of moving objects in complex outdoor scenes. *International Journal of Computer Vision*, 14(1):83–105, 1995.

[193] F. Durst, A. Melling, and J. H. Whitelaw. *Theorie und Praxis der Laser-Doppler-Anemometrie*. Braun, Karlsruhe, 1987.

[194] P. Dutre, E. Lafortune, and Y. D. Willems. Monte Carlo light tracing with direct computation of pixel intensities. In *Compugraphics '93*, pages 128–137, Alvor, 1993.

[195] H. Duvenbeck and A. Schmidt. Darstellung zwei- und dreidimensionaler Strömungen. In H. Jürgens and D. Saupe, editors, *Visualisierung in Mathematik und Naturwissenschaften*, pages 21–38, Berlin, July 1988. Springer.

[196] P.-C. Eccardt, K. Niederer, T. Scheiter, and C. Hierold. Surface micromachined ultrasound transducers in CMOS technology. In *Proceedings IEEE Ultrasonics Symposium*, pages 959–962, 1996.

[197] M. Eck, T. DeRose, T. Duchamp, H. Hoppe, M. Lounsbery, and W. Stuetzle. Multiresolution analysis of arbitrary meshes. In R. Cook, editor, *SIGGRAPH '95 Conference Proceedings*, Annual Conference Series, pages 173–182. Addison Wesley, August 1995.

[198] H. Edelsbrunner. An acyclic theorem for cell complexes in d dimensions. *Combinatorica*, 10(3):251–260, 1990.

[199] H. Edelsbrunner and E. P. Mücke. Three-Dimensional alpha shapes. *ACM Transactions on Graphics*, 13(1):43–72, January 1994.

[200] P. Eisert and B. Girod. Model-based 3D motion estimation with illumination compensation. *International Conference on Image Processing and its Applications*, 1:194–198, July 1997.

[201] P. Eisert and B. Girod. Analyzing facial expressions for virtual conferencing. *IEEE Computer Graphics & Applications*, pages 70–78, September 1998.

[202] P. Eisert and B. Girod. Model-based coding of facial image sequences at varying illumination conditions. In *10th IMDSP Workshop '98*, pages 119–122, 1998.

[203] P. Eisert, E. Steinbach, and B. Girod. Multi-hypothesis volumetric reconstruction of 3D objects from multiple calibrated camera views. In *ICASSP*, March 1999.

[204] P. Ekman and W. V. Friesen. *Facial Action Coding System*. Consulting Psychologists Press, Inc., 1978.

[205] A. Eleftheriadis and A. Jacquin. Automatic face location detection and tracking for model-assisted coding of video teleconferencing sequences at low bit-rates. *Signal Processing: Image Communication*, 7(3):231–248, 1995.

[206] T. T. Elvins. A survey of algorithms for volume visualization. *Computer Graphics*, 26(3):194–201, 1992.

[207] J. Encarnação, W. Straßer, and R. Klein. *Graphische Datenverarbeitung 2*. Oldenbourg, Munich, Germany, 4th edition, 1997.

[208] K. Engel, R. Westermann, and T. Ertl. Isosurface extraction techniques for web-based volume visualization. Technical Report 3, IMMD IX, University of Erlangen-Nürnberg, 1999.

[209] S. Enger, M. Breuer, and B. Basu. Numerical simulation of fluid flow and heat transfer in an industrial Czochralski melt using a parallel-vector supercomputer. In *High Performance Computing in Science and Engineering 1999*, Lecture Notes in Computational Sciences and Engineering. Springer, Berlin, 2000. to appear.

[210] S. M. Ermakov and A. A. Zhiglyavskij. On random search of global extremum. *Probability Theory and Applications*, 28(1):129–136, 1983.

[211] I. Babuska et al. *Accuracy Estimates and Adaptive Refinements in Finite Element Computations*. Wiley, New York, 1986.

[212] P. Ettl. Studien zur hochgenauen Objektvermessung mit dem Kohärenzradar. Master's thesis, Physics Insitute V, University of Erlangen-Nürnberg, 1995.

[213] F. Evans, S. Skiens, and A. Varshney. Optimizing triangle strips for fast rendering. In *IEEE Visualization 1996*, pages 319–326, 1996.

[214] L. Falkenhagen. Depth estimation from stereoscopic image pairs assuming piecewise continuous surfaces. In Y. Paker and S. Wilbur, editors, *Image Processing for Broadcast and Video Production*, Workshops in Computing, pages 115–127. Springer, Hamburg, 1994.

[215] G. Farin. *Curves and Surfaces for Computer Aided Geometric Design*. Acadamic Press, Boston, 1993.

[216] R. Farouki and V. Rajan. On the numerical condition of polynomials in Bernstein form. *Computer Aided Geometric Design*, 4:191–216, 1987.

[217] G. E. Fasshauer and L. L. Schumaker. Minimal energy surfaces using parametric splines. *Computer Aided Geometric Design*, 13:45–79, 1996.

[218] D. D. Faugeras, M. Herbert, P. Mussi, and J. D. Boissonnat. Polyhedral approximation of 3D objects without holes. *Computer Vision, Graphics, and Image Processing*, 26:169–183, February 1984.

[219] O. Faugeras. *Three-Dimensional Computer Vision—A Geometric Viewpoint*. MIT Press, Cambridge, Massachusetts, 1993.

[220] J. Feldmar and N. Ayache. Rigid, affine and locally affine registration of free-form surfaces. *International Journal of Computer Vision*, 18(2):99–119, 1996.

[221] J. H. Ferziger and M. Perić. *Computational Methods for Fluid Dynamics*. Springer, Berlin, 2nd edition, 1999.

[222] A. Fettweis. Multidimensional wave-digital principles: From filtering to numerical integration. In *Proceedings International Conference on Acoustics, Speech, and Signal Processing (ICASSP '94)*, pages VI/173–181. IEEE, April 1994.

[223] P. Fieguth and D. Terzopoulos. Color-based tracking of heads and other mobile objects at video frame rates. In *Proceedings of the 1997 IEEE Computer Society Conference on Computer Vision and Pattern Recognition*, pages 21–27, 1997.

[224] V. Fischer. *Parallelverarbeitung in einem semantischen Netzwerk für die wissensbasierte Musteranalyse*, volume 95 of *DISKI*. infix, Sankt Augustin, 1995.

[225] V. Fischer and H. Niemann. Parallelism in a semantic network for image understanding. In A. Bode and M. Dal Cin, editors, *Parallelrechner: Theorie, Hardware, Software, Anwendungen*, Lecture Notes in Computer Science Nr. 732, pages 203–218. Springer, Berlin, Heidelberg, New York, 1993.

[226] M. S. Floater. Parameterization and smooth approximation of surface triangulations. *Computer Aided Geometric Design*, 14:231–250, 1997.

[227] J. D. Foley, A. van Dam, S. K. Feiner, and J. F. Hughes. *Computer Graphics: Principles and Practice*. Addison-Wesley, 2nd edition, 1990.

[228] T. A. Foley and G. M. Nielson. Knot selection for parametric spline interpolation. In T. Lyche and L. L. Schumaker, editors, *Mathematical Methods in Computer Aided Geometric Design*, pages 261–271. Academic Press, Boston, 1989.

[229] D. R. Forsey and R. H. Bartels. Hierarchical B-spline refinement. In *ACM Computer Graphics (SIGGRAPH '88 Proceedings)*, pages 205–212, 1988.

[230] D. R. Forsey and R. H. Bartels. Surface fitting with hierarchical splines. *ACM Transactions on Graphics*, 14:134–161, 1995.

[231] L. K. Forssell. Visualizing flow over curvilinear grid surfaces using line integral convolution. In Bergeron D. and Kaufman A., editors, *Visualization '94*, pages 240–247, Los Alamitos, California, 1994. IEEE Computer Society Press.

[232] R. Franke and G. M. Nielson. Scattered data interpolation and applications: A tutorial and survey. In H. Hagen and D. Roller, editors, *Geometric Modeling*, pages 131–161. Springer, Berlin, 1991.

[233] R. W. Frischholz. *Beiträge zur automatischen dreidimensionalen Bewegungsanalyse*. PhD thesis, University of Erlangen-Nürnberg, 1998.

[234] T. Frühauf. Interactive visualization of vector data in unstructured volumes. *Computers & Graphics*, 18(1):73–80, 1994.

[235] T. Funkhouser, I. Carlbom, G. Elko, G. Pingali, M. Sondhi, and J. West. A beam tracing approach to acoustic modeling for interactive virtual environments. In *Proceedings SIGGRAPH '98*, 1998.

[236] W. G. Gardener. *3D Audio Using Loudspeakers.* Kluwer Academic Publishers, Boston, 1998.

[237] M. Garland and P. S. Heckbert. Surface simplification using quadric error metrics. In T. Whitted, editor, *SIGGRAPH '97 Conference Proceedings*, Annual Conference Series, pages 209–216. ACM SIGGRAPH, Addison Wesley, August 1997.

[238] A. Gelb. *Applied Optimal Estimation.* The MIT Press, Cambridge, Massachusetts, 1979.

[239] D. B. Gennery. Visual tracking of known 3D objects. *International Journal of Computer Vision*, 7(3):243–270, 1992.

[240] A. Gershun. The light field. *Journal of Mathematics and Physics*, 18:51–151, 1939.

[241] A. Geyer-Schulz. *Fuzzy Rule-Based Expert Systems and Genetic Machine Learning*. Physica, Heidelberg, 1995.

[242] S. Gibson and R. J. Hubbold. Efficient hierarchical refinement and clustering for radiosity in complex environements. *Computer Graphics Forum*, 15(5):297–310, December 1996.

[243] R. H. Gilkey and T. R. Anderson. *Binaural and Spatial Hearing in Real and Virtual Environments.* Lawrence Erlbaum Assoc., Mahwah, USA, 1997.

[244] B. Girod and E. Steinbach. A new method for simultaneous estimation of displacement, depth, and rigid body motion parameters. *IEEE IMDSP*, pages 122–123, March 1996.

[245] S. Girod, E. Keeve, and B. Girod. Soft tissue prediction in orthognatic surgery by 3D CT and 3D laser scanning. *Journal of Oral and Maxillofacial Surgery Suppl.*, 51:167, 1993.

[246] A. Glassner. *Principles of Digital Image Synthesis.* Morgan Kaufmann, 1995.

[247] M. E. Go Ong. *Hierarchical Basis Preconditioning for Second Order Elliptic Problems in Three Dimensions.* PhD thesis, University of California, Los Angeles, 1989.

[248] M. von Golitschek and L. L. Schumaker. Data fitting by penalized least squares. In J. C. Mason, editor, *Algorithms for approximation II*, pages 210–227, Shrivenham, 1988.

[249] G. H. Golub and C. F. van Loan. *Matrix Computations.* The John Hopkins University Press, Baltimore, Maryland, 1989.

[250] J. W. Goodman. Statistical properties of laser speckle patterns. In J. C. Dainty, editor, *Laser Speckle and Related Phenomena.* Springer, Berlin, 1984.

[251] J. W. Goodman. *Introduction to Fourier Optics*. McGraw-Hill, San Francisco, 2nd edition, 1996.

[252] M. Goodwin. *Adaptive Signal Models*. Kluwer Academic Publishers, Boston, 1998.

[253] C. M. Goral, K. E. Torrance, and D. P. Greenberg. Modeling the interaction of light between diffuse surfaces. *Computer Graphics (SIGGRAPH '84 Proceedings)*, 18(3):212–222, July 1984.

[254] S. J. Gortler, R. Grzeszczuk, R. Szeliski, and M.. F. Cohen. The lumigraph. In H. Rushmeier, editor, *SIGGRAPH '96 Conference Proceedings*, Annual Conference Series, pages 43–54. ACM SIGGRAPH, Addison Wesley, August 1996.

[255] S. Gottschalk, M. Lin, and D. Manocha. OBB-tree: A hierarchical structure for rapid interference detection. In *Proceedings of SIGGRAPH '96*, 1996.

[256] N. Greene. Applications of world projections. In *Proceedings of Graphics Interface '86*, pages 108–114, May 1986.

[257] G. Greiner. Surface construction based on variational principles. In P. J. Laurent, A. LeMéhauté, and L. L. Schumaker, editors, *Wavelets, Images, and Surface Fitting*, pages 277–286. AK Peters, Wellesley, 1994.

[258] G. Greiner. Variational design and fairing of spline surfaces. In *Computer Graphics Forum (EUROGRAPHICS '94 Proceedings)*, volume 13, pages 143–154, 1994.

[259] G. Greiner and K. Hormann. Interpolating and approximating scattered 3D data with hierarchical tensor product B-splines. In A. Méhauté, C. Rabut, and L. L. Schumaker, editors, *Surface Fitting and Multiresolution Methods*, pages 163–172. Vanderbilt University Press, 1997.

[260] G. Greiner, J. Loos, and W. Wesselink. Data dependent thin plate energy and its use in interactive surface modeling. In *Computer Graphics Forum (EUROGRAPHICS '96 Proceedings)*, volume 15, pages 175–185, 1996.

[261] G. Greiner and H. P. Seidel. Splines in computer graphics: Polar forms and triangular B-spline surfaces. *Computer Graphics Forum (EUROGRAPHICS '93 Proceedings)*, 1993.

[262] W. E. Grimson. *Object Recognition by Computer: The Role of Geometric Constraints*. MIT Press, Cambridge, Massachusetts, 1990.

[263] R. Grosso and T. Ertl. Progressive iso-surface extraction from hierarchical 3D meshes. *Computers Graphics Forum (EUROGRAPHICS '98)*, 17(3), 1998.

[264] R. Grosso and G. Greiner. Hierarchical meshes for volume data. In *Proceedings CGI*, Hannover, Germany, 1998.

[265] M. Gruber and G. Häusler. Simple, robust and accurate phase-measuring triangulation. *Optik*, 89(3):118–122, 1992.

[266] X. Gu, S. Gortler, H. Hoppe, L. McMillan, B. Brown, and A. Stone. Silhouette mapping. Technical Report TR-1-99, Harvard University, Computer Science Technical Report, March 1999.

[267] T. Guenter. Virim: A massively parallel processor for real-time volume visualization in medicine. In *Proceedings Ninth Eurographics Hardware Workshop*, pages 103–108. Addison-Wesley, 1994.

[268] L. Guibas and J. Stolfi. Primitives for the manipulation of general subdivisions and computation of Voronoi diagrams. *ACM Transactions on Graphics*, 4(2):74–123, April 1985.

[269] I. Guskov, W. Sweldens, and P. Schröder. Multiresolution signal processing for meshes. Technical Report 99-01, Princeton University, Program in Applied and Computational Mathematics, January 1999.

[270] P. Haeberli and M. Segal. Texture mapping as a fundamental drawing primitive. In *Fourth Eurographics Workshop on Rendering*, pages 259–266, June 1993.

[271] H. Hagen. Surface interrogation algorithms. *IEEE Visualization and Computer Graphics*, pages 53–60, 1992.

[272] H. Hagen, G.-P. Bonneau, and S. Hahmann. Variational design and surface interrogation. *Computer Graphics Forum (EUROGRAPHICS '93 Proceedings)*, 13:447–459, 1993.

[273] G. D. Hager and K. Toyama. X vision: Combining image warping and geometric constraints for fast visual tracking. In A. Blake, editor, *Computer Vision—ECCV '96*, Lecture Notes in Computer Science, pages 507–517, Berlin, Heidelberg, New York, London, 1996. Springer.

[274] W. R. Hahn. Optimum signal processing for passive sonar range and bearing estimation. *Journal of the Acoustical Society of America*, 58(1):201–207, 1975.

[275] W. R. Hahn and S. A. Tretter. Optimum processing for delay-vector estimation in passive signal arrays. *IEEE Transactions on Information Theory*, 19(5):608–614, 1973.

[276] M. Halioua, H. Liu, and V. Srinivasan. Automated phase-measuring profilometry of 3D diffuse objects. *Applied Optics*, 23(18):3105–3108, 1984.

[277] R. Hall. *Illumination and Color in Computer Generated Imagery*. Springer, New York, 1989.

[278] M. Halstead, B. Barsky, S. Klein, and R. Mandell. Reconstructing curved surfaces from specular reflection patterns using spline surface fitting of normals.

In H. Rushmeier, editor, *SIGGRAPH '96 Conference Proceedings*, pages 335–342. Addison Wesley, August 1996.

[279] M. Halstead, M. Kass, and T. DeRose. Efficient, fair interpolation using Catmull-Clark surfaces. In *ACM Computer Graphics (SIGGRAPH '93 Proceedings)*, pages 35–44, 1993.

[280] P. Hanrahan. *Radiosity and Realistic Image Synthesis*. Academic Press, 1993.

[281] P. Hanrahan and J. Lawson. A language for shading and lighting calculations. In *Computer Graphics (SIGGRAPH '90 Proceedings)*, pages 289–298, August 1990.

[282] P. Hanrahan, D. Salzman, and L. Aupperle. A rapid hierarchical radiosity algorithm. *Computer Graphics (SIGGRAPH '91 Proceedings)*, 25(4):197–206, 1991.

[283] R. M. Haralick. Performance characterization in computer vision. *Computer Vision, Graphics, and Image Processing*, 60(2):245–249, September 1994.

[284] M. Harbeck. *Objektorientierte linienbasierte Segmentierung von Bildern*. Shaker, Aachen, 1996.

[285] R. L. Hardy. Multiquadric equation of topography and other irregular surfaces. *Journal Geophysical Research*, 76:1905–1915, 1971.

[286] C. Harris. Tracking with rigid models. In A. Blake and A. Yuille, editors, *Active Vision*, pages 59–74. MIT Press, Cambridge, Massachusetts, London, England, 1992.

[287] C. Harris and M. J. Stephens. A combined corner and edge detector. In *Alvey '88*, pages 147–152, 1988.

[288] R. I. Hartley. Estimation of relative camera positions for uncalibrated cameras. In *Proceedings European Conference on Computer Vision (ECCV '92)*, LNCS, pages 579–587. Springer, 1992.

[289] R. I. Hartley. In defense of the eight-point algorithm. *IEEE Transactions on Pattern Analysis and Machine Intelligence*, 19(6):580–593, June 1997.

[290] R. I. Hartley. Lines and points in three views and the trifocal tensor. *International Journal of Computer Vision*, 22(2):125–140, 1997.

[291] H. R. Hashemipour, S. Roy, and A. J. Laub. Decentralized structures for parallel Kalman filtering. *IEEE Transactions on Automatic Control*, 33(1):88–93, 1988.

[292] J. C. Hassab and R. E. Boucher. Optimum estimation of time delay by a generalized correlator. *IEEE Transactions on Acoustics, Speech, and Signal Processing*, 27(4):373–380, 1979.

[293] G. Häusler. Möglichkeiten und Grenzen optischer 3D-Sensoren in der industriellen Praxis. In Koch, Rupprecht, Toedter, and Häusler, editors, *Optische Meßtechnik an diffus reflektierenden Medien*. Expert, 1997.

[294] G. Häusler, G. Ammon, P. Andretzky, S. Blossey, G. Bohn, P. Ettl, H. P. Habermeier, B. Harand, I. Laszlo, and B. Schmidt. New modifications of the coherence radar. In W. Jüptner and W. Osten, editors, *Fringe '97*, Third International Workshop on Automatic Processing of Fringe Pattern, 1997.

[295] G. Häusler, G. Bickel, and M. Maul. Triangulation with expanded range of depth. *Optical Engineering*, 24(6):975–977, 1985.

[296] G. Häusler, P. Ettl, B. Schmidt, M. Schenk, and I. Laszlo. Roughness parameters and surface deformation measured by coherence radar. In *Proceedings of the SPIE*, volume 3407, 1998.

[297] G. Häusler and W. Heckel. Light sectioning with large depth and high resolution. *Applied Optics*, 27:5165–5169, 1988.

[298] G. Häusler, M. B. Hernanz, R. Lampalzer, and H. Schönfeld. 3D real time camera. In W. Jüptner and W. Osten, editors, *Fringe '97*, 3rd International Workshop on Automatic Processing of Fringe Pattern, 1997.

[299] G. Häusler and S. Karbacher. Reconstruction of smoothed polyhedral surfaces from multiple range images. In B. Girod, H. Niemann, and H.-P. Seidel, editors, *3D Image Analysis and Synthesis '97*, pages 191–198, Sankt Augustin, Germany, 1997. infix.

[300] G. Häusler and E. Körner. Imaging with expanded depth of focus. *Zeiss Information*, 98(9), 1986.

[301] G. Häusler, S. Kreipl, R. Lampalzer, A. Schielzeth, and B. Spellenberg. New range sensors at the physical limit of measuring uncertainty. In *Proceedings of the EOS Topical Meeting on Optoelectronic Distance Measurements and Applications*, 1997.

[302] G. Häusler and G. Leuchs. Physikalische Grenzen der optischen Formerfassung mit Licht. *Physikalische Blätter*, 53:417–421, 1997.

[303] G. Häusler and M. Lindner. Coherence radar and spectral radar—new tools for dermatological diagnosis. *Journal of Biomedical Optics*, 3:21–31, 1998.

[304] G. Häusler, H. Schönfeld, and F. Stockinger. Kalibrierung von optischen 3D-Sensoren. *Optik*, 102(3):93–100, May 1996.

[305] P. Havaldar, M.-S. Lee, and G. Medioni. Synthesizing novel views from unregistered 2D images. *Computer Graphics Forum*, 16(1):65–73, 1997.

[306] B. Hayes-Roth. A blackboard architecture for control. *Artificial Intelligence*, 26:251–321, 1985.

[307] X. D. He, K. E. Torrance, F. X. Sillion, and D. P. Greenberg. A comprehensive physical model for light reflection. In *Computer Graphics (SIGGRAPH '91 Proceedings)*, pages 175–186, July 1991.

[308] P. S. Heckbert. Survey of texture mapping. *IEEE Computer Graphics and Applications*, 6(11):56–67, November 1986.

[309] H. C. Hege, T. Höllerer, and D. Stalling. Volume rendering, mathematical foundations and algorithmic aspects. Technical Report TR93-7, Konrad-Zuse-Zentrum für Informationstechnik Berlin, 1993.

[310] W. Heidrich, J. Kautz, P. Slusallek, and H.-P. Seidel. Canned lightsources. In *Rendering Techniques '98 (Proceedings Ninth Eurographics Workshop on Rendering)*, pages 293–300, 1998.

[311] W. Heidrich and H.-P. Seidel. Ray-tracing procedural displacement shaders. In *Graphics Interface '98*, pages 8–16, 1998.

[312] W. Heidrich and H.-P. Seidel. View-independent environment maps. In *Eurographics/SIGGRAPH Workshop on Graphics Hardware*, pages 39–45, 1998.

[313] W. Heidrich, P. Slusallek, and H.-P. Seidel. An image-based model for realistic lens systems in interactive computer graphics. In Wayne A. Davis, Marilyn Mantei, and R. Victor Klassen, editors, *Graphics Interface '97*, pages 68–75. Canadian Information Processing Society, Canadian Human-Computer Communications Society, May 1997.

[314] F. Heitz and P. Bouthemy. Multimodal motion estimation and segmentation using Markov random fields. In *Proceedings of International Conference on Pattern Recognition*, pages 378–383, 1990.

[315] J. L. Helman and L. Hesselink. Automated analysis of fluid flow topology. In *Three-Dimensional Visualization and Display Technologies*, volume 1083, pages 825–855, Washington, January 1989. SPIE.

[316] J. L. Helman and L. Hesselink. Visualizing vector field topology in fluid flows. *IEEE Computer Graphics and Applications*, 11(3):36–46, May 1991.

[317] R. Herpers, H. Kattner, H. Rodax, and G. Sommer. Gaze: An attentive processing strategy to detect and analyze the prominent facial regions. In M. Bichsel, editor, *Proceedings of the International Workshop on Automatic Face- and Gesture-Recognition—IWAFGR '95*, pages 214–220, Zürich, 1995.

[318] R. Herpers, H. Rodax, and G. Sommer. A neural network identifies faces with morphological syndrome. In S. Andreassen, editor, *Artificial Intelligence in Medicine*, pages 481–485. IOS Press, Amsterdam, 1993.

[319] A. Hilton, A. J. Stoddart, J. Illingworth, and T. Windeatt. Marching triangles: Range image fusion for complex object modelling. In *International Conference on Image Processing*, 1996.

[320] A. Hilton, A. J. Stoddart, J. Illingworth, and T. Windeatt. Reliable surface reconstruction from multiple range images. In *Forth European Conference on Computer Vision*, pages 117–126. Springer, 1996.

[321] C. Hirsch. *Numerical Computation of Internal and External Flows*. Wiley series in numerical methods in engineering. John Wiley & Sons, New York, 1990.

[322] M. Hoch, G. Fleischmann, and B. Girod. Modeling and animation of facial expressions based on B-splines. *Visual Computer*, 11:87–95, 1994.

[323] H. Hoppe. *Surface Reconstruction from Unorganized Points*. PhD thesis, University of Washington, 1994.

[324] H. Hoppe. Progressive meshes. In H. Rushmeier, editor, *SIGGRAPH '96 Conference Proceedings*, Annual Conference Series, pages 99–108. Addison Wesley, August 1996.

[325] H. Hoppe, T. DeRose, T. Duchamp, M. Halstead, H. Jin, J. McDonald, J. Schweitzer, and W. Stuetzle. Piecewise smooth surface reconstruction. In *ACM Computer Graphics (SIGGRAPH '94 Proceedings)*, pages 295–302, 1994.

[326] H. Hoppe, T. DeRose, T. Duchamp, J. McDonald, and W. Stuetzle. Surface reconstruction from unorganized points. In *ACM Computer Graphics (SIGGRAPH '92 Proceedings)*, pages 71–78, 1992.

[327] H. Hoppe, T. DeRose, T. Duchamp, J. McDonald, and W. Stuetzle. Mesh optimization. In J. T. Kajiya, editor, *Computer Graphics (SIGGRAPH '93 Proceedings)*, volume 27, pages 19–26, August 1993.

[328] U. Horbach. New techniques for the production of multichannel sound. In *Proceedings 103. AES Convention*. Audio Engineeering Society, 1997.

[329] K. Hormann. Glatte Approximation mit hierarchischen Splineflächen. Master's thesis, IMMD IX, University of Erlangen-Nürnberg, 1997.

[330] B. K. Horn. Extended Gaussian images. *Proceedings IEEE*, 72(12):1671–1686, 1982.

[331] B. K. Horn. Closed-form solution of absolute orientation using unit quaternions. *Journal of the Optical Society of America*, 4:629–642, 1987.

[332] B. K. Horn. Closed-form solution of absolute orientation using orthonormal matrices. *Journal of the Optical Society of America*, 5:1127–1135, 1988.

[333] B. K. Horn. *Robot Vision*. MIT Press, Cambridge, 1998.

[334] B. K. Horn and E. J. Weldon. Direct methods for recovering motion. *International Journal of Computer Vision*, 2:51–76, 1988.

[335] J. Hornegger. *Statistische Modellierung, Klassifikation und Lokalisation von Objekten*. Shaker, Aachen, 1996.

[336] J. Hornegger and H. Niemann. Statistical learning, localization, and identification of objects. In *Proceedings of the Fifth International Conference on Computer Vision (ICCV)*, pages 914–919, Boston, Massachusetts, USA, 1995. IEEE Computer Society Press.

[337] J. Hornegger, H. Niemann, D. Paulus, and G. Schlottke. Object recognition using hidden Markov models. In E. S. Gelsema and L. N. Kanal, editors, *Pattern Recognition in Practice IV: Multiple Paradigms, Comparative Studies and Hybrid Systems*, volume 16 of *Machine Intelligence and Pattern Recognition*, pages 37–44, Amsterdam, June 1994. Elsevier.

[338] S. L. Horowitz and T. Pavlidis. Picture segmentation by a tree traversal algorithm. *J. Assoc. Comput. Mach.*, 23:368–388, 1976.

[339] J. Hoschek. Intrinsic parametrization for approximation. *Computer Aided Geometric Design*, 5:27–31, 1988.

[340] J. Hoschek and W. Dankwort. *Reverse Engineering*. B. G. Teubner, Stuttgart, 1996.

[341] J. Hoschek and D. Lasser. *Fundamentals of Computer Aided Geometric Design*. AK Peters, Wellesley MA, 1993.

[342] G. N. Hounsfield and J. A. Ambrose. Computerized transverse axial scanning (tomography). *British Journal of Radiology*, 46:1016–1022, 1973.

[343] J. W. House. Facial nerve grading systems. *Laryngoscope*, 93:1056–1069, 1983.

[344] W. M. Hsu. Segmented ray casting for data parallel volume rendering. In T. Crockett, C. Hansen, and Whitman S., editors, *1993 Parallel Rendering Symposium*, pages 7–14, New York, 1993. ACM SIGGRAPH.

[345] T. H. Huang and A. N. Netravali. Motion and structure from feature correspondences: A review. *Proceedings of the IEEE*, 82(2):252–268, February 1994.

[346] T. S. Huang, S. Reddy, and K. Aizawa. Human facial motion analysis and synthesis for video compression. *Proceedings of the SPIE*, pages 234–241, 1991.

[347] J. P. Hultquist. Interactive numerical flow visualization using stream surfaces. In *Computing Systems in Engineering*, pages 349–353, 1990.

[348] J. P. Hultquist. Constructing stream surface in steady 3D vector fields. In A. E. Kaufman and G. M. Nielson, editors, *Visualization '92*, pages 171–178. IEEE Computer Society Press, October 1992.

[349] J. Huopaniemi, L. Savioja, and M. Karjalainen. Modeling of reflections and air absorption in acoustical spaces—a digital filter design approach. In *Proceedings IEEE Workshop on Applications of Signal Processing to Audio and Acoustics*, 1997.

[350] H. Hutten. *Biomedizinsiche Technik 1: Diagnostik und bildgebende Verfahren.* Springer, Berlin, 1992.

[351] K. Ikeuchi and S. Sato. Determining reflectance properties of an object using range and brightness images. *IEEE Transactions on Pattern Analysis and Machine Intelligence*, 13(11):1139–1153, 1991.

[352] D. S. Immel, M. F. Cohen, and D. P. Greenberg. A radiosity method for non-diffuse environments. *Computer Graphics (SIGGRAPH '86 Proceedings)*, pages 133–142, August 1986.

[353] V. Interrante and C. Grosch. Visualizing 3D flow. *IEEE Computer Graphics and Applications*, 18(4):47–53, 1998.

[354] S. S. Intille and A. F. Bobick. Disparity-space images and large occlusion stereo. Technical Report no. 220, M.I.T., Media Lab, Vision and Modelling Group, 1993.

[355] M. Irani, B. Rousso, and S. Peleg. Detection and tracking multiple moving objects using temporal integration. In G. Sandini, editor, *Computer Vision—ECCV '92*, Lecture Notes in Computer Science, pages 282–287, Berlin, Heidelberg, New York, London, 1992. Springer.

[356] M. Isard and B. Andrew. CONDENSATION—conditional density propagation for visual tracking. *International Journal of Computer Vision*, 29(1):5–28, 1998.

[357] M. Isard and A. Blake. Contour tracking by stochastic propagation of conditional density. In A. Blake, editor, *Computer Vision—ECCV '96*, Lecture Notes in Computer Science, pages 343–356, Berlin, Heidelberg, New York, London, 1996. Springer.

[358] J. Ivins and J. Porrill. Active region models for segmenting medical images. In *First International Conference on Image Processing*, pages II/227–231, Austin, Texas, 1994.

[359] B. Jähne, H. Haußecker, and P. Geißler. *Handbook of Computer Vision and Applications*, volume I: Sensors and Imaging. Academic Press, Boston, USA, 1999.

[360] A. K. Jain and P. J. Flynn. *Three-Dimensional Object Recognition Systems.* Elsevier, Amsterdam, 1993.

[361] R. Jain and H. H. Nagel. On the analysis of accumulative difference pictures from image sequences of real world scenes. *IEEE Transactions on Pattern Analysis and Machine Intelligence*, 1:206–214, 1979.

[362] T. Jebara, A. Azarbayejani, and A. Pentland. 3D structure from 2D motion. *IEEE Signal Processing Magazine*, pages 66–84, May 1999.

[363] F. V. Jensen. *An Introduction to Bayesian Networks*. UCL Press, London, 1996.

[364] H. W. Jensen. Global illumination using photon maps. In X. Pueyo and P. Schröder, editors, *Rendering Techniques '96 (Proceedings Seventh Eurographics Workshop on Rendering)*, pages 21–30. Springer, June 1996.

[365] B. Joe. Construction of three-dimensional Delaunay triangulations using local transformations. *Computer Aided Geometric Design*, 8:123–142, 1991.

[366] D. H. Johnson and D. E. Dudgeon. *Array Signal Processing*. Prentice Hall, Englewood Cliffs, New Jersey, 1993.

[367] M. Kahrs and K. Brandenburg. *Application of Digital Signal Processing to Audio and Acoustics*. Kluwer Academic Publishers, Boston, 1998.

[368] J. T. Kajiya. Anisotropic reflection models. In *Computer Graphics (SIGGRAPH '85 Proceedings)*, pages 15–21, August 1985.

[369] J. T. Kajiya. The rendering equation. *Computer Graphics (SIGGRAPH '86 Proceedings)*, 20(4):143–150, August 1986.

[370] R. E. Kalman. A new approach to linear filtering and prediction problems. *Journal of Basic Engineering*, pages 35–44, 1960.

[371] M. Kaltenbacher, H. Landes, and R. Lerch. An efficient calculation scheme for the numerical simulation of coupled magnetomechanical systems. *IEEE Transactions on Magnetics*, pages 1646–1649, March 1997.

[372] M. Kaltenbacher, H. Landes, and R. Lerch. A finite-element/boundary-element method for the simulation of coupled electrostatic-mechanical systems. *Journal de Physique III France*, pages 1975–1982, 1997.

[373] M. Kaneko, A. Koike, and Y. Hatori. Coding of facial image sequence based on a 3D model of the head and motion detection. *Journal of Visual Communication and Image Representation*, 2(1):39–54, March 1991.

[374] S. Karbacher. *Rekonstruktion und Modellierung von Flächen aus Tiefenbildern*. PhD thesis, University of Erlangen-Nürnberg, Shaker, Aachen, 1997.

[375] S. Karbacher and G. Häusler. A new approach for modeling and smoothing of scattered 3D data. In R. N. Ellson and J. H. Nurre, editors, *Three-Dimensional Image Capture and Applications*, volume 3313 of *SPIE Proceedings*, pages 168–177, Bellingham, Washington, 1998. The International Society for Optical Engineering.

[376] K. Karhunen. Über lineare Methoden in der Wahrscheinlichkeitsrechnung. *Ann. Acad. Sci. Fenn.*, Ser. A I:37, 1947.

[377] W. Kasprzak and H. Niemann. Adaptive road recognition and ego-state tracking in the presence of obstacles. *International Journal of Computer Vision*, 28(1):5–26, 1998.

[378] M. Kass, A. Witkin, and D. Terzopoulos. Snakes: Active contour models. In *International Journal of Computer Vision*, pages 321–331. Kluwer Academic Publishers, 1988.

[379] A. Kaufman. *Volume Visualization*. IEEE Computer Society Press, 1991.

[380] E. Kaufman and R. Klass. Smoothing surfaces using reflection lines for families of splines. *Computer Aided Design*, 20(6):312–316, 1988.

[381] G. Kay and T. Caelli. Estimating the parameters of an illumination model using photometric stereo. *Graphical Models and Image Processing*, 5(57):365–388, 1995.

[382] E. Keeve, S. Girod, R. Kikinis, and B. Girod. Deformable modeling of facial tissue for craniofacial surgery simulation. *Computer Aided Surgery*, 3(5), 1998.

[383] E. Keeve, S. Girod, P. Pfeifle, and B. Girod. Anatomy-based facial tissue modeling using the finite element method. In *Proceedings IEEE Visualization*, 1996.

[384] E. Keeve, S. Girod, S. Schaller, and B. Girod. Adaptive surface data compression. Technical report, Lehrstuhl für Nachrichtentechnik, University of Erlangen-Nürnberg, 1996.

[385] D. N. Kenwright and D. A. Lane. Optimization of time-dependent particle tracing using tetrahedral decomposition. In G. M. Nielson and Silver D., editors, *Visualization '95*, pages 321–328, Los Alamitos, California, 1995. IEEE Computer Society Press.

[386] G. D. Kerlick. Moving iconic objects in scientific visualization. In A. Kaufman, editor, *Visualization '90*, pages 124–130, San Francisco, California, 1990. IEEE Computer Society Press.

[387] R. Kikinis, H. Cline, D. Altobelli, M. Halle, W. Lorensen, and F. Jolesz. Interactive visualization and manipulation of 3D reconstructions for the planning of surgical procedures. In *Proceedings of Visualization in Biomedical Computing VBC '92*, pages 559–563, 1992.

[388] R. Klass. Correction of local surface irregularities using reflection lines. *Computer Aided Design*, 12:73–77, March 1980.

[389] R. Klein, G. Liebich, and W. Sraßer. Mesh reduction with error control. In *IEEE Visualization '96 (Conference Proceedings)*, pages 311–318, 1996.

[390] W. Klingenberg. *A Course in Differential Geometry*. Springer, Berlin/Heidelberg, 1978.

[391] G. J. Klinker, S. A. Shafer, and T. Kanade. The measurement of highlights in color images. *International Journal of Computer Vision (IJCV)*, 2(1):7–32, 1988.

[392] G. J. Klinker, S. A. Shafer, and T. Kanade. A physical approach to color image understanding. *International Journal of Computer Vision*, 4:7–38, 1990.

[393] G. Knittel and W. Straßer. A compact volume rendering accelerator. In A. Kaufman and W. Krüger, editors, *1994 Symposium on Volume Visualization*, pages 67–74. ACM SIGGRAPH, 1994.

[394] G. Knoll. *Radiation Detection and Measurement*. Wiley, N.Y., 1979.

[395] L. Kobbelt, S. Campagna, and H.-P. Seidel. A general framework for mesh decimation. In *Proceedings of Graphics Interface '98*, pages 43–50, 1998.

[396] L. Kobbelt, S. Campagna, J. Vorsatz, and H.-P. Seidel. Interactive multi-resolution modeling on arbitrary meshes. In M. Cohen, editor, *SIGGRAPH '98 Conference Proceedings*, Annual Conference Series, pages 105–114. ACM SIGGRAPH, Addison Wesley, July 1998.

[397] R. M. Koch. Dynamic 3D scene analysis through synthesis feedback control. *IEEE Transactions on Pattern Analysis and Machine Intelligence*, 15(6):556–568, June 1993.

[398] R. M. Koch, M. H. Gross, D. F. Bueren, G. Frankhauser, Y. Parish, and F. R. Carls. Simulating facial surgery using finite element models. In *SIGGRAPH '96, ACM Computer Graphics*, volume 30, August 1996.

[399] A. Kolb. *Optimierungsansätze bei der Interpolation verteilter Daten*. PhD thesis, University of Erlangen-Nürnberg, August 1995.

[400] A. Kolb, H. Pottmann, and H.-P. Seidel. Fair surface reconstruction using quadratic functionals. In *Computer Graphics Forum (EUROGRAPHICS '95 Proceedings)*, volume 14, pages 469–479, 1995.

[401] D. Koller. *Detektion, Verfolgung und Klassifikation bewegter Objekte in monokularen Bildfolgen am Beispiel von Straßenverkehrsszenen*, volume 13 of *Dissertationen zur künstlichen Intelligenz*. infix, St. Augustin, 1992.

[402] D. Koller, K. Daniilidis, T. Thorhallson, and H. Nagel. Model-based object tracking in traffic scenes. In G. Sandini, editor, *Computer Vision—ECCV '92*, Lecture Notes in Computer Science, pages 437–452, Berlin, Heidelberg, New York, London, 1992. Springer.

[403] J. Konrad and E. Dubois. Bayesian estimation of motion vector fields. *IEEE Transactions on Pattern Analysis and Machine Intelligence*, 14(9):910–927, 1992.

[404] E. P. Krotkov. *Active Computer Vision by Cooperative Focus and Stereo*. Springer, Berlin, Heidelberg, 1989.

[405] E. P. Krotkov and R. Bajcsy. Active vision for reliable ranging: Cooperating focus, stereo and vergence. *International Journal of Computer Vision*, 11(2):187–203, 1993.

[406] N. Krüger and G. Peters. Object recognition with Banana wavelets. In *Proceedings of the European Symposium on Artificial Neural Networks (ESANN97)*, Bruges, 1997.

[407] W. Krüger. The application of transport theory to the visualization of 3D scalar data fields. In A. Kaufman, editor, *Visualization '90*, pages 273–280. IEEE Computer Society Press, 1990.

[408] R. Kumar, P. Anandan, and K. Hanna. Direct recovery of shape from multiple views: a parallax based approach. *International Conference on Pattern Recognition*, pages 685–688, 1994.

[409] R. Kumar and A. Hanson. Robust estimation of camera location and orientation from noisy data having outliers. In *Proceedings IEEE Workshop on Interpretation of 3D Scenes*, pages 52–60, Austin, Texas, ANOV 1989.

[410] D. Kunz, K.-J. Schilling, and T. Vögtle. A new approach for satellite image analysis by means of a semantic network. In W. Förstner and L. Plümer, editors, *Semantic Modeling*, pages 20–36, Basel, 1997. Birkhäuser.

[411] C. Kyriakakis. Fundamental and technological limitations of immersive audio systems. *Proceedings of the IEEE*, 86(5):941–951, 1998.

[412] X. Laboureux, S. Seeger, and G. Häusler. Computation of curvatures from 2.5D raster data. *Chair for Optics, Annual report*, 1999.

[413] P. Lacroute and M. Levoy. Fast volume rendering using a shear-warp factorization of the viewing transformation. In A. S. Glassner, editor, *Computer Graphics Proceedings*, Annual Conference Series, pages 451–457, Los Angeles, California, July 1994. ACM SIGGRAPH, Addison-Wesley Publishing Company, Inc.

[414] E. P. Lafortune, S.-C. Foo, K. E. Torrance, and D. P. Greenberg. Non-linear approximation of reflectance functions. In *Computer Graphics (SIGGRAPH '97 Proceedings)*, pages 117–126, August 1997.

[415] E. P. Lafortune and Y. D. Willems. Bi-directional Path Tracing. In *Proceedings of Third International Conference on Computational Graphics and Visualization Techniques (Compugraphics '93)*, pages 145–153, December 1993.

[416] Y. Lamdan and H. J. Wolfson. Geometric hashing: A general and efficient model-based recognition scheme. In *Second International Conference on Computer Vision*, pages 238–249, 1988.

[417] R. Landsee, F. v. d. Linden, H. Schönfeld, G. Häusler, A. M. Kielbassa, R. J. Radlanski, D. Drescher, and R.-R. Miethke. Die Entwicklung von Datenbanken

zur Unterstützung der Aus-, Fort- und Weiterbildung sowie der Diagnostik und Therapieplanung in der Zahnmedizin — Teil 1. *Kieferorthopädie*, 11:283–290, 1997.

[418] D. A. Lane. Scientific visualization of large-scale unsteady fluid flows. In G. M. Nielson, H. Hagen, and H. Müller, editors, *Scientific Visualization: Overviews, Methodologies, and Techniques*, pages 125–145. IEEE Computer Society Press, Los Alamitos, California, 1997.

[419] S. Lang. *Differential Manifolds*. Addison-Wesley Publishing Company, Inc., Reading, MA, 1972.

[420] W. Larrabee. A finite element model of skin deformation. II: An experimental model of skin deformation. *Laryngoscope*, 96:406–412, 1986.

[421] Y. Lavin, Y. Levy, and L. Hesselink. Singularities in nonuniform tensor fields. In R. Yagel and H. Hagen, editors, *Visualization '97*, pages 59–66, Phoenix, Arizona, 1997. IEEE Computer Society Press.

[422] A. W. Lee, W. Sweldens, P. Schröder, L. Cowsar, and D. Dobkin. MAPS: Multiresolution adaptive parameterization of surfaces. In M. Cohen, editor, *SIGGRAPH '98 Conference Proceedings*, Annual Conference Series, pages 95–104. ACM SIGGRAPH, Addison Wesley, July 1998.

[423] E. T. Lee. Choosing nodes in parametric curve interpolation. *Computer Aided Design*, 21(6):363–370, 1989.

[424] S. Lee, G. Wolberg, and S. Y. Shin. Scattered data interpolation with multilevel B-splines. *IEEE Transactions on Visualization and Computer Graphics*, 3(3):1–17, 1997.

[425] W. C. de Leeuw and R. van Liere. Comparing LIC and spot noise. In D. Ebert, H. Rushmeier, and H. Hagen, editors, *Visualization '98*, pages 359–365, Research Triangle Park, North Carolina, 1998. IEEE Computer Society Press.

[426] W. C. de Leeuw and J. J. van Wijk. A probe for local flow field visualization. In G. M. Nielson and D. Bergeron, editors, *Visualization '93*, pages 39–45, San Jose, California, 1993. IEEE Computer Society Press.

[427] T. Lehmann, W. Oberschelp, E. Pelikan, and R. Repges. *Bildverarbeitung für die Medizin, Grundlagen, Modelle, Methoden, Anwendungen*. Springer, Berlin, 1997.

[428] J. Lengyel and J. Snyder. Rendering with coherent layers. In T. Whitted, editor, *SIGGRAPH '97 Conference Proceedings*, Annual Conference Series, pages 233–242. ACM SIGGRAPH, Addison Wesley, August 1997.

[429] R. Lerch. Finite element analysis of piezoelectric transducers. In *Proceedings IEEE Ultrasonics Symposium*, pages 643–654, 1988.

[430] R. Lerch. Simulations of piezoelectric devices by two- and three-dimensional finite elements. *IEEE Transactions on Ultras., Ferroel. and Freq. Control*, pages 233–247, 1990.

[431] R. Lerch, H. Landes, W. Friedrich, R. Hebel, A. Höß, and H. Kaarmann. Modelling of acoustic antennas with a combined finite-element-boundary-element-method. In *Proceedings IEEE Ultrasonics Symposium*, pages 643–654, 1992.

[432] R. Lerch, H. Landes, and H. Kaarmann. Finite element modelling of the pulse-echo behavior of ultrasound transducers. In *Proceedings IEEE Ultrasonics Symposium*, pages 1021–1025, 1994.

[433] M. Levoy. Display of surfaces from volume data. *IEEE Computer Graphics and Applications*, 8(3):29–37, May 1988.

[434] M. Levoy. Efficient ray tracing of volume data. *ACM Transactions on Graphics*, 9(3):245–261, July 1990.

[435] M. Levoy and P. Hanrahan. Light field rendering. In *Computer Graphics (SIGGRAPH '96 Proceedings)*, pages 31–42, August 1996.

[436] R. R. Lewis. Making shaders more physically plausible. In *Fourth Eurographics Workshop on Rendering*, pages 47–62, June 1993.

[437] H. Li, A. Lundmark, and R. Forchheimer. Image sequence coding at very low bitrates: A review. *IEEE Transactions on Image Processing*, 3(5):589–609, September 1994.

[438] H. Li, P. Roivainen, and R. Forchheimer. 3D motion estimation in model-based facial image coding. *IEEE Transactions on Pattern Analysis and Machine Intelligence*, 15(6):545–555, June 1993.

[439] C.-E. Liedtke, J. Bückner, O. Grau, S. Growe, and R. Tönjes. AIDA: A system for the knowledge based interpretation of remote sensing data. In *Proceedings of the Third International Airborne Remote Sensing Conference and Exhibition*, Kopenhagen, 1997.

[440] M. E. Liggins, C.-Y. Chong, et al. Distributed fusion architectures and algorithms for target tracking. *IEEE Proceedings*, 85(1):97–109, 1997.

[441] T. Lindeberg. Detecting salient blob-like image structures and their scales with a scale-space primal sketch: A method for focus-of-attention. *International Journal of Computer Vision*, 11(3):283–318, 1993.

[442] L. Lippert, M. H. Gross, and C. Kurmann. Compression domain volume rendering for distributed environments. In D. Fellner and L. Szirmay-Kalos, editors, *EUROGRAPHICS '97*, volume 14, pages C95–C107. Eurographics Association, Blackwell Publishers, 1997.

[443] Y. Livnat and C. Hansen. View dependent isosurface extraction. In *IEEE Visualization 1998*, pages 175–181, 1998.

[444] Y. Livnat, H.-W. Shen, and C. R. Johnson. A near optimal isosurface extraction algorithm using span space. *IEEE Transactions on Visualization and Computer Graphics*, 2(1):73–84, 1996.

[445] G. Lohmann and D. von Cramon. Automatic detection and labelling of the human cortical folds in magnetic resonance data sets. In *ECCV '98*, Freiburg, 1998.

[446] H. C. Longuet-Higgins. A computer algorithm for reconstructing a scene from two projections. *Nature*, 293:133–135, ASEP 1981.

[447] H. C. Longuet-Higgins. The reconstruction of a scene from two projections—configurations that defeat the 8-point algorithm. In *Proceedings of the IEEE First Conference on Artificial Intelligence*, pages 395–397, Denver, 1984.

[448] H. C. Longuet-Higgins. The visual ambiguity of a moving plane. *Proceedings of the Royal Society of London*, B 223:165–175, 1984.

[449] J. Loos. *Konstruktion von Flächen mit vorgegebenen Krümmungseigenschaften und Anwendungen in der Augenoptik.* PhD thesis, University of Erlangen-Nürnberg, 1997.

[450] J. Loos, G. Greiner, and H.-P. Seidel. A variational approach to progressive lens design. *Computer Aided Design*, pages 595–602, 1998.

[451] J. Loos, G. Greiner, and H.-P. Seidel. Constructing surface geometry from isophotes and reflection lines. *Computer Aided Geometric Design*, 1999.

[452] W. E. Lorensen and H. E. Cline. Marching cubes: A high resolution 3D surface construction algorithm. In M. C. Stone, editor, *Computer Graphics (SIGGRAPH '87 Proceedings)*, volume 21, pages 163–169, July 1987.

[453] M. Lounsbery. *Multiresolution Analysis for Surfaces of Arbitrary Topological Type*. PhD thesis, University of Washington, Seattle, 1994.

[454] M. Lounsbery, T. D. DeRose, and J. Warren. Multiresolution analysis for surfaces of arbitrary topological type. *ACM Transactions on Graphics*, 16(1):34–73, January 1997.

[455] D. G. Lowe. Robust model-based motion tracking through the integration of search and estimation. *International Journal of Computer Vision*, 8(2):113–122, 1992.

[456] J. Lu and J. Little. Reflectance function estimation and shape recovery from image sequence of a rotating object. In *Proceedings of International Conference on Computer Vision*, pages 80–86, 1995.

[457] B. Ludwig. Everything about surround so far. *Surround 2000*, pages 2–8, November 1999.

[458] Q.-T. Luong and T. Viéville. Canonic representations for the geometries of multiple projective views. In *Proceedings European Conference on Computer Vision (ECCV '94)*, LNCS, pages 589–599. Springer, 1994.

[459] K. L. Ma, J. S. Painter, C. D. Hansen, and M. F. Krogh. A data distributed parallel algorithm for ray-traced volume rendering. In T. Crockett, C. Hansen, and Whitman S., editors, *1993 Parallel Rendering Symposium*, pages 15–22, New York, 1993. ACM SIGGRAPH.

[460] W. Ma and J. P. Kruth. Parameterization of randomly measured points for least squares fitting of B-spline curves and surfaces. *Computer Aided Design*, 27:663–675, 1995.

[461] M. Magnor and B. Girod. Hierarchical coding of light fields with disparity maps. In *ICIP '99*. IEEE Signal Processing Society, 1999.

[462] M. Magnor and B. Girod. Data compression in image-based rendering. *IEEE Transactions on Circuits and Systems for Video Technology, Special Issue on Three-Dimensional Video Technology*, April 2000.

[463] J. Maillot, H. Yahia, and A. Verroust. Interactive texture mapping. In *ACM Computer Graphics (SIGGRAPH '93 Proceedings)*, pages 27–34, 1993.

[464] J. B. Maintz and M. A. Viergever. A survey of medical image registration. *Medical Image Analysis*, 2(1):1–36, 1998.

[465] B. Maitenaz. Image rétinienne donnée par un verre correcteur de puissance progressive. *Revu. Opt. Theor. Instrum.*, 46:233–241, 1967.

[466] R. W. Malz. Codierte Lichtstrukturen für 3D-Meßtechnik und Inspektion. Technical Report 14, University of Stuttgart, 1992.

[467] X. Mao, M. Kikukawa, N. Fujita, and A. Imamiya. Line integral convolution for 3D surfaces. In W. Lefer and M. Grave, editors, *Visualization in Scientific Computing '97*, pages 57–69, Wien, April 1997. Springer.

[468] D. Marr. *Vision: A Computational Investigation into the Human Representation and Processing of Visual Information.* W. H. Freeman and Company, San Francisco, 1982.

[469] D. Marr and E. Hildreth. Theory of edge detection. *Proceedings Royal Society London*, 207:187–217, 1980.

[470] S. R. Marschner. Inverse lighting for photography. In *Fifth Color Imaging Conference*, 1997.

[471] W. P. Mason. *Physical Acoustics*. Academic Press, 1964.

[472] T. Masuda, K. Sakaue, and N. Yokoya. Registration and integration of multiple range images for 3D model construction. In *IEEE Proceedings of ICPR '96*, pages 879–883, 1996.

[473] T. Matsuyama and V. Hwang. *SIGMA: A Knowledge-Based Aerial Image Understanding System*. Plenum Press, New York, 1990.

[474] W. Maurel, Y. Wu, N. M. Thalmann, and D. Thalmann. *Biomechanical Models for Soft Tissue Simulation*. Springer, Berlin, Germany, 1998.

[475] N. L. Max, B. G. Becker, and R. A. Crawfis. Flow volumes for interactive vector field visualization. In G. M. Nielson and D. Bergeron, editors, *Visualization '93*, pages 19–24, San Jose, California, 1993. IEEE Computer Society Press.

[476] L. McMillan and G. Bishop. Plenoptic modeling: An image-based rendering system. In *Computer Graphics (SIGGRAPH '95 Proceedings)*, pages 39–46, August 1995.

[477] R. Mencl. A graph-based approach to surface reconstruction. In *Computer Graphics Forum (EUROGRAPHICS '95 Proceedings)*, volume 14, pages 445–456, 1995.

[478] R. Mencl and H. Müller. Interpolation and approximation of surfaces from three-dimensional scattered data points. In *Computer Graphics Forum (EUROGRAPHICS '96 Proceedings), State of the Art Report (STAR)*, volume 17, 1998.

[479] D. Meyers, S. Skinner, and K. Sloan. Surface from contour: The corresponding and branching problems. *ACM Transactions on Graphics*, 11(3):228–258, 1992.

[480] E. Michaelsen. *Über Koordinatengrammatiken zur Bildverarbeitung und Szenenanalyse*. PhD thesis, University of Erlangen-Nürnberg, 1998.

[481] G. Miller, S. Rubin, and D. Ponceleon. Lazy decompression of surface light fields for precomputed global illumination. In *Rendering Techniques '98 (Proceedings Ninth Eurographics Workshop on Rendering)*, pages 281–292, June 1998.

[482] J. Milnor. *Topology from the Differential Viewpoint*. The University of Virginia, Charlottesville, VA, 1965.

[483] A. Mitiche and P. Bouthemy. Computation and analysis of image motion: A synopsis of current problems and methods. *International Journal of Computer Vision*, 19(1):29–55, 1996.

[484] J.W. Modestino and J. Zhang. A Markov random field model-based approach to image interpretation. *IEEE Transactions on Pattern Analysis and Machine Intelligence*, 14:606–615, 1992.

[485] B. Moghaddam and A. Pentland. An automatic system for model-based coding of faces. Technical Report 317, MIT Media Lab Vismod, 1995.

[486] B. Moghaddam and A. Pentland. Probabilistic visual learning for object representation. *IEEE Transactions on Pattern Analysis and Machine Intelligence*, 19(7):696–710, July 1997.

[487] R. Mohr and B. Triggs. Projective geometry for image analysis. In *Int. Symp. Photogrammetry and Remote Sensing*, July 1996.

[488] P. Moin and J. Kim. The structure of the vorticity field in turbulent channel flow. Part 1. *Journal Fluid Mechanics*, 155:441, 1985.

[489] T. Moons. A guided tour through multiview relations. In *Proceedings SMILE Workshop (post-ECCV '98)*, LNCS. Springer, 1998.

[490] H. P. Moravec. Robot spatial perception by stereoscopic vision and 3D evidence grids. Technical Report CMU-RI-TR-96-34, The Robotics Institue, Carnegie Mellon University, Pittsburgh, Pennsylvania, September 1996.

[491] H. P. Moreton and C. H. Séquin. Functional optimization for fair surface design. In *ACM Computer Graphics (SIGGRAPH '92 Proceedings)*, pages 167–176, 1992.

[492] H. Morneburg. *Bildgebende Systeme für die medizinische Diagnostik*. Publicis MCD, Erlangen, 1995.

[493] MPEG Committee. *ISO/IEC 14496-2, Coding of Audio-Visual Objects: Visual (MPEG-4 video), Committee Draft, Document N1902*, October 1997.

[494] H. Murase and S. K. Nayar. Visual learning and recognition of 3D objects from appearance. *International Journal of Computer Vision*, 14(1):5–24, January 1995.

[495] D. Murray and A. Basu. Motion tracking with an active camera. *IEEE Transactions on Pattern Analysis and Machine Intelligence*, 16(5):449–459, 1994.

[496] H.-H. Nagel. Zur Strukturierung eines Bildfolgen–Auswertesystems. *Informatik Forschung und Entwicklung*, 11:3–11, 1996.

[497] S. K. Nayar, K. Ikeuchi, and T. Kanade. Surface reflection: Physical and geometrical perspectives. *IEEE Transactions on Pattern Analysis and Machine Intelligence*, 13(7):611–634, July 1991.

[498] S. Negahdaripour and B. K. Horn. Direct passive navigation. *IEEE Transactions on Pattern Analysis and Machine Intelligence*, 9(1):168–176, AJAN 1987.

[499] J. Neider, T. Davis, and M. Woo. *OpenGL Programming Guide, Release 1*, 1993.

[500] A. N. Netravali and J. Salz. Algorithms for estimation of three-dimensional motion. *AT&T Technical Journal*, 64(2):335–346, February 1985.

[501] R. Neubauer, M. Ohlberger, M. Rumpf, and R. Schwörer. Efficient visualization of large-scale data on hierarchical meshes. In *Eighth Eurographics Workshop on Visualization in Scientific Computing*, pages 165–174, 1997.

[502] P. J. Neugebauer. Reconstruction of real-world objects via simultaneous registration and robust combination of multiple range images. *International Journal of Shape Modeling*, 3(1 & 2):71–90, 1997.

[503] L. Neumann, W. Purgathofer, R. F. Tobler, A. Neumann, P. Eliás, M. Feda, and X. Pueyo. The stochastic ray method for radiosity. In *Rendering Techniques '95 (Proceedings Sixth Eurographics Workshop on Rendering)*, pages 206–218, Dublin, June 1995. Springer.

[504] J. Nieh and M. Levoy. Volume rendering on scalable shared-memory MIMD architectures. In *1992 Workshop on Volume Visualization*, pages 17–24, Boston, Massachusetts, October 1992. ACM, ACM SIGGRAPH.

[505] G. Nielson and B. Hamann. The asymptotic decider: Resolving the ambiguity in marching cubes. In G. Nielson and Rosenblum. L., editors, *Visualization '91*, pages 83–91. IEEE Computer Society Press, 1991.

[506] G. M. Nielson. Scattered data modeling. *IEEE Computer Graphics & Applications*, pages 60–70, January 1993.

[507] H. Niemann. *Klassifikation von Mustern*. Springer, Berlin, 1983.

[508] H. Niemann. *Pattern Analysis and Understanding*. Springer Series in Information Sciences 4. Springer, Berlin, 2nd edition, 1990.

[509] H. Niemann, H. Bunke, I. Hofmann, G. Sagerer, F. Wolf, and H. Feistel. A knowledge based system for analysis of gated blood pool studies. *IEEE Transactions on Pattern Analysis and Machine Intelligence*, 7:246–259, 1985.

[510] H. Niemann, G. Sagerer, S. Schröder, and F. Kummert. ERNEST: A semantic network system for pattern understanding. *IEEE Transactions on Pattern Analysis and Machine Intelligence*, 9:883–905, 1990.

[511] N. J. Nilsson. *Principles of Artificial Intelligence*. Springer, Berlin, Heidelberg, 1982.

[512] J. Nimeroff, J. Dorsey, and H. Rushmeier. Implementation and analysis of an image-based global illumination framework for animated environments. *IEEE Transactions on Visualization and Computer Graphics*, 2(4):283–298, December 1996.

[513] H. Nowacki. Mathematische Verfahren zum Glätten von Kurven und Flächen. In J. L. Encarnação, J. Hoschek, and J. Rix, editors, *Geometrische Verfahren der Graphischen Datenverarbeitung*, pages 22–45. Springer, 1990.

[514] H. Nowacki and R. Gnatz. *Geometrisches Modellieren (Geometric Modelling)*. Springer, Berlin, 1983.

[515] M. Ohlberger and M. Rumpf. Hierarchical and adaptive visualization on nested grids. *Computing*, 59 (4):269–285, 1997.

[516] J.-R. Ohm. *Digitale Bildcodierung*. Springer, 1995.

[517] E. Oja and J. Parkkinen. On subspace clustering. In *Proceedings International Conference on Acoustics, Speech and Signal Processing*, pages 692–695, San Diego, 1984.

[518] M. Olano and A. Lastra. A shading language on graphics hardware: The PixelFlow shading system. In *Computer Graphics (SIGGRAPH '98 Proceedings)*, pages 159–168, July 1998.

[519] J.-M. Oliva, M. Perrin, and S. Coquillart. 3D reconstruction of complex polyhedral shapes from contours using a simplified generalized Voronoi diagram. *Proceedings Eurographics '96, Computer Graphics Forum*, 15(3):397–408, 1996.

[520] N. Oliver, A. Pentland, and F. Berard. LAFTER: Lips and face real time tracker. In *Proceedings of IEEE Conference on Computer Vision and Pattern Recognition*, pages 123–129, San Juan, Puerto Rico, 1997.

[521] J. O'Rourke. *Computational Geometry in C*. Cambridge University Press, 1993.

[522] J. O'Rourke, H. Booth, and R. Washington. Connect-the-dots: a new heuristic. *Computer Vision, Graphics, and Image Processing*, 39:258–266, 1987.

[523] N. L. Owsley and G. R. Swope. Time delay estimation in a sensor array. *IEEE Transactions on Acoustics, Speech, and Signal Processing*, 29(3):519–523, 1981.

[524] K. Pahlavan. *Active Robot Vision and Primary Ocular Processes*. PhD thesis, CVAP, Stockholm University, 1993.

[525] F. I. Parke. Parameterized models for facial animation. *IEEE Computer Graphics & Applications*, 2(9):61–68, November 1982.

[526] F. I. Parke and K. Waters. *Computer Facial Animation*. AK Peters, Massachusetts, 1996.

[527] D. Paulus, U. Ahlrichs, B. Heigl, J. Denzler, J. Hornegger, and H. Niemann. Active knowledge based scene analysis. In *Computer Vision Systems*, pages 180–199, 1999.

[528] D. Paulus, T. Greiner, and C. Knüvener. Wasserscheidentransformation für Thermographiebilder. In G. Sagerer, S. Posch, and F. Kummert, editors, *Mustererkennung 1995*, pages 355–362, Berlin, September 1995. Springer.

[529] D. Paulus and J. Hornegger. *Applied pattern recognition: A practical introduction to image and speech processing in C++*. Advanced Studies in Computer Science. Vieweg, Braunschweig, 2nd edition, 1998.

[530] D. Paulus and H. Niemann. Iconic-symbolic interfaces. In R. B. Arps and W. K. Pratt, editors, *Image Processing and Interchange: Implementation and Systems*, pages 204–214, San Jose, California, 1992. SPIE.

[531] J. Pearl. *Probabilistic Reasoning in Intelligent Systems: Networks of Plausible Inference*. Morgan Kaufmann, San Mateo, California, 1988.

[532] D. E. Pearson. Texture mapping in model-based image coding. *Signal Processing: Image Communication*, 2(4):377–395, December 1990.

[533] D. E. Pearson. Developments in model-based video coding. *Proceedings of the IEEE*, 83(6):892–906, June 1995.

[534] H. K. Pedersen. Decorating implicit surfaces. In R. Cook, editor, *Computer Graphics (SIGGRAPH '95 Proceedings)*, pages 291–300, 1995.

[535] F. Pedrotti and L. Pedrotti. *Introduction to Optics*. Prentice Hall, 2nd edition, 1993.

[536] M. Pelka, K.-H. Kunzelmann, D. Paulus, and A. Winzen. Automatic digital subtraction radiography using simulated annealing. In *IADR*, Seattle, 1994.

[537] X. Pennec and J. P. Thirion. Validation of 3D registration methods based on points and frames. Technical Report 2470, INRIA, January 1995.

[538] A. Pentland. Finding the illuminant direction. *Journal of the Optical Society of America*, 72(4):170–187, 1982.

[539] A. Pentland. Photometric motion. *IEEE Transactions on Pattern Analysis and Machine Intelligence*, 13(9):879–890, September 1991.

[540] A. Pentland and B. Horowitz. Recovery of nonrigid motion and structure. *IEEE Transactions on Pattern Analysis and Machine Intelligence*, 13(7):730–742, July 1991.

[541] R. Pfeifle. *Approximation und Interpolation mit quadratischen Dreiecks-B-Splines*. PhD thesis, University of Erlangen-Nürnberg, 1995.

[542] R. Pfeifle and H.-P. Seidel. Spherical triangular B-splines with application to data fitting. In *Computer Graphics Forum (EUROGRAPHICS '95 Proceedings)*, volume 14, pages 89–96, 1995.

[543] H. Pfister and A. Kaufman. Cube-4—a scalable architecture for real-time volume rendering. In R. Crawfis and C. Hansen, editors, *Symposium on Volume Visualization*, pages 47–54. ACM SIGGRAPH, 1996.

[544] B. Phong. Illumination for computer generated pictures. *Communications of the ACM*, 18(6):311–317, 1975.

[545] S. D. Pieper. *CAPS: Computer Aided Plastic Surgery*. PhD thesis, MIT, Media Arts and Sciences, Cambrigde, MA, 1991.

[546] G. S. Pingali. Integrated audio-visual processing for object localization and tracking. In *Proceedings of the SPIE*, volume 3310, pages 206–213, 1997.

[547] U. Pinkall and K. Polthier. Computing discrete minimal surfaces and their conjugates. *Experimental Mathematics*, 2(1):15–36, 1993.

[548] T. Poeschl. Detecting surface irregularities using isophotes. *Computer Aided Geometric Design*, 1(2):163–168, 1984.

[549] G. De Poli, A. Piccialli, and C. Roads. *Representation of Musical Signals*. MIT Press, Cambridge, Massachusetts, 1991.

[550] M. Pollefeys, R. Koch, and L. Van Gool. Self-calibration and metric reconstruction in spite of varying and unknown internal camera parameters. In *Proceedings International Conference on Computer Vision (ICCV '98)*, pages 90–95, Bombay, 1998.

[551] J. Ponce, A. Zisserman, and M. Hebert. *Object Representation in Computer Vision*, volume 1144 of *Lecture Notes in Computer Science*. Springer, Heidelberg, 1996.

[552] J. Popović and H. Hoppe. Progressive simplicial complexes. In T. Whitted, editor, *SIGGRAPH '97 Conference Proceedings*, Annual Conference Series, pages 217–224. ACM SIGGRAPH, Addison Wesley, August 1997.

[553] J. Pösl and H. Niemann. Statistical 3D object localization without segmentation using wavelet analysis. In G. Sommer, K. Daniilidis, and J. Pauli, editors, *Computer Analysis of Images and Patterns (CAIP '97, Kiel)*, pages 440–447, Berlin Heidelberg, 1997. Springer.

[554] J. Pösl and H. Niemann. Object localization with mixture densities of wavelet features. In *International Wavelets Conference*, Tangier, Marocco, April 1998. INRIA.

[555] F. H. Post and T. van Walsum. Fluid flow visualization. In H. Hagen, H. Müller, and G. M. Nielson, editors, *Focus on Scientific Visualization*, pages 1–40. Springer, Berlin, 1993.

[556] T. Poston, H. T. Nguyen, P. A. Heng, and T. T. Wong. Skeleton-climbing: Fast isosurfaces with fewer triangles. In *Pacific Graphics 1997*, pages 117–126, 1997.

[557] P. Poulin and A. Fournier. A model for anisotropic reflection. In *Computer Graphics (SIGGRAPH '90 Proceedings)*, volume 24, pages 273–282, August 1990.

[558] H. Prade. A computational approach to approximate and plausible reasoning with applications to expert systems. *IEEE Transactions Pattern Analysis and Machine Intelligence*, 7:260–283, 1985.

[559] M. J. Pratt. Smooth parametric surface approximations to discrete data. *Computer Aided Geometric Design*, 2:165–171, 1985.

[560] W. H. Press, S. A. Teukolsky, W. T. Vetterling, and B. P. Flannery. *Numerical Recipes in C*. Cambridge University Press, 1996.

[561] R. J. Quian, M. I. Sezan, and K. E. Matthews. A robust real-time face tracking algorithm. In *Proceedings of the 1998 IEEE International Conference on Image Processing*, volume 1, pages 131–135, 1998.

[562] F. Quint and M. Stiess. Map-based semantic modeling for the extraction of objects from aerial images. In A. Grün, O. Kübler, and P. Agouris, editors, *Automatic Extraction of Man-Made Objects from Aerial and Space Images*, pages 307–316. Birkhäuser, Basel, 1995.

[563] R. Rabenstein and A. Zayati. A direct method to computational acoustics. In *Proceedings International Conference on Acoustics, Speech, and Signal Processing (ICASSP '99)*. IEEE, 1999.

[564] L. R. Rabiner. Mathematical foundations of hidden Markov models. In H. Niemann, M. Lang, and G. Sagerer, editors, *Recent Advances in Speech Understanding and Dialog Systems*, volume 46 of *NATO ASI Series F*, pages 183–205. Springer, Berlin, 1988.

[565] P. Rademacher and G. Bishop. Multiple-center-of-projection images. In *Computer Graphics (SIGGRAPH '98 Proceedings)*, pages 199–206, July 1998.

[566] B. S. Rao, H. F. Durrant-Whyte, and J. A. Sheen. A fully decentralized multi-sensor system for tracking and surveillance. *International Journal of Robotics Research*, 12(1):20–44, 1993.

[567] C. S. Regazzoni. Distributed extended Kalman filtering network for estimation and tracking of multiple objects. *Electronics Letters*, 30(15):1202–1213, 1994.

[568] C. Rezk-Salama, P. Hastreiter, C. Teitzel, and T. Ertl. Interactive exploration of volume line integral convolution based on 3D-texture mapping. Technical Report 2/1999, IMMD IX, University of Erlangen-Nürnberg, March 1999.

[569] J. M. Richardson and K. A. Marsh. Fusion of multisensor data. *International Journal of Robotics Research*, 7(6):78–96, 1988.

[570] C. Ridder, O. Munkelt, and H. Kirchner. Adaptive background estimation and foreground detection using Kalman-filtering. Technical report, Bavarian Research Center for Knowledge-Based Systems, 1995.

[571] A. Riepl. Interpolation gestreuter Daten mit krümmungsminimierenden Flächen. Master's thesis, IMMD IX, University of Erlangen-Nürnberg, 1995.

[572] B. D. Ripley. *Pattern Recognition and Neural Networks*. Cambridge University Press, Cambridge, 1996.

[573] C. P. Risquet. Visualizing 2D flows: Integrate and draw. In D. Bartz, editor, *Ninth Eurographics Workshop on Visualization in Scientific Computing*, pages 132–142, Blaubeuren, Germany, April 1998.

[574] D. Ritter. *Merkmalsorientierte Ojekterkennung und -lokalisation im 3D-Raum aus einem einzelnen 2D-Grauwertbild und Referenzmodellvermessung mit optischen 3D-Sensoren*. PhD thesis, University of Erlangen-Nürnberg, 1996.

[575] J. W. Roach and J. K. Aggarwal. Determining the movement of objects from a sequence of images. *IEEE Transactions on Pattern Analysis and Machine Intelligence*, 2(6):554–562, 1980.

[576] C. Roads, S. Pope, A. Piccialli, and G. De Poli. *Musical Signal Processing*. Swets & Zeitlinger, Lisse, 1997.

[577] A. Rognone, M. Campani, and A. Verri. Identifying multiple motions from optical flow. *Proceedings Second ECCV*, pages 258–266, AMAY 1992.

[578] R. Ronfard. Region-based strategies for active contour models. *International Journal of Computer Vision*, 13(2):229–251, 1994.

[579] R. Ronfard and J. Rossignac. Full-range approximation of triangulated polyhedra. *Computer Graphics Forum*, 15(3):67–76, August 1996.

[580] L. J. Rosenblum. Research issues in scientific visualization. *IEEE Computer Graphics and Applications*, 14(2):61–85, March 1994.

[581] J. Rossignac and P. Borrel. Multi-resolution 3D approximation for rendering complex scenes. In *Second Conference on Geometric Modelling in Computer Graphics*, pages 453–465, June 1993.

[582] P. J. Rousseeuw and A. M. Leroy. *Robust Regression and Outlier Detection*. John Wiley, 1987.

[583] S. Roy, J. Meunier, and I. J. Cox. Cylindrical rectification to minimize epipolar distortion. *Proceedings of IEEE Conference on Computer Vision and Pattern Recognition*, pages 393–399, 1997.

[584] H. Rushmeier, G. Taubin, and A. Guéziec. Applying shape from lighting variation to bump map capture. In *Rendering Techniques '97 (Proceedings Eighth Eurographics Workshop on Rendering)*, pages 35–44. Springer Wien, 1997.

[585] S. Russell and P. Norvig. *Artificial Intelligence—A Modern Approach*. Prentice Hall Series in Artificial intelligence. Prentice Hall, Englewood Cliffs, New Jersey, 1995.

[586] M. Rutishauser, M. Stricker, and M. Trobina. Merging range images of arbitrarily shaped objects. In *Proceedings of IEEE Conference on Computer Vision and Pattern Recognition*, pages 573–580, 1994.

[587] M. Rydfalk. *CANDIDE: A Parameterized Face.* PhD thesis, Linköping University, 1978.

[588] P. A. Sabella. A rendering algorithm for visualizing 3D scalar fields. *Computer Graphics*, 22(4):51–58, August 1988.

[589] M. A. Sabin. Contouring—the state of the art. In R. A. Earnshaw, editor, *Fundamental Algorithms for Computer Graphics*, pages 411–482. Springer, Berlin, 1985.

[590] A. Sadarjoen, T. van Walsum, A. Hin, and F. H. Post. Particle tracing algorithms for 3D curvilinear grids. In *Fifth Eurographics Workshop on Visualization in Scientific Computing*, 1994.

[591] G. Sagerer and H. Niemann. *Semantic Networks for Understanding Scenes.* Advances in Computer Vision and Machine Intelligence. Plenum Press, New York and London, 1997.

[592] V. Salari and I. K. Sethi. Feature point correspondence in the presence of occlusion. *IEEE Transactions on Pattern Analysis and Machine Intelligence*, pages 87–91, AJAN 1990.

[593] R. Salzbrunn. *Wissensbasierte Erkennung und Lokalisierung von Objekten.* PhD thesis, University of Erlangen-Nürnberg, 1995.

[594] A. Samal and P. Iyengar. Automatic recognition and analysis of human faces and facial expressions: A survey. *Pattern Recognition*, 8:65–77, 1992.

[595] N. S. Sapidis. *Designing Fair Curves and Surfaces.* SIAM, Philadelphia, 1994.

[596] B. Sarkar and C.-H. Menq. Parameter optimization in approximating curves and surfaces to measurement data. *Computer Aided Geometric Design*, 8:267–290, 1991.

[597] Y. Sato and K. Ikeuchi. Reflectance analysis for 3D computer graphics model generation. *Graphical Models and Image Processing*, 58(5):437–451, 1996.

[598] Y. Sato, M. D. Wheeler, and K. Ikeuchi. Object shape and reflectance modeling from observation. *ACM Computer Graphics (SIGGRAPH '97 Proceedings)*, pages 379–387, 1997.

[599] L. Savioja, J. Backman, A. Järvinen, and T. Takala. Waveguide mesh method for low-frequency simulation of room acoustics. In *Proceedings International Congress on Acoustics (ICA '95)*, pages 637–641, 1995.

[600] L. Savioja and V. Välimäki. Improved discrete-time modeling of multidimensional wave propagation using the interpolated digital waveguide mesh. In *Proceedings International Conference on Acoustics, Speech, and Signal Processing (ICASSP '97)*, pages 459–462, 1997.

[601] H. S. Sawhney. 3D geometry from planar parallax. *Proceedings of IEEE Conference on Computer Vision and Pattern Recognition*, pages 929–934, 1994.

[602] G. P. Scavone. Digital waveguide modeling of the non-linear excitation of single reed woodwind instruments. In *Proceedings International Computer Music Conference*, 1995.

[603] M. Schäfer, M. Höfken, and F. Durst. Detailed LDV measurements for visualization of the flow field within a stirred-tank reactor equipped with a Rushton turbine. *Trans IChemE*, 75:729–736, 1997.

[604] M. Schäfer, M. Yianneskis, P. Wächter, and F. Durst. Trailing vortices around a 45° pitched-blade impeller. *AIChE Journal*, 44(6):1233–1246, 1998.

[605] S. Schaller. Ein Beitrag zur Implementierung eines interaktiven Operationsplanungssystems — Reduzierung von Polygonnetzen zur interaktiven Manipulation und Simulation von verformbaren Objektoberflächen in computergraphischen Anwendungen. Master's thesis, University of Erlangen-Nürnberg, February 1995.

[606] L. L. Scharf. *Statistical Signal Processing—Detection, Estimation, and Time Series Analysis*. Addison-Wesley, 1991.

[607] T. Schetelig and R. Rabenstein. Simulation of three-dimensional sound propagation with multidimensional wave digital filters. In *Proceedings International Conference on Acoustics, Speech, and Signal Processing (ICASSP '98)*, pages 3537–3540. IEEE, 1998.

[608] G. Scheuermann, H. Hagen, and H. Krüger. Vector field topology with Clifford algebra. In N. M. Thalmann and V. Skala, editors, *WSCG '98, The Sixth International Conference in Central Europe on Computer Graphics and Visualization '98*, volume II, pages 347–353, Plzen, Czech Republic, February 1998. University of West Bohemia Press.

[609] G. Scheuermann, H. Hagen, H. Krüger, M. Menzel, and A. Rockwood. Visualization of higher order singularities in vector fields. In R. Yagel and H. Hagen, editors, *Visualization '97*, pages 67–74, Phoenix, Arizona, 1997. IEEE Computer Society Press.

[610] B. Schiele. *Object Recognition using Multidimensional Receptive Field Histograms (English translation)*. PhD thesis, Institut National Polytechnique de Grenoble, Grenoble Cedex, 1997.

[611] A. Schilling, G. Knittel, and W. Straßer. Texram: A smart memory for texturing. *IEEE Computer Graphics and Applications*, 16(3):32–41, May 1996.

[612] H. Schirmacher, W. Heidrich, and H.-P. Seidel. Adaptive acquisition of lumigraphs from synthetic scenes. In P. Brunet and R. Scopigno, editors, *Eurographics Rendering Workshop*, New York City, NY, 1999. Eurographics, Springer Wien.

[613] C. Schlick. A customizable reflectance model for everyday rendering. In *Fourth Eurographics Workshop on Rendering*, pages 73–83, June 1993.

[614] C. Schlick. A survey of shading and reflectance models. *Computer Graphics Forum*, 13(2):121–132, June 1994.

[615] K. Schluens and M. Teschner. Analysis of 2D color spaces for highlight elimination in 3D shape reconstruction. In *Proceedings ACCV*, volume 2, pages 801–805, 1995.

[616] N. Schön. Erfassung von 3D-Objekten mit Farbtextur. Master's thesis, Physics Insitute V, University of Erlangen-Nürnberg, 1998.

[617] H. Schönfeld, G. Häusler, and S. Karbacher. Reverse engineering using optical 3D sensors. In *Three-Dimensional Image Capture and Applications*, volume 3313 of *Proceedings of SPIE*. R. N. Ellson and J. H. Nurre, 1998.

[618] P. Schröder and J. B. Salem. Fast rotation of volume data on data parallel architectures. In G. M. Nielson and L. Rosenblum, editors, *Visualization '91*, San Diego, California, October 1991. IEEE Computer Society Press.

[619] P. Schröder and G. Stoll. Data parallel volume rendering as line drawing. In *1992 Workshop on Volume Visualization*, pages 25–32, Boston, Massachusetts, October 1992. ACM SIGGRAPH.

[620] W. J. Schroeder. Polygon reduction techniques. In *IEEE Visualization '95. Advanced Techniques for Scientific Visualization*, 1995.

[621] W. J. Schroeder. A topology modifying progressive decimation algorithm. In R. Yagel and H. Hagen, editors, *IEEE Visualization '97*, pages 205–212. IEEE, November 1997.

[622] W. J. Schroeder, K. Martin, and B. Lorensen. *The Visualization Toolkit*. Prentice Hall, Upper Saddle River, New Jersey, 2nd edition, 1998.

[623] W. J. Schroeder, C. R. Volpe, and W. E. Lorensen. The stream polygon: A technique for 3D vector field visualization. In G. M. Nielson and L. Rosenblum, editors, *Visualization '91*, pages 126–132, San Diego, California, October 1991. IEEE Computer Society Press.

[624] W. J. Schroeder, J. A. Zarge, and W. E. Lorensen. Decimation of triangle meshes. In E. E. Catmull, editor, *Computer Graphics (SIGGRAPH '92 Proceedings)*, volume 26, pages 65–70, July 1992.

[625] E. G. Schukat-Talamazzini. *Automatische Spracherkennung*. Vieweg, Wiesbaden, 1995.

[626] L. L. Schumaker. Fitting surfaces to scattered data. In G. G. Lorentz, C. K. Chui, and L. L. Schumaker, editors, *Approximation Theory II*, pages 203–268. Academic Press, Boston, 1976.

[627] L. L. Schumaker. *Spline Functions: Basic Theory*. John Wiley & Sons, New York, 1981.

[628] T. W. Sederberg and S. R. Parry. Free-form deformation of solid geometric models. *ACM Computer Graphics (SIGGRAPH '86 Proceedings)*, 20(4):151–159, August 1986.

[629] M. Segal, C. Korobkin, R. van Widenfelt, J. Foran, and P. Haeberli. Fast shadow and lighting effects using texture mapping. *Computer Graphics (SIGGRAPH '92 Proceedings)*, 26(2):249–252, July 1992.

[630] H.-P. Seidel. Geometrische Grundlagen des Computer Aided Geometric Design. In *Geometrie und ihre Anwendungen*, pages 201–246. Hanser, 1994.

[631] I. K. Sethi and R. Jain. Finding trajectories of feature points in a monocular image sequence. *IEEE Transactions on Pattern Analysis and Machine Intelligence*, pages 56–73, 1987.

[632] M. I. Sezan and R. L. Lagendijk. *Motion Analysis and Image Sequence Processing*. Kluwer Academic Publishers, 1993.

[633] SGI. *OpenGL on Silicon Graphics Systems*. Silicon Graphics Inc., Mountain View, California, 1996.

[634] J. W. Shade, S. J. Gortler, L.-W. He, and R. Szeliski. Layered depth images. In *Computer Graphics (SIGGRAPH '98 Proceedings)*, pages 231–242, July 1998.

[635] S. A. Shafer. Using color to separate reflection components. *COLOR research and application*, 10(4):210–218, 1985.

[636] C. E. Shannon. A mathematical theory of communication. *The Bell System Technical Journal*, 27, 1948.

[637] H. Shariat and K. Price. Motion estimation using more than two images. In W. Martin and J. Aggarwal, editors, *Motion Understanding: Robot and Human Vision*, pages 143–188. Kluwer Academic Publishers, Boston, 1988.

[638] R. Shekkar, W. Fayyad, and J. Fredrick. Octree-based decimation of marching cubes surface. In *Proceedings Visualization '96*, pages 335–342. IEEE Computer Society Press, 1996.

[639] H. Shen and C. Johnson. Sweeping simplices: A fast iso-surface extraction algorithm for unstructured grids. In G. M. Nielson and D. Silver, editors, *Visualization '95*, pages 143–150. IEEE Computer Society Press, 1995.

[640] H.-W. Shen, C. Hansen, Y. Livnat, and C. R. Johnson. Isosurfacing in span space with utmost efficiency (ISSUE). In *Proceedings IEEE Visualization '96*, pages 287–294, 1996.

[641] D. Shepard. A two dimensional interpolation function for irregularly spaced data. In *Proceedings of ACM 23rd National Conference*, pages 517–524, 1968.

[642] J. Shi, A. Zhang, J. Encarnação, and M. Göbel. A modified radiosity algorithm for integrated visual and auditory rendering. *Computers & Graphics*, 17(6):633–642, 1993.

[643] P. Shirley. Hybrid Radiosity/Monte Carlo methods. SIGGRAPH '94 course notes on Advanced Radiosity, 1994.

[644] P. Shirley and A. Tuchman. A polygonal approximation to direct scalar volume rendering. *San Diego Workshop on Volume Visualization, Computer Graphics*, 24(5):63–70, December 1988.

[645] K. Shoemake. Animating rotation with quaternion curves. *ACM Computer Graphics (SIGGRAPH '85 Proceedings)*, 19(3):245–254, July 1985.

[646] F. Sillion. Clustering and volume scattering for hierarchical radiosity calculations. In *Photorealistic Rendering Technique (Proceedings Fifth Eurographics Workshop on Rendering)*, pages 105–120, Darmstadt, June 1994. Springer.

[647] F. Sillion, G. Drettakis, and C. Soler. A clustering algorithm for radiance calculation in general environments. In *Rendering Techniques '95 (Proceedings Sixth Eurographics Workshop on Rendering)*, pages 196–205. Springer, August 1995.

[648] F. Sillion and C. Puech. A general two-pass method integrating specular and diffuse reflection. *Computer Graphics (SIGGRAPH '89 Proceedings)*, 23(3):335–344, July 1989.

[649] C. Silva, L. Hong, and A. Kaufman. Flow surface probes for vector field visualization. In G. M. Nielson, H. Hagen, and H. Müller, editors, *Scientific Visualization: Overviews, Methodologies, and Techniques*, pages 295–310. IEEE Computer Society Press, Los Alamitos, California, 1997.

[650] C. T. Silva, J. S. Mitchell, and P. L. Williams. An exact interactive time visibility ordering algorithm for polyhedra cell complexes. In W. E. Lorensen and R. Yagel, editors, *1998 Symposium on Volume Visualization*, pages 87–94. IEEE Computer Society Press and ACM Press, 1998.

[651] L. De Silva, K. Aizawa, and M. Hatori. Detection and tracking of facial features. In *SPIE Visual Communications and Image Processing '95 (VCIP '95)*, volume 2501, pages 2501/1161–1172. The International Society for Optical Engineering, May 1995.

[652] P. Slusallek, W. Heidrich, C. Vogelgsang, M. Ott, and H.-P. Seidel. Radiance maps: An image-based approach to global illumination. Technical Report 23/1997, IMMD IX, University of Erlangen-Nürnberg, 1997.

[653] A. R. Smith. A pixel is not a little square. Technical report, Microsoft Research, 1995.

[654] B. G. Smith. Geometrical shadowing of a random rough surface. *IEEE Transactions on Antennas and Propagation*, 15(5):668–671, September 1967.

[655] J. O. Smith. Physical modeling using digital waveguides. *Computer Music Journal*, 16(4):74–91, 1992.

[656] S. M. Smith and J. M. Brady. ASSET-2: Real-time motion segmentation and shape tracking. *IEEE Transactions on Pattern Analysis and Machine Intelligence*, 17(8):814–820, 1995.

[657] B. Smits, J. Arvo, and D. Greenberg. A clustering algorithm for radiosity in complex environments. *Computer Graphics (SIGGRAPH '94 Proceedings)*, pages 435–442, July 1994.

[658] B. Smits, J. Arvo, and D. Salesin. An importance driven radiosity algorithm. *Computer Graphics (SIGGRAPH '92 Proceedings)*, 26(2):273–282, July 1992.

[659] U. Soergel. Rekonstruktion einer dreidimensionalen Szene aus Videosequenzen. Master's thesis, IMMD V, University of Erlangen-Nürnberg, 1996.

[660] J. F. Sowa. *Principles of Semantic Networks*. Morgan Kaufmann, San Mateo, California, 1991.

[661] O. Staadt and M. Gross. Progressive tetrahedralization. In *IEEE Visualization (Conference Proceedings)*, pages 397–402, 1998.

[662] D. Stalling and H.-C. Hege. Fast and resolution independent line integral convolution. In *Computer Graphics Proceedings*, Annual Conference Series, pages 249–256, Los Angeles, California, August 1995. ACM SIGGRAPH, Addison-Wesley Publishing Company, Inc.

[663] D. Stalling, M. Zöckler, and H.-C. Hege. Fast display of illuminated field lines. *IEEE Transactions on Visualization and Computer Graphics*, 3(2):118–128, April 1997.

[664] M. Stamminger, A. Scheel, X. Granier, F. Perez-Cazorla, G. Drettakis, and F. Sillion. Efficient glossy global illumination with interactive viewing. In *Proceedings Graphics Interface '99*, 1999.

[665] M. Stamminger, P. Slusallek, and H.-P. Seidel. Bounded clustering—finding good bounds on clustered light transport. In *Proceedings Pacific Graphics '98*, 1998.

[666] J. Stauder. Estimation of point light source parameters for object-based coding. In *Signal Processing: Image Communication*, pages 355–379, 1995.

[667] C. Stein, B. G. Becker, and N. L. Max. Sorting and hardware assisted rendering for volume visualization. In A. Kaufman and W. Krüger, editors, *1994 Symposium on Volume Visualization*, pages 83–90. ACM SIGGRAPH, 1994.

[668] E. Steinbach, S. Chaudhuri, and B. Girod. Robust estimation of three-dimensional motion and structure of multiple objects from image sequences. *3D Image Analysis and Synthesis*, pages 53–59, November 1996.

[669] E. Steinbach, S. Chaudhuri, and B. Girod. Robust estimation of multi-component motion in image sequences using the epipolar constraint. *Proceedings International Conference on Acoustics, Speech and Signal Processing*, pages 2689–2692, April 1997.

[670] E. Steinbach and B. Girod. Estimation of rigid body motion and scene structure from image sequences using a novel epipolar transform. *Proceedings International Conference on Acoustics, Speech and Signal Processing*, pages 1911–1914, 1996.

[671] E. Stennert, C. H. Limberg, and K. P. Frentrup. Parese- und Defektheilungs-Index. *HNO*, 25:238–245, 1977.

[672] A. J. Stoddart and K. Brunnström. Free-form surface matching using mean field theory. In *British Machine Vision Conference*, pages 33–42, Edinburgh, UK, 1996.

[673] J. Stolk and J. J. van Wijk. Surface-particles for 3D flow visualization. In F. H. Post and A. J. S. Hin, editors, *Advances in Scientific Visualization*. Springer, 1992.

[674] E. J. Stollnitz, T. D. DeRose, and D. H. Salesin. *Wavelets for Computer Graphics*. Morgan Kaufmann Publishers, Inc., 1996.

[675] W. Straßer and H.-P. Seidel. *Theory and Practice of Geometric Modeling*. Springer, 1989.

[676] D. E. Sturim, M. S. Brandstein, and H. F. Silverman. Tracking multiple talkers using microphone-array measurements. In *Proceedings International Conference on Acoustics, Speech and Signal Processing*, volume 1, pages 371–374, 1997.

[677] M. Subbarao and A. M. Waxman. Closed form solution to image flow equations for planar surfaces in motion. *Computer Vision, Graphics and Image Processing*, 36:208–228, 1986.

[678] G. Subsol, J. P. Thirion, and N. Ayache. A scheme for automatically building three-dimensional morphometric anatomical atlases: application to a skull atlas. *Medical Image Analysis*, 2(1):37–60, 1998.

[679] M. J. Swain and D. H. Ballard. Color indexing. *International Journal of Computer Vision*, 7(1):11–32, November 1991.

[680] M. J. Swain and M. Stricker. Promising directions in active vision. Technical Report CS 91-27, University of Chicago, 1991.

[681] R. Szeliski and S. Lavallée. Matching 3D anatomical surfaces with non-rigid deformations using octree-splines. *International Journal of Computer Vision*, 18(2):171–186, 1996.

[682] R. Szeliski and H.-Y. Shum. Creating full view panoramic image mosaics and environment maps. In *Computer Graphics (SIGGRAPH '97 Proceedings)*, pages 251–258, August 1997.

[683] L. Szirmay-Karlos. Stochastic Methods in Global Illumination - State of the Art Report. Technical Report TR-186-2-98-23, Department of Control Engineering and Information Technology, Technical University of Budapest, August 1998.

[684] H. Tao and T. Huang. Deriving facial articulation models from image sequences. In *Proceedings IEEE International Conference on Image Processing, ICIP '98, Chicago*, pages III 158–161. IEEE, October 1998.

[685] M. J. Tarr and M. J. Black. A computational and evolutionary perspective on the role of representation in vision. *Computer Vision, Graphics, and Image Processing*, 60(1):65–73, 1994.

[686] G. Taubin. Curve and surface smoothing without shrinkage. In *Fifth International Conference on Computer Vision*, Conference Proceedings, pages 852–857. IEEE Computer Society Press, June 1995.

[687] G. Taubin. A signal processing approach to fair surface design. In R. Cook, editor, *SIGGRAPH '95 Conference Proceedings*, Annual Conference Series, pages 351–358. Addison Wesley, August 1995.

[688] C. Teitzel, R. Grosso, and T. Ertl. Efficient and reliable integration methods for particle tracing in unsteady flows on discrete meshes. In W. Lefer and M. Grave, editors, *Visualization in Scientific Computing '97*, pages 31–41, Wien, April 1997. Springer.

[689] C. Teitzel, R. Grosso, and T. Ertl. Line integral convolution on triangulated surfaces. In N. M. Thalmann and V. Skala, editors, *WSCG '97, The Fifth International Conference in Central Europe on Computer Graphics and Visualization '97*, volume III, pages 572–581, Plzen, Czech Republic, February 1997. University of West Bohemia Press.

[690] M. A. Tekalp. *Digital Video Processing*. Prentice-Hall, 1995.

[691] D. Terzopoulos and D. Metaxas. Dynamic 3D models with local and global deformations: Deformable superquadrics. *IEEE Transactions on Pattern Analysis and Machine Intelligence*, 13(7):703–714, July 1991.

[692] D. Terzopoulos and K. Waters. Analysis and synthesis of facial image sequences using physical and anatomical models. *IEEE Transactions on Pattern Analysis and Machine Intelligence*, 15(6):569–579, June 1993.

[693] M. Teschner, S. Girod, and B. Girod. Efficient and robust soft tissue prediction in craniofacial surgery simulation using individual patient's data sets. In *Proceedings CARS '99*, 1999.

[694] W. M. Theimer and H. A. Mallot. Phase-based binocular vergence control and depth reconstruction using active vision. *Computer Vision, Graphics, and Image Processing*, 60(3):343–358, 1995.

[695] J. P. Thirion. New feature points based on geometric invariants for 3D image registration. *International Journal of Computer Vision*, 18(2):121–137, 1996.

[696] A. Tirumalai, B. Schunk, and R. Jain. Robust dynamic stereo for incremental disparity map refinement. In *Proceedings International Workshop on Robust Computer Vision*, Seattle, WA, 1990.

[697] S. Tölg. *Strukturuntersuchungen zur Informationsverarbeitung in neuronaler Architektur am Beispiel der Modellierung von Augenbewegungen für aktives Sehen*. VDI Verlag, Düsseldorf, 1992.

[698] S. Tominaga and B. A. Wandell. Standard surface-reflectance model and illuminant estimation. *Journal of the Optical Society of America*, 6(4):576–584, 1989.

[699] P. H. Torr and D. W. Murray. Statistical detection of independent movement from a moving camera. In D. Hogg and R. Boyle, editors, *British Machine Vision Conference 1992*, pages 79–88. Springer, 1992.

[700] P. H. Torr and A. Zisserman. Robust parameterization and computation of the trifocal tensor. Technical report, Robotics Research Group, Department of Engineering Science, Oxford University, 1997.

[701] K. E. Torrance and E. M. Sparrow. Theory for off-specular reflection from roughened surfaces. *Journal of the Optical Society of America*, 57(9):1105–1114, September 1967.

[702] K. E. Torrance, E. M. Sparrow, and R. C. Birkebak. Polarization, directional distribution, and off-specular peak phenomena in light reflected from roughened surfaces. *Journal of the Optical Society of America*, 56(7):916–925, July 1966.

[703] T. Totsuka and M. Levoy. Frequency domain volume rendering. *Computer Graphics*, 27(4):271–78, August 1993.

[704] L. Trautmann and R. Rabenstein. Digital sound synthesis based on transfer function models. In *Proceedings 1999 IEEE Workshop on Applications of Signal Processing to Audio and Acoustics (WASPAA '99)*, pages 83–86, New Paltz, New York, October 1999.

[705] B. Triggs. Matching constraints and the joint image. In *Proceedings Fifth International Conference on Computer Vision (ICCV '95)*, pages 338–343, Cambridge, MA, 1995.

[706] B. Triggs. Autocalibration and the absolute quadric. In *Int. Conf. Computer Vision and Pattern Recognition*, pages 609–614, Puerto Rico, June 1997.

[707] I. J. Trotts, B. Hamann, K. I. Joy, and D. F. Wiley. Simplification of tetrahedral meshes. In D. Ebert, H. Rushmeier, and H. Hagen, editors, *Visualization '98*, pages 287–295, Research Triangle Park, North Carolina, 1998. IEEE Computer Society Press.

[708] E. Trucco and A. Verri. *Introductory Techniques for 3D Computer Vision*. Prentice Hall, New York, 1998.

[709] R. Y. Tsai. A versatile camera calibration technique for high-accuracy 3D machine vision metrology using off-the-shelf TV cameras and lenses. *IEEE Journal of Robotics and Automation*, RA-3(4):323–344, August 1987.

[710] R. Y. Tsai and T. S. Huang. Estimating three-dimensional motion parameters of a rigid planar patch. *IEEE Transactions on Acoustics, Speech and Signal Processing*, 29(6):1147–1152, December 1981.

[711] R. Y. Tsai and T. S. Huang. Uniqueness and estimation of three-dimensional motion parameters of rigid objects with curved surfaces. *IEEE Transactions on Pattern Analysis and Machine Intelligence*, 6(1):13–27, 1984.

[712] G. Tsang, S. Ghali, E. Fiume, and A. Venetsanopoulos. A novel parameterization of the light field. In H. Niemann, H.-P. Seidel, and B. Girod, editors, *IMDSP '98 Workshop Proceedings*, pages 319–322. IEEE Signal Processing Society, July 1998.

[713] J. K. Tsotsos. On the relative complexity of active vs. passive visual search. *International Journal of Computer Vision*, 7(2):127–141, 1992.

[714] J. K. Tsotsos. There is no one way to look at vision. *Computer Vision, Graphics, and Image Processing*, 60(1):95–97, 1994.

[715] G. Turk. Re-tiling polygonal surfaces. In E. E. Catmull, editor, *Computer Graphics (SIGGRAPH '92 Proceedings)*, volume 26, pages 55–64, July 1992.

[716] G. Turk and M. Levoy. Zippered polygon meshes from range images. In A. Glassner, editor, *Proceedings of SIGGRAPH '94*, Annual Conference Series, pages 311–318. ACM SIGGRAPH, July 1994.

[717] M. A. Turk and A. Pentland. Face recognition using eigenfaces. In *Proceedings IEEE Computer Society Conference on Computer Vision and Pattern Recognition*, pages 586–590, Hawaii, June 1992.

[718] S. K. Ueng, K. Sikorski, and M. Kwan-Liu. Fast algorithms for visualizing fluid motion in steady flow on unstructured grids. In G. M. Nielson and D. Silver, editors, *Visualization '95*, pages 313–320. IEEE Computer Society Press, 1995.

[719] A. Vahedian, M. Frater, et al. Estimation of speaker position using audio information. In *Proceedings of IEEE TENCON '97*, volume 1, pages 181–184, 1997.

[720] V. Välimäki, J. Huopaniemi, and M. Karjalainen. Physical modeling of plucked string instruments with application to real-time sound synthesis. *Journal Audio Engineering Society*, 44(5):331–353, 1996.

[721] V. Välimäki and T. Takala. Virtual musical instruments—natural sound using physical models. *Organised Sound*, 1(2):75–86, 1996.

[722] M. Vannier, J. Marsh, and J. Warren. Three-dimensional computer graphics for craniofacial surgical planning and evaluation. *Computer Graphics*, 17(3):262–273, 1983.

[723] V. N. Vapnik. *The Nature of Statistical Learning Theory*. Springer, Heidelberg, 1996.

[724] VDI/VDE. *Handbuch Meßtechnik II*, volume VDI/VDE 2628. VDI/VDE Gesellschaft Meß- und Automatisierungstechnik (GMA), 1985.

[725] E. Veach and L. J. Guibas. Bidirectional estimators for light transport. In *Rendering Techniques '94 (Proceedings Fifth Eurographics Workshop on Rendering)*, 1994.

[726] R. C. Veltkamp. *Closed Object Boundaries from Scattered Points*, volume 885 of *Lecture Notes in Computer Science*. Springer, Berlin, Heidelberg, New York, 1994.

[727] E. Verheijen. *Sound Reproduction by Wave Field Synthesis*. PhD thesis, Delft University of Technology, 1997.

[728] T. Vetter and V. Blanz. Estimating coloured 3D face models from single images: An example based approach. *ECCV*, 1998.

[729] G. Vezina, P. Fletcher, and P. Robertson. Volume rendering on the MasPar MP-1. In *1992 Workshop on Volume Visualization*, pages 3–8, Boston, Massachusetts, October 1992. ACM SIGGRAPH.

[730] L. Vincent and P. Soille. Watersheds in digital spaces: An efficient algorithm based on immersion simulations. *IEEE Transactions on Pattern Analysis and Machine Intelligence (PAMI)*, 15(6):583–598, 1991.

[731] P. Viola and W. M. Wells III. Alignment by maximization of mutual information. In *Fifth International Conference on Computer Vision*, Cambridge, Massachusetts, USA, 1995. IEEE.

[732] T. Wada, H. Ukida, and T. Matsuyama. Shape from shading with interreflections under proximal light source, 1995.

[733] J. R. Wallace, M. F. Cohen, and D. P. Greenberg. A two-pass solution to the rendering equation: A synthesis of ray tracing and radiosity methods. *Computer Graphics (SIGGRAPH '87 Proceedings)*, 21(4):311–320, July 1987.

[734] H. Wang and P. Chu. Voice source localization for automatic camera pointing system in videoconferencing. In *Proceedings of the 1997 IEEE International Conference on Acoustics, Speech, and Signal Processing*, volume 1, pages 187–190, Munich, 1997.

[735] L. Wang and J. J. Clark. Shape from active shadow motion. *SPIE Conference on Intelligent Robots and Computer Vision: Active Vision and 3D Methods*, 1993.

[736] W. Wang and J. H. Duncan. Recovering the three-dimensional motion and structure of multiple moving objects from binocular image flows. *Computer Vision and Image Understanding*, 63(3):430–446, May 1996.

[737] Y. Watanabe and Y. Suenaga. A trigonal prism-based method for hair image generation. *IEEE Computer Graphics & Applications*, 12(1):47–53, January 1992.

[738] A. M. Waxman, B. Kamgar-Parsi, and M. Subbarao. Closed-form solutions to image flow equations for 3D structure and motion. *International Journal of Computer Vision*, 1:239–258, 1987.

[739] K. Wechsler, M. Breuer, and F. Durst. Steady and unsteady computations of turbulent flows induced by a $4/45°$ pitched-blade impeller. *Journal of Fluids Engineering*, 121:318–329, 1999.

[740] R. Wegenkittl and E. Gröller. Fast oriented line integral convolution for vector field visualization via the internet. In R. Yagel and H. Hagen, editors, *Visualization '97*, pages 309–316, Phoenix, Arizona, 1997. IEEE Computer Society Press.

[741] K. Weiler. Edge-based data structures for solid modeling in curved-surface enviroments. *IEEE Computer Graphics & Applications*, pages 21–40, January 1985.

[742] G. Welch and G. Bishop. An introduction to Kalman filter. Technical report, Department of Computer Science, University of North Carolina, Chapel Hill, NC, 1998.

[743] W. Welch and A. Witkin. Variational surface modeling. In *ACM Computer Graphics (SIGGRAPH '92 Proceedings)*, pages 157–166, 1992.

[744] W. J. Welsh, S. Searsby, and J. B. Waite. Model-based image coding. *British Telecom Technology Journal*, 8(3):94–106, July 1990.

[745] J. Weng, M. Ahuja, and T. S. Huang. Motion and structure from two perspective views: Algorithms, error analysis and error estimation. *IEEE Transactions on Pattern Analysis and Machine Intelligence*, 11(5):451–476, May 1989.

[746] J. Weng, M. Ahuja, and T. S. Huang. Optimal motion and structure estimation. *IEEE Transactions on Pattern Analysis and Machine Intelligence*, 15(9):864–884, September 1993.

[747] W. Wesselink. *Variational Modeling of Curves and Surfaces.* PhD thesis, University of Technology, Eindhoven, 1996.

[748] R. Westermann. Compression domain rendering of time-resolved volume data. In G. M. Nielson and D. Silver, editors, *Visualization '95*, pages 168–175. IEEE Computer Society Press, 1995.

[749] R. Westermann and T. Ertl. Efficiently using graphics hardware in volume rendering applications. In *Computer Graphics Proceedings*, Annual Conference Series, pages 169–177, Orlando, Florida, 1998. ACM SIGGRAPH.

[750] R. Westermann, L. Kobbelt, and T. Ertl. Real-time exploration of regular volume data by adaptive reconstruction of iso-surfaces. *The Visual Computer Journal*, 1999.

[751] L. Westover. Footprint evaluation for volume rendering. *Computer Graphics*, 24(4):367–376, August 1990.

[752] D. Wetzel and H. Niemann. A robust cognitive approach to traffic scene analysis. In *Second IEEE Workshop on Applications of Computer Vision*, pages 65–72, Sarasota, Florida, 1994.

[753] J. J. van Wijk. Spot noise—texture synthesis for data visualization. In *Computer Graphics Proceedings*, volume 25 of *Annual Conference Series*, pages 309–318, Las Vegas, July 1991. ACM SIGGRAPH, Addison-Wesley Publishing Company, Inc.

[754] J. J. van Wijk. Implicit stream surfaces. In G. M. Nielson and D. Bergeron, editors, *Visualization '93*, pages 245–260, Los Alamitos, California, 1993. IEEE Computer Society Press.

[755] J. Wilhelms and A. van Gelder. A coherent projection approach for direct volume rendering. *Computer Graphics*, 25(4):275–284, July 1991.

[756] J. Wilhelms and A. van Gelder. Octrees for faster iso-surface generation. In *ACM Transactions on Graphics*, pages 201–227, 1992.

[757] L. Williams. Casting curved shadows on curved surfaces. In *Computer Graphics (SIGGRAPH '78 Proceedings)*, pages 270–274, August 1978.

[758] L. Williams. Pyramidal parametrics. In *Computer Graphics (SIGGRAPH '83 Proceedings)*, pages 1–11, July 1983.

[759] T. Wilson and C. Sheppard. *Theory and Practice of Scanning Microscopy.* Academic Press, London, 1994.

[760] G. Winkler. *Image Analysis, Random Fields, and Dynamic Monte Carlo Methods*, volume 27 of *Applications of Mathematics*. Springer, Heidelberg, 1995.

[761] A. Witkin and W. Welch. Fast animation and control of nonrigid structures. *ACM Computer Graphics (SIGGRAPH '90 Proceedings)*, 24(4):243–252, August 1990.

[762] L. E. Wixson. Exploiting world structure to efficiently search for objects. Technical Report Number 434, University of Rochester, 1992.

[763] S. Wolf, M. Müller, W. Schneider, C. Haid, and M. Wigand. Facial nerve function after transtemporal removal of acoustic neurinomas: Results, time course or function and rehabilitation. In M. Samii, editor, *Skull Base Surgery*, pages 894–897. Springer, Hannover, 1992.

[764] T.-T. Wong, P.-A. Heng, S.-H. Or, and W.-Y. Ng. Image-based rendering with controllable illumination. In *Rendering Techniques '97 (Proceedings Eighth Eurographics Workshop on Rendering)*, pages 13–22, June 1997.

[765] R. J. Woodham, Y. Iwahori, and R. A. Barman. Photometric stereo: Lambertian reflectance and light sources with unknown direction and strength. Technical Report 91-18, University of British Columbia, 1991.

[766] G. Xu and Z. Zhang. *Epipolar Geometry in Stereo, Motion and Object Recognition—A Unified Approach*, volume 6 of *Computational Imaging and Vision*. Kluwer Academic Publishers, Dordrecht, 1996.

[767] R. Yagel and Z. Shi. Accelerating volume animation by space leaping. In *Proceedings Visualization '93*, pages 62–69, Los Alamitos, 1993. IEEE Computer Society Press.

[768] T. Yasuda, Y. Hashimoto, S. Yokoi, and J. Toriwaki. Computer system for craniofacial surgical planning based on CT images. *IEEE Transactions on Medical Imaging*, 9(3):270–280, September 1990.

[769] M. Young. *Optik, Laser, Wellenleiter*. Springer, Berlin, Germany, 4th edition, 1997.

[770] W. Yu, K. Daniilidis, and G. Sommer. Rotated wedge averaging method for junction characterization. In *Proceedings of IEEE Conference on Computer Vision and Pattern Recognition*, pages 390–395, Santa Barbara, California, USA, 1998.

[771] A. L. Yuille. Deformable templates for face recognition. *Journal of Cognitive Neuroscience*, 3(1):59–70, 1991.

[772] Q. Zaidi. Identification of illuminant and object colors: heuristic-based algorithms. *Journal of the Optical Society of America*, 15(7):1767–1776, 1998.

[773] Y. J. Zhang. A survey on evaluation methods for image segmentation. *Pattern Recognition*, 29(8):1335–1346, 1996.

[774] Z. Zhang. Iterative point matching for registration of free-form curves and surfaces. *International Journal of Computer Vision*, 13(1):119–152, 1994.

[775] Z. Zhang, R. Deriche, O. Faugeras, and Q.-T. Luong. A robust technique for matching two uncalibrated images through the recovery of the unknown epipolar geometry. *Artificial Intelligence Journal*, 78:87–119, October 1995.

[776] Z. Zhang, O. Faugeras, and R. Deriche. An effective technique for calibrating a binocular stereo through projective reconstruction using both a calibration object and the environment. *Videre: Journal of Computer Vision Research*, 1(1), 1997.

[777] Q. Zheng and R. Chellappa. Estimation of illuminant direction, albedo, and shape from shading. *IEEE Transactions on Pattern Analysis and Machine Intelligence*, 13(7):680–702, 1991.

[778] L. J. Ziomek. *Fundamentals of Acoustic Field Theory and Space-Time Signal Processing*. CRC Press, Boca Raton, 1995.

[779] M. Zöckler, D. Stalling, and H.-C. Hege. Interactive visualization of 3D-vector fields using illuminated stream lines. In R. Yagel and G. M. Nielson, editors, *Visualization '96*, pages 107–113, San Francisco, California, 1996. IEEE Computer Society Press.

Index